VOLUME FIVE HUNDRED AND THIRTY FIVE

METHODS IN ENZYMOLOGY

Endosome Signaling Part B

METHODS IN ENZYMOLOGY

VOLUME FIVE HUNDRED AND THIRTY FIVE

METHODS IN ENZYMOLOGY

Endosome Signaling Part B

Edited by

P. MICHAEL CONN

Senior Vice President for Research
Associate Provost
Professor of Internal Medicine and Cell Biology
Texas Tech University Health Sciences Center
Lubbock, TX 79430, USA

AMSTERDAM • BOSTON • HEIDELBERG • LONDON
NEW YORK • OXFORD • PARIS • SAN DIEGO
SAN FRANCISCO • SINGAPORE • SYDNEY • TOKYO
Academic Press is an imprint of Elsevier

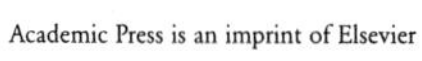

Academic Press is an imprint of Elsevier
525 B Street, Suite 1800, San Diego, CA 92101-4495, USA
225 Wyman Street, Waltham, MA 02451, USA
Radarweg 29, PO Box 211, 1000 AE Amsterdam, The Netherlands
The Boulevard, Langford Lane, Kidlington, Oxford, OX5 1GB, UK
32 Jamestown Road, London NW1 7BY, UK

First edition 2014

ISBN: 978-0-12-397925-4
ISSN: 0076-6879

Printed and bound in United States of America
14 15 16 17 11 10 9 8 7 6 5 4 3 2 1

CONTENTS

CONTRIBUTORS

Veronica Aran
Division of Cellular and Molecular Physiology, Institute of Translational Medicine, University of Liverpool, Liverpool, United Kingdom

François Authier
Service information scientifique et technique (IST) de l'Inserm, Délégation régionale Inserm Paris V, Paris, France

Masahiro Azuma
Department of Microbiology and Immunology, Hokkaido University Graduate School of Medicine, Kita-ku, Sapporo, Japan

Tadashi Baba
Department of Microbiology and Infection Control Science, Juntendo University Graduate School of Medicine, Tokyo, Japan

Rachel Barrow
Centre for Tumour Biology, Barts Cancer Institute – A Cancer Research UK Centre of Excellence, Queen Mary University of London, John Vane Science Centre, London, United Kingdom

John J.M. Bergeron
Department of Medicine, and Department of Cell Biology, McGill University, Montreal, Quebec, Canada

Rose Cairns
Faculty of Pharmacy, University of Sydney, Sydney, New South Wales, Australia

Eric C. Chang
Lester and Sue Smith Breast Center, Department of Molecular and Cellular Biology, Baylor College of Medicine, Houston, Texas, USA

Xue Chen
Atopy (Allergy) Research Center, Juntendo University Graduate School of Medicine, Tokyo, Japan, and Department of Dermatology, Peking University People's Hospital, Beijing, China

Helen R. Clark
Virginia Bioinformatics Institute, and Department of Biochemistry, Virginia Tech, Blacksburg, Virginia, USA

Bernard Desbuquois
Inserm U567, Institut Cochin, CNRS UMR 8104, Université Paris-Descartes, Paris, France

Gianni M. Di Guglielmo
Department of Physiology and Pharmacology, Western University, London, Ontario, Canada

Nikolai Engedal
Prostate Cancer Research Group, Centre for Molecular Medicine Norway, Nordic EMBL Partnership, University of Oslo and Oslo University Hospital, Oslo, Norway

Carlos Enrich
Departament de Biologia Cellular, Immunologia i Neurociències, IDIBAPS, Facultat de Medicina, Universitat de Barcelona, Barcelona, Spain

Marco Falasca
Inositide Signalling Group, Centre for Diabetes, Blizard Institute, Barts and The London School of Medicine and Dentistry, Queen Mary University of London, London, United Kingdom

Gareth W. Fearnley
Endothelial Cell Biology Unit, School of Molecular & Cellular Biology, University of Leeds, Leeds, United Kingdom

Theodore Fotsis
Laboratory of Biological Chemistry, Medical School, University of Ioannina, and Department of Biomedical Research, Foundation for Research & Technology – Hellas, Institute of Molecular Biology & Biotechnology, University Campus of Ioannina, Ioannina, Greece

Jürgen Fritsch
Institute of Immunology, Christian-Albrechts-University of Kiel, Kiel, Germany

Kenji Funami
Department of Microbiology and Immunology, Hokkaido University Graduate School of Medicine, Kita-ku, Sapporo, Japan

Mariona Gelabert-Baldrich
Departament de Biologia Cellular, Immunologia i Neurociències, IDIBAPS, Facultat de Medicina, Universitat de Barcelona, Barcelona, Spain

Thomas Grewal
Faculty of Pharmacy, University of Sydney, Sydney, New South Wales, Australia

Neil Grimsey
Department of Pharmacology, School of Medicine, University of California, La Jolla, California, USA

Mutsuko Hara
Atopy (Allergy) Research Center, Juntendo University Graduate School of Medicine, Tokyo, Japan

Michael A. Harrison
School of Biomedical Sciences, University of Leeds, Leeds, United Kingdom

Tristan A. Hayes
Virginia Bioinformatics Institute, and Department of Biological Sciences, Virginia Tech, Blacksburg, Virginia, USA

Carina Hellberg
University of Birmingham, School of Biosciences, Birmingham, United Kingdom

Maria Hernandez-Valladares
Division of Cellular and Molecular Physiology, Institute of Translational Medicine, University of Liverpool, Liverpool, United Kingdom

Keiichi Hiramatsu
Department of Microbiology and Infection Control Science, Juntendo University Graduate School of Medicine, Tokyo, Japan

Monira Hoque
Faculty of Pharmacy, University of Sydney, Sydney, New South Wales, Australia

Lukas A. Huber
Biocenter, Division of Cell Biology, Innsbruck Medical University, Innsbruck, Austria

Shigaku Ikeda
Atopy (Allergy) Research Center, Juntendo University Graduate School of Medicine, Tokyo, Japan, and Department of Dermatology and Allergology, Juntendo University Graduate School of Medicine, Tokyo, Japan

Roshanak Irannejad
Department of Psychiatry and, Department of Cellular & Molecular Pharmacology, University of California School of Medicine, San Francisco, California, USA

Kamil Jastrzębski
Laboratory of Cell Biology, International Institute of Molecular and Cell Biology, Warsaw, Poland

Carine Joffre
Centre for Tumour Biology, Barts Cancer Institute – A Cancer Research UK Centre of Excellence, Queen Mary University of London, John Vane Science Centre, London, United Kingdom, and UNITE 830 INSERM, Institut Curie Centre de Recherche, Paris Cedex 05, France

Shiv D. Kale
Virginia Bioinformatics Institute, Virginia Tech, Blacksburg, Virginia, USA

Seiji Kamijo
Atopy (Allergy) Research Center, Juntendo University Graduate School of Medicine, Tokyo, Japan

Junko Kawasaki
Atopy (Allergy) Research Center, and Department of Dermatology and Allergology, Juntendo University Graduate School of Medicine, Tokyo, Japan

Stéphanie Kermorgant
Centre for Tumour Biology, Barts Cancer Institute – A Cancer Research UK Centre of Excellence, Queen Mary University of London, John Vane Science Centre, London, United Kingdom

Hirokazu Kinoshita
Atopy (Allergy) Research Center, and Department of Dermatology and Allergology, Juntendo University Graduate School of Medicine, Tokyo, Japan

Eleftherios Kostaras
Laboratory of Biological Chemistry, Medical School, University of Ioannina, and Department of Biomedical Research, Foundation for Research & Technology – Hellas, Institute of Molecular Biology & Biotechnology, University Campus of Ioannina, Ioannina, Greece

Sarah J. Kotowski
Department of Psychiatry and, Department of Cellular & Molecular Pharmacology, University of California School of Medicine, San Francisco, California, USA

Antony M. Latham
Endothelial Cell Biology Unit, School of Molecular & Cellular Biology, University of Leeds, Leeds, United Kingdom

Tuan Anh Le
Atopy (Allergy) Research Center; Department of Dermatology and Allergology, Juntendo University Graduate School of Medicine, Tokyo, Japan, and Department of Dermatology and Allergology, Institute of Clinical Medical and Pharmaceutical Sciences 108, Hanoi, Vietnam

Huilan Lin
Department of Pharmacology, School of Medicine, University of California, La Jolla, California, USA

Tania Maffucci
Inositide Signalling Group, Centre for Diabetes, Blizard Institute, Barts and The London School of Medicine and Dentistry, Queen Mary University of London, London, United Kingdom

Bénédicte Manoury
INSERM, Unité 1013, and Université Paris Descartes, Sorbonne Paris Cité, Faculté de médecine, Paris, France

Misako Matsumoto
Department of Microbiology and Immunology, Hokkaido University Graduate School of Medicine, Kita-ku, Sapporo, Japan

Sarah McLean
Department of Anatomy and Cell Biology, Department of Physiology and Pharmacology, Western University, London, Ontario, Canada

Marta Miaczynska
Laboratory of Cell Biology, International Institute of Molecular and Cell Biology, Warsaw, Poland

Ian G. Mills
Prostate Cancer Research Group, Centre for Molecular Medicine Norway, Nordic EMBL Partnership, University of Oslo; Department of Cancer Prevention, Institute of Cancer Research; Department of Urology, Oslo University Hospital, Oslo, Norway, and Uro-Oncology Research Group, Cambridge Research Institute, University of Cambridge, Cambridge, United Kingdom

Ludovic Ménard
Centre for Tumour Biology, Barts Cancer Institute – A Cancer Research UK Centre of Excellence, Queen Mary University of London, John Vane Science Centre, London, United Kingdom

Jessica G. Moreland
Division of Critical Care, Department of Pediatrics and the Inflammation Program, The University of Iowa, Iowa City, Iowa, USA

Shunsuke Mori
Department of Oncogene Research, Research Institute for Microbial Diseases, Osaka University, Suita, Osaka, Japan

Carol Murphy
Department of Biomedical Research, Foundation for Research & Technology – Hellas, Institute of Molecular Biology & Biotechnology, University Campus of Ioannina, Ioannina, Greece

Shigeyuki Nada
Department of Oncogene Research, Research Institute for Microbial Diseases, Osaka University, Suita, Osaka, Japan

Adam F. Odell
Endothelial Cell Biology Unit, School of Molecular & Cellular Biology, University of Leeds, Leeds, United Kingdom

Hideoki Ogawa
Atopy (Allergy) Research Center, Juntendo University Graduate School of Medicine, Tokyo, Japan, and Department of Dermatology and Allergology, Juntendo University Graduate School of Medicine, Tokyo, Japan

Masato Okada
Department of Oncogene Research, Research Institute for Microbial Diseases, Osaka University, Suita, Osaka, Japan

Ko Okumura
Atopy (Allergy) Research Center, Juntendo University Graduate School of Medicine, Tokyo, Japan

A. Paige Davis Volk
Division of Critical Care, Department of Pediatrics and the Inflammation Program, The University of Iowa, Iowa City, Iowa, USA

Nina Marie Pedersen
Centre for Cancer Biomedicine, Faculty of Medicine, University of Oslo, Oslo, Norway

Albert Pol
Departament de Biologia Cellular, Immunologia i Neurociències, IDIBAPS, Facultat de Medicina, Universitat de Barcelona, and ICREA, Institució Catalana de Recerca Avançada, Barcelona, Spain

Sreenivasan Ponnambalam
Endothelial Cell Biology Unit, School of Molecular & Cellular Biology, University of Leeds, Leeds, United Kingdom

Barry I. Posner
Department of Medicine, and Department of Cell Biology, McGill University, Montreal, Quebec, Canada

Ian A. Prior
Division of Cellular and Molecular Physiology, Institute of Translational Medicine, University of Liverpool, Liverpool, United Kingdom

Elżbieta Purta
Laboratory of Bioinformatics and Protein Engineering, International Institute of Molecular and Cell Biology, Warsaw, Poland

Carles Rentero
Departament de Biologia Cellular, Immunologia i Neurociències, IDIBAPS, Facultat de Medicina, Universitat de Barcelona, Barcelona, Spain

Łukasz Sadowski
Laboratory of Cell Biology, International Institute of Molecular and Cell Biology, Warsaw, Poland

Julia M. Scheffler
Biocenter, Division of Cell Biology, Innsbruck Medical University, Innsbruck, Austria

Natalia Schiefermeier
Biocenter, Division of Cell Biology, Innsbruck Medical University, Innsbruck, Austria

Stefan Schütze
Institute of Immunology, Christian-Albrechts-University of Kiel, Kiel, Germany

Tsukasa Seya
Department of Microbiology and Immunology, Hokkaido University Graduate School of Medicine, Kita-ku, Sapporo, Japan

Sudha K. Shenoy
Department of Medicine, Duke University Medical Center, Durham, North Carolina, USA

Gina A. Smith
Endothelial Cell Biology Unit, School of Molecular & Cellular Biology, University of Leeds, Leeds, United Kingdom

Harald Stenmark
Centre for Cancer Biomedicine, Faculty of Medicine, University of Oslo, Oslo, Norway

Yusuke Takahashi
Department of Oncogene Research, Research Institute for Microbial Diseases, Osaka University, Suita, Osaka, Japan

Toshiro Takai
Atopy (Allergy) Research Center, Juntendo University Graduate School of Medicine, Tokyo, Japan

Megumi Tatematsu
Department of Microbiology and Immunology, Hokkaido University Graduate School of Medicine, Kita-ku, Sapporo, Japan

Vladimir Tchikov
Institute of Immunology, Christian-Albrechts-University of Kiel, Kiel, Germany

Francesc Tebar
Departament de Biologia Cellular, Immunologia i Neurociències, IDIBAPS, Facultat de Medicina, Universitat de Barcelona, Barcelona, Spain

Mira Tohmé
INSERM, Unité 1013; Université Paris Descartes, Sorbonne Paris Cité, Faculté de médecine, and INSERM, Unité 932, Institut Curie, Paris, France

Darren C. Tomlinson
Biomedical Health Research Centre & Astbury Centre for Structural Molecular Biology, University of Leeds, Leeds, United Kingdom

JoAnn Trejo
Department of Pharmacology, School of Medicine, University of California, La Jolla, California, USA

Hiroko Ushio
Atopy (Allergy) Research Center, Juntendo University Graduate School of Medicine, Tokyo, Japan

Mark von Zastrow
Department of Psychiatry and, Department of Cellular & Molecular Pharmacology, University of California School of Medicine, San Francisco, California, USA

Anh Tuan Vu
Atopy (Allergy) Research Center; Department of Dermatology and Allergology, Juntendo University Graduate School of Medicine, Tokyo, Japan, and Quyhoa National Leprosy-Dermatology Hospital, Quynhon, Vietnam

Stephen B. Wheatcroft
Division of Diabetes and Cardiovascular Research, Faculty of Medicine & Health, University of Leeds, Leeds, United Kingdom

Yang Xie
Atopy (Allergy) Research Center, Juntendo University Graduate School of Medicine, Tokyo, Japan, and Department of Dermatology, The 3rd Affiliated Hospital of Sun Yat-sen University, Guangzhou, China

Ahmed Zahraoui
Phagocytosis and Bacterial Invasion Laboratory, INSERM U.1016-CNRS UMR8104, Institut Cochin, Université Paris Descartes, Paris, France

Ze-Yi Zheng
Lester and Sue Smith Breast Center, Department of Molecular and Cellular Biology, Baylor College of Medicine, Houston, Texas, USA

PREFACE

Endosomes are membrane-bound compartments that transport internalized material from the plasma membrane to the lysosome and elsewhere. These compartments, often about 500 nm, but ranging in size, have the capability to sort molecules, routing some contents to the lysosomes for degradation, and recycling other materials back to the plasma membrane. The Golgi apparatus also provides molecules to the endosome, some of which are delivered to lysosomes and others are recycled back to the Golgi. Because of this ability to differentially deliver molecules, the endosome is viewed as a presorting structure.

Endosomes are categorized by size, enzymatic content, morphology, and by other criteria such as the length of time it takes internalized material to reach them. Endosomes may provide platforms for cross talk between signaling systems, and this consideration has provided them elite status among cellular components that contribute to signaling.

This volume provides descriptions of the range of methods used to analyze and evaluate these important compartments. The authors explain how these methods are able to provide important biological insights in the context of particular models.

Authors were selected based on both their research contributions and on their ability to describe their methodological contributions in a clear and reproducible way. They have been encouraged to make use of graphics, comparisons to other methods, and to provide tricks and approaches not revealed in prior publications that make it possible to adapt their methods to other systems.

The editor wants to express appreciation to the contributors for providing their contributions in a timely fashion, to the senior editors for guidance, and to the staff at Academic Press for helpful input.

P. Michael Conn
Lubbock, TX, USA

SECTION I

SIGNALING

CHAPTER ONE

Assessment of Insulin Proteolysis in Rat Liver Endosomes: Its Relationship to Intracellular Insulin Signaling

François Authier*,[1], Bernard Desbuquois†

*Service information scientifique et technique (IST) de l'Inserm, Délégation régionale Inserm Paris V, Paris, France
†Inserm U567, Institut Cochin, CNRS UMR 8104, Université Paris-Descartes, Paris, France
[1]Corresponding author: e-mail address: francois.authier@inserm.fr

Contents

Abstract

Insulin binding to insulin receptor (IR) at the cell surface results in the activation of IR kinase and initiates the translocation of insulin–IR complexes to clathrin-coated pits and to early endosomes containing internalized but still active receptors. In liver parenchyma, several mechanisms are involved in the regulation of endosomal IR tyrosine kinase activity. Two of these regulatory mechanisms are at the level of intraendosomal

Methods in Enzymology, Volume 535
ISSN 0076-6879
http://dx.doi.org/10.1016/B978-0-12-397925-4.00001-8

ligand. First, a progressive decrease in endosomal pH mediated by the vacuolar H^+-ATPase proton pump promotes dissociation of the insulin–IR complex. Second, free dissociated insulin is degraded by a soluble endosomal acidic insulinase, which has been identified as aspartic acid protease cathepsin D. This enzyme catalyzes the cleavage of insulin at the Phe^{B24}–Phe^{B25} bond, generating a major clipped molecule, A^{1-21}–B^{1-24} insulin, that can no longer bind to IR within endosomes. Concomitant with, or shortly after, the tyrosine-phosphorylated IR is deactivated by two independent processes: its rapid dephosphorylation by endosome-associated phosphotyrosine phosphatase(s) and its association with the molecular adaptor Grb14, with resulting inhibition of IR catalytic activity. By mediating the removal and degradation of circulating insulin, as well as the deactivation of the activated IR, internalization of the insulin-receptor complex into endosomes represents a major mechanism involved in the negative regulation of insulin signaling.

1. INTRODUCTION

The first step in insulin action and degradation in cells is its interaction with a specific receptor localized at the cell surface. The insulin receptor (IR) is a heterotetrameric glycoprotein composed of two extracellular α-subunits, which contain the insulin binding site, and two transmembrane β-subunits, which possess in their cytoplasmic domain a tyrosine kinase activity. Upon insulin binding, at least six tyrosine residues in the β-subunits are phosphorylated and the tyrosine kinase is activated, leading to the recruitment and phosphorylation of two main families of substrates, the IRS (IR substrates) and Shc (Src homology 2 domain containing) proteins. These in turn recruit through their phosphotyrosine (PY) residues effector proteins containing SH2 domains, leading to the activation of the PI3K (phosphoinositide 3-kinase)/Akt (protein kinase B) and Ras/Erk (extracellular signal-regulated kinase) signal transduction pathways.

Concomitantly to insulin binding, both insulin and the activated IR are rapidly internalized into endosomes. Within the acidic endosomal lumen, the insulin-receptor complex dissociates; insulin is then degraded by endosomal protease(s), whereas the receptor is dephosphorylated by protein tyrosine phosphatases and is to a large extent recycled to the cell surface. Although both ligand proteolysis and receptor dephosphorylation are clearly involved in the termination of insulin signaling, the identification of several insulin-signaling proteins in IR-containing endosomes and/or their insulin-induced translocation to endosomes (for review, see Sorkin & Von Zastrow, 2009) raises the possibility that insulin signaling may also occur, or continue, at this locus after IR internalization. On the other hand, the recruitment to the internalized receptor of at least two negative regulators of IR kinase activity, the molecular adaptor Grb14, and the protein tyrosine phosphatase PTP1B would facilitate termination of signaling.

This chapter presents biochemical procedures used to monitor the degradation of insulin in liver endosomes, to map the sites of cleavage within the insulin chains, and to identify the responsible endosomal proteases. Procedures used to assess the *in vivo* effects of insulin and slowly processed insulin analogs on tyrosine phosphorylation of the endosomal IR and IR-mediated association of Grb14 to endosomes are also described.

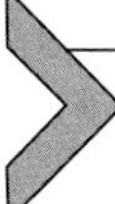

2. Assays for Insulin Degradation

2.1. Trichloroacetic acid precipitation assay

The standard assay for insulin proteolysis has been made using the radiolabeled peptide [^{125}I]-insulin by monitoring the increase in soluble counts after precipitation of radioactive insulin with cold trichloroacetic acid (TCA) (Authier, Danielsen, Kouach, Briand, & Chauvet, 2001; Authier, Métioui, Fabrega, Kouach, & Briand, 2002). The limitation of this type of assay is that it greatly underestimates actual degradation and measures the proteolytic activity under conditions far from the initial reaction rate. For valid results with TCA, proteolytic activity must be kept within the linear portion of the assay by shortening the incubation time and/or diluting the sample. Moreover, radiolabeled insulin may lack many structural features of the natural insulin substrates that might be important in directing protease interaction and cleavage. Finally, all of the early insulin intermediates displaying a molecular mass higher than 3125 Da are not TCA-soluble (Authier et al., 2001). Consequently, the TCA solubility of radioactive products reflects additional and/or further processing of transient insulin intermediates and their late transformation into short end products and/or free amino acids (Authier et al., 2002).

The loss of TCA-precipitable radioactivity during [^{125}I]-insulin proteolysis can be measured as follows:

1. Add 2 ml ice-cold TCA (6%) and 1 mg BSA (bovine serum albumin) to hydrolysate samples (0.2 ml) containing [^{125}I]-insulin (~20 fmol).
2. Incubate 2 h at 4 °C.
3. Centrifuge at 10,000 × g_{av} for 20 min at 4 °C.
4. Decant supernatant.

5. Quantify the radioactivity associated with supernatants and pellets using a γ-counter.
6. Calculate insulin degradation as the percentage of total [^{125}I]-insulin degraded within the linear range of the degradation curve.

Other assays such as receptor binding or immunoprecipitation have been described and may have specialized uses (Desbuquois, Janicot, & Dupuis, 1990).

2.2. Reverse-phase HPLC assay using radiolabeled insulin

Reverse-phase (RP) HPLC is by far the most sensitive assay able to detect a single peptide bond cleavage. With HPLC procedures, radiolabeled insulin may be used as the protease substrate. Monoiodinated insulins (especially [^{125}I]Tyr^{A14} insulin and [^{125}I]Tyr^{A19} insulin) have allowed the first identification of radioactive insulin intermediates produced *in vivo* within rat hepatic endosomes (Seabright & Smith, 1996). This can be accomplished as follows:

1. Load the radioactive hydrolytic sample (diluted in 15% acetic acid) onto a RP-HPLC column (microBondapak C18 column, 0.39 × 30 cm, 10^{-5} m particle size) connected to a single pump liquid chromatograph.
2. Chromatograph the radioactive samples using a mixture of 0.1% trifluoroacetic acid (TFA) in water (solvent A) and 0.1% TFA in acetonitrile (solvent B) with a flow rate of 1 ml/min and an isocratic elution of 28% solvent B (70 min).
3. Monitor eluates online for radioactivity using a γ-detector.
4. Collect the major radioactive components in the eluates and freeze-dry the samples.
5. Submit resuspended samples to automated Edman degradation as described by Clot et al. (1990).
6. Determine the cycle at which the radioactivity ([^{125}I]-Tyr) is released and identify subsequently the amino-terminal residue of radiolabeled intermediate.

The radiosequencing procedure gives only limited information about the cleavage sites of insulin since the peptide bonds cleaved beyond the radiolabeled tyrosine (Tyr^{A14}, Tyr^{A19}, Tyr^{B16}, or Tyr^{B26}) are not identified.

2.3. Reverse-phase HPLC assay using native insulin

The most appropriate substrate for assaying insulin proteolysis is native insulin that contains the physiological prerequisites for protease recognition.

Using native HI as the substrate and RP-HPLC connected to mass spectrometer to purify and characterize insulin intermediates, the following protocol can be used:

1. Load the hydrolytic sample (diluted in 15% acetic acid) onto a RP-HPLC column (microBondapak C18 column, 0.39 × 30 cm, 10^{-5} m particle size) connected to a double pump liquid chromatograph.
2. Chromatograph the samples using a mixture of 0.1% TFA in water (solvent A) and 0.1% TFA in acetonitrile (solvent B) with a flow rate of 1 ml/min and the following elution: two sequential linear gradients of 0–15% solvent B (5 min) and 15–39% solvent B (32 min) followed by an isocratic elution of 39% solvent B (13 min).
3. Monitor eluates online for absorbance at 214 nm with a spectrophotometer.
4. Collect the major components in the eluates.
5. Dilute components at a final concentration of 5–10 pmol/μl in 20% CH_3CN containing 0.1% HCOOH.
6. Analyze components using ion spray mass spectrometry as described by Authier et al. (2001) to determine the molecular mass of peptide products and cleavage sites in the insulin molecule.

3. ASSAY FOR ENDOSOMAL PROTEOLYSIS OF INSULIN

3.1. *In vivo* studies

With earlier observations that TCA-soluble radioactive products of [^{125}I]-insulin were found in rat hepatic endosomes various times after injection of radiolabeled insulin to rats (Hamel, Posner, Bergeron, Frank, & Duckworth, 1988), the hypothesis that the endosomal apparatus represent a physiological site of degradation of internalized insulin was elaborated. This was supported by subsequent studies using HPLC analysis of radiolabeled endosomal contents various times after injection of mono-iodinated insulin isomers (Clot et al., 1990; Hamel et al., 1988; Seabright & Smith, 1996). This technique has enabled an examination of the kinetics of insulin degradation intermediates produced by the action of endosomal proteases and allowed the detection of three major radioactive degradation products containing an intact A-chain. At a later stage, insulin is degraded to its constituent amino acids within endosomes, and iodotyrosine is the ultimate radiolabeled degradation product able to cross the endosomal membrane and escape the endosome (Seabright & Smith, 1996).

The following protocol can be used:

1. Inject [^{125}I]Tyr^{A14} insulin (5–20 pmol in 0.4 ml of 0.15 *M* NaCl) into the penile vein of fasted rats (body weight 180–220 g) under light anesthesia with ether.
2. Kill rats 2–90 min following insulin injection.
3. Remove the liver and homogenize the tissue in ice-cold isotonic 0.25 *M* sucrose (about 5 ml/g liver) using a Potter-Elvehjem homogenizer (six up-and-down strokes of a motor-driven Teflon pestle).
4. Prepare the endosomal fraction (EN) from the microsomal (P) or combined light mitochondrial and microsomal (L + P) fractions by discontinuous sucrose density gradient centrifugation as described in Desbuquois, Béréziat, Authier, Girard, and Burnol (2008) and Desbuquois et al. (1990). All fractionation procedures, unless otherwise is indicated, are carried out at 4 °C.
5. Collect the EN (density, 1.03–1.16 g/cm^3) at the 0.25–1.0 *M* sucrose interface.
6. Measure insulin proteolysis using the TCA precipitation assay of radioactive insulin by following the procedure described in Section 2.1.
7. Measure insulin proteolysis and characterize radioactive insulin degradation intermediates using the HPLC assay of radioactive insulin by following the procedure described in Section 2.2.

3.2. Cell-free endosome studies

The endosomal proteolysis of internalized insulin was accurately characterized using intact cell-free endosomes. EN were isolated after intravenous injection of [^{125}I] insulin into rats (when most of the internalized hormone should be located within hepatic endosomes; see procedure described in Section 3.1, Steps 1–5) and subsequently incubated in isotonic buffered medium (Desbuquois et al., 1990; Doherty, Kay, Lai, Posner, & Bergeron, 1990). In these *in vitro* conditions, the endosomes remain intact, maintain intraluminal proteolytic capacity, and degrade previously internalized insulin. The degradation of insulin in cell-free endosomes is maximal at pH 5.5 and functionally linked to ATP-dependent endosomal acidification. Thus, ATP stimulates endosomal insulinase activity at neutral pH by promoting endosome acidification as judged by acridine orange uptake and fluorescence of internalized fluorescein-labeled dextran (Desbuquois et al., 1990). The ability of ATP to stimulate endosomal degradation of insulin does not occur if the isotonic medium was depleted of Cl$^-$ (as expected from the electrogenic properties of the vacuolar proton pump) or supplemented with weak bases (chloroquine), proton ionophore, or inhibitor of the vacuolar type H^+-ATPase. It was also proposed that inhibiting the dissociation of insulin from its receptor in endosomes reduces insulin degradation, suggesting that free luminal insulin may represent the preferred substrate for the endosomal acidic insulinase(s). These conclusions were supported by studies of chloroquine injection into rats that was accompanied by a net increase in the insulin content (reduced degradation) within endosomal vesicles (Desbuquois et al., 1990). Finally, endosomal insulin degradation intermediates generated in cell-free endosomes under conditions that promote endosomal acidification are similar to those identified *in vivo* (Clot et al., 1990; Seabright & Smith, 1996).

The following protocol can be used:

1. Prepare EN containing internalized radioactive insulin by following the procedure described in Section 3.1 up to Step 5 (rats should be killed at an early postinjection time of 2–10 min).
2. Suspend EN at 1 mg/ml in 0.15 *M* KCl containing 25 m*M* citrate-phosphate buffer, pH 4.5–7.5, in the presence or absence of various compounds (ATP, protease inhibitors, weak bases, proton ionophores, proton pump inhibitors, polypeptides, etc.).
3. Incubate samples at 37 °C.
4. Measure insulin proteolysis using the TCA precipitation assay of radioactive insulin by following the procedure described in Section 2.1.
5. Measure insulin proteolysis and characterize radioactive insulin degradation intermediates using the HPLC assay of radioactive insulin by following the procedure described in Section 2.2.

3.3. *In vitro* studies

A pure *in vitro* assay based on measurement of the initial degradation step of iodinated (Authier et al., 2001; Authier, Rachubinski, Posner, & Bergeron, 1994) or native insulin (Authier et al., 2002) by purified endosomal lysates allowed for the rigorous screening of endosomal insulinase activity (endosomal distribution, affinity for insulin peptides, sensitivity to pH, protease inhibitors and ions, etc.). The endosomal insulinase activity displayed a broad acidic optimal pH of 4–5.5 (Authier, Rachubinski, et al., 1994). Hypotonic shock released greater than 80% of endosomal acidic insulinase activity indicating that acidic insulinase is a luminal soluble entity (Authier et al., 2001). It was strongly inhibited (>95%) by pepstatin A, an inhibitor of aspartic acid protease. It displayed a 20–40 μ*M* IC_{50} value for native insulin, an IC_{50} parameter close to the endosomal concentration of free insulin (related to the small volume of endosomes, $\sim10^{-17}$ l; Doherty et al., 1990) (Table 1.1).

Table 1.1 Insulin-degrading activities associated with hepatic endosomes

	Cathepsin D	Insulin-degrading enzyme	Neutral Arg aminopeptidase
Class of protease	Aspartic acid protease	Metalloprotease	Aminopeptidase and metalloprotease
Optimum pH	4	7	7
Molecular weight	45 kDa (rat), 31 kDa (human)	110 kDa	n.d.
Proteolytic activity	Endopeptidase	Endopeptidase	Arg aminopeptidase
Proform in endosomes	Yes (rat, 68 kDa; human, 45 kDa)	No	n.d.
Endosomal distribution	Lumen	Lumen? Cytosolic side of the endosomal membrane?	n.d.
Inhibitors	Pepstatin A	Chelating reagents	Bestatin and chelating reagents
Specificity for insulin peptides	Human insulin (HI) [Arg^{A0}]-HI (monoarginyl insulin) [Arg^{B31}–Arg^{B32}]-HI (diarginyl insulin) [His^{A8},His^{B4},Glu^{B10}, His^{B27}]-HI (H2 analog) [Glu^{A13},Glu^{B10}]-HI [Asp^{B10}]-HI	Human insulin (HI) [His^{A8},His^{B4}, Glu^{B10},His^{B27}]-HI (H2 analog) [Glu^{A13},Glu^{B10}]-HI [Asp^{B10}]-HI	[Arg^{A0}]-HI (monoarginyl insulin)
Primary cleavages within insulin peptides	Phe^{B24}–Phe^{B25}, Phe^{B25}–Tyr^{B26}	Phe^{B24}–Phe^{B25}, Phe^{B25}–Tyr^{B26}	Arg^{A0}–Gly^{A1}
Affinity binding for insulin peptides	HI: ~20 μ*M* Monoarginyl insulin: ~30 μ*M* Diarginyl insulin: ~10 μ*M* H2 analog: ~20 μ*M* [Glu^{A13},Glu^{B10}]-HI: ~20 μ*M* [Asp^{B10}]-HI: ~10 μ*M*	HI: ~0.05 μ*M*	n.d.

Table 1.1 Insulin-degrading activities associated with hepatic endosomes—cont'd

	Cathepsin D	Insulin-degrading enzyme	Neutral Arg aminopeptidase
References	Authier et al. (2001), Authier, Di Guglielmo, Danielsen, and Bergeron (1998), Authier, Merlen, Amessou, and Danielsen (2004), and Authier et al. (2002)	Hamel, Mahoney, and Duckworth (1991)	Kouach, Desbuquois, and Authier (2009)

The major biochemical characteristics of well-defined (cathepsin D), undefined (neutral Arg aminopeptidase), or potential (insulin-degrading enzyme) endosomal insulinase activities are presented. The identity of neutral Arg aminopeptidase is still unknown (Kouach et al., 2009). The role of insulin-degrading enzyme in endosomal proteolysis of internalized insulin remains controversial (Authier, Posner, Bergeron, 1994; Authier et al., 1995; Authier, Posner, & Bergeron, 1996a, 1996b).
n.d., not determined.

Endosomal lysates processed nonradiolabeled insulin to the major inactive A^{1-21}–B^{1-24} insulin intermediate resulting from a major cleavage at residues Phe^{B24}–Phe^{B25} bond (Authier et al., 2002) (see Section 4).

The following protocol can be used:

1. Prepare EN from untreated rats by following the procedure described in Section 3.1 (Steps 3–5).
2. Freeze/thaw EN in 5 m*M* Na phosphate, pH 7.4, and disrupt the fraction in the same hypotonic medium with a small Dounce homogenizer (15 strokes with the tight Type A pestle).
3. Centrifuge at $300{,}000 \times g_{av}$ for 30 min and collect the supernatant (soluble endosomal extract, ENs).
4. Incubate ENs (~1 μg) for varying lengths of time at 37 °C with 10 μg insulin peptides in 20 μl of 25 m*M* citrate-phosphate buffer, pH 4, containing 6 m*M* $CaCl_2$ in the presence or absence of protease inhibitors or other compounds.
5. Stop the proteolytic reaction by the addition of acetic acid (15%). Measure insulin proteolysis and characterize unlabeled insulin degradation intermediates using the HPLC assay by following the procedure described in Section 2.3.
6. Alternatively or in combination with Steps 4 and 5, substitute native insulin for $[^{125}I]Tyr^{A14}$ insulin (~20 fmol) and measure insulin proteolysis using the TCA precipitation assay of radioactive insulin by following the procedure described in Section 2.1.

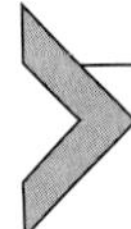

4. SITES OF CLEAVAGE AND DEGRADATION PRODUCTS OF INSULIN WITHIN ENDOSOMES

Despite the limitation of the use of radiolabeled insulin (see in the preceding text) an accurate examination of the major radiolabeled degradation peptides has been possible using monoiodinated insulin isomers ([^{125}I] Tyr^{A14} insulin) combined with RP-HPLC and radiosequencing procedures (Clot et al., 1990; Hamel et al., 1988; Seabright & Smith, 1996). Using this approach, endosome-associated degradation products extracted at an early time of endocytosis displayed intact A-chain and cleavages in the B-chain at Phe^{B24}–Phe^{B25}, Gly^{B23}–Phe^{B24}, Tyr^{B16}–Leu^{B17}, and Ala^{B14}–Leu^{B15} bonds (Fig. 1.1A). Kinetic analyses of *in vivo* endosomal intermediates have allowed the determination of an ordered sequence involving an initial cleavage in the B-chain at Phe^{B24}–Phe^{B25} bond, which produces des (hexapeptide)insulin and the carboxy-terminal B-chain Phe^{B25}–Thr^{B30}.

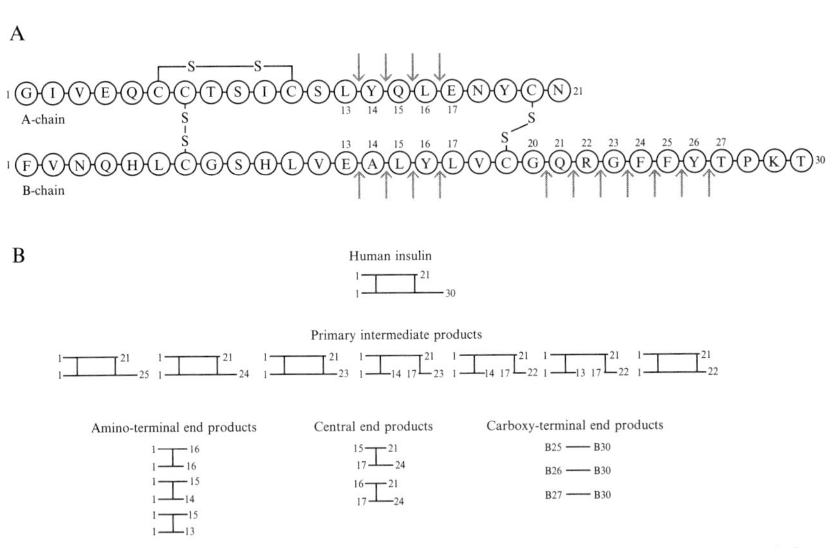

Figure 1.1 Cleavage sites and structure of the cleavage products generated from human insulin within hepatic endosomes. (A) Schematic representation of the primary structure of human insulin (HI). Amino acid residues are depicted in a single-letter code. The arrows show cleavage sites generated upon incubation of HI with a soluble endosomal extract (ENs) isolated from rat liver, partially purified endosomal acidic insulinase (EAI) activity or cathepsin D. (B) Structure of HI intermediates generated by hepatic endosomes, partially purified EAI activity or cathepsin D. *Adapted from Authier et al. (2001, 2002).* (For color version of this figure, the reader is referred to the online version of this chapter.)

This proteolytic process has been confirmed by a complete identification of the early insulin intermediates, which has been achieved using a RP-HPLC/mass spectrometry assay based on measurement of the initial degradation step of native human insulin (HI) by endosomal lysates (Authier et al., 2001, 2002). Proteolysis of native HI by endosomal acidic insulinase (EAI) results in one cleavage in the A-chain and eight in the B-chain. The endosomal protease cleaves the A-chain at the Gln^{A15}–Leu^{A16} bond, while the B-chain cleavages occur in two main regions of the polypeptide, one comprising the carboxy-terminal part Arg^{B22}–Thr^{B27} (five sites) and the second appearing in the central region Glu^{B13}–Leu^{B17} (three sites) (Authier et al., 2001). The kinetic data have shown that the degradation steps proceed *via* the following ordered sequential pathway: two primary end products of HI result from proteolytic cleavages occurring at residues Phe^{B24}–Phe^{B25} (major pathway) and Phe^{B25}–Phe^{B26} (minor pathway) (Fig. 1.1). This is followed by the sequential removal of the carboxy-terminal Phe^{B24} and Gly^{B23} with the concomitant release of residues Ala^{B14}, Leu^{B15}, and Tyr^{B16} from the central region of the B-chain. The A-chain is then processed by a cleavage occurring at the Gln^{A15}–Leu^{A16} bond (Fig. 1.1).

Analysis of the nine cleavage sites showed a preference for hydrophobic and aromatic amino acid residues on both the carboxy and amino sides of a cleaved peptide bond. A crucial role for the carboxy-terminal region of the B-chain in the high-affinity recognition of EAI was supported by the competition studies, which revealed that the B-chain alone and the Arg^{B22}–Ala^{B30} peptide displayed a 30-fold increase in the affinity for EAI and a two- to sixfold increase in the rate of hydrolysis compared with native insulin (Authier et al., 2001). Within this carboxy-terminal region of the insulin B-chain, the aromatic locus Phe^{B24}–Phe^{B25}–Tyr^{B26} appear to confer high-affinity binding of insulin to EAI (Authier et al., 2001, 2002). Interestingly, the same region of insulin has been implicated in the high-affinity binding of the hormone to IR. However, other structural determinants required for binding to the IR and EAI should be distinct since insulin analogs that display an increase in binding affinity to IR showed variable rates of processing from high to low (Authier et al., 1998, 2001). Moreover, the insulin B-chain, which displays a complete loss of IR affinity, had one of the highest affinities for EAI and the highest rate of proteolysis by the protease. Clearly, there is a fundamental difference between the mechanisms by which the endosomal protease and the IR recognize the insulin molecule.

5. ENDOSOMAL INSULINASES

5.1. Cathepsin D

The pepstatin A-sensitive cathepsin D is an aspartic acid endopeptidase that is localized in the endolysosomal compartment of cells and induces the proteolytic cleavage of a wide variety of endosomal substrates at acidic pH, such as polypeptide hormones, proteins, and plant or bacterial toxins (Authier et al., 1996a; El Hage, Lorin, Decottignies, Djavaheri-Mergny, & Authier, 2010; El Hage, Merlen, Fabrega, & Authier, 2007; Merlen et al., 2005). In conjunction with the thiol-dependent cysteine protease cathepsin B, cathepsin D has been shown to be responsible for the majority of proteolytic activities within endosomes (Authier et al., 1996a). Its role in the endosomal processing of internalized insulin has been demonstrated through a series of biochemical (using rat liver endosomes) and morphological (using immunofluorescence of primary hepatocyte culture) investigations (Authier et al., 2002). Using confocal immunofluorescence microscopy, a partial colocalization of internalized insulin with cathepsin D, and cathepsin D with the early endocytic marker EEA1, has been demonstrated (Authier et al., 2002), confirming the physiological presence of both the protease and substrate at the same cellular locus. Sequences cleaved by cathepsin D contain apolar or hydrophobic residues situated on both sides of the cleavage site, with the Phe–Phe bond strongly favored and, in most cases, a hydrophobic residue in position P_1 (Table 1.1). This cleavage specificity conforms well to the cleavage pattern for HI that was generated using the endosomal proteolytic activity (see Section 4) (Authier et al., 2001, 2002).

5.2. Neutral Arg aminopeptidase

An endosomal Arg convertase, sensitive to bestatin and potentially related to aminopeptidase B-type enzyme, has been involved in the endosomal removal of Arg residues from internalized monoarginyl insulin prior to organelle acidification (Table 1.1; Kouach et al., 2009). The action of neutral Arg aminopeptidase is highly selective towards the [Arg^{A0}]-human insulin (HI) peptide, a proinsulin intermediate containing an additional Arg at the amino-terminal insulin A-chain (Table 1.1). Thus, no degradation products were observed with other proinsulin intermediates ([Arg^{B31}–Arg^{B32}] insulin nor with HI itself (which contains Arg residue at the B^{22} position)) (Fig. 1.1).

5.3. Insulin-degrading enzyme

It has been proposed that ATP-dependent acidification of endosomes was not required for the initial step of insulin degradation and that insulin-degrading enzyme (IDE), a neutral endopeptidase, may operate in the endosomal apparatus before the progressive decrease in endosomal pH (Hamel et al., 1991). IDE is an evolutionarily conserved neutral thiol-dependent metallo-endopeptidase of 110 kDa that was first postulated to be responsible for insulin proteolysis *in vivo* almost 55 years ago (Table 1.1; Authier et al., 1996b). The major arguments for the implication of IDE in endosomal proteolysis of insulin are (a) the extremely low K_m value of IDE for insulin (originally supporting the hypothesis that the enzyme is specific to insulin), (b) the primary sites of cleavage of internalized insulin are nearly comparable to those seen with purified IDE, and (c) a small level of IDE was detected by immunoblotting in liver EN (Table 1.1). However, in this study (Hamel et al., 1991), no comparison was made with IDE present in other liver subcellular fractions suggesting that the endosome-associated IDE may have derived from organelle contamination (Authier, Rachubinski, et al., 1994). Whatever the physiological significance of the endosomal pool of IDE described in Hamel et al. (1991), significant progress in understanding the physiological enzymes for insulin degradation has toned down this postulated role, mainly based on the fact that (a) overexpression of IDE has not unanimously been found to modify the rate of insulin degradation in intact cells and (b) IDE displays a major dual cytoplasmic and peroxisomal location and is not readily available for internalized insulin that is mainly located within endosomes (Authier et al., 1995; Authier, Cameron, & Taupin, 1996). Thus, the potential biological significance of the proteolytic activity of IDE towards insulin *in vivo* has not yet been established.

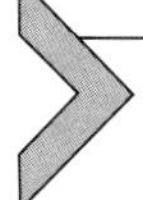

6. EFFECTS OF INSULIN AND SLOWLY PROCESSED INSULIN ANALOGS ON INSULIN SIGNALING IN LIVER ENDOSOMES

6.1. Endosomal IR tyrosine phosphorylation

The tyrosine phosphorylation state of the IR reflects a balance between its intrinsic tyrosine kinase activity and the action of protein tyrosine phosphatases, which result, respectively, in receptor autophosphorylation and dephosphorylation. Early assays of IR phosphorylation and dephosphorylation involved receptor ^{32}P labeling procedures combined with receptor immunoprecipitation. This allowed the first demonstration of *in vitro* (Khan et al., 1989) and *in vivo* (Burgess et al., 1992) insulin-induced tyrosine phosphorylation of the IR in liver endosomes, as well as *in vitro* dephosphorylation of the phosphorylated IR by endosomal PY phosphatases (Faure, Baquiran, Bergeron, & Posner, 1992). These observations were soon confirmed by immunoblotting procedures using anti-IR and anti-PY antibodies, which allow an easier and safer assessment of the tyrosine phosphorylation state of the IR.

As reported previously (Balbis, Baquiran, Bergeron, & Posner, 2000; Burgess et al., 1992) and confirmed here (Fig. 1.2), *in vivo* insulin treatment resulted in a rapid but transient increase in total and tyrosine-phosphorylated IR content in liver ENs, with a maximum by 2 min. The PY content of the IR as determined from densitometric analyses of IR and phospho-IRβ autoradiographic signals first abruptly increased and then more slowly decreased (Burgess et al., 1992). Relative to the plasma membrane-associated receptor, the endosomal receptor showed a lesser increase in PY content in response to insulin but a paradoxical, greater increase in autophosphorylating and exogenous IR kinase activities, which was attributed to a dephosphorylation-dependent kinase activation (Burgess et al., 1992).

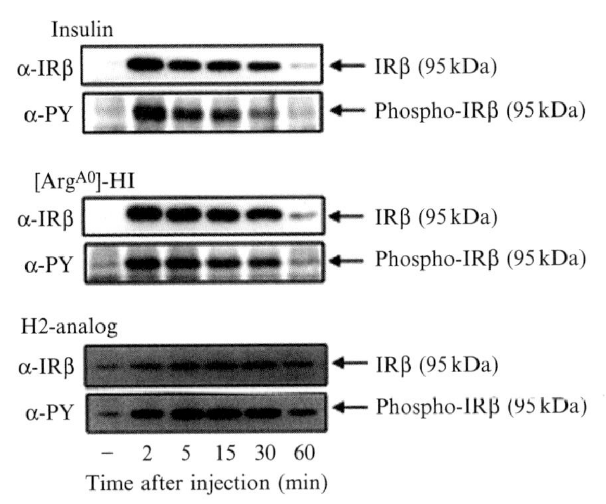

Figure 1.2 *In vivo* effects of insulin, [Arg^{A0}]-HI, and H2 analog on IR and phosphorylated IR association with liver endosomes. Endosomes were isolated from rats killed at the indicated times after injection of insulin, [Arg^{A0}]-HI, or H2 analog (30 μg each). They were evaluated for their content in IR and phospho-IR β-subunit by immunoblotting using polyclonal anti-IR and monoclonal anti-PY antibodies, respectively. Endosomes from control rats were prepared from saline-injected rats. *Adapted from Authier et al. (2004) and Kouach et al. (2009).*

Although the activated IR is known to remain tyrosine-phosphorylated in the absence of insulin, following *in vivo* activation endosomal IR dephosphorylation parallels endosomal insulin proteolysis. Furthermore, comparative studies with receptor-saturating doses of insulin and insulin analogs indicate that these two events are functionally related. Thus, the high-affinity insulin analog [His^{A8}, His^{B4}, Glu^{B10}, HisB^{24}] insulin (H2 analog) (Authier et al., 1998, 2001, 2004) and the low-affinity insulin precursor proinsulin (Desbuquois, Chauvet, Kouach, & Authier, 2003), which are both more slowly processed than insulin in endosomes, increase IR tyrosine phosphorylation to a higher level and/or for a longer duration than insulin, as does [Arg^{A0}]-HI, which is converted to insulin within endosomes prior to degradation (Kouach et al., 2009). Consistent with these observations, *in vivo* administration of the cathepsin D inhibitor pepstatin A (Authier et al., 2002) and the acidotropic agent chloroquine (Bevan et al., 1997), which both inhibit endosomal insulin degradation, also extends the lifetime of the tyrosine-phosphorylated IR in endosomes.

We describe here the immunoblotting procedure used to assess the endosomal content of IR and tyrosine-phosphorylated IR in rats treated with insulin, H2-insulin analog, and [Arg^{A0}]-HI. As shown on Fig. 1.2, relative to insulin, the two latter analogs increase the residence time of the IR in endosomes and induce a higher and more prolonged tyrosine phosphorylation of IR-β-subunit.

6.1.1 Procedure

1. Inject insulin, H2 analog, or [Arg^{A0}]-HI (30 μg in 0.4 ml of 0.15 *M* NaCl) into the penile vein of fasted rats (body weight 180–220 g) under light anesthesia with ether.
2. Kill rats, remove and homogenize the liver, and prepare EN containing internalized ligand by following the procedure described in Section 3.1 (Steps 2–5). Protease (1 m*M* benzamidine and 0.5 m*M* PMSF) and phosphatase (2.5 m*M* NaF and 1 m*M* Na_3VO_4) inhibitors may be included in the homogenization and fractionation media.
3. Suspend EN in Laemmli sample buffer (62.5 m*M* Tris–HCl, pH 6.8, 2% SDS, 10% glycerol, and 2% β-mercaptoethanol), heat at 95 °C for 5–15 min and subject to SDS-PAGE on 8–10% acrylamide gels.
4. Transfer electrophoresed samples to nitrocellulose membrane (0.45 μm) for 60 min at 380 mA in buffer containing 25 m*M* Tris base and 192 m*M* glycine.

5. Block the membranes by an overnight incubation with 5% skim milk (for IR immunoblotting) or 2% BSA (for PY immunoblotting) in 10 m*M* Tris–HCl, pH 7.5, 300 m*M* NaCl, and 0.05% Tween 20 (TNT (Tris/NaCl/Tween) buffer).
6. Incubate the membranes with affinity-purified polyclonal antibody to IR (Upstate Biotechnology Inc. or Santa Cruz Biotechnology) diluted 1:1000 in 5% skim milk or monoclonal anti-PY antibody (clone 4G10, Upstate Biotechnology Inc.) diluted 1:2500 in TNT buffer for 1–2 h at room temperature.
7. Wash the membranes three times with 0.5% skim milk (IR immunoblotting) or 0.2% BSA (PY immunoblotting) in TNT buffer over a period of 1 h at room temperature.
8. Incubate the washed membranes with HRP (horseradish peroxidase)-conjugated antirabbit IgG (IR immunoblotting) or goat antimouse IgG (PY immunoblotting) for 1 h at room temperature.
9. Reveal immune complexes using an enhanced chemiluminescence detection kit (Pierce) and expose membranes to Kodak X-Omat films.
10. Quantify autoradiographic signals by scanning laser densitometry and calculate the ratio phosphorylated IR-β/IR-β.

6.2. Translocation of insulin-signaling proteins to endosomes

Major insulin signaling proteins, including the IR substrates IRS1 and IRS2, the p85 and p110 subunits of phosphoinositide 3-kinase (PI3K), and the Akt/PKB serine kinase have been shown by immunoprecipitation and immunoblotting to be compartmentalized in liver and to undergo insulin-dependent recruitment to plasma membrane and endosomal compartments (Balbis et al., 2000; Drake, Balbis, Wu, Bergeron, & Posner, 2000; and references therein). Besides the plasma membrane, endosomes have been identified as a major site at IRS proteins undergo insulin-dependent tyrosine phosphorylation and association with the p85 subunit of PI3K, and at which Akt/PKB is serine-phosphorylated and activated (Balbis et al., 2000). In addition to IR substrates, two negative regulators of IR catalytic activity known to interact with the phosphorylated IR in intact cells, the protein tyrosine phosphatase PTP1B (Li et al., 2006) and the molecular adaptor Grb14 (Desbuquois et al., 2008), have been identified in liver endosomes. However, although involved in IR dephosphorylation *in vivo*, PTP1B does not appear to be responsible for *in vitro* dephosphorylation of the IR in hepatic endosomes (Li et al., 2006).

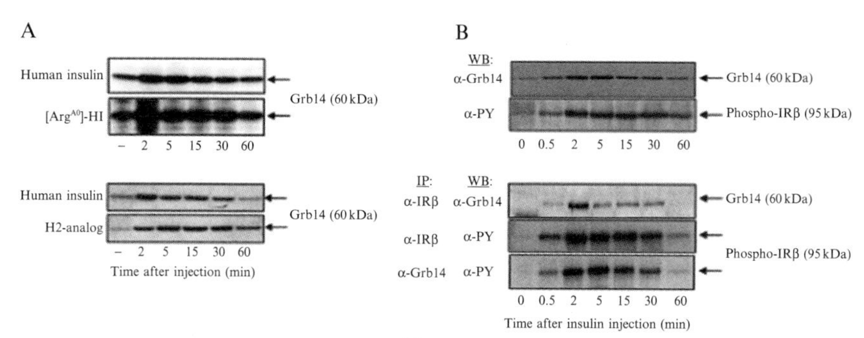

Figure 1.3 *In vivo* effects of insulin, [Arg^{A0}]-HI, and H2 analog on Grb14 translocation to liver endosomes and assessment of insulin-dependent association of Grb14 with phosphorylated IR. (A) Endosomes were isolated at the indicated times after *in vivo* administration of insulin and insulin analogs (30 μg each) and evaluated for content of Grb14 by immunoblotting using polyclonal anti-Grb14 antibody. (B) Endosomes were isolated from rats killed at the indicated times after insulin injection (30 μg). Endosomal lysates were prepared and subjected to immunodetection using polyclonal anti-Grb14 and monoclonal anti-PY antibodies, either directly (WB) or after previous immunoprecipitation (IP) using monoclonal anti-IR-β (α-IRβ) or polyclonal anti-Grb14 antibody. *Panel (A) data adapted from Desbuquois et al. (2008) and Kouach et al. (2009); panel (B) data adapted from Desbuquois et al. (2008).*

We describe here the procedures used to assess the translocation of Grb14 to liver endosomes following administration of insulin, H2 insulin analog, and [Arg^{A0}]-HI to rats (Fig. 1.3A) and the insulin-induced interaction of Grb14 with the tyrosine-phosphorylated IR (Fig. 1.3B). In these procedures, endosomal lysates are subjected to immunoblotting using anti-Grb14 antibodies and to immunoprecipitation using anti-Grb14 and anti-IR antibodies, following which phospho-IRβ and Grb14 in immunoprecipitates are detected using anti-PY and anti-Grb14 antibodies, respectively. As shown on Fig. 1.3, insulin, the H2 analog, and [Arg^{A0}]-HI induce each a time-dependent translocation of Grb14 with endosomes, and consistent with their sustained effects on IR phosphorylation, the two latter analogs induce a greater and more sustained association of Grb14 with endosomes than does insulin. In addition, at least with insulin, an association of Grb14 with the phosphorylated IR is detectable.

6.2.1 Procedure

1. Inject rats with insulin, H2 analog, or [Arg^{A0}]-HI (30 μg in 0.15 *M* NaCl) as described in Section 6.1 (Step 1).
2. Kill rats, remove and homogenize the liver, and prepare EN by following the procedure described in Section 3.1 (Steps 2–5).

3. For direct immunoblot analysis of Grb14, first proceed as described for IR immunoblotting in Section 6.1 (Steps 3–4). Then, block the membranes by overnight incubation with 5% skim milk in TNT buffer at 4 °C, incubate the membranes with affinity-purified rabbit anti-Grb14 antibody diluted 1:2000 in milk/TNT buffer for 2–3 h at room temperature, wash the membranes in TNT buffer three times over 1 h at room temperature, incubate the washed membranes with HRP-conjugated antirabbit IgG in milk/TNT buffer for 1 h at room temperature, and reveal immune complexes as described in Section 6.1 (Step 9). For Grb14 and IR immunoprecipitation, proceed as described in the succeeding text (Steps 4–8).
4. Dilute EN in 25 m*M* HEPES buffer, pH 7.6, containing 0.5–1% Triton X-100, 1 m*M* Na_3VO_4, 2.5 m*M* NaF, 1 m*M* PMSF, 1 m*M* benzamidine, 1 μg/ml pepstatin A, 2 μg/ml leupeptin, and 0.25 mg/ml aprotinin. After 20 min at 4 °C, centrifuge the medium at 150,000 × g_{av} for 60 min.
5. Incubate supernatants with monoclonal anti-IR-β antibody (Santa Cruz Biotechnology Inc.) diluted 1:100 or polyclonal anti-Grb14 antibody diluted 1:200 for 16 h at 4 °C.
6. Add protein G–Sepharose (IR immunoprecipitation) or protein A–agarose (Grb14 immunoprecipitation) and incubate for 2 h at 4 °C with rotatory shaking.
7. Wash four times Sepharose or agarose beads by low-speed centrifugation in 25 m*M* HEPES buffer, 100 m*M* NaCl, 1 m*M* Na_3VO_4, and 0.1% Triton X-100.
8. Resuspend beads in Laemmli sample buffer and subject immunoprecipitates to SDS-PAGE and immunoblotting using anti-Grb14 (IR-β immunoprecipitates) or anti-IR-β and anti-PY (Grb14 immunoprecipitates) as described in the preceding text (Step 3) and in Section 6.1 (Steps 3–10).

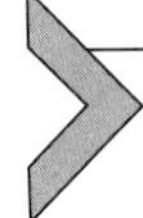

7. INVOLVEMENT OF IR ENDOCYTOSIS IN POSITIVE REGULATION OF INSULIN SIGNALING

The studies described in the preceding text clearly show that cathepsin D-induced insulin proteolysis (Baass, Di Guglielmo, Authier, Posner, & Bergeron, 1996), in conjunction with PY phosphatase-dependent IR dephosphorylation, limit the temporal window of activation of the internalized IR in endosomes and thus modulate negatively insulin signaling. There is also evidence that Grb14 recruited by the activated IR in liver endosomes exerts an inhibitory effect on IR catalytic activity at this locus (Desbuquois et al., 2008). However, under certain circumstances, IR endocytosis and endosome-associated events may be involved in an opposite, positive regulation of insulin signaling. For instance, inhibition of clathrin-dependent endocytosis attenuates insulin-stimulated Shc phosphorylation and ERK1/2 activation in hepatoma cells and 3T3-L1 adipocytes (Ceresa, Kao, Santeler, & Pessin, 1998), and depletion of the endosomal protein WDFY2 impairs insulin-induced Akt activation in 3T3-L1 adipocytes (Walz et al., 2010). Also, *in vivo* studies have shown that the PY phosphatase inhibitor bisperoxovanadium in the presence of a selective activation of hepatic endosomal signaling induces a hypoglycemic action (Bevan et al., 1995).

REFERENCES

Authier, F., Bergeron, J. J. M., Ou, W. J., Rachubinski, R. A., Posner, B. I., & Walton, P. A. (1995). Degradation of the cleaved leader peptide of thiolase by a peroxisomal proteinase. In: *Proceedings of the National Academy of Sciences of the United States of America, 92*, 3859–3863.

Authier, F., Cameron, P. H., & Taupin, V. (1996). Association of insulin-degrading enzyme with a 70 kDa cytosolic protein in hepatoma cells. *The Biochemical Journal, 319*, 149–158.

Authier, F., Danielsen, G. M., Kouach, M., Briand, G., & Chauvet, G. (2001). Identification of insulin domains important for binding to and degradation by endosomal acidic insulinase. *Endocrinology, 142*, 276–289.

Authier, F., Di Guglielmo, G. M., Danielsen, G. M., & Bergeron, J. J. M. (1998). Uptake and metabolic fate of [HisA8, HisB4, GluB10, HisB27]insulin in rat liver in vivo. *The Biochemical Journal, 332*, 421–430.

Authier, F., Merlen, C., Amessou, M., & Danielsen, G. M. (2004). Use of high affinity insulin analogues to assess the functional relationships between insulin receptor trafficking, mitogenic signaling and mRNA expression in rat liver. *Biochimie, 86*, 157–166.

Authier, F., Métioui, M., Fabrega, S., Kouach, M., & Briand, G. (2002). Endosomal proteolysis of internalized insulin at the C-terminal region of the B chain by cathepsin D. *The Journal of Biological Chemistry, 277*, 9437–9446.

Authier, F., Posner, B. I., & Bergeron, J. J. M. (1994). Hepatic endosomes are a major physiological locus of insulin and glucagon degradation *in vivo*. In A. Ciechanover, & A. L. Schwartz (Eds.), *Modern Cell Biology Series: Vol. 12. Cellular proteolytic system* (pp. 89–113). New York: Wiley-Liss.

Authier, F., Posner, B. I., & Bergeron, J. J. M. (1996a). Endosomal proteolysis of internalized proteins. *FEBS Letters, 389*, 55–60.

Authier, F., Posner, B. I., & Bergeron, J. J. M. (1996b). Insulin-degrading enzyme. *Clinical and Investigative Medicine, 19*, 149–160.

Authier, F., Rachubinski, R. A., Posner, B. I., & Bergeron, J. J. M. (1994). Endosomal proteolysis of insulin by an acidic thiol metalloprotease unrelated to insulin-degrading enzyme. *The Journal of Biological Chemistry, 269*, 3010–3016.

Baass, P. C., Di Guglielmo, G. M., Authier, F., Posner, B. I., & Bergeron, J. J. M. (1996). Compartmentalized signal transduction by receptor tyrosine kinases. *Trends in Cell Biology, 5*, 465–470.

Balbis, A., Baquiran, G., Bergeron, J. J. M., & Posner, B. I. (2000). Compartmentalization and insulin-induced translocations of insulin receptor substrates, phosphatidylinositol 3-kinase, and protein kinase B in rat liver. *Endocrinology, 141*, 4041–4049.

Bevan, A. P., Burgess, J. W., Drake, P. G., Shaver, A., Bergeron, J. J. M., & Posner, B. I. (1995). Selective activation of the rat hepatic endosomal insulin receptor kinase. Role for the endosome in insulin signaling. *The Journal of Biological Chemistry, 270*, 10784–10791.

Bevan, A. P., Krooke, A., Tikerpae, J., Seabright, P. J., Siddle, K., & Smith, G. D. (1997). Chloroquine extends the lifetime of the activated insulin receptor in endosomes. *The Journal of Biological Chemistry, 278*, 26833–26840.

Burgess, J. W., Wada, I., Ling, N., Khan, M. N., Bergeron, J. J. M., & Posner, B. I. (1992). Decrease in β-subunit phosphotyrosine correlates with internalization and activation of the endosomal insulin receptor kinase. *The Journal of Biological Chemistry, 267*, 10077–10086.

Ceresa, B. P., Kao, A. W., Santeler, S. R., & Pessin, J. F. (1998). Inhibition of clathrin-mediated endocytosis attenuates specific insulin receptor signal transduction pathways. *Molecular and Cellular Biology, 18*, 3862–3870.

Clot, J. P., Janicot, M., Fouque, F., Desbuquois, B., Haumont, P. Y., & Lederer, F. (1990). Characterization of insulin degradation products generated in liver endosomes: In vivo and in vitro studies. *Molecular and Cellular Endocrinology, 10*, 175–185.

Desbuquois, B., Béréziat, V., Authier, F., Girard, J., & Burnol, A. F. (2008). Compartmentalization and in vivo insulin-induced translocation of the insulin-signaling inhibitor Grb14 in rat liver. *The FEBS Journal, 275*, 4363–4377.

Desbuquois, B., Chauvet, G., Kouach, M., & Authier, F. (2003). Cell itinerary and metabolic fate of proinsulin in rat liver: In vivo and In vitro studies. *Endocrinology, 144*, 5308–5321.

Desbuquois, B., Janicot, M., & Dupuis, A. (1990). Degradation of insulin in isolated liver endosomes is functionally linked to ATP-dependent endosomal acidification. *European Journal of Biochemistry, 193*, 501–512.

Doherty, J. J., II, Kay, D. G., Lai, W. H., Posner, B. I., & Bergeron, J. J. M. (1990). Selective degradation of insulin within rat liver endosomes. *The Journal of Cell Biology, 110*, 35–42.

Drake, P. G., Balbis, A., Wu, J., Bergeron, J. J. M., & Posner, B. I. (2000). Association of phosphatidylinositol with the insulin receptor: Compartmentation in rat liver. *American Journal of Physiology Endocrinology and Metabolism, 279*, E266–E274.

El Hage, T., Lorin, S., Decottignies, P., Djavaheri-Mergny, M., & Authier, F. (2010). Proteolysis of Pseudomonas exotoxin A within hepatic endosomes by cathepsins B and D produces ADP-ribosylating and apoptotic fragments. *The FEBS Journal, 277*, 3735–3749.

El Hage, T., Merlen, C., Fabrega, S., & Authier, F. (2007). Role of receptor-mediated endocytosis, endosomal acidification and cathepsin D in cholera toxin cytotoxicity. *The FEBS Journal, 274*, 2614–2629.

Faure, R., Baquiran, G., Bergeron, J. J. M., & Posner, B. I. (1992). The dephosphorylation of insulin and epidermal growth factor receptors. Role of endosome-associated phosphotyrosine phosphatases. *The Journal of Biological Chemistry, 267*, 11215–11221.

Hamel, F. G., Mahoney, M. J., & Duckworth, W. C. (1991). Degradation of intraendosomal insulin by insulin-degrading enzyme without acidification. *Diabetes, 40*, 436–443.

Hamel, F. G., Posner, B. I., Bergeron, J. J. M., Frank, B. H., & Duckworth, W. C. (1988). Isolation of insulin degradation products from endosomes derived from intact rat liver. *The Journal of Biological Chemistry, 263*, 6703–6708.

Khan, M. N., Baquiran, G., Brule, C., Burgess, J., Foster, B., Bergeron, J. J. M., et al. (1989). Internalization and activation of the rat liver insulin receptor kinase in vivo. *The Journal of Biological Chemistry, 264*, 12391–12940.

Kouach, M., Desbuquois, B., & Authier, F. (2009). Endosomal proteolysis of internalised [Arg^{A0}]-human insulin at neutral pH generates the mature insulin peptide in rat liver *in vivo*. *Diabetologia, 52*, 2621–2632.

Li, C., Baquiran, G., Gu, F., Tremblay, M. L., Fazel, A., Bergeron, J. J. M., et al. (2006). Insulin receptor kinase-associated phosphotyrosine phosphatases in hepatic endosomes: Assessing the role of phosphotyrosine phosphatase-1B. *Endocrinology, 147*, 912–918.

Merlen, C., Fayol-Messaoudi, D., Fabrega, S., El Hage, T., Servin, A., & Authier, F. (2005). Proteolytic activation of internalized cholera toxin within hepatic endosomes by cathepsin D. *The FEBS Journal*, *272*, 4385–4397.

Seabright, P. J., & Smith, G. D. (1996). The characterization of endosomal insulin degradation intermediates and their sequence of production. *The Biochemical Journal*, *320*, 947–956.

Sorkin, A., & Von Zastrow, M. (2009). Endocytosis and signaling: Intertwining molecular networks. *Nature Reviews Molecular Cell Biology*, *10*, 609–622.

Walz, H. A., Shi, X., Chouinard, M., Bue, C. A., Navaroli, D. M., Hayakawa, A., et al. (2010). Isoform-specific regulation of Akt signaling by the endosomal protein WDFY2. *The Journal of Biological Chemistry*, *285*, 14101–14108.

CHAPTER TWO

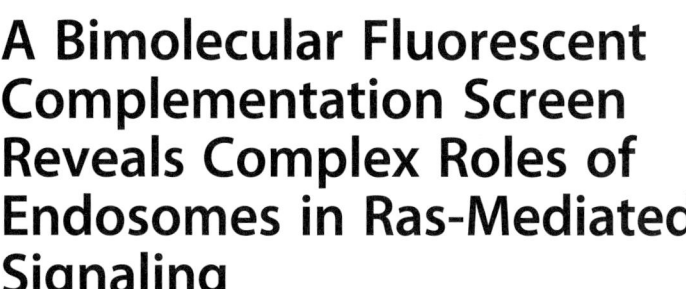

A Bimolecular Fluorescent Complementation Screen Reveals Complex Roles of Endosomes in Ras-Mediated Signaling

Ze-Yi Zheng, Eric C. Chang[1]

Lester and Sue Smith Breast Center, Department of Molecular and Cellular Biology, Baylor College of Medicine, Houston, Texas, USA

[1]Corresponding author: e-mail address: echang1@bcm.tmc.edu

Contents

Abstract

While Ras GTPases are best known for mediating growth factor signaling on the plasma membrane, these proteins also have surprisingly complex activities in the endosome.

Methods in Enzymology, Volume 535
ISSN 0076-6879
http://dx.doi.org/10.1016/B978-0-12-397925-4.00002-X

Assisted by a method called bimolecular fluorescent complementation (BiFC), which can detect weak and transient protein–protein interactions and reveal where the binding takes place in live cells, we have identified three effectors, Cdc42, CHMP6, and VPS4A that interact with Ras proteins in endosomes. These effectors are all necessary for Ras-induced transformation, suggesting that for Ras proteins to efficiently induce tumor formation, they must also activate effectors in cytoplasm, such as those in endosomes. Here, we describe how BiFC can be used to detect and screen for Ras effectors and for readily revealing where in the cell the binding occurs.

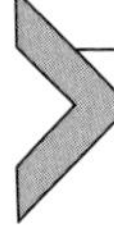

1. INTRODUCTION

1.1. Ras protein and growth factor signaling

Ras proteins cycle between the GDP- and GTP-bound states to function as binary switches in order to mediate a wide range of signaling events. The best-known activity of the Ras proteins is mediating extracellular proliferative signals controlled by receptor tyrosine kinases (RTKs) in response to growth factors, such as EGF (epidermal growth factor). Deregulations in these signaling pathways have been linked to the development of many human cancers (Pylayeva-Gupta, Grabocka, & Bar-Sagi, 2011). Evidently, the growth factor signaling initiates at the plasma membrane to activate Ras proteins; however, what happens next to these Ras proteins on the plasma membrane is poorly understood.

1.2. RTK and Ras internalization

Internalizing receptors with seven transmembrane helices via endocytosis is a well-established mechanism for attenuating signals transduced by trimeric G proteins. As it turns out, receptor internalization appears to be a very common mechanism in reducing signaling outputs at the plasma membrane, as many RTKs, such as EGFRs (EGF receptors), are also internalized (Sigismund et al., 2012).

While Ras proteins are well known for lipidations (farnesylation and palmitoylation) at the C-terminus to influence how they associate with various membrane compartments, in a surprising study by Jura, Scotto-Lavino, Sobczyk, and Bar-Sagi (2006), H- and N-Ras proteins have been found to be also ubiquitylated via a K-63 linkage. This mode of ubiquitylation is often required for endocytosis and endosomal sorting, and indeed, we and others have found that these two Ras proteins frequently localize to the endosomes (Cheng & Chang, 2011;

Cheng et al., 2011; Fehrenbacher, Bar-Sagi, & Philips, 2009; Zheng, Cheng, et al., 2012; Zheng, Xu, Bar-Sagi, & Chang, 2012).

While many Ras proteins enter the endosomes, the complex fates of these internalized Ras proteins are just beginning to resolve. When the lysine residues in the Ras proteins are mutated (Jura et al., 2006; Zheng, Cheng, et al., 2012) or when an ubiquitin ligase responsible for Ras ubiquitylation (Rabex-5) is repressed (Xu, Lubkov, Taylor, & Bar-Sagi, 2010; Yan, Jahanshahi, Horvath, Liu, & Pfleger, 2010), Ras signaling outputs are apparently enhanced. Conversely, when Rabex-5 is overexpressed, the Ras signaling outputs are inhibited. These results suggest that Ras internalization can lead to attenuation of Ras-mediated signaling at the plasma membrane. Furthermore, one of the Ras effectors is RIN1 (Han et al., 1997), a guanine nucleotide exchange factor for Rab5, which controls endocytosis (Tall, Barbieri, Stahl, & Horazdovsky, 2001). Thus, it seems that Ras proteins can promote their own internalization by activating RIN1 in an apparent negative feedback loop. While this model explains why some of the Ras proteins are internalized, it does not appear to explain observations made by us (see in the succeeding text) and by others that endosomal Ras proteins are in fact functional and stimulate a specific set of effectors to induce transformation.

1.3. Ras interaction with Cdc42 in the endosome is required for efficient transformation

We and others have initially used the fission yeast *Schizosaccharomyces pombe* as a model system to study Ras functions (Chang et al., 1994; Chang & Philips, 2006; Fukui & Yamamoto, 1988; Onken, Wiener, Philips, & Chang, 2006). *S. pombe* has a single Ras protein, Ras1, that controls two spatially segregated pathways. Ras1 controls a protein kinase called Byr2, which activates a MAP kinase module at the plasma membrane to mediate signaling carried by the mating pheromone. Furthermore, Ras1 activates Cdc42 in the endomembrane via its quinine nucleotide exchange factor Scd1/Ral1 to mediate cell polarity to maintain an elongated cell morphology.

In a more recent study, we investigated whether the Ras–GEF–Cdc42 pathway has been conserved evolutionarily in mammalian cells (Cheng & Chang, 2011; Cheng et al., 2011). Indeed, we demonstrated that H-Ras, in a GTP-dependent manner, binds Cdc42 in the endosome, and Cdc42 is required for cellular transformation induced by oncogenic H-Ras. Activation of Ras in turn induces activation of Cdc42 apparently by recruiting and/or activating several guanine nucleotide exchange factors for Cdc42.

1.4. Using BiFC to seek and study new Ras effectors

To our knowledge, we were the first to demonstrate that Ras and Cdc42 can form a complex. We believe that many, including us, have tried using more conventional methods, such as immunoprecipitation, to measure the binding and found that it was not easy to do so because the binding is weak and transient. We overcame this difficulty by employing a technology called bimolecular fluorescent complementation (BiFC) (Hu, Chinenov, & Kerppola, 2002).

In BiFC (Hu et al., 2002), one protein is fused to an N-terminal fragment of YFP (Yn, Fig. 2.1A), while its binding partner is fused to an C-terminal fragment of YFP (Yc, Fig. 2.1A). When the proteins of interest form a complex, the fragments of the fluorescent protein also refold to restore fluorescence, which can be readily detected by both FACS (Fig. 2.1B) and fluorescence microscopy (Fig. 2.1C). The BiFC method can detect weak and transient binding because it can "capture" a steady pool of proteins

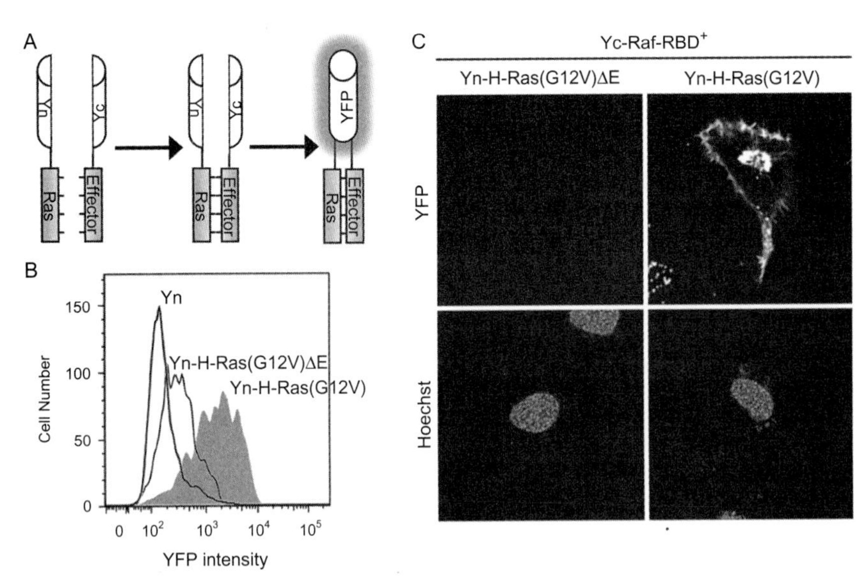

Figure 2.1 Detecting H-Ras and Raf-RBD interaction by BiFC. (A) A schematic representation of the BiFC system—see text for details. (B) Yc-tagged Raf-RBD was cotransfected with indicated Yn-tagged constructs into HT1080 cells. The transfected cells were cultured for another 36 h before being examined by FACS analysis for YFP. (C) The same cells as in (B) were examined by confocal microscopy for the reconstituted YFP signal. Cells were also stained with Hoechst 33258 to mark nuclei. Cells expressing the Yn control yielded the same results as those with H-Ras(G12V)ΔE (not shown).

that bind one another, even though dynamic and/or transient protein binding often has a high off-rate at steady state. Moreover, as the two fragments of a fluorescent protein refold, extensive H-bonding between them further stabilizes the protein complex. In addition to being very sensitive, the BiFC system can readily reveal where the binding occurs in live cells. This latter property is particularly critical for the study of compartmentalized Ras signaling.

1.5. Promoting Ras recycling and transformation by components of the ESCRT-III complex

Since there is clear evidence that Ras proteins can interact with Cdc42 in a cell compartment-specific manner, we argue that it is important to understand this process thoroughly by systematically determining where in the cell a particular Ras–effector pair interacts. We are particularly interested in knowing how often a given Ras protein interacts with effectors in the cytoplasm, a cell compartment that is the least understood in the study of Ras signaling. In a recent proof-of-principle study, we screened human cDNA libraries using BiFC to identify Ras effectors that interact with a Ras protein selectively in the cytoplasm, but not on the plasma membrane (Zheng, Cheng, et al., 2012). Our screen uncovered a known Ras effector, A-Raf, plus over 20 new effectors with an intriguing enrichment in proteins known to mediate intracellular trafficking. Two of them, in particular, CHMP6/VPS20 and VPS4A, are well known to interact with the ESCRT-III (Endosomal Sorting Complex Required for Transport-III) complex (Raiborg & Stenmark, 2009). These two proteins were chosen for detail analysis.

A key role of the ESCRT-III complex is to promote protein sorting by controlling scission of the ubiquitylated cargo-containing buds in the endosomes to form intralumenal vesicles, which is a critical step of multivesicular body biogenesis (Hurley & Hanson, 2010). In our study, CHMP6 and VPS4A act like typical Ras effectors, in that they bind H-Ras directly in a GTP-dependent manner. More importantly, when their levels are reduced, Ras-induced cell transformation is attenuated. This phenomenon correlates with the loss of EGFR recycling, suggesting that Ras can act through these two effectors to promote RTK recycling back to the plasma membrane. The binding to CHMP6 and VPS4A also appears to require Ras ubiquitylation, suggesting that Ras proteins themselves act as cargoes when passing through the endosomes. To investigate this possibility further, we examined biochemically the portion of Ras in the cytoplasm versus that on the plasma membrane and found that the latter was reduced when the expression of either CHMP6 or VPS4A was repressed. We then carried out photobleaching experiment and the data show that reducing either CHMP6 or VPS4A decreases the movement of Ras from the cytoplasm to the plasma membrane. These data collectively support the model that CHMP6 and VPS4A, if not the whole ESCRT-III complex, are responsible for recycling components of the Ras pathway (e.g., Ras proteins and RTKs) back to the plasma membrane to continue supporting the signal for growth. Thus, while internalized Ras proteins can remain dormant, some of them can also activate a recycling mechanism, presumably during persistent stimulation by the growth factors, in an apparent positive feedback loop for sustained and prolonged growth factor signaling (Zheng, Cheng, et al., 2012; Zheng, Xu, et al., 2012). In the succeeding text, we will describe in detail the methodology of using BiFC to study Ras–effector interactions.

2. DETECT RAS–EFFECTOR INTERACTIONS BY BiFC

2.1. General considerations for using BiFC to study Ras

While a large and growing list of fluorescent proteins have been reported to work efficiently in the BiFC system (Kodama & Hu, 2012), YFP and its derivatives remain popular choices because they have relative high quantum yields and complementation efficiencies among the GFP variants. In addition, several mutations have been introduced into the original YFP to accelerate maturation, which leads to the production of brighter YFP proteins (Kodama & Hu, 2012). We have chosen the "Venus" version of the YFP (Nagai et al., 2002), which contains a F46L mutation in the backbone of the "super"-EYFP (SEYFP) (Sawano & Miyawaki, 2000). The Venus is not only brighter than SEYFP but also more resistant to acid and Cl^-, which are common in many cellular compartments. In our study (Chen, Liu, & Songyang, 2007), the Yn is an N-terminal fragment of Venus, truncated at residue 155, a common split site for BiFC (Kodama & Hu, 2012). Since the majority of the "enhancement" mutations are in the N-terminus and included in the Yn, for Yc, we just use that of the EYFP (residues 156–239) because it was already made (Chen et al., 2007).

Ras proteins undergo important posttranslational modifications at the C-terminus to influence where they localize in the cell (Ahearn, Haigis, Bar-Sagi, & Philips, 2012). The C-terminus is also cleaved by a protease. We thus believe that it is better to tag Ras proteins at the N-terminus. As would be expected, in the BiFC system, the binding to the effector is also stronger with constitutively active (oncogenic) Ras proteins. We examined two of the most common oncogenic substitutions, G12V and Q61L, in H-Ras and found that the former binds slightly more strongly to the effectors. Seeking the appropriate negative controls for the BiFC system is critical and not trivial because when Yn and Yc are expressed at high enough levels in the same compartment, they too could interact to produce detectable fluorescence (Kodama & Hu, 2012). We created a Ras mutant as negative control, H-RasΔE(G12V), in which all the amino acid residues in the effector-binding region (residues 33–40) of H-Ras(G12V) were mutated to alanines. This works much better than the N17 dominant-negative Ras proteins, which returns higher "background." We note that even the binding to H-RasΔE(G12V) is not "zero" with the effectors we have tested, and we have thus also examined GST (see in the succeeding text) as a negative control. To reduce background due to overexpression as a result of transient transfection, we have created cell lines in which fusion proteins are stably expressed at appropriate levels.

Before the screen, in addition to Cdc42, we have examined two known Ras effectors, Raf and RIN1, by BiFC as positive controls. While full-length Raf can readily bind H-Ras(G12V), a truncated form of Raf, Raf-RBD, containing just the Ras-binding domain, binds even more strongly to Ras. Thus, for general studies of Ras–effector interactions, Raf-RBD is an ideal positive control. As discussed in the preceding text, we believe that Ras proteins should be tagged at the N-terminus, and in the BiFC system, they were tagged with Yn. Conversely, Yc was fused at the C-terminus to the effectors. Since the authentic N- and C-termini of the YFP are maintained (and exposed) in the correct orientations in the fusion proteins, we postulate that the binding between Ras and the effector would not substantially interfere with the refolding of the YFP. We stress that we have not vigorously tested this hypothesis, and some of the isolated effectors (e.g., TESC) can bind Ras no matter whether they are N- and C-terminally tagged (Zheng, Cheng, et al., 2012). cDNAs expressing the described fusion proteins were carried by pBabe vectors, modified by us to make it easier for constructing BiFC fusion proteins (Chen et al., 2007).

Finally, a variety of cell lines have been used successfully for the BiFC experiments (Kerppola, 2006). We have chosen the human fibrosarcoma HT1080 cells in part due to their relatively low autofluorescence. For example, when examining Yn-H-RasΔE(G12V) and Yc-Raf-RBD by FACS, only 0.6% of HT1080 cells showed detectable fluorescence, 20 $\times$ lower than that in MCF-7 cells. In addition, HT1080 cells endogenously overexpress an oncogenic N-Ras, which is at least partly responsible for the transforming phenotype in these cells (Zheng, Cheng, et al., 2012). As such, one can use the same cells to functionally validate the roles of novel Ras-binding proteins that act in the Ras pathways operative in these cells. One caveat with HT1080 is that they are prone to dying when transfected with Lipofectamine 2000.

2.2. Detect Ras–effector binding by FACS

1. Seed HT1080 cells in six-well plates and grow to 30–40% confluence.
2. Cotransfect 1 μg pBabe-Yc-Raf-RBD with 1 μg of either pBabe-Yn-H-Ras(G12V) or pBabe-Yn-H-RasΔE(G12V) to each well using 6 μl Lipofectamine 2000 reagent (Life Technologies, Carlsbad, CA). pBabe-Yc-Raf-RBD and pBabe-Yn empty vectors were also cotransfected as an additional negative control.
3. Incubate these cells at 37 °C with CO_2 for overnight.
4. Trypsinize and pellet the cells by centrifugation at $150 \times g$ at room temperature for 5 min.
5. Resuspend the cells in 300 μl of DMEM supplemented with 10% FBS and 4 μg/ml propidium iodide, which was used during FACS to exclude dead cells, and transfer them to 4 °C for 30 min.
6. Examine these cells by FACS for YFP. As shown in Fig. 2.1B, the YFP signal in Raf-RBD cells expressing H-Ras(G12V) can be readily separated from those carrying the negative controls.

2.3. Detect Ras–effector binding by microscopy

1. Seed HT1080 cells in 24-well glass bottom plates (MatTek), and grow them to 30–40% confluence.
2. Cotransfect 0.2 μg pBabe-Yc-Raf-RBD with 0.2 μg pBabe-Yn-H-Ras(G12V) or pBabe-Yn-H-RasΔE(G12V) vector to each well using 1.2 μl Lipofectamine 2000.
3. Incubate these cells at 37 °C with CO_2 for overnight.
4. Replace the transfection media with phenol red-free DMEM supplemented with 2% FBS.
5. Examine the cells on a Leica TCS SP5 confocal microscope with a 63×/1.4 oil objective lens. The reconstituted YFP signals (Fig. 2.1C) were imaged as described previously (Zheng, Cheng, et al., 2012).

3. SCREEN FOR RAS-BINDING PROTEINS BY BiFC

3.1. Create "baits" and negative controls

We used Yn-H-Ras(G12V) as the bait. In addition to Yn-H-RasΔE (G12V), we also used Yn-GST as a negative control, in case some effectors still retain a weak binding to H-RasΔE(G12V). To reduce background and improve reproducibility, we found it helpful when the bait is stably expressed as follows:

1. Seed HEK293FT cells (less than 30 passages) in DMEM supplemented with 10% FBS in 10 cm Petri dishes. Grow the cells to 60–70% confluence.
2. Transfect 6 μg pBabe-Yn-H-Ras(G12V), pBabe-Yn-H-RasΔE (G12V), or pBabe-Yn-GST plasmid individually with 6 μg retroviral packaging pAmpho vector into HEK293FT cells using 36 μl Lipofectamine 2000.
3. Incubate the cells in a 37 °C CO_2 incubator overnight.
4. Carefully aspirate the transfection medium, and replace it with 5 ml DMEM 10% FBS.
5. Incubate the cells in at 37 °C with CO_2 for 48 h.
6. Harvest the viruses by filtering the media through 0.45 μm PES syringe-top filter.
7. Mix the filtered virus-containing medium (~5 ml) with 2.5 ml fresh DMEM 10% FBS and 4 μg/ml polybrene and then added to the pres-eeded HT1080 cells (20–30% confluence) in 10 cm Petri dishes.
8. Incubate HT1080 cells at 37 °C with CO_2 for 36 h.
9. Replace the infection medium with DMEM 10% FBS and 500 μg/ml G418.
10. Incubate the cells at 37 °C with CO_2 for 2 weeks. Replace with fresh G418-containing growth medium every 3 days.
11. Pick single colonies and expand the cells.
12. Harvest the cells by a cell scraper with a lysis buffer containing 50 m*M* Tris–HCl (pH7.4), 200 m*M* NaCl, 2.5 m*M* $MgCl_2$, 10% glycerol, 1% NP-40, 1 m*M* PMSF, and a proteinase inhibitor cocktail (Roche Applied Science, Indianapolis, IN).
13. Examine the expression level of Yn-H-Ras(G12V), Yn-H-RasΔE (G12V), or Yn-GST in HT1080 clones by Western blots.
14. Select and pool three clones expressing medium levels of tested proteins.

3.2. Create cDNA libraries

We used human ORFeome v1.1 containing 8,100 ORFs as the resource of cDNAs (Rual et al., 2004). To rule out the possibility that the binding between Ras and some effectors may be sensitive to whether the effectors are tagged at the C- or N-terminus, we built two libraries, nYc and cYc (Zheng, Cheng, et al., 2012), in which the cDNAs were fused to the 5′ and 3′ of the coding sequence of Yc by the Gateway cloning system.

3.3. Initial screen for binding partners by FACS

1. Seed 5×10^5 of the "bait" cells (HT1080 cells stably expressing Yn-H-Ras(G12V)) to 10-cm Petri dishes; 24 h later, transfect 4 μg library DNAs using 16 μl Lipofectamine 2000.
2. For the negative control, transfect pBabe-Yc-Raf-RBD into HT1080-Yn-H-RasΔE(G12V) cells. This can be done at a smaller scale since only a small amount of cells are needed for gating.
3. Incubate these cells at 37 °C with CO_2 for 24 h.
4. Trypsinize and pellet the cells by centrifugation at $150 \times g$ at room temperature for 5 min.
5. Resuspend the cells in DMEM supplemented with 10% FBS and 4 μg/ml propidium iodide to a density of 1×10^6 cells/ml. Incubate at 4 °C for 30 min.
6. To set the gate, analyze the control cells (carrying the Yc-Raf-RBD/Yn-H-RasΔE(G12V) pair) first for YFP signal by FACS. Collect all the cells carrying the cDNA that are outside the boundary set by the negative control. We note that this is intended to maximize the yield, and the other assays discussed in the succeeding text are designed to more stringently exclude false-positives.
7. Centrifuge the isolated cells at $150 \times g$ at room temperature for 5 min.
8. Resuspend the cell pellet with 300 μl DNA extraction buffer containing 50 m*M* Tris–HCl (pH 8.0), 100 m*M* EDTA (pH 8.0), 100 m*M* NaCl, 1% SDS, and 0.2 mg/ml protease K.
9. Incubate the cells at 55 °C for 1 h.
10. Add 0.5 mg/ml RNase A to the cells, and incubate them at room temperature for another 1 h.
11. Extract the DNA by phenol/chloroform/isoamyl alcohol (25:24:1), and then precipitate the DNA by 1/10 volume of 3 N sodium acetate, 3 volume of ethanol, and 50 μg/ml of glycogen. Wash with 70% ethanol, and finally, allow it to be air-dried at room temperature.

12. Dissolve the DNA in 10 μl TE buffer (pH 8.0), and transform the isolate DNAs into high-efficiency 10-beta bacterial cells (New England Biolabs, Ipswich, MA).
13. Pick colonies and extract the plasmid DNAs by QIAprep 96 Turbo Miniprep Kit (Qiagen).

3.4. Follow-up screens to confirm the binding

1. Seed 3×10^4/well HT1080 cells stably expressing Yn-tagged H-Ras(G12V), H-RasΔE(G12V), or GST into 24-well plates.
2. Grow the cells in at 37 °C with CO_2 for 24 h.
3. Transfect 0.2 μg of each isolated cDNA clone using 0.8 μl Lipofectamine 2000 to each well of cells. Transfect pBabe-Yc and pBabe-RIN1-Yc individually into all the three cell lines (Step 1, above) as the controls.
4. Incubate the cells in a 37 °C CO_2 incubator for overnight.
5. Trypsinize the cells, and pellet them by centrifugation at 150 × *g* at room temperature for 5 min.
6. Resuspend the cells in 200 μl DMEM supplemented with 10% FBS and 4 μg/ml propidium iodide. Incubate at 4 °C for 30 min.
7. Examine the cells for YFP by FACS.
8. To readily exclude nonspecific binding, we estimated the binding strength between each protein pair by calculating the geometric mean of the reconstituted YFP intensity using FlowJo. We then divided the "binding" to the bait with that to either GST or H-RasΔE(G12V) (see Supplemental figure S2 in Zheng, Cheng, et al. (2012), leading to the final selection of 64 cDNA clones with a relative binding that is >1 with *both* negative controls. To further exclude false positive clones in a more stringent manner, we performed two experiments to identify Ras-binding proteins that also affect Ras functions.

3.5. Functional validation

Oncogenic Ras proteins induce cellular transformation; we thus measured whether overexpressing the isolated cDNAs can alter the ability of NIH3T3 cells carrying H-Ras(G12V) to form colonies in soft agar (Zheng, Cheng, et al., 2012). In addition, we carried out an Elk1-mediated luciferase assay to measure gene expression as induced by the Ras-Raf-MAP kinase pathway. Since we conducted these studies using standard protocols, which are published, we will not describe them here in detail. We should point out, however, that out of the final 26 clones, only one can enhance these Ras activities, while the rest inhibit them. At first glance, this seems rather counterintuitive. In the case of CHMP6 and VPS4A, it is evident that overexpressing these two genes created dominant-negative effects on the Ras pathways; thus, we expect that overexpressing many other isolated cDNAs may similarly act in a dominant-negative fashion. Of course, another possibility is that some of the isolated cDNAs control Ras activities that are different from those tested here. Regardless, it is critical to select Ras activities that are most relevant to your interest when designing such a screen. While the observed dominant-negativity associated with gene overexpression has caveats, it is convenient for narrowing down the list of potential Ras effectors. Ultimately, gene silencing is likely to be more informative in determining how isolated components work in a particular Ras pathway.

4. CONCLUDING REMARKS

While Ras proteins are well known to mediate growth factor signaling at the plasma membrane, we and others have discovered that Ras proteins can signal from endosomes. In particular, the results described in the preceding text strongly suggest that Ras proteins can activate the ESCRIT-III complex to induce recycling of components of the Ras pathway to enhance growth factor signaling. Besides Cdc42, CHMP6, and VPS4A, PI3K, one of the most cancer-relevant Ras effectors, has also been shown to interact with Ras proteins in the endosome (Tsutsumi, Fujioka, Tsuda, Kawaguchi, & Ohba, 2009). Therefore, better understanding of the roles of Ras in endosomes is likely to uncover key mechanisms by which Ras proteins are involved in tumorigenesis.

It is perhaps not surprising that Ras proteins move dynamically in the cell and along the way interact with many effectors. BiFC has offered us a great tool to identify these interacting proteins because of its unique property to secure protein–protein interactions that are otherwise transient and weak. Despite this, in reflection, the BiFC system can be improved in several important ways with improved reagents and technology.

The single most labor-intensive (and costly) step in our screen was to sort out individual cDNA clones from the pool. We have first tried to sort out individual YFP^+ cells but found that only a very few cells survived the sorting, and they frequently carrying the cDNAs expressing Raf. Thus, this approach may work if your genes of interest promote cell survival after sorting. We have also sought to use the pooled isolated cDNAs to conduct a second round of screening using FACS with the intention to "enrich" positive clones. However, we found that some positive clones, such as A-Raf, that were present in the initial pool were in fact lost in the second screen. We speculate that the BiFC signals are greatly influenced by copy number of the vector, which is hard to control during transfection. This problem can be addressed using cDNA libraries that are printed in 96-well format and by screening through a robotic automated FACS system. Finally, more Ras effectors can be found if the BiFC libraries have better coverage of the human genome. We note that since the completion of our study, over 90% of human cDNAs are now available (Team et al., 2009).

ACKNOWLEDGMENTS

Z. Z. was supported by a postdoctoral fellowship from the Susan G. Komen for the Cure Foundation. E. C. C. is supported by grants from NIH (CA90464, GM81627, CA125123, and P30-CA58183) and by the Nancy Owens Memorial Foundation and the Mary Kay Ash Foundation.

REFERENCES

Ahearn, I. M., Haigis, K., Bar-Sagi, D., & Philips, M. R. (2012). Regulating the regulator: Post-translational modification of RAS. *Nature Reviews Molecular Cell Biology*, *13*, 39–51.

Chang, E. C., Barr, M., Wang, Y., Jung, V., Xu, H., & Wigler, H. M. (1994). Cooperative interaction of *S. pombe* proteins required for mating and morphogenesis. *Cell*, *79*, 131–141.

Chang, E. C., & Philips, M. R. (2006). Spatial segregation of Ras signaling: New evidence from fission yeast. *Cell Cycle*, *5*, 1936–1939.

Chen, L. Y., Liu, D., & Songyang, Z. (2007). Telomere maintenance through spatial control of telomeric proteins. *Molecular and Cellular Biology*, *27*, 5898–5909.

Cheng, C. M., & Chang, E. C. (2011). Busy traveling Ras. *Cell Cycle*, *10*, 1180–1181.

Cheng, C. M., Li, H., Gasman, S., Huang, J., Schiff, R., & Chang, E. C. (2011). Compartmentalized Ras proteins transform NIH 3T3 cells with different efficiencies. *Molecular and Cellular Biology*, *31*, 983–997.

Fehrenbacher, N., Bar-Sagi, D., & Philips, M. (2009). Ras/MAPK signaling from endomembranes. *Molecular Oncology*, *3*, 297–307.

Fukui, Y., & Yamamoto, M. (1988). Isolation and characterization of Schizosaccharomyces pombe mutants phenotypically similar to ras1. *Molecular & General Genetics*, *215*, 26–31.

Han, L., Wong, D., Dhaka, A., Afar, D., White, M., Xie, W., et al. (1997). Protein binding and signaling properties of RIN1 suggest a unique effector function. In: *Proceedings of the National Academy of Sciences of the United States of America*, *94*, 4954–4959.

Hu, C. D., Chinenov, Y., & Kerppola, T. K. (2002). Visualization of interactions among bZIP and Rel family proteins in living cells using bimolecular fluorescence complementation. *Molecular Cell*, *9*, 789–798.

Hurley, J. H., & Hanson, P. I. (2010). Membrane budding and scission by the ESCRT machinery: It's all in the neck. *Nature Reviews Molecular Cell Biology*, *11*, 556–566.

Jura, N., Scotto-Lavino, E., Sobczyk, A., & Bar-Sagi, D. (2006). Differential modification of Ras proteins by ubiquitination. *Molecular Cell*, *21*, 679–687.

Kerppola, T. K. (2006). Design and implementation of bimolecular fluorescence complementation (BiFC) assays for the visualization of protein interactions in living cells. *Nature Protocols, 1*, 1278–1286.

Kodama, Y., & Hu, C. D. (2012). Bimolecular fluorescence complementation (BiFC): A 5-year update and future perspectives. *BioTechniques, 53*, 285–298.

Nagai, T., Ibata, K., Park, E. S., Kubota, M., Mikoshiba, K., & Miyawaki, A. (2002). A variant of yellow fluorescent protein with fast and efficient maturation for cell-biological applications. *Nature Biotechnology, 20*, 87–90.

Onken, B., Wiener, H., Philips, R. M., & Chang, E. C. (2006). Compartmentalized signaling of Ras in fission yeast. *Proceedings of the National Academy of Sciences of the United States of America, 103*, 9045–9050.

Pylayeva-Gupta, Y., Grabocka, E., & Bar-Sagi, D. (2011). RAS oncogenes: Weaving a tumorigenic web. *Nature Reviews Cancer, 11*, 761–774.

Raiborg, C., & Stenmark, H. (2009). The ESCRT machinery in endosomal sorting of ubiquitylated membrane proteins. *Nature, 458*, 445–452.

Rual, J. F., Hirozane-Kishikawa, T., Hao, T., Bertin, N., Li, S., Dricot, A., et al. (2004). Human ORFeome version 1.1: A platform for reverse proteomics. *Genome Research, 14*, 2128–2135.

Sawano, A., & Miyawaki, A. (2000). Directed evolution of green fluorescent protein by a new versatile PCR strategy for site-directed and semi-random mutagenesis. *Nucleic Acids Research, 28*, E78.

Sigismund, S., Confalonieri, S., Ciliberto, A., Polo, S., Scita, G., & Di Fiore, P. P. (2012). Endocytosis and signaling: Cell logistics shape the eukaryotic cell plan. *Physiological Reviews, 92*, 273–366.

Tall, G. G., Barbieri, M. A., Stahl, P. D., & Horazdovsky, B. F. (2001). Ras-activated endocytosis is mediated by the Rab5 guanine nucleotide exchange activity of RIN1. *Developmental Cell, 1*, 73–82.

Team, M. G. C. P., Temple, G., Gerhard, D. S., Rasooly, R., Feingold, E. A., Good, P. J., et al. (2009). The completion of the Mammalian Gene Collection (MGC). *Genome Research, 19*, 2324–2333.

Tsutsumi, K., Fujioka, Y., Tsuda, M., Kawaguchi, H., & Ohba, Y. (2009). Visualization of Ras-PI3K interaction in the endosome using BiFC. *Cellular Signalling, 21*, 1672–1679.

Xu, L., Lubkov, V., Taylor, L. J., & Bar-Sagi, D. (2010). Feedback regulation of Ras signaling by Rabex-5-mediated ubiquitination. *Current Biology, 20*, 1372–1377.

Yan, H., Jahanshahi, M., Horvath, E. A., Liu, H. Y., & Pfleger, C. M. (2010). Rabex-5 ubiquitin ligase activity restricts Ras signaling to establish pathway homeostasis in Drosophila. *Current Biology, 20*, 1378–1382.

Zheng, Z. Y., Cheng, C. M., Fu, X. R., Chen, L. Y., Xu, L., Terrillon, S., et al. (2012). CHMP6 and VPS4A mediate the recycling of Ras to the plasma membrane to promote growth factor signaling. *Oncogene, 31*, 4630–4638.

Zheng, Z. Y., Xu, L., Bar-Sagi, D., & Chang, E. C. (2012). Escorting Ras. *Small GTPases, 3*, 236–239.

CHAPTER THREE

TGFβ in Endosomal Signaling

Sarah McLean[*,†], Gianni M. Di Guglielmo[†,1]
[*]Department of Anatomy and Cell Biology, Western University, London, Ontario, Canada
[†]Department of Physiology and Pharmacology, Western University, London, Ontario, Canada
[1]Corresponding author: e-mail address: john.diguglielmo@schulich.uwo.ca

Contents

Abstract

The transforming growth factor beta (TGFβ) signaling pathway is important for normal cell homeostasis and has critical roles in apoptosis, cell-cycle arrest, and cellular differentiation (reviewed in Massague, 2008). In the classical TGFβ pathway, the endosomal trafficking of receptors has a direct outcome on signal transduction—receptors internalized via clathrin-mediated endocytosis enter the early endosome and propagate signaling, while those internalized via membrane rafts are targeted for degradation. Recently, there have been a number of articles that have identified TGFβ receptor-binding proteins that direct receptor endocytosis and/or intracellular trafficking and affect signal output (Atfi et al., 2007; Bauge, Girard, Leclercq, Galera, & Boumediene, 2012; Bizet et al., 2011, 2012; Chen et al., 2007; Gunaratne, Benchabane, & Di Guglielmo, 2012; Hao et al., 2011; McLean, Bhattacharya, & Di Guglielmo, 2013; Zhao et al., 2012). Given the importance of TGFβ receptor trafficking to signaling outcome, this chapter will focus on strategies to isolate membrane rafts and techniques to follow the trafficking of cell-surface TGFβ receptors and provide examples of functional readouts to assess TGFβ signal transduction.

Methods in Enzymology, Volume 535
ISSN 0076-6879
http://dx.doi.org/10.1016/B978-0-12-397925-4.00003-1

1. INTRODUCTION

The transforming growth factor beta (TGFβ) signaling pathway is a cell-type and context-dependent pathway that has pleiotropic effects. It was initially thought that TGFβ signal transduction simply occurred from the cell surface and was mediated by the Smad family of transcription factors. In this linear fashion to propagate TGFβ signaling, TβRIII presents TGFβ to TβRII (Lopez-Casillas, Wrana, & Massague, 1993). The binding of TGFβ to TβRII recruits TβRI to the ligand–receptor complex. TβRI is then phosphorylated by TβRII, activating its kinase activity (Wrana, Attisano, Wieser, Ventura, & Massague, 1994). TβRI then recruits Smad2 and phosphorylates its MH2 domain (Macias-Silva et al., 1996; Nakao, Imamura, et al., 1997; Nakao, Roijer, et al., 1997). Following Smad2 phosphorylation, Smad4 is recruited to form a heteromeric complex (Nakao, Imamura, et al., 1997) and the Smad complex then translocates to the nucleus and interacts with transcriptional coactivators and corepressors to induce cell-specific transcriptional programs (Shi & Massague, 2003).

While the role of Smads in signal propagation has been well established, it is now understood that TGFβ signal transduction does not simply occur in a linearly manner from the cell membrane. TGFβ receptor internalization and trafficking play key roles in regulating signaling outcome (Le Roy & Wrana, 2005). TGFβ receptors internalized by clathrin-mediated endocytosis traffic to the early endosome where they interact with SARA (Smad anchor for receptor activation) to propagate Smad-mediated transcription (Di Guglielmo, Le Roy, Goodfellow, & Wrana, 2003; Hayes, Chawla, & Corvera, 2002; Mitchell, Choudhury, Pagano, & Leof, 2004; Runyan, Schnaper, & Poncelet, 2005). However, receptors internalized by membrane raft-dependent, clathrin-independent mechanisms traffic to the caveolin-1-positive vesicle and are prevented from signal propagation and are targeted for degradation (Di Guglielmo et al., 2003; Kavsak et al., 2000; Ogunjimi et al., 2005).

2. ENDOCYTOSIS

Endocytosis refers to the process whereby cell-surface-associated molecules enter the cell without passing through the plasma membrane. Internalization of cell-surface receptors is important in the control of signal transduction, functioning to downregulate either signaling or trafficking receptors to specific endocytic compartments. There are several methods of endocytosis of cell-surface receptors including membrane raft-dependent endocytosis, caveolin-dependent endocytosis, Arf6-dependent endocytosis, and clathrin-mediated endocytosis (reviewed in Doherty & McMahon, 2009). The focus of this chapter will be clathrin-mediated endocytosis and membrane raft-dependent endocytosis as they are implicated in the TGFβ pathway (Di Guglielmo et al., 2003).

Clathrin-mediated endocytosis is a highly conserved mechanism implicated in the internalization of many receptor types. Clathrin-mediated endocytosis occurs when clathrin from the cytosol is recruited to the plasma membrane and aggregates to form pits (Doherty & McMahon, 2009). Protein motifs of cargo play a role in the development of clathrin-coated pits, as dileucine and tyrosine motifs in the cytoplasmic domains of receptors are detected by adaptor protein 2 and promote clathrin polymerization (Marks, Woodruff, Ohno, & Bonifacino, 1996; Sorkin, 2004). The polymerization of clathrin into lattices increases plasma membrane curvature (Doherty & McMahon, 2009). Upon sufficient membrane curvature, the GTPase dynamin forms a helix around the neck of the clathrin-coated pit and with GTP hydrolysis promotes scission of the clathrin-coated pit from the plasma membrane (Praefcke & McMahon, 2004). These excised pits then form vesicles, known as endosomes, and lose their clathrin coat. Endosomes may be routed to the cell membrane for recycling or mature and go on to form other compartments.

Clathrin-independent endocytosis through membrane rafts is also a common mechanism for the uptake of cell-surface receptors. The membrane raft model proposes that cholesterol–sphingolipid–protein complexes form in the plasma membrane to make a tightly packed, liquid-ordered phase mediating endocytosis and signal transduction (Simons & Ikonen, 1997). The lipid composition of membrane rafts is distinct from the rest of the plasma membrane. Membrane rafts are enriched in cholesterol and sphingolipids and are therefore more rigid and less fluid than the surrounding plasma membrane (Staubach & Hanisch, 2011). Membrane rafts have been shown to be especially important in the endocytosis of proteins with glycophosphatidylinositol-binding domains (Simons & Gerl, 2010).

Intriguingly, a role for membrane raft-mediated signal transduction has been identified for TGFβ signaling. At the cell surface, TGFβ receptor complexes can access both clathrin-coated pits and membrane rafts (Di Guglielmo et al., 2003). Inhibition of clathrin-coated pit internalization through the use of a dominant-negative Eps15 mutant shifts receptors into membrane raft fractions. Similarly, inhibition of membrane raft formation through cholesterol depletion shifts receptors back into nonmembrane raft fractions (Di Guglielmo et al., 2003). As previously mentioned, TGFβ receptors internalized via clathrin-mediated endocytosis access the early endosome, which propagates TGFβ signal transduction through the recruitment of Smads (Di Guglielmo et al., 2003; Mitchell et al., 2004). Membrane raft-mediated endocytosis, however, promotes ubiquitin-dependent receptor degradation (Di Guglielmo et al., 2003). Therefore, endocytic route plays a powerful role in dictating TGFβ receptor intracellular trafficking and signal transduction.

2.1. Isolation of membrane rafts

The study of membrane rafts has been limited at times due to various methods used to isolate rafts. Indeed, several techniques have been described that isolate membrane rafts based on their detergent insolubility and/or buoyant density (Garner, Smith, & Hooper, 2008; Radeva & Sharom, 2004). The tight packing of lipids in the liquid-ordered phase of membrane rafts prevents detergent incorporation and therefore disruption (Xu et al., 2001). Following detergent extraction of membrane rafts, cell lysates are frequently subjected to sucrose-density ultracentrifugation, as the enrichment of membrane rafts with cholesterol and sphingolipids increases their buoyancy relative to the rest of the plasma membrane. Different compositions of membrane rafts can be obtained depending on the type of detergent and the duration of extraction (Lingwood & Simons, 2007). Membrane rafts may also be isolated using the detergent-free method below that may decrease some of the extraction-based issues of membrane raft isolation (Di Guglielmo et al., 2003; Luga et al., 2009; McCabe & Berthiaume, 2001; McLean & Di Guglielmo, 2010).

2.1.1 Detergent-free membrane raft isolation protocol

1. Cells should be grown to near-confluence for the isolation of membrane rafts. As the membrane raft content of different cell types can vary, it is recommended to grow several 100 mm dishes to confluence per condition (e.g., four dishes of HEK 293T cells). *Note*: This protocol is set up to use a Beckman SW40 rotor, which accommodates six buckets, each capable of holding approximately 13 mL ultracentrifuge tubes. Therefore, up to six different conditions can be carried out in one experiment.

2. Prepare the following stock solutions to be kept at 4 °C:
   i. 1 *M* Na_2CO_3, pH 11
   ii. 0.5 *M* MES (2-(*N*-morpholino)ethanesulfonic acid) pH 6.5
   iii. 5 *M* NaCl
   iv. 80% sucrose in 25 m*M* MES and 150 m*M* NaCl
3. Prepare the following solutions and keep on ice.

   Lysis buffer: 6.5 mL Na_2CO_3 pH 11 + protease inhibitors + npH_2O to 13 mL final volume.

   30% sucrose solution: 11.25 mL 80% sucrose/MES/NaCl + 7.5 mL 1 *M* Na_2CO_3 pH 11 + 0.94 mL 0.5 *M* MES pH 6.5 + 0.56 mL 5 *M* NaCl + 9.75 mL H_2O to give a total volume of 30 mL.

   5% sucrose solution: 1.875 mL 80% sucrose/MES/NaCl + 7.5 mL 1 *M* Na_2CO_3 pH 11 + 1.4 mL 0.5 *M* MES pH 6.5 + 0.843 mL 5 *M* NaCl + 18.4 mL H_2O to give a total volume of 30 mL.
4. Gently rinse cells twice with ice-cold PBS and aspirate PBS completely before lysing.
5. If using four plates/condition, add 2 mL of lysis buffer to the first plate, scrape cells, and transfer to the second plate. Scrape the second plate and transfer to the third plate. Continue this process until all four plates per given condition have been lysed. Transfer the lysate (~2.5 mL) to a 15 mL conical tube. The cell lysate should be very viscous.
6. Repeat the previously mentioned lysis procedure for each of the other conditions and place into separate 15 mL conical tubes.
7. Homogenize the cell lysate with three 10-s bursts in a Polytron tissue grinder (settings 7–8). It is recommended that the samples be kept on ice between homogenizations. Following homogenization, sonicate samples with three 20-s bursts of sonication on ice. The lysate should no longer be viscous.
8. Transfer 2 mL of the homogenate to the bottom of a 13 mL Beckman centrifuge polyallomer tube.
9. Adjust the sucrose concentration of the homogenates to 40% sucrose by the addition of 2 mL of 80% sucrose/MES/NaCl buffer. Mix well with a transfer pipette. Do not invert the tube as the sucrose will coat the sides of the tube and disrupt the formation of subsequent sucrose cushions.
10. Very gently apply 4 mL of 30% sucrose buffer on top of the 40% homogenate with a syringe (18 gauge needle). While applying, move the needle up the side of the tube and follow the 30% sucrose cushion.

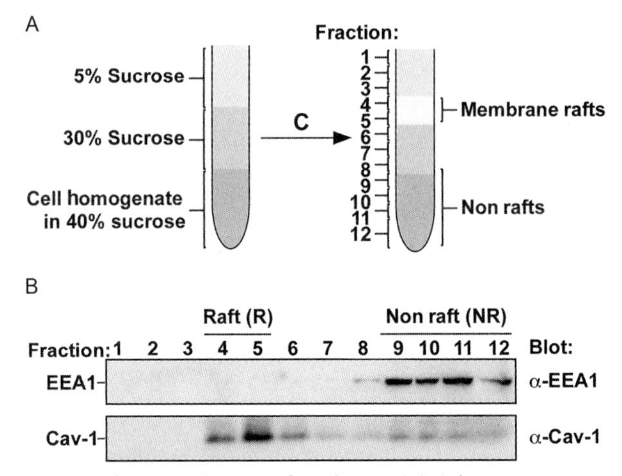

Figure 3.1 *Detergent-free membrane raft isolation.* (A) Schematic representing membrane raft preparation. Prior to centrifugation, cell homogenates are sonicated and adjusted to 40% sucrose. They are then overlaid with 30% and 5% sucrose gradients. Following overnight ultracentrifugation (C; 200,000 × g_{av}), 12 fractions are collected and subjected to SDS-PAGE and immunoblotting. (B) Blots were probed for markers of the early endosome (EEA1) and membrane raft (caveolin-1 and Cav-1). (For color version of this figure, the reader is referred to the online version of this chapter.)

11. Repeat Step 10 with 5% sucrose buffer on top of the 30% sucrose cushion. You should now be able to see three distinct phases in the tube (Fig. 3.1). The bottom 4 mL may look slightly pink due to residual medium in the homogenate.
12. Centrifuge for 16 h at 4 °C at 40,000 rpm (200,000 × g_{av}) in an SW40 rotor (Beckman Instruments) with low brake during the deceleration portion of centrifugation.
13. Collect 12 1-mL fractions from the top of the tube going downward and transfer into 1.5 mL tubes. The rafts are at the interface between the 5% and 30% cushions and should appear as a white pellicle. This pellicle should collect in fractions 4–6. The nonraft fractions (i.e., the majority of cellular proteins/compartments) are at the bottom of the tube in fractions 8–12.
14. Test a small aliquot (50–100 μL) of each fraction by SDS-PAGE and immunoblotting to determine the fractions enriched in raft and non-raft components. For example, caveolin-1 can be used as a marker for membrane raft fractions and EEA1 (early endosome antigen 1) can be used as a marker for nonraft fractions (Fig. 3.1).

3. THE ROLE OF THE EARLY ENDOSOME IN TGFβ SIGNAL TRANSDUCTION

While the classic paradigm of signal transduction suggests that following receptor endocytosis signal transduction is terminated, it has been shown in many different systems that signaling continues following receptor internalization into endosomes (reviewed in Murphy, Padilla, Hasdemir, Cottrell, & Bunnett, 2009).

As previously described, clathrin-mediated endocytosis of TGFβ receptors targets their localization to the early endosome (Di Guglielmo et al., 2003; Mitchell et al., 2004), which enhances TGFβ signaling. An elegant study performed by Runyan and colleagues illustrated that internalization of the TGFβ receptor complex is essential for maximal signal transduction (Runyan et al., 2005). The authors illustrated that inhibition of clathrin-mediated endocytosis did not greatly prevent the ability of the receptor complex to phosphorylate Smad2, but endocytic inhibition *did* prevent nuclear translocation of Smad2, thus preventing TGFβ-dependent transcription (Runyan et al., 2005), suggesting a spatial component to TGFβ signaling. One key player in the spatial control of TGFβ signaling is SARA. SARA contains an FYVE domain (Fab1, YOTB, Vac1, and EEA1), a common motif in early endosomal proteins, which has been shown to bind to phosphatidylinositol-3-phosphate (Lawe, Patki, Heller-Harrison, Lambright, & Corvera, 2000). SARA also binds Smad2 via MH2 domains and preferentially binds the nonphosphorylated forms of the Smads (Tsukazaki, Chiang, Davison, Attisano, & Wrana, 1998). It has been proposed that SARA functions to link the TGFβ receptors with Smad2. Once Smad2 has been phosphorylated by the receptor complex, Smad2 dissociates from SARA and binds Smad4, translocating to the nucleus and initiating TGFβ-driven transcription (Tsukazaki et al., 1998).

Membrane raft endocytosis of TGFβ receptors results in receptors being targeted to the caveolin-1-positive vesicle. Unlike the early endosome, the caveolin-1-positive vesicle promotes association of Smad7, not Smad2, with the receptor complex (Di Guglielmo et al., 2003). Smad7 belongs to the inhibitory Smads and counteracts the canonical TGFβ pathway (Yan, Liu, & Chen, 2009). The antagonistic role of Smad7 is mediated by two mechanisms. Firstly, Smad7 is able to interact with activated TβRI and therefore sterically inhibits the association of TβRI with Smad2, preventing the subsequent activation of Smad2 and its association with Smad4 (Hayashi et al., 1997; Nakao, Afrakhte, et al., 1997). Secondly, Smad7 acts as an adaptor between the TGFβ complex and an ubiquitin regulatory factor, Smurf2 (Kavsak et al., 2000). Smurf2 is an E3 ubiquitin ligase that is localized primarily in the nucleus (Kavsak et al., 2000). Upon TGFβ stimulation, however, Smurf2 translocates to the cytoplasm and forms a stable interaction with Smad7 and the receptor complex (Kavsak et al., 2000). Smurf2 then ubiquitinates the receptor complex, targeting it for degradation via proteasomal and lysosomal pathways (Kavsak et al., 2000).

3.1. Fluorescence techniques to follow cell-surface receptor endocytosis

Since the trafficking of TGFβ receptors following endocytosis plays a critical role in TGFβ signal transduction, visualizing the intracellular trafficking of TGFβ receptors may help elucidate the signaling outcome.

To visualize receptor trafficking, one can use an antibody-chase approach followed by immunofluorescence microscopy. Due to the relative low abundance of TGFβ receptors at the cell surface, this technique works best in cells overexpressing TGFβ receptors. Although transiently expressing cells may be used for this technique, it is recommended to use cells stably expressing extracellularly tagged receptors (such as FLAG-, HA-, or MYC-tagged receptors). Antibodies directed towards the extracellular tags can be used in instances where antibodies that recognize the extracellular portion of receptors are not available. It is also important to use adherent cells, as the washing steps in this method may cause the lifting of less adherent cell types. Alternatively, one can increase the adhesiveness of the cells to the glass coverslips by first coating the coverslips with poly-L-lysine.

1. Trypsinize and plate cells on glass coverslips in a 12-well dish.
2. Twenty-four to forty-eight hours following plating, begin the antibody-chase technique. At this point, cells should be 50–60% confluent.
3. Place plates on ice for 10 min to prevent receptor internalization.
4. Gently rinse cells twice in ice-cold 0.5% bovine serum albumin (BSA)/ Krebs Ringer HEPES (KRH) buffer.
5. Dilute the antibodies that will recognize the extracellular portion of the receptors (e.g., TβRII, HA, and FLAG) at a concentration of 1:200 in 0.5% BSA/KRH buffer and add to cells. In order to avoid receptor clustering, antibody Fab fragments should be used for this step.

6. Incubate at 4 °C for 2 h on a gently rocking platform.
7. Following incubation, place cells on ice and rinse gently with cold 0.5% BSA/KRH three times. Dilute fluorophore-conjugated secondary antibody at a dilution of 1:250 in 0.5% BSA/KRH and add to the cells. All subsequent steps must now be completed in the DARK. Incubate the cells at 4 °C for 1 h on a gently rocking platform.
8. Following the incubation with secondary antibody, rinse cells three times with 0.5% BSA/KRH. The cells are either immediately fixed (this represents receptors at the cell surface) or incubated with media (containing 10% final FBS) at 37 °C for an additional time (e.g., 30 min and 60 min). The incubation at 37 °C permits the cell-surface-labeled receptors to internalize.
9. The fixation procedure is as follows:
   i. Gently rinse coverslips five times with PBS.
   ii. Fix cells by adding 4% paraformaldehyde/PBS and incubate for 10 min at room temperature.
   iii. Gently rinse five times with PBS at room temperature.
10. Permeabilization procedure:
    i. Add 0.25% TX-100/PBS and incubate for 5 min at room temperature.
    ii. Gently rinse five times with PBS at room temperature.
11. Coverslips now must be blocked for at least 1 h at room temperature in 10% fetal bovine serum (FBS)/PBS in order to prevent nonspecific binding of antibody.
12. Coverslips are now incubated with primary antibodies that detect markers of subcellular compartments such as the early endosome (using EEA1 or Rab5) or the caveolin-1-positive vesicles. Generally, a dilution of 1:100 in 10% FBS/PBS is recommended, but this may depend on the commercial sources of these antibodies. Ensure that the species of origin for the antibody is different from the antibody recognizing the receptor. The coverslips can then be incubated for 2–3 h at room temperature or at 4 °C overnight.
13. Following incubation with primary antibody, rinse cells five times with PBS at room temperature.
14. Add fluorophore-conjugated secondary antibody (with a different wavelength than the fluorophore used to detect the receptors) and incubate for 45 min at room temperature.
15. Gently rinse five times with PBS at room temperature.

16. Stain the coverslips with a dye that recognizes nuclear structures (e.g., DAPI) using the manufacturer's protocols. Some mounting reagents contain DAPI or Hoechst stains. If this is the case, skip to Step 18.
17. Rinse with PBS at room temperature.
18. Use mounting media to mount slides and analyze using standard or confocal immunofluorescence microscopy (Fig. 3.2A).

3.2. Biotinylation of TGFβ

In addition to using antibody to label cell-surface receptors, it is also possible to use biotinylated TGFβ followed by fluorophore-conjugated streptavidin to visualize the trafficking of TGFβ receptors.

Biotinylation of TGFβ

1. For this procedure, it is desirable to have an excess of biotin per TGFβ molecule. This protocol uses a ratio of 30 molecules of biotin per TGFβ molecule.
2. Dissolve 5 μg TGFβ in 10 μL 0.1% acetonitrile/trifluoroacetic acid and add to 90 μL of 125 μ*M* sodium phosphate.
3. Make a 225 μ*M* stock of EZ-link-NHS-biotin (Thermo, Inc.) in npH_2O. Add 25 μL of this stock to the 100 μL of TGFβ/sodium phosphate.
4. After mixing biotin with the TGFβ, incubate at room temperature for 3 h.
5. To purify the biotinylated TGFβ from free biotin, it will be necessary to filter the mixture through a PD10 desalting column.
6. Prepare column wash buffer: 0.1% BSA in 75 m*M* NaCl, pH to 2.4 using HCl.
7. Equilibrate the column with 10–15 mL of column wash buffer then let the column empty.
8. Add biotinylated TGFβ/biotin mixture to the column and let it flow through.
9. Fill the column with column wash buffer (~6 mL).
10. Collect the eluent from the column in 0.5 mL fractions (12 fractions total).
11. Test samples for biotinylated TGFβ by spotting a small aliquot (5 μL) from each fraction on nitrocellulose paper and let dry.
12. Following standard procedures from immunoblotting protocols, block and incubate the nitrocellulose strip with HRP-conjugated streptavidin followed by chemiluminescence (ECL) reagents to visualize and

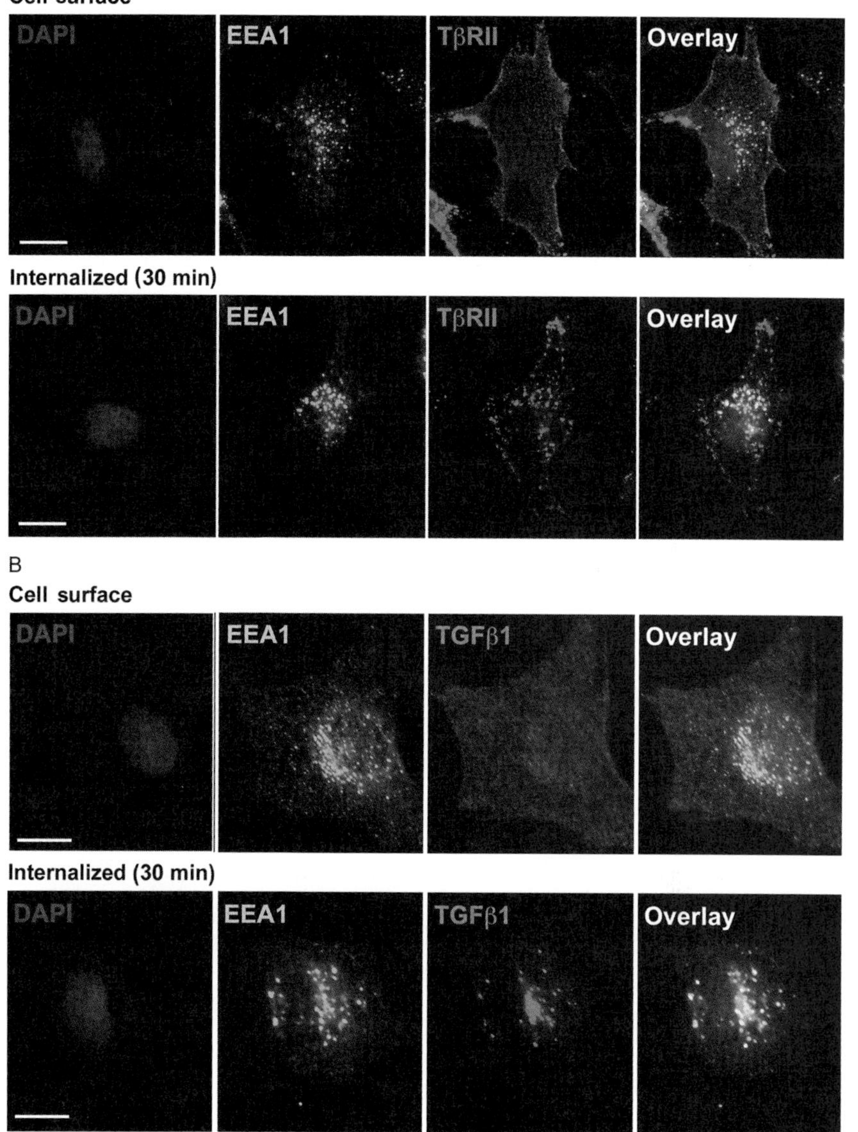

Figure 3.2 *TGFβ receptor trafficking to the early endosome.* (A) Mv1Lu cells stably overexpressing HA–TβRII were labeled at 4 °C with anti-HA antibodies. Following incubation of Cy3-labeled anti-rabbit antibodies at 4 °C, cells were fixed (cell surface) or allowed to internalize for 30 min at 37 °C to permit receptor internalization. Standard immunofluorescence staining was used to visualize EEA1, a marker for the early endosome, and nuclei (DAPI staining). (B) Mv1Lu cells stably overexpressing HA–TβRII were labeled at 4 °C with biotinylated TGFβ1. Following incubation of Cy3-labeled streptavidin at 4 °C, cells were fixed or incubated at 37 °C as described in panel A. Standard immunofluorescence staining was used to visualize EEA1 and nuclei (DAPI staining). Receptor complex localization with the early endosomal marker results in a yellow overlay. (For interpretation of the references to color in this figure legend, the reader is referred to the online version of this chapter.)

identify the fractions that contain biotinylated TGFβ (typically, fractions 6–9).

13. To visualize biotinylated TGFβ-bound receptor internalization by immunofluorescence microscopy, use the antibody technique described earlier (Section 3.1, Steps 1–18) but replace the antibody that recognizes the TGFβ receptors (Step 5) with biotinylated TGFβ and the fluorophore-labeled antibody (step 7) with fluorophore-labeled streptavidin (Fig. 3.2B).

4. READOUTS OF TGFβ SIGNAL TRANSDUCTION

One complexity of studying TGFβ signaling is the varying cell- and context-dependent outcomes that it elicits. For example, while TGFβ stimulates cell-cycle arrest and apoptosis in epithelial cells, it stimulates migration and proliferation in fibroblasts (Massague & Gomis, 2006). However, in both instances, the Smad family of intracellular signaling molecules mediates the TGFβ-dependent effects. Therefore, Smad phosphorylation could be used as an initial readout to evaluate the effects of trafficking on TGFβ signal transduction.

Here, we outline a quick method to evaluate TGFβ receptor activation. This method assesses the level of phosphorylated Smad2 relative to "total" levels of Smad2. Other complementary approaches to assess Smad activation include assessing a Smad-inducible luciferase construct (ARE-lux) (McLean et al., 2013) and assessing the nuclear translocation of Smad2 (To et al., 2008). For a comprehensive study of TGFβ signaling outcome assays, readers are encouraged to view the paper by Halder, Beauchamp, and Datta (2005).

4.1. Smad2 phosphorylation

Assessing phosphorylated Smad2 levels is an efficient way to evaluate TGFβ signal transduction. Smad2 is phosphorylated by TβRI in response to the activation of TβRI by ligand-bound TβRII. In the absence of ligand, there is very little phosphorylation of Smad2. The phosphorylated serine residues of Smad2 serve as a docking site for Smad4 and promote the dissociation of Smad2 from TβRI and the formation of a heteromeric complex with Smad4 (Macias-Silva et al., 1996; Nakao, Roijer, et al., 1997). Smad2 is generally located cytoplasmically in the absence of ligand but upon ligand stimulation translocates to the nucleus with Smad4, which in the absence of ligand is found distributed equally between the nucleus and the cytoplasm

(Schmierer & Hill, 2007). Therefore, the phosphorylation of Smad2 acts as a limiting factor for Smad-dependent transcription, as only phosphorylated Smad2 can interact with Smad4.

1. Prior to performing the phosphorylated Smad2 assay, cells should be at approximately 80% confluence.
2. Rinse cells gently with PBS and incubate in low-serum (0.2% FBS) or serum-free media for at least 4 h (preferably overnight). This step removes exogenous TGFβ found in serum and suppresses background Smad2 phosphorylation.
3. In order to induce Smad2 phosphorylation, treat cells with 50–250 pM TGFβ (Fig. 3.3). Incubate one plate in the absence of ligand (control), and incubate the remaining plates for 0.5–1 h in low-serum media containing TGFβ. Following incubation, immediately lyse cells in 0.5% TX-100/Tris-buffered saline/0.2 m*M* EDTA buffer containing protease and phosphatase inhibitors.

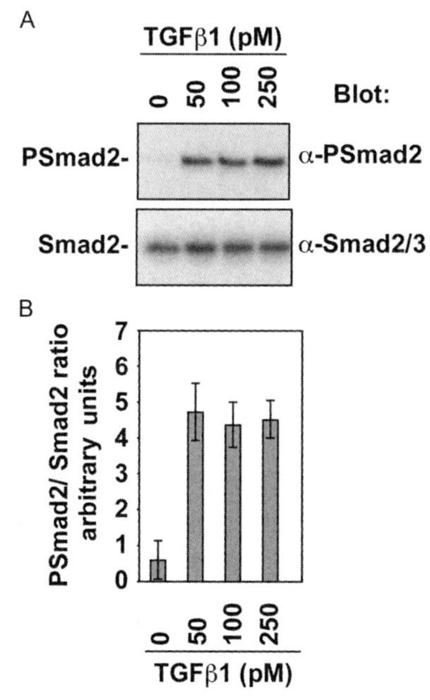

Figure 3.3 Smad2 phosphorylation assay. (A) A549 cells were serum-starved and treated with increasing concentrations of TGFβ1 for 1. Following lysis, cells were subjected to SDS-PAGE and immunoblotting for phosphorylated Smad2 (PSmad) or Smad 2. (B) Three separate experiments as described in panel A were carried out and the amounts of PSmad2 and Smad2 were quantified using Quantity One software and plotted as the ratio of PSmad2/Smad2. The mean (arbitrary units) ± SD is shown.

4. Process for standard SDS-PAGE followed by immunoblotting for phosphorylated Smad2 and total Smad2 using commercially available antibodies. Use densitometry to quantitatively evaluate the induction of PSmad2 (Fig. 3.3).

ACKNOWLEDGMENTS

This manuscript was supported by a grant from the Canadian Institutes of Health Research (MOP-93625) awarded to G. M. D. G.

REFERENCES

Atfi, A., Dumont, E., Colland, F., Bonnier, D., L'Helgoualc'h, A., Prunier, C., et al. (2007). The disintegrin and metalloproteinase ADAM12 contributes to TGF-beta signaling through interaction with the type II receptor. *The Journal of Cell Biology*, *178*(2), 201–208.

Bauge, C., Girard, N., Leclercq, S., Galera, P., & Boumediene, K. (2012). Regulatory mechanism of transforming growth factor beta receptor type II degradation by interleukin-1 in primary chondrocytes. *Biochimica et Biophysica Acta*, *1823*(5), 983–986.

Bizet, A. A., Liu, K., Tran-Khanh, N., Saksena, A., Vorstenbosch, J., Finnson, K. W., et al. (2011). The TGF-beta co-receptor, CD109, promotes internalization and degradation of TGF-beta receptors. *Biochimica et Biophysica Acta*, *1813*(5), 742–753.

Bizet, A. A., Tran-Khanh, N., Saksena, A., Liu, K., Buschmann, M. D., & Philip, A. (2012). CD109-mediated degradation of TGF-beta receptors and inhibition of TGF-beta responses involve regulation of SMAD7 and Smurf2 localization and function. *Journal of Cellular Biochemistry*, *113*(1), 238–246.

Chen, C. L., Liu, I. H., Fliesler, S. J., Han, X., Huang, S. S., & Huang, J. S. (2007). Cholesterol suppresses cellular TGF-beta responsiveness: Implications in atherogenesis. *Journal of Cell Science*, *120*(Pt. 20), 3509–3521.

Di Guglielmo, G. M., Le Roy, C., Goodfellow, A. F., & Wrana, J. L. (2003). Distinct endocytic pathways regulate TGF-beta receptor signalling and turnover. *Nature Cell Biology*, *5*(5), 410–421.

Doherty, G. J., & McMahon, H. T. (2009). Mechanisms of endocytosis. *Annual Review of Biochemistry*, *78*, 857–902.

Garner, A. E., Smith, D. A., & Hooper, N. M. (2008). Visualization of detergent solubilization of membranes: Implications for the isolation of rafts. *Biophysical Journal*, *94*(4), 1326–1340.

Gunaratne, A., Benchabane, H., & Di Guglielmo, G. M. (2012). Regulation of TGFbeta receptor trafficking and signaling by atypical protein kinase C. *Cellular Signalling*, *24*(1), 119–130.

Halder, S. K., Beauchamp, R. D., & Datta, P. K. (2005). A specific inhibitor of TGF-beta receptor kinase, SB-431542, as a potent antitumor agent for human cancers. *Neoplasia*, *7*(5), 509–521.

Hao, X., Wang, Y., Ren, F., Zhu, S., Ren, Y., Jia, B., et al. (2011). SNX25 regulates TGF-beta signaling by enhancing the receptor degradation. *Cellular Signalling*, *23*(5), 935–946.

Hayashi, H., Abdollah, S., Qiu, Y., Cai, J., Xu, Y. Y., Grinnell, B. W., et al. (1997). The MAD-related protein Smad7 associates with the TGFbeta receptor and functions as an antagonist of TGFbeta signaling. *Cell*, *89*(7), 1165–1173.

Hayes, S., Chawla, A., & Corvera, S. (2002). TGF beta receptor internalization into EEA1-enriched early endosomes: Role in signaling to Smad2. *The Journal of Cell Biology*, *158*(7), 1239–1249.

Kavsak, P., Rasmussen, R. K., Causing, C. G., Bonni, S., Zhu, H., Thomsen, G. H., et al. (2000). Smad7 binds to Smurf2 to form an E3 ubiquitin ligase that targets the TGF beta receptor for degradation. *Molecular Cell*, *6*(6), 1365–1375.

Lawe, D. C., Patki, V., Heller-Harrison, R., Lambright, D., & Corvera, S. (2000). The FYVE domain of early endosome antigen 1 is required for both phosphatidylinositol 3-phosphate and Rab5 binding. Critical role of this dual interaction for endosomal localization. *The Journal of Biological Chemistry*, *275*(5), 3699–3705.

Le Roy, C., & Wrana, J. L. (2005). Clathrin- and non-clathrin-mediated endocytic regulation of cell signalling. *Nature Reviews Molecular Cell Biology*, *6*(2), 112–126.

Lingwood, D., & Simons, K. (2007). Detergent resistance as a tool in membrane research. *Nature Protocols*, *2*(9), 2159–2165.

Lopez-Casillas, F., Wrana, J. L., & Massague, J. (1993). Betaglycan presents ligand to the TGF beta signaling receptor. *Cell*, *73*(7), 1435–1444.

Luga, V., McLean, S., Le Roy, C., O'Connor-McCourt, M., Wrana, J. L., & Di Guglielmo, G. M. (2009). The extracellular domain of the TGFbeta type II receptor regulates membrane raft partitioning. *The Biochemical Journal*, *421*(1), 119–131.

Macias-Silva, M., Abdollah, S., Hoodless, P. A., Pirone, R., Attisano, L., & Wrana, J. L. (1996). MADR2 is a substrate of the TGFbeta receptor and its phosphorylation is required for nuclear accumulation and signaling. *Cell*, *87*(7), 1215–1224.

Marks, M. S., Woodruff, L., Ohno, H., & Bonifacino, J. S. (1996). Protein targeting by tyrosine- and di-leucine-based signals: Evidence for distinct saturable components. *The Journal of Cell Biology*, *135*(2), 341–354.

Massague, J. (2008). TGFbeta in cancer. *Cell*, *134*(2), 215–230.

Massague, J., & Gomis, R. R. (2006). The logic of TGFbeta signaling. *FEBS Letters*, *580*(12), 2811–2820.

McCabe, J. B., & Berthiaume, L. G. (2001). N-terminal protein acylation confers localization to cholesterol, sphingolipid-enriched membranes but not to lipid rafts/caveolae. *Molecular Biology of the Cell*, *12*(11), 3601–3617.

McLean, S., Bhattacharya, M., & Di Guglielmo, G. M. (2013). betaarrestin2 interacts with TbetaRII to regulate Smad-dependent and Smad-independent signal transduction. *Cellular Signalling*, *25*(1), 319–331.

McLean, S., & Di Guglielmo, G. M. (2010). TGF beta (transforming growth factor beta) receptor type III directs clathrin-mediated endocytosis of TGF beta receptor types I and II. *The Biochemical Journal*, *429*(1), 137–145.

Mitchell, H., Choudhury, A., Pagano, R. E., & Leof, E. B. (2004). Ligand-dependent and -independent transforming growth factor-beta receptor recycling regulated by clathrin-mediated endocytosis and Rab11. *Molecular Biology of the Cell*, *15*(9), 4166–4178.

Murphy, J. E., Padilla, B. E., Hasdemir, B., Cottrell, G. S., & Bunnett, N. W. (2009). Endosomes: A legitimate platform for the signaling train. In: *Proceedings of the National Academy of Sciences of the United States of America*, *106*(42), 17615–17622.

Nakao, A., Afrakhte, M., Moren, A., Nakayama, T., Christian, J. L., Heuchel, R., et al. (1997). Identification of Smad7, a TGFbeta-inducible antagonist of TGF-beta signalling. *Nature*, *389*(6651), 631–635.

Nakao, A., Imamura, T., Souchelnytskyi, S., Kawabata, M., Ishisaki, A., Oeda, E., et al. (1997). TGF-beta receptor-mediated signalling through Smad2, Smad3 and Smad4. *The Embo Journal*, *16*(17), 5353–5362.

Nakao, A., Roijer, E., Imamura, T., Souchelnytskyi, S., Stenman, G., Heldin, C. H., et al. (1997). Identification of Smad2, a human Mad-related protein in the transforming growth factor beta signaling pathway. *The Journal of Biological Chemistry*, *272*(5), 2896–2900.

Ogunjimi, A. A., Briant, D. J., Pece-Barbara, N., Le Roy, C., Di Guglielmo, G. M., Kavsak, P., et al. (2005). Regulation of Smurf2 ubiquitin ligase activity by anchoring the E2 to the HECT domain. *Molecular Cell*, *19*(3), 297–308.

Praefcke, G. J., & McMahon, H. T. (2004). The dynamin superfamily: Universal membrane tubulation and fission molecules? *Nature Reviews Molecular Cell Biology*, *5*(2), 133–147.

Radeva, G., & Sharom, F. J. (2004). Isolation and characterization of lipid rafts with different properties from RBL-2H3 (rat basophilic leukaemia) cells. *The Biochemical Journal*, *380*(Pt. 1), 219–230.

Runyan, C. E., Schnaper, H. W., & Poncelet, A. C. (2005). The role of internalization in transforming growth factor beta1-induced Smad2 association with Smad anchor for receptor activation (SARA) and Smad2-dependent signaling in human mesangial cells. *The Journal of Biological Chemistry*, *280*(9), 8300–8308.

Schmierer, B., & Hill, C. S. (2007). TGFbeta-SMAD signal transduction: Molecular specificity and functional flexibility. *Nature Reviews Molecular Cell Biology*, *8*(12), 970–982.

Shi, Y., & Massague, J. (2003). Mechanisms of TGF-beta signaling from cell membrane to the nucleus. *Cell*, *113*(6), 685–700.

Simons, K., & Gerl, M. J. (2010). Revitalizing membrane rafts: New tools and insights. *Nature Reviews Molecular Cell Biology*, *11*(10), 688–699.

Simons, K., & Ikonen, E. (1997). Functional rafts in cell membranes. *Nature*, *387*(6633), 569–572.

Sorkin, A. (2004). Cargo recognition during clathrin-mediated endocytosis: A team effort. *Current Opinion in Cell Biology*, *16*(4), 392–399.

Staubach, S., & Hanisch, F. G. (2011). Lipid rafts: Signaling and sorting platforms of cells and their roles in cancer. *Expert Review of Proteomics*, *8*(2), 263–277.

To, C., Kulkarni, S., Pawson, T., Honda, T., Gribble, G. W., Sporn, M. B., et al. (2008). The synthetic triterpenoid 2-cyano-3,12-dioxooleana-1,9-dien-28-oic acid-imidazolide alters transforming growth factor beta-dependent signaling and cell migration by affecting the cytoskeleton and the polarity complex. *The Journal of Biological Chemistry*, *283*(17), 11700–11713.

Tsukazaki, T., Chiang, T. A., Davison, A. F., Attisano, L., & Wrana, J. L. (1998). SARA, a FYVE domain protein that recruits Smad2 to the TGFbeta receptor. *Cell*, *95*(6), 779–791.

Wrana, J. L., Attisano, L., Wieser, R., Ventura, F., & Massague, J. (1994). Mechanism of activation of the TGF-beta receptor. *Nature*, *370*(6488), 341–347.

Xu, X., Bittman, R., Duportail, G., Heissler, D., Vilcheze, C., & London, E. (2001). Effect of the structure of natural sterols and sphingolipids on the formation of ordered sphingolipid/sterol domains (rafts). Comparison of cholesterol to plant, fungal, and disease-associated sterols and comparison of sphingomyelin, cerebrosides, and ceramide. *The Journal of Biological Chemistry*, *276*(36), 33540–33546.

Yan, X., Liu, Z., & Chen, Y. (2009). Regulation of TGF-beta signaling by Smad7. *Acta Biochimica et Biophysica Sinica*, *41*(4), 263–272.

Zhao, B., Wang, Q., Du, J., Luo, S., Xia, J., & Chen, Y. G. (2012). PICK1 promotes caveolin-dependent degradation of TGF-beta type I receptor. *Cell Research*, *22*(10), 1467–1478.

CHAPTER FOUR

Annexins and Endosomal Signaling

Francesc Tebar*, Mariona Gelabert-Baldrich*, Monira Hoque†, Rose Cairns†, Carles Rentero*, Albert Pol*,‡, Thomas Grewal†, Carlos Enrich*,1

*Departament de Biologia Cellular, Immunologia i Neurociències, IDIBAPS, Facultat de Medicina, Universitat de Barcelona, Barcelona, Spain
†Faculty of Pharmacy, University of Sydney, Sydney, New South Wales, Australia
‡ICREA, Institució Catalana de Recerca Avançada, Barcelona, Spain
[1]Corresponding author: e-mail address: enrich@ub.edu

Contents

Abstract

Cell signaling and endocytosis are intimately linked in eukaryotic cells. Signaling receptors at the cell surface enter the endocytic pathway and continue to activate downstream effectors in endosomal compartments. This spatiotemporal regulation of signal transduction provides opportunity for signal diversity and a cell-specific machinery of scaffolding/targeting proteins contributes to establish compartment-specific signaling complexes. Members of the annexin (Anx) protein family, in particular AnxA1, AnxA2, and AnxA6, appear to target their interaction partners to specific membrane microdomains to contribute to the formation of compartment-specific signaling platforms along the endocytic pathway. A major challenge to understand the impact of scaffolding/targeting proteins on spatiotemporal signal transduction along endocytic pathways is the identification, isolation, and functional analysis of low-abundance

Methods in Enzymology, Volume 535
ISSN 0076-6879
http://dx.doi.org/10.1016/B978-0-12-397925-4.00004-3

signal-transducing protein complexes in endocytic compartments. Here, we describe methods to isolate endosomes and to target signaling molecules to endosomes. Applying these methodologies to suitable animal or cell models will enable the dissection of signal transduction in the endocytic compartment in the presence or absence of annexins.

1. INTRODUCTION

After internalization, endocytosed molecules are routed to early endosomes (EEs). In this compartment, the internalized material can be recycled back to the plasma membrane (PM) or remain there during their maturation to late endosomes (LEs) or multivesicular bodies (MVBs). LE/MVBs eventually fuse with lysosomes, where endocytosed molecules are degraded. The organelles of the endocytic pathway are endowed with SNARE and motor proteins, Rab GTPases, and phosphoinositides that confer spatiotemporal functionality to the system. Endosomal membranes are enriched in compartment-specific molecular markers such as Rab5, early endosome antigen 1 (EEA1), and phosphatidylinositol 3-phosphate (PI3P) in EE, Rab7, and phosphatidylinositol 3, 5-bisphosphate (PI3, $5P_2$) in LE or LAMP proteins in lysosomes (Sigismund et al., 2012).

The endocytic compartment represents a functional continuity of the PM, which allows signaling events to continue once generated at the cell surface, making EE and LE to represent specific signaling platforms. By interaction with the cytoskeleton, endosomes allow the traffic of signaling molecules to different cellular locations and enable the degradation of signaling complexes through directing them to lysosomes.

Over the last decade, the importance of endosomal signaling for growth factor/tyrosine kinase receptors (EGFR, TGFβR, Met, Notch, and TNFR) and G-protein-coupled receptors has been well established. Receptor signaling in endosomal compartments is associated with the presence of small Ras GTPases to control cellular outputs like growth, proliferation, and cell mobility (Flinn, Yan, Goswami, Parker, & Backer, 2010; Gould & Lippincott-Schwartz, 2009; Kermorgant & Parker, 2008; Lobert & Stenmark, 2011; Ohashi et al., 2011; Palamidessi et al., 2008; Platta & Stenmark, 2011; Sorkin & von Zastrow, 2009; Taub, Teis, Ebner, Hess, & Huber, 2007).

Along these lines, we and others have studied the EGFR signaling pathway initiated at the PM and active in endosomes. Accordingly, many activated signaling proteins of the EGFR pathway have been detected in endosomes, for example, Shc, Grb2, Sos, Ras, and the kinases Raf-1, Mek, and Erk1/2 (Balbis, Parmar, Wang, Baquiran, & Posner, 2007; Lu et al., 2009; Moreto et al., 2008; Pol, Calvo, & Enrich, 1998; Sorkin & von Zastrow, 2009; Teis et al., 2006). It is generally believed that this reflects sustained signaling of internalized receptors (Mor & Philips, 2006), but more importantly, the multiple and differential location of signal-transducing molecules is crucial to diversify cellular signaling output (Calvo, Agudo-Ibanez, & Crespo, 2010).

Despite the improved knowledge on compartment-specific signaling, there are still fundamental questions about the role and specificity of signaling molecules in endosomes. Alternatively, besides providing a signaling continuity derived from receptor activation at the cell surface, targeting of signaling molecules specifically to endosomes could elucidate its activation and signaling exclusively from endosomes and discriminate this from signals derived from the PM.

In this context, scaffolding/targeting proteins are essential to organize and stabilize the formation of signal-transducing complexes in distinct cellular compartments (Kheifets & Mochly-Rosen, 2007; Palfy, Remenyi, & Korcsmaros, 2012; White, Erdemir, & Sacks, 2012). Their ability to specifically interact with certain signaling molecules provides a platform to recruit regulators/effectors and modulate signaling amplitude and kinetics in a cell- and stimulus-specific manner. Scaffolding/targeting proteins regulating signaling and trafficking of activated growth factor receptors in EE and LE include several annexins, in particular AnxA1, AnxA2, AnxA6, and AnxA8 (Grewal & Enrich, 2009). Annexins are a dynamic, multifunctional, and evolutionary conserved superfamily of proteins and are characterized by their ability to interact with biological membranes in a calcium-dependent manner (Gerke, Creutz, & Moss, 2005).

The current understanding of the contribution of these annexins in the regulation of spatiotemporal signaling along the endocytic pathway is mostly derived from studies examining the EGFR/Ras/MAPK signaling cascade. It would go beyond the scope of this chapter to recapitulate all these findings and we refer the reader to reviews from our laboratory and others summarizing the role of annexins in cellular signaling and the endocytic pathway in more detail (Futter & White, 2007; Gerke et al., 2005; Grewal & Enrich, 2009). In brief, AnxA1 was one the first EGFR tyrosine kinase substrates identified and regulates EGF-induced MVB biogenesis to facilitate lysosomal degradation of EGFR (White, Bailey, Aghakhani, Moss, & Futter, 2006).

The role of AnxA2 is more complex and involves the sorting of EGFR to recycling endosomes or MVB (Morel, Parton, & Gruenberg, 2009; Zobiack, Rescher, Ludwig, Zeuschner, & Gerke, 2003), possibly via interaction with cholesterol- and PI(4,5)P2-rich domains and the actin cytoskeleton and interaction and modulation of signaling output of regulators/ effectors such as src, PKC, SHP2, Pyk2, cdc42, and Rho (Hayes, Rescher, Gerke, & Moss, 2004). AnxA8 is found in LE/MVB and its ability to bind phosphatidylinositols and F-actin might affect final steps in EGFR degradation (Goebeler, Poeter, Zeuschner, Gerke, & Rescher, 2008). Our laboratory has focused on the role of AnxA6, which is a targeting protein for p120GAP and PKCα, both negative regulators of EGFR and Ras (Enrich et al., 2011). The ability of annexins to act as membrane organizers and scaffolding proteins is exemplified by AnxA6, which is a highly dynamic protein involved in several cellular events linked to membrane transport. This includes the reorganization of the actin cytoskeleton at the PM (Monastyrskaya et al., 2009), the delivery of cholesterol from LE to the Golgi and PM (Cubells et al., 2007), the cholesterol-dependent recruitment of cytosolic phospholipase A_2 to the Golgi for Golgi vesiculation (Cubells et al., 2008), and t-SNARE trafficking and assembly in the secretory pathway (Reverter et al., 2011).

AnxA6 is synthesized as a cytosolic protein but is predominantly targeted to the PM upon cell activation. This possibly includes the association of AnxA6 with cholesterol-enriched membrane microdomains (lipid rafts, both caveolae and noncaveolae). Furthermore, AnxA6 is a major component of rat liver endosomes (Jackle et al., 1994; Pol, Ortega, & Enrich, 1997). Together with cell culture studies, these findings linked AnxA6 with low-density lipoprotein (LDL) receptor-mediated endocytosis (Grewal et al., 2000; Kamal, Ying, & Anderson, 1998) and the delivery of LDL-containing LE to lysosomes for degradation (Pons et al., 2000). LDL-induced translocation of AnxA6 to LE (Grewal et al., 2000) correlates with the recruitment of AnxA6 to cholesterol-enriched LE, establishing a cholesterol-dependent membrane-binding capacity of AnxA6 in LE (de Diego et al., 2002). In addition, AnxA6 interacts with EGFR, Ras, Raf-1, MAPK, and effectors/regulators of the EGFR/Ras pathway at the cell surface and endosomes (Grewal & Enrich, 2006; Koese et al., 2012; Vila de Muga et al., 2009). In line with studies on other scaffolding proteins, expression levels and subcellular localization of annexins seem to modulate membrane transport and signal complex formation to differentially determine signaling output at the PM or endosomes.

One of the challenges in dissecting signaling events in different subcellular compartments is the identification and isolation of low-abundance signal protein complexes and to provide material for further functional assays. In the following sections, we describe methods to isolate endosomes and to target signaling molecules to endosomes. In combination with suitable animal and cell models, this can serve to dissect the endocytic compartment and determine the contribution of annexins in the signaling output of intracellular signaling pathways.

2. ISOLATION OF ENDOCYTIC COMPARTMENTS

In this section, we describe procedures to isolate endosomes from rat or mice liver and cell cultures, such as Chinese hamster ovary (CHO) cells, respectively. A number of methods have been developed for the isolation of endosomes largely free of contamination from other intracellular membranes. The similarity in the equilibrium densities of endocytic vesicles, Golgi membranes, and other smooth membranes has led to the development of methods for specifically modifying the density of endocytic vesicles based on their ligand content (Belcher et al., 1987; Bergeron, Searle, Khan, & Posner, 1986; Debanne, Evans, Flint, & Regoeczi, 1982; Evans & Flint, 1985; Mueller & Hubbard, 1986).

For rat or mice liver, we used the method developed by Belcher et al. (1987), which distinguishes three endosomal fractions on the basis of their distinct morphologies and their association with, or absence of, various radioligands at different time points with "early" (compartment of uncoupling receptors and ligands (CURLs)) and "late" (MVB) endosomes and a third endosomal fraction, the receptor-recycling compartment (RRC), representing recycling and transcytotic structures (Enrich, Jackle, & Havel, 1996). Although this procedure is based on the shift density that endocytic structures undergo after loading with LDL in estradiol-treated rats, these fractions have been proven most valuable to characterize signaling platforms in endosomes. Indeed, one- and two-dimensional gel electrophoresis to investigate the protein composition of the three isolated endosome fractions identified a differential distribution of various receptors and ligands, including growth factor receptors and signaling proteins. In summary, rat and mouse liver endocytic compartments consist of specific and defined stations, with CURL being the central and most important location for sorting and RRC being involved in recycling and transcytosis (Enrich, Pol, Calvo, Pons, & Jackle, 1999; Pol et al., 1997).

2.1. Isolation of endocytic fractions from rat liver

Three rats are treated for 3 days with 17-α-ethinyl estradiol, dissolved in propylene glycol (1 mg/ml) and injected (5 mg/kg) subcutaneously, to induce a high rate of uptake of LDL into the liver. Rats are then anesthetized with 4% isoflurane, and human LDL (5 mg of protein) is injected into the femoral vein (Belcher et al., 1987).

1. After 15 min, livers are flushed via the portal vein with 0.15 *M* NaCl and 0.1% EDTA. Then, the livers are removed, minced with scalpels, and using a Dounce homogenizer, homogenized (1:3 wt/vol) in 0.25 *M* sucrose with protease inhibitors (1 m*M* PMSF, 100 μ*M* leupeptin, 150 μ*M* aprotinin, and 1 μ*M* pepstatin).
2. The homogenate is centrifuged for 10 min at 500 × *g* in an SW28 rotor (Beckman Coulter).
3. The supernatant is centrifuged at 4800 × *g* for 20 min in an SW28 rotor (Beckman Coulter).
4. Finally, the supernatant is centrifuged at 16,600 × *g* for 20 min in a type 30 rotor (Beckman Coulter).
5. This third supernatant (64 ml) is diluted with 30 ml of isotonic Percoll solution at pH 7.4 (27 ml Percoll plus 3 ml 2.5 *M* sucrose, 9:1 vol/vol), layered in tubes each containing a marker density of 1.062 g/ml (Pharmacia, Uppsala, Sweden) and centrifuged for 45 min at 40,700 × *g* in a type 30 rotor (Beckman Coulter).
6. The fraction over the marker density is collected, 2 volumes of 0.9% NaCl are added, and the fraction is then layered on 2 ml of 2.5 *M* sucrose and centrifuged at 24,250 × *g* for 45 min in an SW28 rotor (Beckman Coulter).
7. The white crude endosome band is harvested and 2.5 *M* sucrose is added to a final density of 1.15 g/ml. The 1.15 g/ml portion is loaded at the bottom of the tubes containing a discontinuous sucrose gradient of densities 1.032, 1.074, 1.11, and 1.139 g/ml. Tubes are centrifuged for 170 min at 140,000 × *g* in an SW28 rotor (Beckman Coulter). Three distinct populations are obtained at the density interfaces: MVB at 1.032/1.074 g/ml, CURL at 1.074/1.110 g/ml, and RRC at 1.110/1.130 g/ml.
8. Each fraction is collected and ice-cold water is added to generate isotonic fractions. Finally, the isotonic fractions are pelleted by centrifugation at 71,000 × *g* for 30 min in a 50 Ti rotor (Beckman Coulter), resuspended in 0.9% NaCl, and stored at −80 °C.

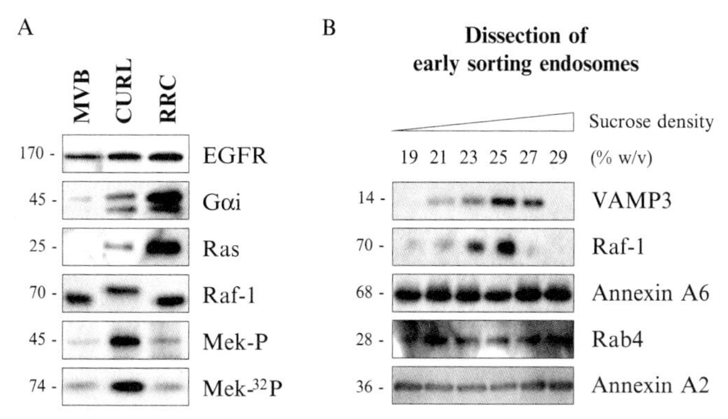

Figure 4.1 *Signaling in isolated rat liver endosomes.* (A) Differential distribution of signaling proteins (EGFR, Ras, Raf-1, and Mek) and G-proteins (Gαi) in three endosomal fractions (MVB, CURL, and RRC) from rat liver by Western blotting. The electrophoretical mobility of Raf-1 differs in the various fractions suggesting posttranslational modification. The last panel shows Raf-1 kinase activity as judged by GST-Mek phosphorylation from Raf-1 immunoprecipitates purified from MVB, CURL, and RRC (see Section 3.2). (B) Biochemical dissection of CURL. Isolated early sorting endosomes (CURL) were loaded at the bottom of a multistep sucrose gradient (from 19% to 29% w/v) and centrifuged for 2 h and 50 min at 78,000 × *g* in an SW28 rotor. Samples from this gradient (2 ml) were pelleted, resuspended in 0.9% NaCl, TCA-precipitated, electrophoresed (4 μg/lane), and transferred to Immobilon-P membranes. Western blotting was used to analyze the distribution of proteins along the gradient.

If further separation is needed, the interface corresponding to CURL or RRC can be mixed with heavy sucrose (2.5 *M*) and loaded at the bottom of a discontinuous sucrose gradient with 19%, 21%, 23%, 25%, 27%, and 29% sucrose (w/v) for CURL or 29%, 31%, 33%, 35%, 37%, and 39% for RRC. Then, the tubes are centrifuged at 136,000 × *g* for 2 h 50 min in a Beckman SW28 rotor. Following centrifugation, the interfaces are collected, pelleted, resuspended in 0.5 ml 0.9% NaCl, and stored at −80 °C. Figure 4.1A shows a representative Western blot profile of signaling proteins in the three endosomal fractions. Further subfractionation of the CURL fraction is shown in Fig. 4.1B.

2.1.1 Characterization of the recycling endocytic compartment (RRC)

In view of the complexity of the endocytic compartment, it is not surprising that isolation and biochemical characterization of this pleiomorphic organelle from tissue homogenates has been a major task. In respect to compartmentalized signaling, recycling endosomes have received increased attention in recent years, as growth factor receptors and signaling proteins can be diverted to this compartment for transport back to the cell surface. For instance, EGFR recycling was previously considered a default pathway, but we showed that this involves a complex and PKCδ-dependent regulation of actin dynamics (Llado et al., 2004, 2008).

It was assumed that tubules contained in RRC have their origin in the membrane appendages emanating from CURL and MVB, as structures involved in the recycling of receptors (Belcher et al., 1987; Geuze et al., 1984). In principle, RRC is a subcellular fraction that comprises all those morphological-described endocytic structures, tubules, and vesicles, which do not contain LDL (after LDL loading) and therefore float at the same density in the sucrose gradient ($d = 1.13$ g/ml, 35.5% sucrose). However, these endocytic membranes possibly originate from different cellular compartments.

Thus, whereas RRCr (recycling) and RRCt (transcytosis) might come from different regions of the membranous tubular extensions of CURL (and MVB), RRCc (caveolae) originates from the caveolae-enriched PM domains at the sinusoidal PM (a subcellular well-characterized fraction denominated caveolin-enriched PM fraction) (Pol, Calvo, Lu, & Enrich, 1999).

2.2. Isolation of endocytic fractions from CHO cells

Several methods have been developed for the isolation of endocytic fractions from cells in culture and even procedures to obtain enriched populations of early and late endocytic structures. However, none of these procedures allow the degree of enrichment obtained compared to those prepared from the liver, at least in terms of morphology of the isolated fractions. Also, and very importantly, neither of the methods using cells in culture are able to obtain a reasonably pure and enriched fraction representative of the recycling endocytic compartment (RRC). In the following sections, we describe those fractionation procedures optimized for CHO cells. Over the years, this protocol has also been used to isolate endocytic fractions from other cell lines, such as baby hamster kidney cells (Jost, Zeuschner, Seemann, Weber, & Gerke, 1997; Morel et al., 2009). However, given the cell-specific differences in endosome morphology and endocytic transport, detailed analysis and characterization of isolated fractions using established early/late and lysosomal markers is required.

2.2.1 Early and late endosomes

Subcellular endosomal fractions are prepared according to the protocol described by Gorvel, Chavrier, Zerial, and Gruenberg (1991).

1. CHO cells are grown in Ham's F-12 containing 10% fetal bovine serum (FBS), 2 m*M* glutamine, 100 U/ml penicillin, and 100 mg/ml streptomycin. $4–6 \times 10^7$ CHO cells are used for each gradient. If needed, increased amounts of enlarged LE can be obtained upon treatment with U18666A (4.5 μg/ml for 16 h) (Cubells et al., 2007; de Diego et al., 2002) or bafilomycin A1 (50 n*M* for 16 h) (Lu et al., 2009). U18666A treatment or LDL loading (0.5 mg/ml for 120 min) can be utilized to recruit or increase AnxA6, and possibly other annexins, in LE fractions (de Diego et al., 2002; Grewal et al., 2000).
2. Cells are put on ice, washed two times with cold PBS, and collected in 5 ml of homogenization buffer (250 m*M* sucrose, 3 m*M* imidazole, pH 7.4, and protease inhibitors: 1 m*M* PMSF, 1 m*M* aprotinin, 20 μ*M* leupeptin, and 5 m*M* Na_3VO_4).
3. Cells are pelleted at $500 \times g$ in a bench centrifuge, resuspended in 2 ml of homogenization buffer, and homogenized by 10 passages through a 22 gauge needle. Complete homogenization is confirmed under the phase microscope.
4. The homogenate is centrifuged for 15 min at 4 °C at $1000 \times g$ in a bench centrifuge.
5. The postnuclear supernatant (PNS) is brought to a final 40.2% sucrose (w/v) concentration by adding 62% sucrose (3 m*M* imidazole, pH 7.4) and loaded (4 ml) at the bottom of an SW41 centrifugation tube (Beckman Ultraclear). Then, 35% sucrose (3 ml), 25% sucrose (3 ml), and finally 3 ml of homogenization buffer are poured stepwise on top of the PNS.
6. The gradient is centrifuged for 90 min at $210{,}000 \times g$ and 4 °C in a swing out Beckman SW41 Ti rotor.
7. After centrifugation, 1 ml fractions are collected from top to bottom. Aliquots of each fraction are trichloroacetic acid-precipitated to determine the distribution of markers by Western blotting.

3. TARGETING Raf-1 SIGNALING TO EARLY ENDOSOMES

EGFR activates different signaling pathways that can trigger a variety of cellular responses. The Ras/Raf-1/Mek/Erk (MAPK) pathway is an excellent example and regulates fundamental processes such as proliferation, differentiation, migration, and apoptosis. Upon ligand binding, EGFR activates Ras GTPases, which transmit the extracellular signals to Raf kinases. Raf-1 plays a central role in EGFR-mediated activation of the MAPK pathway and is recruited to the PM through binding to activated Ras. Once translocated to the PM, Raf-1 is activated by sequential dephosphorylation and phosphorylation events. Several phosphatases, kinases, and scaffold proteins have now been identified to finely coordinate and guarantee the proper spatiotemporal activation of Raf-1 and subsequently Mek and ERK (McKay & Morrison, 2007). In particular, the phosphorylation of Ser338 strongly correlates with the activated state of Raf-1.

As outlined earlier, Ras/Raf-1 and MAPK signaling has also been observed in EE, but the low abundance of these signaling proteins in endocytic compartments has yet limited the analysis of endosomal Raf-1 signaling. To analyze Raf-1 activation specifically in EE, we have developed a method to target Raf-1 to this compartment. Proteins like EEA1 or Hrs (hepatocyte growth factor-regulated tyrosine kinase substrate) are recruited to endosomes through their FYVE domains, which interact specifically with PI3P on early endosomal membranes. This was demonstrated in elegant studies by the Stenmark laboratory, showing that the fusion of 2xFYVE domains with the green fluorescent protein (GFP) can serve as an *in vivo* marker of the EE compartment (Gillooly et al., 2000; Pattni, Jepson, Stenmark, & Banting, 2001; Petiot, Faure, Stenmark, & Gruenberg, 2003). Figure 4.2 shows GFP–2xFYVE localization in EEA1-positive endosomes (EE), but not in LBPA-positive LE. Based on the ability of the FYVE domain to target EE and to overcome the technical difficulty to analyze Raf-1 signaling in this compartment, we therefore cloned an expression vector encoding Raf-1 fused to CFP–2xFYVE. To avoid interference of CFP–2xFYVE with the signaling capacity of Raf-1, the fluorescently tagged FYVE domain was fused to the Raf-1 N-terminus. Therefore, the human Raf-1 cDNA (kindly provided by Dr. Richard Marais, Institute of Cancer Research, London, United Kingdom) was subcloned into a CFP-containing Living Colors expression vector (Clontech). Then, the PCR-amplified 2xFYVE motif was cloned in between CFP and Raf-1 using BsrG1 restriction sites, generating a fusion protein with CFP N-terminal of the chimeric 2xFYVE-Raf-1 protein.

3.1. Expression of CFP–2xFYVE-Raf-1 and Raf-1 activation analysis

1. COS-1 cells are grown in DMEM, containing 10% FBS, 1 m*M* pyruvic acid, antibiotic (50 U/ml penicillin and 50 mg/ml streptomycin), 2 m*M* glutamine, and 1% nonessential amino acids. Cells are plated on coverslips up to 50% confluence for transient transfections.

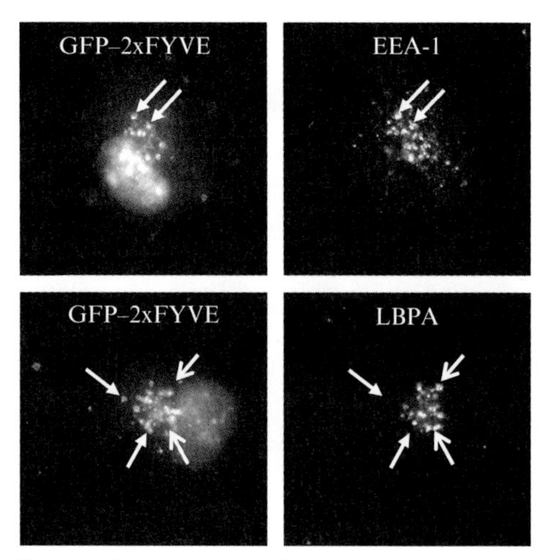

Figure 4.2 *GFP–2xFYVE localizes in early endosome compartment.* COS-1 cells grown on coverslips were transfected with GFP–2xFYVE. After 24 h, cells were fixed with 4% paraformaldehyde, and immunocytochemistry with monoclonal anti-EEA1 (BD Transduction Laboratories) (upper panels) or monoclonal anti-LBPA (Echelon Inc.) (lower panels) and the corresponding Alexa555 (Molecular Probes, Invitrogen) secondary antibody was performed. Images were acquired through GFP and CY3 channels with the epifluorescence Axiovert 200M microscope (Zeiss). Arrows in panel A indicate colocalization between GFP–2xFYVE and EEA1. Arrows in panel B show no colocalization between GFP–2xFYVE and LBPA.

2. Cells are transfected with CFP–2xFYVE-Raf-1 using the Effectene kit (Qiagen) and analyzed 24 h thereafter.
3. Cells are starved for 2 h and then stimulated with EGF (100 ng/ml) in DMEM, 0.1% BSA for 0, 5, and 15 min.
4. Cells are washed twice in PBS and fixed with freshly prepared 4% paraformaldehyde (Electron Microscopy Sciences) for 15 min at RT.
5. Detection of Ser338-phosphorylated Raf-1 (activated Raf-1) in endosomes can be performed using immunofluorescence microscopy with the anti-PSer338-Raf-1 monoclonal antibody (Upstate, Millipore). Therefore, fixed cells are mildly permeabilized with PBS, 0.1% BSA, and 0.1% saponin, for 5 min. After two washes with PBS, cells are incubated with blocking solution (PBS and 1% BSA) for 5 min at RT. Then, coverslips are incubated with primary antibody (anti-PSer338-Raf-1) diluted in PBS, 0.1% BSA, and 0.02% saponin for 1 h, washed extensively, and thereafter incubated with the adequate Cy5-labeled secondary antibody (Jackson, ImmunoResearch; Europe Ltd.) for 30 min. After

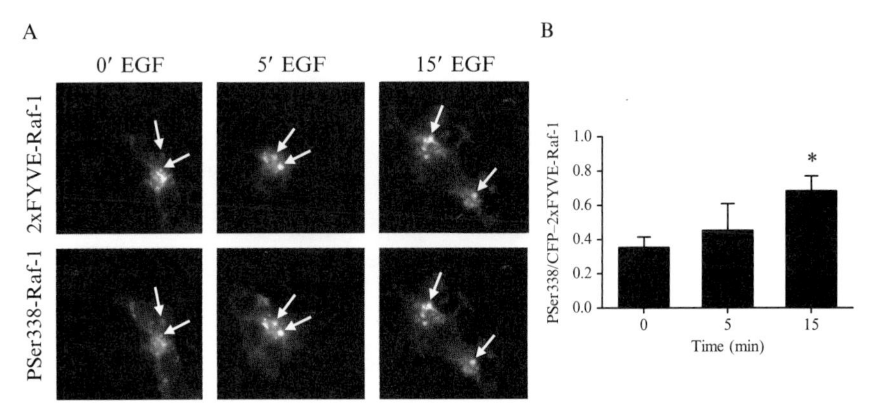

Figure 4.3 *Raf-1 targeted to early endosomes is activated by EGF.* (A) COS-1 cells expressing CFP–2xFYVE-Raf1 were serum-starved for 2 h and then treated with EGF (100 ng/ml) for 0–15 min as indicated. Fixation and immunocytochemistry with the anti-PSer338-Raf-1 monoclonal antibody followed by the corresponding Cy5 secondary antibody was performed as described (see Section 3.1). Images were acquired through the cyan and Cy5 channels with the epifluorescence Axiovert 200M microscope (Zeiss) (arrows indicate endosomal Raf-1 localization). (B) Graph showing the ratio of active PSer338-Cy5-Raf-1 (Cy5) and total CFP–2xFYVE-Raf-1 (cyan) in endosomes after 0–15 min EGF treatment. Statistical significances of differences between control and EGF treatment were determined using the Student's *t* test. Data are means ± SEM; $^{*}p<0.05$.

staining, the coverslips are washed with PBS, rinsed in H_2O, and mounted in Mowiol (Calbiochem, Merck).

6. Images are acquired through the cyan and Cy5 channels with any appropriate imaging system, in our case the epifluorescence Axiovert 200M microscope (Zeiss) equipped with HQ camera (Photometrics) (Fig. 4.3A). Final analysis of all images is performed using SlideBook software (Intelligent Imaging Innovations, Inc). To analyze Raf-1 activation, the intensities in endosomes of the PSer338-Cy5 of Raf-1 (Cy5 channel) and the CFP–2xFYVE-Raf-1 (cyan channel) can be calculated and compared over time (Fig. 4.3B). The ratio of intensities calculated for PSer338-Cy5 (active) and CFP–2xFYVE-Raf-1 (total) then serves as an indicator for Raf-1 activation. The results presented here demonstrate that endosomal-targeted Raf-1 was effectively phosphorylated upon EGF incubation, in particular at later time points ($t=15$ min).

While this method appears as an attractive approach to compare the signaling capacity of Raf-1 with and without annexins in this compartment, we would like to emphasize that special caution should be taken when molecules are directed to endosomes using the 2xFYVE targeting sequence. Overexpression of 2xFYVE may result in blocking/saturation of the

PI3P-binding sites for endogenous FYVE-containing proteins such as EEA1 and Hrs. In fact, although fluid-phase transport to LE was not compromised, receptors like EGFR were trapped in EE in cells overexpressing GFP–2xFYVE (Petiot et al., 2003).

To study the impact of annexins on Raf-1 signaling in EE, overexpression and knockdown models for AnxA1, AnxA2, AnxA6, and AnxA8 in cell culture and mouse models are now available. However, all of these annexins are also found in other cellular compartments, such as the PM and LE, which might reduce the efficacy to specifically address their impact on Raf-1 signaling in EE. In the final of chapter four, we will provide some suggestions to overcome this problem.

3.2. Raf-1 immunoprecipitation and kinase assay

To measure Raf-1 activity in isolated endosomal fractions (Marais, Light, Paterson, & Marshall, 1995), and to possibly identify and characterize the impact of annexins on Raf-1 containing signal-transducing complexes, the following procedures are carried out:

1. Immunoprecipitations are performed as described (Morrison, 1995). 60 μg of liver endosomal fractions is solubilized in 300 μl of RIPA buffer (20 m*M* Tris–HCl, pH 8, 137 m*M* NaCl, 10% glycerol, 1% Nonidet P-40, 0.1% SDS, 0.5% sodium deoxycholate, 2 m*M* EDTA) containing 1 m*M* PMSF, 1 m*M* aprotinin, 20 μ*M* leupeptin, and 5 m*M* sodium vanadate. Provided fractionation procedures for cell cultures are scaled-up appropriately to increase protein yield, this method can also be utilized for the immunoprecipitation of Raf-1 from EE and LE fractions from cell cultures.
2. To immunoprecipitate Raf-1 from fractions, 2.5 μg of anti-Raf-1 monoclonal antibody or 2.5 μg of a nonrelated monoclonal antibody (control) is first prebound to 20 μl protein G-Sepharose beads (Sigma Biochemical) in 1 ml of RIPA buffer for 1 h at RT. After washing the anti-Raf-coated beads in RIPA buffer twice, the endosomal fractions are added and incubated for 2 h at 4 °C.
3. The immunoprecipitated complexes can then be analyzed for coprecipitated signaling proteins (e.g., Western blotting or more sophisticated -omics procedures) or be examined for Raf-1 activity. For the latter, samples are washed three times in 1 ml of cold NP-40 lysis buffer (20 m*M* Tris–HCl, pH 8, 137 m*M* NaCl, 10% glycerol, 1% Nonidet P-40, 2 m*M* EDTA) containing the protease inhibitor mixture (as in

the preceding text), resuspended, and incubated for 20 min at 25 °C in 40 μl of freshly made kinase buffer containing 30 m*M* HEPES–Na, pH 7.4, 7 m*M* $MnCl_2$ (made fresh), 5 m*M* $MgCl_2$, 100 m*M* NaCl, 1 m*M* DTT, 15 μ*M* ATP, 20 μCi of [γ-^{32}P] ATP (3000 Ci/mmol), and 40 mg/ml pGST-Mek (Millipore) inactive fusion protein.

4. The reaction is terminated by the addition of 20 μl ice-cold stop buffer (30 m*M* Tris, 6 m*M* EDTA, 0.3% mercaptoethanol, 0.1% Triton X-100, 5 m*M* NaF, 0.2 m*M* Na_3VO_4).
5. The samples are then electrophoresed on SDS-polyacrylamide gels and gels are stained with Coomassie blue, dried, and exposed to X-ray films at −80 °C.
6. Alternatively, samples are loaded onto P81 sheets, washed three times (20 min each) in 75 m*M* orthophosphoric acid, and counted for ^{32}P incorporation.

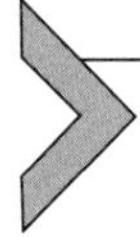

4. MONITORING ENDOSOMAL SIGNALING BY FLUORESCENCE RESONANCE ENERGY TRANSFER MICROSCOPY

In order to build a model for the understanding of signaling networks, it is essential to obtain spatiotemporal activities for each of the signaling molecules. Of special interest are those techniques that allow analysis of signaling events in living cells.

In recent years, a number of tools have been developed based on the principle of Förster (or fluorescence) resonance energy transfer (FRET). Intermolecular and intramolecular FRET between two spectrally overlapping GFP variants fused to two different host proteins or at two different sites within the same protein offers a unique opportunity to monitor real-time protein–protein interactions or protein conformational changes.

We recently used FRET analysis based on the sensitized emission method to measure the recruitment of Raf-1 to K-Ras at the PM (Moreto et al., 2009). It is evident that these protocols can be extended to CFP–2xFYVE-Raf-1-expressing cells to analyze endosomal Raf-1 signaling using model systems with upregulated or depleted annexins. In addition, annexins directly interact with signaling proteins of the Ras/Raf-1/MAPK signaling cascade. To get more insights into the factors that regulate these interactions, we recently analyzed the Ca^{2+}-inducible binding of AnxA6 to active, but not inactive, H-Ras using "conventional" FRET technology (see the preceding text) and an independent approach,

FRET acceptor photobleaching (Vila de Muga et al., 2009). The protocol for the latter is given later and suitable for inter- and intramolecular interactions.

4.1. FRET acceptor photobleaching

1. For the FRET acceptor photobleaching method using CFP as donor fluorochrome paired with YFP as acceptor fluorochrome (Karpova et al., 2003), a Leica TCS SL laser scanning confocal spectral microscope (Leica Microsystems) equipped with a DMIRE2 inverted microscope, argon laser, and a ×63 PLAN APO oil immersion objective (NA 1.32) or similar can be used. For CFP and YFP visualization, images are acquired in a line-by-line sequential mode using 458 and 514 nm laser lines, double dichroic 458/514 nm mirror, emission detection ranges 465–510 and 525–650 nm, respectively, and the confocal pinhole set to 2 airy units. To minimize the effect of photobleaching, images are collected at low laser intensity. Acceptor photobleaching is performed by irradiating half of the cell in a region of interest (zoom 6.5, pixel size 72 × 72 nm) with the 514 nm laser line set to maximum intensity for six rounds with a line average of two. Under these conditions, more than 80% of acceptor is generally bleached.
2. Apparent FRET efficiency ($FRET_{eff}$) is calculated by normalizing the difference of the donor post- (I_{post}) and prebleach (I_{pre}) intensity by the donor postbleach intensity in the bleached region:

$$FRET_{eff} = \left(\frac{I_{post} - I_{pre}}{I_{post}}\right) \times 100$$

As internal negative control, an unbleached region in the same cell has to be measured. FRET efficiency is expressed as the mean ± s.d. for $n > 5$ cells for each group. Image analysis can be performed using the Image Processing Leica Confocal Software (FRET wizard) and ImageJ (Schneider, Rasband, & Eliceiri, 2012; Vila de Muga et al., 2009) or similar.

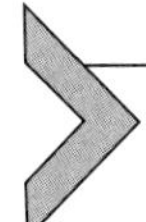

5. TARGETING ANNEXINS TO ENDOSOMES AND OTHER CELLULAR COMPARTMENTS

Every annexin is generally found in multiple locations in each cell, making it difficult to specifically analyze the contribution or participation of endosomal annexins in signaling events and protein–protein interactions in this compartment. In addition, the transient and reversible nature of the Ca^{2+}-dependent membrane association of annexins complicates investigations aiming to address their scaffolding function in certain cellular sites. We therefore recently developed a model system for a constitutive membrane association of AnxA6 and A1 (Monastyrskaya et al., 2009). Using the PM-anchoring sequences of H- and K-Ras, we targeted both annexins to the PM independently of Ca^{2+}. Similarly, other researchers have developed other PM anchors or ER- and Golgi-targeting sequences (Matallanas et al., 2006). As shown for CFP–2FYVE-Raf-1 (see the preceding text), the fusion of the 2FYVE domain to fluorescently tagged AnxA1, A2, A6, and A8 could serve as an approach to target annexins to the endosomal compartment. Given the low abundance of annexins in these compartments, even upon cell stimulation, these experimental approaches will ultimately help to improve our understanding of the contribution of scaffold/targeting proteins in compartmentalized signaling.

6. SUMMARY

Although originally regarded as a route for the degradation or recycling of membrane receptors, endosomal structures are now credited to be a crucial site for signal transduction. Studies using isolated fractions, *in situ* probes that identify activities in life cells or by targeting specific molecules, will provide new insights into the signaling dynamics of the endocytic compartment. The techniques presented in this chapter should be useful in future investigations to identify and characterize new details of the machinery along the vesicle and tubulovesicular membranes of the endocytic compartment.

ACKNOWLEDGMENTS

We would like to thank all members of our laboratories, past and present, for their invaluable contributions. C. R., A. P., and C. E. acknowledge funding from Consolider-Ingenio (CSD2009-00016); the work in this laboratory is supported by grants from Plan Nacional from the Spanish Ministerio de Economía y Competitividad (BFU2012-36272 to C. E., BFU2011-23745 to A. P. and BFU2012-38259 to F. T.), Fundació Marató TV3, and Generalitat de Catalunya (AGAUR). T. G. is supported by the National Health and Medical Research Council of Australia (NHMRC) and the University of Sydney (2010-02681). M. G.-B. is thankful to IDIBAPS fellowship.

REFERENCES

Balbis, A., Parmar, A., Wang, Y., Baquiran, G., & Posner, B. I. (2007). Compartmentalization of signaling-competent epidermal growth factor receptors in endosomes. *Endocrinology, 148*, 2944–2954.

Belcher, J. D., Hamilton, R. L., Brady, S. E., Hornick, C. A., Jaeckle, S., Schneider, W. J., et al. (1987). Isolation and characterization of three endosomal fractions from the liver of estradiol-treated rats. In: *Proceedings of the National Academy of Sciences of the United States of America, 84*, 6785–6789.

Bergeron, J. J., Searle, N., Khan, M. N., & Posner, B. I. (1986). Differential and analytical subfractionation of rat liver components internalizing insulin and prolactin. *Biochemistry, 25*, 1756–1764.

Calvo, F., Agudo-Ibanez, L., & Crespo, P. (2010). The Ras-ERK pathway: Understanding site-specific signaling provides hope of new anti-tumor therapies. *BioEssays: News and Reviews in Molecular, Cellular and Developmental Biology, 32*, 412–421.

Cubells, L., Vila de Muga, S., Tebar, F., Bonventre, J. V., Balsinde, J., Pol, A., et al. (2008). Annexin A6-induced inhibition of cytoplasmic phospholipase A2 is linked to caveolin-1 export from the Golgi. *The Journal of Biological Chemistry, 283*, 10174–10183.

Cubells, L., Vila de Muga, S., Tebar, F., Wood, P., Evans, R., Ingelmo-Torres, M., et al. (2007). Annexin A6-induced alterations in cholesterol transport and caveolin export from the Golgi complex. *Traffic, 8*, 1568–1589.

Debanne, M. T., Evans, W. H., Flint, N., & Regoeczi, E. (1982). Receptor-rich intracellular membrane vesicles transporting asialotransferrin and insulin in liver. *Nature, 298*, 398–400.

de Diego, I., Schwartz, F., Siegfried, H., Dauterstedt, P., Heeren, J., Beisiegel, U., et al. (2002). Cholesterol modulates the membrane binding and intracellular distribution of annexin 6. *The Journal of Biological Chemistry, 277*, 32187–32194.

Enrich, C., Jackle, S., & Havel, R. J. (1996). The polymeric immunoglobulin receptor is the major calmodulin-binding protein in an endosome fraction from rat liver enriched in recycling receptors. *Hepatology, 24*, 226–232.

Enrich, C., Pol, A., Calvo, M., Pons, M., & Jackle, S. (1999). Dissection of the multifunctional "receptor-recycling" endocytic compartment of hepatocytes. *Hepatology, 30*, 1115–1120.

Enrich, C., Rentero, C., de Muga, S. V., Reverter, M., Mulay, V., Wood, P., et al. (2011). Annexin A6-Linking Ca(2+) signaling with cholesterol transport. *Biochimica et Biophysica Acta, 1813*, 935–947.

Evans, W. H., & Flint, N. (1985). Subfractionation of hepatic endosomes in Nycodenz gradients and by free-flow electrophoresis. Separation of ligand-transporting and receptor-enriched membranes. *The Biochemical Journal, 232*, 25–32.

Flinn, R. J., Yan, Y., Goswami, S., Parker, P. J., & Backer, J. M. (2010). The late endosome is essential for mTORC1 signaling. *Molecular Biology of the Cell, 21*, 833–841.

Futter, C. E., & White, I. J. (2007). Annexins and endocytosis. *Traffic, 8*, 951–958.

Gerke, V., Creutz, C. E., & Moss, S. E. (2005). Annexins: Linking Ca2+ signalling to membrane dynamics. *Nature Reviews Molecular Cell Biology, 6*, 449–461.

Geuze, H. J., Slot, J. W., Strous, G. J., Peppard, J., von Figura, K., Hasilik, A., et al. (1984). Intracellular receptor sorting during endocytosis: Comparative immunoelectron microscopy of multiple receptors in rat liver. *Cell, 37*, 195–204.

Gillooly, D. J., Morrow, I. C., Lindsay, M., Gould, R., Bryant, N. J., Gaullier, J. M., et al. (2000). Localization of phosphatidylinositol 3-phosphate in yeast and mammalian cells. *The EMBO Journal, 19*, 4577–4588.

Goebeler, V., Poeter, M., Zeuschner, D., Gerke, V., & Rescher, U. (2008). Annexin A8 regulates late endosome organization and function. *Molecular Biology of the Cell*, *19*, 5267–5278.

Gorvel, J. P., Chavrier, P., Zerial, M., & Gruenberg, J. (1991). Rab5 controls early endosome fusion in vitro. *Cell*, *64*, 915–925.

Gould, G. W., & Lippincott-Schwartz, J. (2009). New roles for endosomes: From vesicular carriers to multi-purpose platforms. *Nature Reviews Molecular Cell Biology*, *10*, 287–292.

Grewal, T., & Enrich, C. (2006). Molecular mechanisms involved in Ras inactivation: The annexin A6-p120GAP complex. *BioEssays: News and Reviews in Molecular, Cellular and Developmental Biology*, *28*, 1211–1220.

Grewal, T., & Enrich, C. (2009). Annexins—Modulators of EGF receptor signalling and trafficking. *Cellular Signalling*, *21*, 847–858.

Grewal, T., Heeren, J., Mewawala, D., Schnitgerhans, T., Wendt, D., Salomon, G., et al. (2000). Annexin VI stimulates endocytosis and is involved in the trafficking of low density lipoprotein to the prelysosomal compartment. *The Journal of Biological Chemistry*, *275*, 33806–33813.

Hayes, M. J., Rescher, U., Gerke, V., & Moss, S. E. (2004). Annexin-actin interactions. *Traffic*, *5*, 571–576.

Jackle, S., Beisiegel, U., Rinninger, F., Buck, F., Grigoleit, A., Block, A., et al. (1994). Annexin VI, a marker protein of hepatocytic endosomes. *The Journal of Biological Chemistry*, *269*, 1026–1032.

Jost, M., Zeuschner, D., Seemann, J., Weber, K., & Gerke, V. (1997). Identification and characterization of a novel type of annexin-membrane interaction: Ca^{2+} is not required for the association of annexin II with early endosomes. *Journal of Cell Science*, *110*(Pt. 2), 221–228.

Kamal, A., Ying, Y., & Anderson, R. G. (1998). Annexin VI-mediated loss of spectrin during coated pit budding is coupled to delivery of LDL to lysosomes. *The Journal of Cell Biology*, *142*, 937–947.

Karpova, T. S., Baumann, C. T., He, L., Wu, X., Grammer, A., Lipsky, P., et al. (2003). Fluorescence resonance energy transfer from cyan to yellow fluorescent protein detected by acceptor photobleaching using confocal microscopy and a single laser. *Journal of Microscopy*, *209*, 56–70.

Kermorgant, S., & Parker, P. J. (2008). Receptor trafficking controls weak signal delivery: A strategy used by c-Met for STAT3 nuclear accumulation. *The Journal of Cell Biology*, *182*, 855–863.

Kheifets, V., & Mochly-Rosen, D. (2007). Insight into intra- and inter-molecular interactions of PKC: Design of specific modulators of kinase function. *Pharmacological Research: The Official Journal of the Italian Pharmacological Society*, *55*, 467–476.

Koese, M., Rentero, C., Kota, B. P., Hoque, M., Cairns, R., Wood, P., et al. (2012). Annexin A6 is a scaffold for PKCalpha to promote EGFR inactivation. *Oncogene*, *32*, 2858–2872.

Llado, A., Tebar, F., Calvo, M., Moreto, J., Sorkin, A., & Enrich, C. (2004). Protein kinaseCdelta-calmodulin crosstalk regulates epidermal growth factor receptor exit from early endosomes. *Molecular Biology of the Cell*, *15*, 4877–4891.

Llado, A., Timpson, P., Vila de Muga, S., Moreto, J., Pol, A., Grewal, T., et al. (2008). Protein kinase Cdelta and calmodulin regulate epidermal growth factor receptor recycling from early endosomes through Arp2/3 complex and cortactin. *Molecular Biology of the Cell*, *19*, 17–29.

Lobert, V. H., & Stenmark, H. (2011). Cell polarity and migration: Emerging role for the endosomal sorting machinery. *Physiology (Bethesda)*, *26*, 171–180.

Lu, A., Tebar, F., Alvarez-Moya, B., Lopez-Alcala, C., Calvo, M., Enrich, C., et al. (2009). A clathrin-dependent pathway leads to KRas signaling on late endosomes en route to lysosomes. *The Journal of Cell Biology, 184*, 863–879.

Marais, R., Light, Y., Paterson, H. F., & Marshall, C. J. (1995). Ras recruits Raf-1 to the plasma membrane for activation by tyrosine phosphorylation. *The EMBO Journal, 14*, 3136–3145.

Matallanas, D., Sanz-Moreno, V., Arozarena, I., Calvo, F., Agudo-Ibanez, L., Santos, E., et al. (2006). Distinct utilization of effectors and biological outcomes resulting from site-specific Ras activation: Ras functions in lipid rafts and Golgi complex are dispensable for proliferation and transformation. *Molecular and Cellular Biology, 26*, 100–116.

McKay, M. M., & Morrison, D. K. (2007). Integrating signals from RTKs to ERK/MAPK. *Oncogene, 26*, 3113–3121.

Monastyrskaya, K., Babiychuk, E. B., Hostettler, A., Wood, P., Grewal, T., & Draeger, A. (2009). Plasma membrane-associated annexin A6 reduces Ca2+ entry by stabilizing the cortical actin cytoskeleton. *The Journal of Biological Chemistry, 284*, 17227–17242.

Mor, A., & Philips, M. R. (2006). Compartmentalized Ras/MAPK signaling. *Annual Review of Immunology, 24*, 771–800.

Morel, E., Parton, R. G., & Gruenberg, J. (2009). Annexin A2-dependent polymerization of actin mediates endosome biogenesis. *Developmental Cell, 16*, 445–457.

Moreto, J., Llado, A., Vidal-Quadras, M., Calvo, M., Pol, A., Enrich, C., et al. (2008). Calmodulin modulates H-Ras mediated Raf-1 activation. *Cellular Signalling, 20*, 1092–1103.

Moretó, J., Vidal-Quadras, M., Pol, A., Santos, E., Grewal, T., Enrich, C., et al. (2009). Differential involvement of H- and K-Ras in Raf-1 activation determines the role of calmodulin in MAPK signaling. *Cellular Signalling, 21*, 1827–1836.

Morrison, D. K. (1995). Activation of Raf-1 by Ras in intact cells. *Methods in Enzymology, 255*, 301–310.

Mueller, S. C., & Hubbard, A. L. (1986). Receptor-mediated endocytosis of asialoglycoproteins by rat hepatocytes: Receptor-positive and receptor-negative endosomes. *The Journal of Cell Biology, 102*, 932–942.

Ohashi, E., Tanabe, K., Henmi, Y., Mesaki, K., Kobayashi, Y., & Takei, K. (2011). Receptor sorting within endosomal trafficking pathway is facilitated by dynamic actin filaments. *PLoS One, 6*, e19942.

Palamidessi, A., Frittoli, E., Garre, M., Faretta, M., Mione, M., Testa, I., et al. (2008). Endocytic trafficking of Rac is required for the spatial restriction of signaling in cell migration. *Cell, 134*, 135–147.

Palfy, M., Remenyi, A., & Korcsmaros, T. (2012). Endosomal crosstalk: Meeting points for signaling pathways. *Trends in Cell Biology, 22*, 447–456.

Pattni, K., Jepson, M., Stenmark, H., & Banting, G. (2001). A PtdIns(3)P-specific probe cycles on and off host cell membranes during Salmonella invasion of mammalian cells. *Current Biology, 11*, 1636–1642.

Petiot, A., Faure, J., Stenmark, H., & Gruenberg, J. (2003). PI3P signaling regulates receptor sorting but not transport in the endosomal pathway. *The Journal of Cell Biology, 162*, 971–979.

Platta, H. W., & Stenmark, H. (2011). Endocytosis and signaling. *Current Opinion in Cell Biology, 23*, 393–403.

Pol, A., Calvo, M., & Enrich, C. (1998). Isolated endosomes from quiescent rat liver contain the signal transduction machinery. Differential distribution of activated Raf-1 and Mek in the endocytic compartment. *FEBS Letters, 441*, 34–38.

Pol, A., Calvo, M., Lu, A., & Enrich, C. (1999). The "early-sorting" endocytic compartment of rat hepatocytes is involved in the intracellular pathway of caveolin-1 (VIP-21). *Hepatology, 29*, 1848–1857.

Pol, A., Ortega, D., & Enrich, C. (1997). Identification of cytoskeleton-associated proteins in isolated rat liver endosomes. *The Biochemical Journal, 327*(Pt. 3), 741–746.

Pons, M., Ihrke, G., Koch, S., Biermer, M., Pol, A., Grewal, T., et al. (2000). Late endocytic compartments are major sites of annexin VI localization in NRK fibroblasts and polarized WIF-B hepatoma cells. *Experimental Cell Research, 257*, 33–47.

Reverter, M., Rentero, C., de Muga, S. V., Alvarez-Guaita, A., Mulay, V., Cairns, R., et al. (2011). Cholesterol transport from late endosomes to the Golgi regulates t-SNARE trafficking, assembly, and function. *Molecular Biology of the Cell, 22*, 4108–4123.

Schneider, C. A., Rasband, W. S., & Eliceiri, K. W. (2012). NIH Image to ImageJ: 25 years of image analysis. *Nature Methods, 9*, 671–675.

Sigismund, S., Confalonieri, S., Ciliberto, A., Polo, S., Scita, G., & Di Fiore, P. P. (2012). Endocytosis and signaling: Cell logistics shape the eukaryotic cell plan. *Physiological Reviews, 92*, 273–366.

Sorkin, A., & von Zastrow, M. (2009). Endocytosis and signalling: Intertwining molecular networks. *Nature Reviews Molecular Cell Biology, 10*, 609–622.

Taub, N., Teis, D., Ebner, H. L., Hess, M. W., & Huber, L. A. (2007). Late endosomal traffic of the epidermal growth factor receptor ensures spatial and temporal fidelity of mitogen-activated protein kinase signaling. *Molecular Biology of the Cell, 18*, 4698–4710.

Teis, D., Taub, N., Kurzbauer, R., Hilber, D., de Araujo, M. E., Erlacher, M., et al. (2006). p14-MP1-MEK1 signaling regulates endosomal traffic and cellular proliferation during tissue homeostasis. *The Journal of Cell Biology, 175*, 861–868.

Vila de Muga, S., Timpson, P., Cubells, L., Evans, R., Hayes, T. E., Rentero, C., et al. (2009). Annexin A6 inhibits Ras signalling in breast cancer cells. *Oncogene, 28*, 363–377.

White, I. J., Bailey, L. M., Aghakhani, M. R., Moss, S. E., & Futter, C. E. (2006). EGF stimulates annexin 1-dependent inward vesiculation in a multivesicular endosome subpopulation. *The EMBO Journal, 25*, 1–12.

White, C. D., Erdemir, H. H., & Sacks, D. B. (2012). IQGAP1 and its binding proteins control diverse biological functions. *Cellular Signalling, 24*, 826–834.

Zobiack, N., Rescher, U., Ludwig, C., Zeuschner, D., & Gerke, V. (2003). The annexin 2/S100A10 complex controls the distribution of transferrin receptor-containing recycling endosomes. *Molecular Biology of the Cell, 14*, 4896–4908.

CHAPTER FIVE

Analysis, Regulation, and Roles of Endosomal Phosphoinositides

Tania Maffucci, Marco Falasca[1]

Inositide Signalling Group, Centre for Diabetes, Blizard Institute, Barts and The London School of Medicine and Dentistry, Queen Mary University of London, London, United Kingdom

[1]Corresponding author: e-mail address: m.falasca@qmul.ac.uk

Contents

Abstract

Phosphoinositides (PIs) are minor lipid components of cellular membranes that play critical roles in membrane dynamics, trafficking, and cellular signaling. Among the seven naturally occurring PIs, the monophosphate phosphatidylinositol 3-phosphate (PtdIns3P) and the bisphosphate phosphatidylinositol 3,5-bisphosphate [PtdIns(3,5)P_2] have been mainly associated with endosomes and endosomal functions. Metabolic labeling and HPLC analysis revealed that a bulk of PtdIns3P is constitutively present

Methods in Enzymology, Volume 535
ISSN 0076-6879
http://dx.doi.org/10.1016/B978-0-12-397925-4.00005-5

in cells, making it the only detectable product of the enzymes phosphoinositide 3-kinases in unstimulated, normal cells. The use of specific tagged-PtdIns3P-binding domains later demonstrated that this constitutive PtdIns3P accumulates in endosomes where it critically regulates trafficking and membrane dynamics.

1. INTRODUCTION

Phosphoinositides (PIs) are phospholipids comprising a water-soluble head group (*myo*-inositol) linked by a glycerol moiety to two fatty acid chains, usually a saturated C_{18} residue (stearoyl) in the 1-position and a tetra-unsaturated C_{20} residue (arachidonoyl) in the 2-position (Michell, 2008). The founding member of the PIs family is the unphosphorylated phosphatidylinositol (PtdIns), primarily synthesized in the endoplasmic reticulum and then delivered to other membranes by vesicular transport or via cytosolic PtdIns transfer protein (Di Paolo & De Camilli, 2006). PIs derive from phosphorylation of the hydroxyls at positions 3-, 4- and 5- within the *myo*-inositol headgroup of PtdIns and the different possible combinations generate seven distinct derivatives, which can be interconverted into each other by the action of specific kinases or phosphatases. PIs play many intracellular roles either as component of cellular membranes or by regulating the activity of targeted proteins temporally and/or spatially. Because of their lipid tail, PIs are obligatory membrane-bound; therefore, they can mark specific membrane compartments or subdomains within a membrane. On the other hand, PIs also possess a soluble headgroup, which allows them to bind to cytosolic proteins and to mediate the association of these proteins to specific membranes (spatial regulation). In addition, since some PIs are only synthesized upon cellular stimulation in normal cells, they can also regulate activation of target proteins temporally (Maffucci, 2012).

2. ENDOSOMAL PIs

Two PIs have been mainly associated with endosomes and endosomal functions: the monophosphate phosphatidylinositol 3-phosphate (PtdIns3P) and the bisphosphate phosphatidylinositol 3,5-bisphosphate [PtdIns(3,5)P_2] (Nicot & Laporte, 2008).

2.1. Phosphatidylinositol 3-phosphate

PtdIns3P comprises 0.1–0.5% of all PIs (Falasca & Maffucci, 2006) and almost 0.002% of total membrane lipids (Stephens, McGregor, &

Hawkins, 2000). PtdIns3P is the most abundant among the PIs phosphorylated at position 3 in resting mammalian cells and its intracellular levels are regulated by a coordinated action of kinases and phosphatases (Falasca & Maffucci, 2006, 2009). Synthesis of PtdIns3P is catalyzed by members of the family of enzymes phosphoinositide 3-kinase (PI3K). Three classes of PI3Ks exist (Falasca & Maffucci, 2007; Vanhaesebroeck, Guillermet-Guibert, Graupera, & Bilanges, 2010) and synthesis of PtdIns3P occurs *in vivo* through phosphorylation of PtdIns by class II or class III PI3K isoforms. No evidence so far has supported a role for class I PI3Ks in direct synthesis of PtdIns3P from PtdIns *in vivo*.

Class III comprises only one PI3K isoform, vacuolar protein sorting 34 (Vps34), which is the most ancient forms of PI3Ks and the only one conserved from lower eukaryotes to plants and mammals (Backer, 2008; Engelman, Luo, & Cantley, 2006). It is still not clear whether the activity of Vps34 is modulated by cellular stimulation (Vanhaesebroeck et al., 2010) although it has been suggested that the basal activation of hVps34 may be regulated by nutrients (Byfield, Murray, & Backer, 2005). In contrast to the other PI3Ks, which can produce three 3-phosphorylated PIs at least *in vitro* (PtdIns3P, phosphatidylinositol 3,4-bisphosphate, and phosphatidylinositol 3,4,5-trisphosphate), hVps34 can only generate PtdIns3P (Backer, 2008). Class II comprises three PI3K isoforms, PI3K-C2α (Domin et al., 1997), PI3K-C2β (Brown, Ho, Weber-Hall, Shipley, & Fry, 1997), and PI3K-C2γ (Misawa et al., 1998; Ono et al., 1998; Rozycka et al., 1998). Several studies have now demonstrated that at least some members of the class II PI3K subfamily catalyze synthesis of PtdIns3P *in vivo* (Falasca & Maffucci, 2012; Mazza & Maffucci, 2011). PI3K-C2α regulates an insulin-dependent, plasma membrane (PM)-associated pool of PtdIns3P in muscle cells (Falasca et al., 2007; Maffucci, Brancaccio, Piccolo, Stein, & Falasca, 2003) and it generates PtdIns3P in large dense-core vesicles (LDCVs) of PC12 upon stimulation of exocytosis (Wen et al., 2008). Recently, studies using MEFs upon induction of PI3K-C2α knockout (Yoshioka et al., 2012) and endothelial cells (Biswas et al., 2013) have confirmed PtdIns3P as the main specific product of PI3K-C2α. Similarly, PI3K-C2β regulates lysophosphatidic acid-induced and sphingosine-1-phosphate-dependent synthesis of PtdIns3P in cancer (Maffucci et al., 2005) and endothelial cells (Tibolla et al., 2013), respectively. It has also been proposed that PI3K-C2β regulates a nuclear pool of PtdIns3P during cell cycle progression (Visnjić et al., 2003). As far as we know, no study so far has investigated whether PI3K-C2γ also catalyzes PtdIns3P synthesis *in vivo*.

Levels of PtdIns3P are further regulated through dephosphorylation at position 3 by several phosphatases, mainly belonging to the family of myotubularins (MTMs; Robinson & Dixon, 2006). Fourteen MTMs have been detected in humans, nearly half of which are predicted to be catalytically inactive. Recombinant myotubularin MTM1 and myotubularin-related (MTMR) 1, 2, 3, 4, 6, and 7 were initially reported to selectively dephosphorylate PtdIns3P although it was later shown that some of these (such as MTM1 and MTMR2, 3, and 6) can also dephosphorylate PtdIns(3,5)P_2 (Maffucci, 2012).

2.2. Phosphatidylinositol 3,5-bisphosphate

PtdIns(3,5)P_2 represents 0.0001% of total membrane lipids (Stephens et al., 2000) and it is usually only 0.1% or less of total cellular PIs in unstressed mammalian cells (Michell, Heath, Lemmon, & Dove, 2006). It is generally accepted that PtdIns(3,5)P_2 derives from phosphorylation of PtdIns3P by a member of the family of type III PIPKs (Dove, Dong, Kobayashi, Williams, & Michell, 2009), mainly PI kinase for five positions containing a FYVE finger (PIKfyve) whose activation involves formation of a complex with the activator ArPIKfyve (Shisheva, 2008). PtdIns(3,5)P_2 dephosphorylation can be mediated by different phosphatases including the 5-phosphatase FIG4/Sac3 and the scaffolding protein VAC14 although Sac3 can also associate with the complex PIKfyve/ArPIKfyve and promote maximal PtdIns(3,5)P_2 synthesis. It has also been suggested that inactive MTMs can dimerize with active MTMs and change their specificity towards this PI rather than PtdIns3P (Lecompte, Poch, & Laporte, 2008). Other 5-phosphatases have been proposed to be able to dephosphorylate PtdIns(3,5)P_2 including 72 kDa 5-phosphatase/INPP5E or SHIP2 (Ooms et al., 2009), but it is not known whether the endogenous enzymes have a role in modulating the levels of PtdIns(3,5)P_2 *in vivo*.

3. ANALYSIS OF PtdIns3P LEVELS

3.1. Metabolic labeling and HPLC analysis

The classical experimental procedure to quantitatively measure the intracellular levels of PtdIns3P involves labeling of cells with *myo*-[^{3}H] inositol, lipid extraction, and HPLC analysis. Alternatively, cells can be labeled with [^{32}P] orthophosphate.

3.1.1 Cellular labeling

Incorporation of *myo*-[^{3}H] inositol is performed in inositol-free media for the time required to reach isotopic equilibrium, which can vary between cellular types and therefore needs to be optimized for the specific cell system to be used. Labeling can be performed in the presence or absence of serum depending on the cell types and the efficiency of incorporation. Similarly, optimization is required to determine the amount of *myo*-[^{3}H] inositol, which allows detection of the PIs of interest. In our studies, we typically used 5 μCi *myo*-[^{3}H] inositol/ml (Falasca et al., 2007; Maffucci et al., 2005). The following protocol refers to cells plated in a 6-well plate.

3.1.2 Lipid extraction

Following incorporation, cells are lysed and lipids are extracted by phase separation using a mix of water/hydrochloric acid/methanol/chloroform. [^{3}H]PIs containing two acyl chains accumulate in the organic phase, whereas the water-soluble [^{3}H]inositol phosphates accumulate in the aqueous phase.

1. Lyse cells with 500 μl ice-cold HCl 1 *M* + 1 m*M* tetrabutylammonium hydrogen sulfate/well, scrape, and collect the lysates in glass Chromacol vials.
2. Wash each well with 670 μl ice-cold CH_3OH, collect, and transfer to the corresponding vial.
3. Prepare a mix of $CHCl_3$:PIs from bovine brain (1.33 ml $CHCl_3$ + 2 μg PIs/sample).
4. Add 1.332 ml of mix/sample.
5. Vortex.
6. Centrifuge (1500 rpm, 5 min, room temperature).
7. Separate aqueous (upper) from organic (lower) phase.
8. Prepare washing mix (3.545 ml HCl 1 *M* + 18.75 μl tetrabutylammonium hydrogen sulfate 1 *M* + 187.5 μl EDTA 0.5 *M*, 5 ml CH_3OH, and 10 ml $CHCl_3$).
9. Add 1 ml of aqueous washing phase (upper) to the sample organic phase.
10. Add 1.5 ml of organic (lower) washing phase to the sample aqueous phase.
11. Vortex.
12. Centrifuge (1500 rpm, 5 min, room temperature).
13. Separate aqueous (upper) from organic (lower) phase.

3.1.3 PIs deacylation

Extracted PIs are then deacylated using a mix of monomethylamine/methanol/water/butanol to avoid that acyl chains affect the binding to the HPLC column.

1. Dry samples (speed vac).
2. Add 500 μl of monomethylamine solution (CH_3NH_2:CH_3OH:H_2O:C_4H_9OH 5:4:3:1).
3. Vortex.
4. Incubate at 53 °C, 45 min.
5. Cool down.
6. Centrifuge samples (2500 rpm, 1 min, room temperature).
7. Dry samples (speed vac) for at least 1 h.

3.1.4 Glycerophosphoinositide extraction

A mix butanol/petroleum ether/ethyl acetate is used to separate the organic phase containing the acyl chains from the aqueous phase containing the [^{3}H] glycerophosphoinositides (gPIs).

1. Prepare a mix C_4H_9OH:C_6H_{14}:$C_4H_8O_2$ (20:4:1).
2. Add 600 μl mix/sample.
3. Add 500 μl of distilled water/sample.
4. Vortex.
5. Centrifuge (2500 rpm, 1 min, room temperature).
6. Discard upper phase.
7. Add 600 μl mix/sample.
8. Vortex.
9. Centrifuge (2500 rpm, 1 min, room temperature).
10. Discard upper phase.
11. Dry samples (speed vac).
12. Resuspend pellet of gPIs in 600 μl H_2O.

3.1.5 HPLC analysis

gPIs are separated by HPLC using PartiSphere™ 5 μm strong anion exchange column (25 × 4.6 mm). We optimized a protocol to better separate PtdIns3P using a nonlinear gradient of 1 m*M* EDTA (buffer A) and 1.3 *M* $(NH_4)_2HPO_4$ + 1 m*M* EDTA (pH 3.8) (buffer B): 0–1 min, 0% buffer B; 1–40 min, 0–5% buffer B; 40–41 min, 5–15% buffer B; 41–75 min, 15–24% buffer B; 75–76 min, 24–33% buffer B; 76–95 min, 33–60% buffer B; 95–96 min, 60–100% buffer B; 96–100 min, 100% buffer B;

100–101 min, 100% to 0% buffer B; and 101–121 min, 0% buffer B wash. Fractions are collected (1 ml/min) and radioactivity determined using appropriate scintillation liquids.

3.1.6 Pro&Con

This method allows an accurate and quantitative analysis of PtdIns3P levels in cell lines and in some primary cells. One of the main limitations of this approach is the fact that many primary cells cannot be maintained in culture for a sufficient length of time to allow equilibrium radiolabeling. In addition, this technique cannot be used to directly quantify PtdIns3P levels in samples such as biopsies. Because of the many steps of the experimental procedures, it can also be less sensitive to detect trace amounts of PIs or to monitor subtle changes in specific intracellular compartments.

3.2. Liquid chromatography–mass spectrometry

An experimental protocol to analyze all PIs in total lipid extract was developed a few years ago based on liquid chromatography coupled directly to a mass spectrometry detector (Pettitt, Dove, Lubben, Calaminus, & Wakelam, 2006). This approach provides a very sensitive analysis of all PIs without prior fractionation and modification. However, the experimental procedure and protocol requires specific instrument and expertise.

3.3. Novel assay to measure PtdIns3P levels

Recently, a novel mass assay has been described to quantify PtdIns3P, which can potentially overcome the difficulties of primary cells radiolabeling and also allow the quantification of PtdIns3P in various biological samples such as biopsies. The method is based on the use of recombinant PYKfyve and [γ-^{32}P]ATP and the quantification of radiolabeled PtdIns(3,5)P_2 from PtdIns3P (Chicanne et al., 2012). In this protocol, lipids are extracted from the sample of interest and resolved by Thin Layer Chromatography (TLC). PIs are then scraped off, extracted from silica, and then combined with phosphatidylethanolamine to allow vesicle formation. Lipid kinase assay is then performed using recombinant GST-PIKfyve and [γ-^{32}P]ATP and phosphorylated lipids are extracted and separated by TLC. The radioactive bisphosphates are then scraped off, deacylated, and analyzed by HPLC.

4. MONITORING PtdIns3P INTRACELLULAR LOCALIZATION

4.1. PtdIns3P-binding domains

The identification of protein domains specifically able to bind PtdIns3P has allowed the development of protocols to visualize the specific intracellular compartment where this PI is synthesized. Fluorescent-tagged PtdIns3P-binding domains followed by confocal microscopy analysis and GST-tagged domains and electron microscopy analysis have been extensively used to assess PtdIns3P cellular localization.

4.1.1 Fab1/YOTB/Vac1/EEA1 domain

The best characterized and most used PtdIns3P-binding domain is the Fab1/YOTB/Vac1/EEA1 (FYVE) domain, a zinc fingers module of about 60–70 amino acids (Stenmark & Aasland, 1999) consisting of two double-stranded antiparallel β-sheets and a small C-terminal α-helix. The structure is held together by two tetrahedrally coordinated Zn^{2+} ions (Lemmon, 2008), which are critical for preservation of the structure and function of the domain (Kutateladze, 2006). Specificity of FYVE domain for PtdIns3P seems to derive from hydrogen bonds with the 4-, 5- and 6-hydroxyl groups and with the 1- and 3-phosphates (Lemmon, 2008). Nonspecific electrostatic interaction between basic residues within the FYVE domain and negatively charged phospholipids within the membrane, such as phosphatidylserine and phosphatidic acid, in early endosomes can facilitate membrane recruitment of FYVE domain-containing proteins (Kutateladze, 2006). Dimerization is often required for proper endosomal targeting. Green Fluorescent Protein (GFP)-tagged tandem FYVE domains (GFP–2XFYVE) of specific proteins, such as early endosome antigen 1 (EEA1) and hepatocyte growth factor-regulated tyrosine kinase substrate (Hrs), have been used to study PtdIns3P intracellular localization. For instance, this tool allowed the visualization of a pool of PtdIns3P in endosomes (Gillooly et al., 2000) and it also allowed to identify the *de novo* synthesis of PtdIns3P at the PM (Maffucci, Brancaccio, et al., 2003; Maffucci et al., 2005) or in LDCV (Wen et al., 2008). More recently, a distinct approach has been used to control dimerization of the FYVE domain probe intracellularly (Hayakawa et al., 2004; Stuffers et al., 2010). Specifically, the monomeric FYVE domain of Hrs was fused to the rapamycin-binding protein FKBP (Fv). The resulting GFP-Fv-FYVE can homodimerize in the presence of a rapamycin derivative, allowing controlled dimerization and endosomal targeting of the probe and more specific analysis of PtdIns3P endosomal localization (Stuffers et al., 2010).

4.1.2 Phox homology domain

Phox homology (PX) domains are regions of 130 amino acids named after the two phagocyte NADPH oxidase subunits $p40^{phox}$ and $p47^{phox}$ (Kutateladze, 2007). PX domains show a highly conserved three-dimensional structure consisting of three-stranded β-sheet and a subdomain of three to four α-helices. All PX domains found in *S. cerevisiae* bind PtdIns3P (Lemmon, 2008) although PX domains able to bind PtdIns(3,5)P_2 and PtdIns(3,4)P_2 have also been found in mammals. Nevertheless, it is still generally accepted that PX domains preferentially bind PtdIns3P (Lemmon, 2008). There are only few examples of PX domains able to bind PtdIns3P with high affinity and whose interaction with this PI is sufficient to target them to membranes. GFP-PX domains were used together with GFP-FYVE to visualize changes in the levels and subcellular localization of PtdIns3P during phagosome formation (Ellson et al., 2001).

4.1.3 Pleckstrin homology domain

Pleckstrin homology (PH) domains are modules of about 100 amino acids (Maffucci & Falasca, 2001). Although very different in their primary structure, all PH domains possess a similar tertiary structure, consisting of a 7-stranded β-sandwich structure formed by two near-orthogonal β-sheets (Lemmon, 2008). Most PH domains show either low specificity or low affinity (or both) for PIs and require additional mechanisms to guarantee the specific targeting of the host protein (Maffucci & Falasca, 2001). Examples of PtdIns3P-binding PH domains include the amino terminal PH domain of phospholipase C (PLC)β1 (Razzini, Brancaccio, Lemmon, Guarnieri, & Falasca, 2000) and PH domain of insulin receptor substrate (IRS) 3 (Maffucci, Razzini, et al., 2003). In its tagged form, the PLCβ1 PH domain first allowed us to visualize a pool of PtdIns3P specifically localized at the PM, which is important for PLCβ1 activation (Razzini et al., 2000). Similarly, comparison of the intracellular localization of tagged full-length IRS3, the isolated PH domain, and mutants unable to bind PtdIns3P led us to the identification of a nuclear pool of PtdIns3P (Maffucci, Razzini, et al., 2003), later also observed using the PtdIns3P-binding PH domain of casein kinase 2-interacting protein-1 (Safi et al., 2004).

4.2. Immunofluorescence microscopy analysis

PtdIns3P-binding PH domain has been used both in their GFP-tagged and GST-tagged forms. GFP-tagged domains allow not only direct immunofluorescence analysis on fixed cells but also *in vivo* imaging of PtdIns3P dynamics.

4.2.1 Sample preparation

1. Plate cells on glass coverslips at least one day before starting the experiment. If cells are stably transfected with the specific GFP-tagged or GST-tagged PtdIns3P-binding domain, go to Step 3.
2. For transient experiment, transfect cells with cDNA expressing the tagged PtdIns3P-binding domain of choice. Several transfection reagents can be used according to the specific cell lines and this needs to be optimized in order to achieve a high efficiency of transfection.
3. Overexpressed proteins are usually easily detectable 24 h post-transfection. If no cellular stimulation is required, coverslips can be fixed at this stage. Alternatively, cells can be serum-starved for an appropriate length of time (depending on the cell type, usually from 6 to 24 h), stimulated as required, and then fixed.
4. Wash coverslips three times with PBS.
5. Fix coverslips with paraformaldehyde 4% in PBS for 15–30 min at room temperature.
6. For GST domains expressing cells and to perform colocalization analysis with specific intracellular compartments markers (for instance, EEA1 for early endosomes), permeabilize cells with 0.1–0.25% Triton X-100 for 5 min at room temperature.
7. Block coverslips with 1% bovine serum albumin (BSA) in PBS for 30 min at room temperature.
8. Incubate coverslips with the required primary antibodies diluted in PBS/BSA for at least 1 h. Dilution depends on the specific primary antibody and needs to be optimized.
9. Wash coverslips three times with PBS.
10. Incubate with the required specific fluorescent secondary antibodies (for instance, Alexa 488- or Alexa 594-conjugated antibodies) for 1 h.
11. Wash coverslips two times with PBS.
12. If required, incubate with 4′,6-diamidino-2-phenylindole to stain nuclei in PBS for 5 min.
13. Wash coverslips three times with PBS and once with H_2O.
14. Mount coverslips using an antiphotobleaching agent.

4.3. Immunofluorescence analysis: Pro&Con

Coverslips can then be analyzed using an inverted fluorescence microscope. For a more precise, accurate analysis, the use of confocal microscope is suggested. Quantification of the degree of colocalization of PtdIns3P and the distinct intracellular compartments markers can be performed (Wen et al., 2008). Quantification of the number of 2XFYVE-positive vesicles/cell has also been performed in some studies (Biswas et al., 2013; Yoshioka et al., 2012). Similarly, changes in fluorescence intensity of 2XFYVE-positive vesicles have been quantified (Wen et al., 2008). Despite this, it must be noted that the immunofluorescence staining remains mainly a qualitative assay. It is not clear whether these domains are indeed able to bind all PtdIns3P intracellular pools: for instance, no nuclear localization has been detected using tagged-2XFYVE domains, indicating that these probes may selectively recognize distinct PtdIns3P pools. Another complication in data interpretation derives from the fact that overexpression of these domains can affect and impair the normal intracellular functions regulated by PtdIns3P.

4.4. Immunofluorescence using anti-PtdIns3P antibody

Specific anti-PI antibodies have been developed and used in some studies to visualize the PIs. For instance, a couple of studies suggested that amino acids are able to stimulate PtdIns3P synthesis *in vivo* based on the use of an anti-PtdIns3P antibody and indirect analysis of fluorescence (Gulati et al., 2008; Nobukuni et al., 2005). This approach was also used to detect a perinuclear PtdIns3P in zebrafish myofibers (Dowling et al., 2009), but it is still not widely accepted.

4.5. Immunoelectron microscopy analysis

An alternative method to detect the specific intracellular compartment and structure where PtdIns3P accumulates is based on transfection of cells with GST-tagged FYVE domains and immunoelectron microscopy analysis using anti-GST antibody and protein A-gold (Gillooly et al., 2000; Wen et al., 2008).

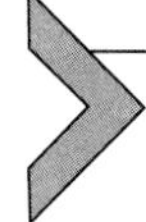

5. ENDOSOMAL PtdIns3P

5.1. Regulation of the endosomal pool of PtdIns3P

Different pools of PtdIns3P have been detected intracellularly, a "constitutive" pool, already detectable in resting cells, and regulated pools, generated upon cellular stimulation (Falasca & Maffucci, 2009). Originally, studies showed that the "constitutive" PtdIns3P is specifically localized in endosomes, mostly on early endosomal membranes and on the internal membranes of multivesicular endosomes (Gillooly et al., 2000). Spatiotemporal analysis of PtdIns3P endosomal distribution, obtained using the "controlled dimerization" of FYVE domain, has revealed that the GFP-Fv-FYVE domain associates very rapidly with EEA1-positive, early endosomes compartment and only later with CD63-containing late endosomes (Stuffers et al., 2010). It is generally accepted that hVps34 is responsible for regulation of this PtdIns3P pool. However, the observation that downregulation of hVps34 affects late endosomal structure but not early endosome morphology and trafficking pathways (Johnson, Overmeyer, Gunning, & Maltese, 2006) first suggested that hVps34 might not be the only enzyme responsible for the synthesis of the endosomal PtdIns3P. Indeed, it was later shown that PtdIns3P can be generated in early endosomes through an enzymatic cascade involving class I PI3K and the sequential action of PI 5- and PI 4-phosphatases (Shin et al., 2005). More recently, it has been reported that downregulation of PI3K-C2α in endothelial cells decreases the number of PtdIns3P-enriched endosomes (Biswas et al., 2013; Yoshioka et al., 2012), suggesting that class II PI3K(s) can also be involved in regulation of endosomal PtdIns3P. Finally, the endosomal pool of PtdIns3P is also regulated by phosphatases, specifically MTM1, which localizes to Rab5-positive early endosomes (Cao, Laporte, Backer, Wandinger-Ness, & Stein, 2007) where it regulates PtdIns3P levels (Cao, Backer, Laporte, Bedrick, & Wandinger-Ness, 2008), and MTMR2, which localizes to Rab7-positive late endosomes (Cao et al., 2007) where it can regulate late endosomal PtdIns3P levels (Cao et al., 2008).

5.2. Cellular functions regulated by the endosomal pool of PtdIns3P

The endosomal pool of PtdIns3P controls membrane transport and membrane dynamics mostly by recruiting key proteins containing FYVE, PX, and PH domains (Lindmo & Stenmark, 2006) including EEA1, which has a critical role in endosome fusion, and Hrs, which regulates the first steps of receptor sorting and internalization within the multivesicular bodies (MVBs). Endosomal pool of PtdIns3P can also contribute to growth factor signaling through regulation of maturation of an early endocytic intermediate (Zoncu et al., 2009).

It is now becoming increasingly clear that the endosomal PtdIns3P has also a role in the control of autophagy. PtdIns3P has a well-established and key role in various steps of the autophagy process, including autophagosome biogenesis, maturation, and intracellular transport (Dall'Armi, Devereaux, & Di Paolo, 2013). Specifically related to the endosomal PtdIns3P pool is the control of the maturation steps, which involve fusion of the autophagosome with the endosomes, generating the amphisomes that then fuse with the lysosomes (Dall'Armi et al., 2013).

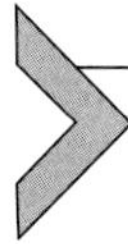

6. PHOSPHATIDYLINOSITOL 3,5-BISPHOSPHATE

6.1. Intracellular levels

PtdIns(3,5)P_2 was first identified in a study using [^{3}H]inositol-labeled *S. cerevisiae* grown almost to isotopic equilibrium in an inositol-free and/or low-phosphate minimal medium and HPLC analysis (Dove et al., 1997). Subsequent studies using this technique were performed in yeasts (Dove et al., 2002; Phelan, Millson, Parker, Piper, & Cooke, 2006) and mammalian cell lines (Jefferies et al., 2008; Sbrissa & Shisheva, 2005).

6.2. Intracellular localization

Because specific PtdIns(3,5)P_2-binding domains have not been described, information on the intracellular localization of this PI has mostly been derived from the localization of PIKfyve. The endogenous enzyme localizes to discrete peripheral punctae resembling endosomes in 3T3L1 adipocytes (Shisheva, 2008). Endosomal localization (both on early and late endosomes) and localization in MVBs and at the *trans*-Golgi network was reported for the overexpressed wild-type PIKfyve. However, it has also been suggested that the localization is strongly dependent on the cell type, the level of protein expressed, and on the ratio between PtdIns(3,5)P_2 and PtdIns3P (Shisheva, 2008).

6.3. Intracellular roles

PtdIns(3,5)P_2 has a well-established role in membrane trafficking and it is currently believed that specific cargo molecules can stimulate synthesis of this PI to regulate their correct endosome–lysosome trafficking (Dove et al., 2009). Accumulation of swollen intracellular vacuoles, enlarged late endosomes, and abnormal MVB has been observed in cultured cells, in animal models, and in human tissues as a result of defects in PtdIns(3,5)P_2

synthesis. Recent data have also revealed a critical role for PtdIns(3,5)P_2 in autophagy in the central nervous system (Ferguson, Lenk, & Meisler, 2009, 2010).

ACKNOWLEDGMENTS

Work in our laboratory is supported by Pancreatic Cancer Research Fund (M. F.), Bowel and Cancer Research (M. F.), Prostate Cancer UK (M. F.), Diabetes UK (T. M.) and Barts and The London Charity-Cancer Fund (T. M.).

REFERENCES

Backer, J. M. (2008). The regulation and function of Class III PI3Ks: Novel roles for Vps34. *The Biochemical Journal*, *410*, 1–17.

Biswas, K., Yoshioka, K., Asanuma, K., Okamoto, Y., Takuwa, N., Sasaki, T., et al. (2013). Essential role of class II phosphatidylinositol-3-kinase-C2α in sphingosine 1-phosphate receptor-1-mediated signaling and migration in endothelial cells. *The Journal of Biological Chemistry*, *288*, 2325–2339.

Brown, R. A., Ho, L. K. F., Weber-Hall, S. J., Shipley, J. M., & Fry, M. J. (1997). Identification and cDNA cloning of a novel mammalian C2 domain-containing phosphoinositide 3-kinase, HsC2-PI3K. *Biochemical and Biophysical Research Communications*, *233*, 537–544.

Byfield, M. P., Murray, J. T., & Backer, J. M. (2005). hVps34 is a nutrient-regulated lipid kinase required for activation of p70 S6 kinase. *The Journal of Biological Chemistry*, *280*, 33076–33082.

Cao, C., Backer, J. M., Laporte, J., Bedrick, E. J., & Wandinger-Ness, A. (2008). Sequential actions of myotubularin lipid phosphatases regulate endosomal PI(3)P and growth factor receptor trafficking. *Molecular Biology of the Cell*, *19*, 3334–3346.

Cao, C., Laporte, J., Backer, J. M., Wandinger-Ness, A., & Stein, M. P. (2007). Myotubularin lipid phosphatase binds the hVps15/hVps34 lipid kinase complex on endosomes. *Traffic*, *8*, 1052–1067.

Chicanne, G., Severin, S., Boscheron, C., Terrisse, A. D., Gratacap, M. P., Gaits-Iacovoni,-F., et al. (2012). A novel mass assay to quantify the bioactive lipid PtdIns3P in various biological samples. *The Biochemical Journal*, *447*, 17–23.

Dall'Armi, C., Devereaux, K. A., & Di Paolo, G. (2013). The role of lipids in the control of autophagy. *Current Biology*, *23*, R33–R45.

Di Paolo, G., & De Camilli, P. (2006). Phosphoinositides in cell regulation and membrane dynamics. *Nature*, *443*, 651–657.

Domin, J., Pages, F., Volinia, S., Rittenhouse, S. E., Zvelebil, M. J., Stein, R. C., et al. (1997). Cloning of a human phosphoinositide 3-kinase with a C2 domain that displays reduced sensitivity to the inhibitor wortmannin. *The Biochemical Journal*, *326*, 139–147.

Dove, S. K., Cooke, F. T., Douglas, M. R., Sayers, L. G., Parker, P. J., & Michell, R. H. (1997). Osmotic stress activates phosphatidylinositol-3,5-bisphosphate synthesis. *Nature*, *390*, 187–192.

Dove, S. K., Dong, K., Kobayashi, T., Williams, F. K., & Michell, R. H. (2009). Phosphatidylinositol 3,5-bisphosphate and Fab1p/PIKfyve underPPIn endo-lysosome function. *The Biochemical Journal*, *419*, 1–13.

Dove, S. K., McEwen, R. K., Mayes, A., Hughes, D. C., Beggs, J. D., & Michell, R. H. (2002). Vac14 controls PtdIns(3,5)P(2) synthesis and Fab1-dependent protein trafficking to the multivesicular body. *Current Biology*, *12*, 885–893.

Dowling, J. J., Vreede, A. P., Low, S. E., Gibbs, E. M., Kuwada, J. Y., Bonnemann, C. G., et al. (2009). Loss of myotubularin function results in T-tubule disorganization in zebrafish and human myotubular myopathy. *PLoS Genetics*, *5*, e1000372.

Ellson, C. D., Anderson, K. E., Morgan, G., Chilvers, E. R., Lipp, P., Stephens, L. R., et al. (2001). Phosphatidylinositol 3-phosphate is generated in phagosomal membranes. *Current Biology*, *11*, 1631–1635.

Engelman, J. A., Luo, J., & Cantley, L. C. (2006). The evolution of phosphatidylinositol 3-kinases as regulators of growth and metabolism. *Nature Reviews Genetics*, *7*, 606–619.

Falasca, M., Hughes, W. E., Dominguez, V., Sala, G., Fostira, F., Fang, Q. M., et al. (2007). The role of phosphoinositide 3-kinase C2alpha in insulin signalling. *The Journal of Biological Chemistry*, *282*, 28226–28236.

Falasca, M., & Maffucci, T. (2006). Emerging roles of phosphatidylinositol-3-monophosphate as a dynamic lipid second messenger. *Archives of Physiology and Biochemistry*, *112*, 274–284.

Falasca, M., & Maffucci, T. (2007). Role of class II phosphoinositide 3-kinase in cell signalling. *Biochemical Society Transactions*, *35*, 211–214.

Falasca, M., & Maffucci, T. (2009). Rethinking phosphatidylinositol 3-monophosphate. *Biochimica et Biophysica Acta*, *1793*, 1795–1803.

Falasca, M., & Maffucci, T. (2012). Regulation and cellular functions of class II phosphoinositide 3-kinases. *The Biochemical Journal*, *443*, 587–601.

Ferguson, C. J., Lenk, G. M., & Meisler, M. H. (2009). Defective autophagy in neurons and astrocytes from mice deficient in PI(3,5)P2. *Human Molecular Genetics*, *18*, 4868–4878.

Ferguson, C. J., Lenk, G. M., & Meisler, M. H. (2010). PtdIns(3,5)P2 and autophagy in mouse models of neurodegeneration. *Autophagy*, *6*, 170–171.

Gillooly, D. J., Morrow, I. C., Lindsay, M., Gould, R., Bryant, N. J., Gaullier, J. M., et al. (2000). Localization of phosphatidylinositol 3-phosphate in yeast and mammalian cells. *The EMBO Journal*, *19*, 4577–4588.

Gulati, P., Gaspers, L. D., Dann, S. G., Joaquin, M., Nobukuni, T., Natt, F., et al. (2008). Amino acids activate mTOR complex 1 via Ca2+/CaM signaling to hVps34. *Cell Metabolism*, *7*, 456–465.

Hayakawa, A., Hayes, S. J., Lawe, D. C., Sudharshan, E., Tuft, R., Fogarty, K., et al. (2004). Structural basis for endosomal targeting by FYVE domains. *The Journal of Biological Chemistry*, *279*, 5958–5966.

Jefferies, H. B., Cooke, F. T., Jat, P., Boucheron, C., Koizumi, T., Hayakawa, M., et al. (2008). A selective PIKfyve inhibitor blocks PtdIns(3,5)P(2) production and disrupts endomembrane transport and retroviral budding. *EMBO Reports*, *9*, 164–170.

Johnson, E. E., Overmeyer, J. H., Gunning, W. T., & Maltese, W. A. (2006). Gene silencing reveals a specific function of hVps34 phosphatidylinositol 3-kinase in late versus early endosomes. *Journal of Cell Science*, *119*, 1219–1232.

Kutateladze, T. G. (2006). Phosphatidylinositol 3-phosphate recognition and membrane docking by the FYVE domain. *Biochimica et Biophysica Acta*, *1761*, 868–877.

Kutateladze, T. G. (2007). Mechanistic similarities in docking of the FYVE and PX domains to phosphatidylinositol 3-phosphate containing membranes. *Progress in Lipid Research*, *46*, 315–327.

Lecompte, O., Poch, O., & Laporte, J. (2008). PtdIns5P regulation through evolution: Roles in membrane trafficking? *Trends in Biochemical Sciences*, *33*, 453–460.

Lemmon, M. A. (2008). Membrane recognition by phospholipid-binding domains. *Nature Reviews Molecular Cell Biology*, *9*, 99–111.

Lindmo, K., & Stenmark, H. (2006). Regulation of membrane traffic by phosphoinositide 3-kinases. *Journal of Cell Science*, *119*, 605–614.

Maffucci, T. (2012). An introduction to phosphoinositides. *Current Topics in Microbiology and Immunology*, *362*, 1–42.

Maffucci, T., Brancaccio, A., Piccolo, E., Stein, R. C., & Falasca, M. (2003). Insulin induces phosphatidylinositol-3-phosphate formation through TC10 activation. *The EMBO Journal, 22*, 4178–4189.

Maffucci, T., Cooke, F. T., Foster, F. M., Traer, C. J., Fry, M. J., & Falasca, M. (2005). Class II phosphoinositide 3-kinase defines a novel signaling pathway in cell migration. *The Journal of Cell Biology, 169*, 789–799.

Maffucci, T., & Falasca, M. (2001). Specificity in pleckstrin homology (PH) domain membrane targeting: A role for a phosphoinositide-protein co-operative mechanism. *FEBS Letters, 506*, 173–179.

Maffucci, T., Razzini, G., Ingrosso, A., Chen, H., Iacobelli, S., Sciacchitano, S., et al. (2003). Role of pleckstrin homology domain in regulating membrane targeting and metabolic function of insulin receptor substrate 3. *Molecular Endocrinology, 17*, 1568–1579.

Mazza, S., & Maffucci, T. (2011). Class II phosphoinositide 3-kinase C2alpha: What we learned so far. *International Journal of Biochemistry and Molecular Biology, 2*, 168–182.

Michell, R. H. (2008). Inositol derivatives: Evolution and functions. *Nature Reviews Molecular Cell Biology, 9*, 151–161.

Michell, R. H., Heath, V. L., Lemmon, M. A., & Dove, S. K. (2006). Phosphatidylinositol 3,5-bisphosphate: Metabolism and cellular functions. *Trends in Biochemical Sciences, 31*, 52–63.

Misawa, H., Ohtsubo, M., Copeland, N. G., Gilbert, D. J., Jenkins, N. A., & Yoshimura, A. (1998). Cloning and characterization of a novel class II phosphoinositide 3-kinase containing C2 domain. *Biochemical and Biophysical Research Communications, 244*, 531–539.

Nicot, A. S., & Laporte, J. (2008). Endosomal phosphoinositides and human diseases. *Traffic, 9*, 1240–1249.

Nobukuni, T., Joaquin, M., Roccio, M., Dann, S. G., Kim, S. Y., Gulati, P., et al. (2005). Amino acids mediate mTOR/raptor signaling through activation of class 3 phosphatidylinositol 3OH-kinase. *Proceedings of the National Academy of Sciences of the United States of America, 102*, 14238–14243.

Ono, F., Nakagawa, T., Saito, S., Owada, Y., Sakagami, H., Goto, K., et al. (1998). A novel class II phosphoinositide 3-kinase predominantly expressed in the liver and its enhanced expression during liver regeneration. *The Journal of Biological Chemistry, 273*, 7731–7736.

Ooms, L. M., Horan, K. A., Rahman, P., Seaton, G., Gurung, R., Kethesparan, D., et al. (2009). The role of the inositol polyphosphate 5-phosphatases in cellular function and human disease. *The Biochemical Journal, 419*, 29–49.

Pettitt, T. R., Dove, S. K., Lubben, A., Calaminus, S. D., & Wakelam, M. J. (2006). Analysis of intact phosphoinositides in biological samples. *Journal of Lipid Research, 47*, 1588–1596.

Phelan, J. P., Millson, S. H., Parker, P. J., Piper, P. W., & Cooke, F. T. (2006). Fab1p and AP-1 are required for trafficking of endogenously ubiquitylated cargoes to the vacuole lumen in S. cerevisiae. *Journal of Cell Science, 119*, 4225–4234.

Razzini, G., Brancaccio, A., Lemmon, M. A., Guarnieri, S., & Falasca, M. (2000). The role of the pleckstrin homology domain in membrane targeting and activation of phospholipase Cbeta(1). *The Journal of Biological Chemistry, 275*, 14873–14881.

Robinson, F. L., & Dixon, J. E. (2006). Myotubularin phosphatases: Policing 3-phosphoinositides. *Trends in Cell Biology, 16*, 403–412.

Rozycka, M., Lu, Y. J., Brown, R. A., Lau, M. R., Shipley, J. M., & Fry, M. J. (1998). cDNA cloning of a third human C2-domain-containing class II phosphoinositide 3-kinase, PI3K-C2gamma, and chromosomal assignment of this gene (PIK3C2G) to 12p12. *Genomics, 54*, 569–574.

Safi, S., Vandromme, M., Caussanel, S., Valdacci, L., Baas, D., Vidal, M., et al. (2004). Role for the pleckstrin homology domain-containing protein CKIP-1 in phosphatidylinositol 3-kinase-regulated muscle differentiation. *Molecular and Cellular Biology, 24*, 1245–1255.

Sbrissa, D., & Shisheva, A. (2005). Acquisition of unprecedented phosphatidylinositol 3,5-bisphosphate rise in hyperosmotically stressed 3T3-L1 adipocytes, mediated by ArPIKfyve-PIKfyve pathway. *The Journal of Biological Chemistry*, *280*, 7883–7889.

Shin, H. W., Hayashi, M., Christoforidis, S., Lacas-Gervais, S., Hoepfner, S., Wenk, M. R., et al. (2005). An enzymatic cascade of Rab5 effectors regulates phosphoinositide turnover in the endocytic pathway. *The Journal of Cell Biology*, *170*, 607–618.

Shisheva, A. (2008). PIKfyve: Partners, significance, debates and paradoxes. *Cell Biology International*, *32*, 591–604.

Stenmark, H., & Aasland, R. (1999). FYVE-finger proteins—Effectors of inositol lipid. *Journal of Cell Science*, *112*, 4175–4183.

Stephens, L., McGregor, A., & Hawkins, P. (2000). Phosphoinositide 3-kinases: Regulation by cell-surface receptors and functions of 3-phosphorylated lipids. In S. Cockcroft (Ed.), *Biology of phosphoinositides*. Oxford, UK: Oxford Univ. Press.

Stuffers, S., Malerød, L., Schink, K. O., Corvera, S., Stenmark, H., & Brech, A. (2010). Time-resolved ultrastructural detection of phosphatidylinositol 3-phosphate. *The Journal of Histochemistry and Cytochemistry*, *58*, 1025–1032.

Tibolla, G., Piñeiro, R., Chiozzotto, D., Mavrommati, I., Wheeler, A. P., Norata, G. D., et al. (2013). Class II phosphoinositide 3-kinases contribute to endothelial cells morphogenesis. *PLoS One*, *8*, e53808.

Vanhaesebroeck, B., Guillermet-Guibert, J., Graupera, M., & Bilanges, B. (2010). The emerging mechanisms of isoform-specific PI3K signalling. *Nature Reviews Molecular Cell Biology*, *11*, 329–341.

Visnjić, D., Curić, J., Crljen, V., Batinić, D., Volinia, S., & Banfić, H. (2003). Nuclear phosphoinositide 3-kinase C2beta activation during G2/M phase of the cell cycle in HL-60 cells. *Biochimica et Biophysica Acta*, *631*, 61–71.

Wen, P. J., Osborne, S. L., Morrow, I. C., Parton, R. G., Domin, J., & Meunier, F. A. (2008). Ca^{2+}-regulated pool of phosphatidylinositol-3-phosphate produced by phosphatidylinositol 3-kinase C2alpha on neurosecretory vesicles. *Molecular Biology of the Cell*, *12*, 5593–5603.

Yoshioka, K., Yoshida, K., Cui, H., Wakayama, T., Takuwa, N., Okamoto, Y., et al. (2012). Endothelial PI3K-C2α, a class II PI3K, has an essential role in angiogenesis and vascular barrier function. *Nature Medicine*, *18*, 1560–1569.

Zoncu, R., Perera, R. M., Balkin, D. M., Pirruccello, M., Toomre, D., & De Camilli, P. (2009). A phosphoinositide switch controls the maturation and signaling properties of APPL endosomes. *Cell*, *136*, 1110–1121.

CHAPTER SIX

Mild Fixation and Permeabilization Protocol for Preserving Structures of Endosomes, Focal Adhesions, and Actin Filaments During Immunofluorescence Analysis

Julia M. Scheffler, Natalia Schiefermeier, Lukas A. Huber[1]

Biocenter, Division of Cell Biology, Innsbruck Medical University, Innsbruck, Austria
[1]Corresponding author: e-mail address: lukas.a.huber@i-med.ac.at

Contents

Abstract

Intracellular membrane trafficking is a highly dynamic process to sort proteins into either the recycling or degradation pathway. The late endosome is a major component of this endosomal biogenesis toward degradation by the lysosome. The endocytotic system is spread throughout the cytoplasm, and vesicle motility is achieved by multiple proteins including Rabs, motor proteins, and cytostructural elements. The subcellular localization of the late endosome is distributed from the accumulation in the perinuclear region toward the cell periphery. Using immunofluorescence methods combined with live-cell microscopy, we want to show that the preservation of the peripheral late endosomal compartment can be successfully achieved by two different techniques. On one hand, we compare two different widely used permeabilization methods: Triton X-100 and saponin. Comparing live-cell microscopic pictures of the same cell with immunofluorescences after fixation and permeabilization revealed

Methods in Enzymology, Volume 535
ISSN 0076-6879
http://dx.doi.org/10.1016/B978-0-12-397925-4.00006-7

improved results by the use of saponin. On the other hand, we present here a protocol of mild fixation to preserve peripheral structures like focal adhesion in combination with endosomes and actin filaments.

1. INTRODUCTION

The primary task of the endocytic system is to separate proteins, which will recycle to distinct subcellular locations from those that undergo degradation. By this, the cell can regulate its surface composition and multiple further biological functions. The presorting of cargo in the early endosomal compartment can initiate either recycling or degradation of those proteins. Early endosomes undergo a gradual maturation into late endosomes. First, multivesicular bodies (MVBs) are formed by inward budding of the limiting membrane into the lumen. Then, those intraluminal vesicles are degraded when MVBs fuse to lysosomes, which are specialized compartments for the degradation of endocytosed and intracellular material (Piper & Katzmann, 2007; Saftig & Klumperman, 2009). The distinct spatial distribution of endosomes and vesicles throughout the cell is coordinated by the so-called Rabs. These small GTPases are membrane organizers that together with their effectors control vesicle formation, transport, and tethering to the target compartment. Together with motor proteins, they control the active transport of vesicles along cytoskeletal elements (Zerial & McBride, 2001). The visualization of the endosomal compartment of cells via fluorescent labeling is a widely used technique. However, how fixation and permeabilization affect the *in vivo* situation of the endosomal compartment and especially of the subpopulation of peripheral late endosomes is not clear. Standard protocols for endosomal immunofluorescence staining propose fixations with formaldehyde, which then subsequently requires a permeabilization of the cellular membrane to allow antibody access. The most commonly used reagents for permeabilization are saponins and nonionic detergents like Triton X-100. Saponins are amphipathic glycosides removing selectively cholesterol, while nonionic detergents unselectively solubilize lipids of membranes. Ultrastructural analysis revealed that treatment with Triton X-100 after formaldehyde fixation induced a decrease in cytoplasmic density and an apparent loss of organelles (Schnell, Dijk, Sjollema, & Giepmans, 2012). Another way of detecting endosomes is the visualization of these organelles in living cells using fluorescent proteins.

However, fixation of fluorescent proteins can result in highly variable outcome. For example, methanol fixation was previously shown to cause extraction of soluble green fluorescent protein (GFP), while short fixation with paraformaldehyde (PFA) preserves the fluorescence of GFP (Wang, Miller, Shaw, & Shaw, 1996). Another fluorescent protein, cyan fluorescent protein (CFP), was reported to change significantly its spectral properties after aldehyde fixation (Domin, Lan, & Kaminski, 2004). Careful analysis of different fixation protocols to monitor the localization of the epidermal growth factor receptor (EGFR) fused to the GFP protein was performed by Brock, Hamelers, and Jovin (1999). The authors observed that PFA fixation preserved the localization and fluorescence of the EGFR–GFP fusion protein. The methanol fixation, on the other hand, completely removed EGFR–GFP from cells. Moreover, they also demonstrated that the Mowiol-mounting media, routinely used in many laboratories, may cause redistribution of the GFP signal in cells (Brock et al., 1999). Therefore, it is important to carefully study redistribution of fluorescent proteins in different fixation and mounting conditions. Further, careful fixation of cells expressing sensitive fluorescent proteins may be combined with immunofluorescence protocols to allow colocalization studies.

Here, our focus lies on the preservation of peripheral endosomes. By a combination of live-cell imaging, followed by immunofluorescent stainings, we demonstrate that a mild permeabilization with saponin results in superior staining quality of peripheral endosomes. We also introduce a new technique of *in vivo* fixation to visualize GFP and mCherry fluorescent proteins localized to late endosomes and focal adhesion. This technique was combined with immunofluorescence to monitor colocalization of endosomes, focal adhesion, and actin filaments in the periphery of the cell.

2. MOLECULAR TOOLS

Rab7 is a late endosomal marker, which is crucial for proper aggregation and fusion of late endocytic structures in the perinuclear region and by this coordinates endo- and lysosomal biogenesis (Zerial & McBride, 2001). The late endosome also serves as a convergence point for mTor and MAP-kinase signaling, which is achieved by the so-called LAMTOR complex consisting of p18, p14, MP1, HBXIP, and C7orf59 (Bar-Peled, Schweitzer, Zoncu, & Sabatini, 2012; Nada et al., 2009; Sancak et al., 2010; Teis, Wunderlich, & Huber, 2002). Since the small GTPase Rab7 and the LAMTOR complex are localized to late endosomes (Fig. 6.1), we have chosen LAMTOR 1/2/3 (p18, p14, and MP1) to visualize those organelles.

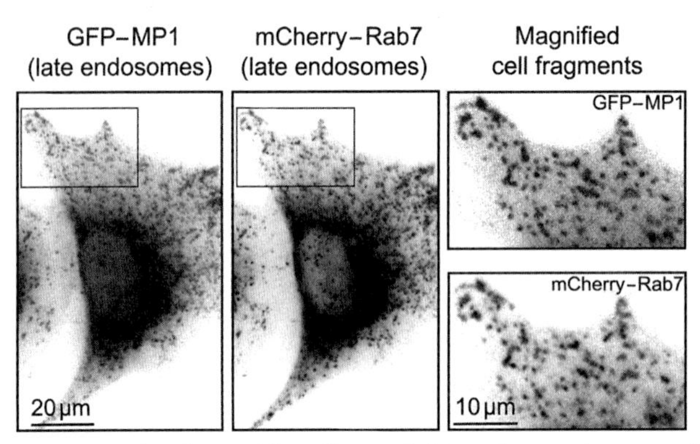

Figure 6.1 GFP–MP1 colocalizes with mCherry–Rab7 on late endosomes. Depicted are HeLa cells cotransfected with GFP–MP1 and mCherry–Rab7. Images were taken from paraformaldehyde-fixed cells. On the right, magnifications of the same cell show a colocalization of GFP–MP1 and mCherry–Rab7.

3. SAPONIN TREATMENT ENHANCES THE PRESERVATION OF PERIPHERAL ENDOSOMES AFTER FIXATION

To compare different permeabilization methods, the *in vivo* situation of endosomal positioning and trafficking must be analyzed. For that, we made use of mouse embryonic fibroblasts (MEF) expressing GFP–p14 to visualize late endosomes. The comparison of pictures from the same cell taken before and after fixation and permeabilization can elucidate the methodological impacts, which may disturb the cellular architecture:

1. The stable cell line $p14^{-/-}$ GFP-p14 MEF was generated as described previously (Obexer, Geiger, Ambros, Meister, & Ausserlechner, 2007; Stasyk et al., 2010) and cultured in high-glucose DMEM supplemented with 10 m*M* HEPES (pH 7.3), 100 IU/ml penicillin, 100 μg/ml streptomycin, and 10% FCS at 37°C, in 5% CO_2, and in 95% humidity. The media and reagents for tissue culture were purchased from Gibco–BRL (Life Technologies).
2. Plate cells on μ-dish (35 mm, high, standard bottom, ibidi) and grow until 90–100% confluent.
3. Change the medium for live-cell microscopy to phenol red-free DMEM (Sigma-Aldrich), supplemented with 10% FCS (Gibco–BRL (Life Technologies)). Wash cells several times with the phenol red-free DMEM.

4. Perform scratches with a yellow tip on the monolayer. Let the cells at the scratch form a migrating front for 1–2 h.
5. Mark the location of the scratches on the lid and also mark the orientation of the dish on the microscope (Fig. 6.2).
6. Take images of cells expressing GFP–p14 endosomes and mark their location at the scratch on the lid (Fig. 6.2). The pictures were taken on an Axiovert 200M microscope equipped with a heatable 1.4NA 63 × oil objective, a CO_2 chamber, a 37 °C heating plate (Carl Zeiss MicroImaging GmbH), and a CoolSnapHQ2 CCD camera (Photometrics). The acquisition was controlled by VisiView software (Visitron Systems).
7. Fix cells in prewarmed 4% PFA (Sigma) in phosphate-buffered salt solution (PBS, 137 m*M* NaCl, 12 m*M* phosphate, 2.7 m*M* KCl, pH 7.4) for 20 min at room temperature (RT). Wash cells several times with PBS.
8. Permeabilize cells with either
   a. 0.2% or 0.5% Triton X-100 (Thermo Scientific) in PBS for 2 min or
   b. 0.025% or 0.05% saponin (Sigma) during blocking, primary, and secondary antibody incubation.
9. Block cells with 10% goat serum in PBS at RT for 30 min.
10. Incubate cells with the antibody against p18 (Atlas Antibodies) 1:200 in blocking buffer at RT for 2 h. Afterward, wash the cells at least three times with PBS.

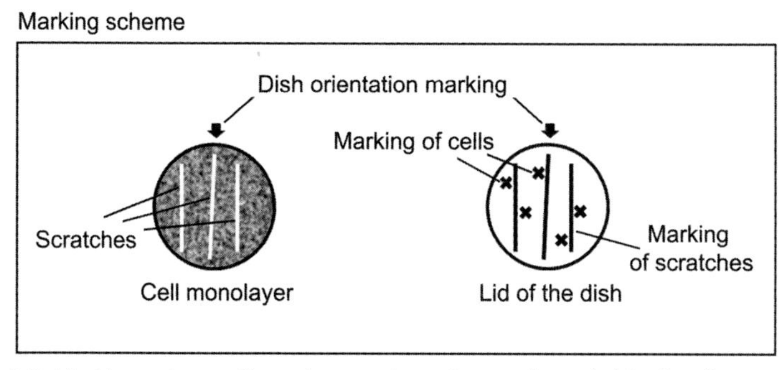

Figure 6.2 Marking scheme. Scratches on the cell monolayer (white lines) are copied as black lines on the lid. Orientation of the dish on the microscope must be marked (arrow). Location of cells taken for live-cell pictures must be marked (cross).

11. Incubate the cells for 30 min with the Alexa Fluor 568 conjugated anti-rabbit secondary antibody (Molecular Probes, Invitrogen) 1:1000 in PBS.
12. Mount the cells with Mowiol (Calbiochem) and a round 15 mm coverslip.
13. Put the dish in the same orientation as before on the microscope and search with the help of the markings on the lid for the cell taken before fixation (Fig. 6.2). Record pictures of GFP–p14 and p18 (Fig. 6.3).

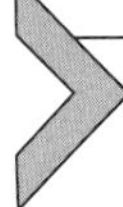

4. MILD FIXATION ALLOWS PRESERVING THE MCHERRY AND GFP LOCALIZED TO FOCAL ADHESIONS AND LATE ENDOSOMES

The visualization of different proteins in the cell using the expression of fluorescent proteins may help to avoid generation of artifacts caused by different fixation conditions. It is important to carefully study distribution of fluorescent proteins *in vivo* in comparison to different fixation conditions. mCherry-paxillin localizes to focal adhesions *in vivo* (Efimov et al., 2008), and GFP–MP1 (LAMTOR3) was chosen to visualize late endosomes:

1. Culture NIH3T3 fibroblasts in growth medium (DMEM, 10% FCS, 1% antibiotics), at 37 °C, and in 5% CO_2.
2. Grow cells until 85% confluence in a 35 mm plastic petri dish. Cotransfect cells with mCherry-paxillin and GFP–MP1 using Opti-MEM (Invitrogen) and Lipofectamine™ 2000 reagents (Invitrogen) according to the manufacturer's instructions. Five hours after transfection, wash cells with growth medium and let them rest in the incubator for approximately 2 h.
3. Trypsinize cells and transfer them in growth medium (1:3) into three glass-bottom μ-dishes (35 mm, high, glass-bottom, ibidi).
4. Start fixation 24–36 h after transfection. Replace growth medium with 1 ml of fresh growth medium (37 °C). Keep cells in the incubator until PFA is prepared.
5. Prepare 15 °C cold 8% PFA in cytoskeletal buffer (CB, 10 m*M* Mes, 150 m*M* NaCl, 5 m*M* EGTA, 5 m*M* glucose, 5 m*M* $MgCl_2$, pH 6.1).
6. Perform fixation at RT. Take cells out of incubator and very slowly add dropwise 1 ml of PFA directly on the cells with generous shaking. Due to the lower temperature, PFA will go to the bottom of the dish. When this procedure is performed correctly, it will result in clear distinction

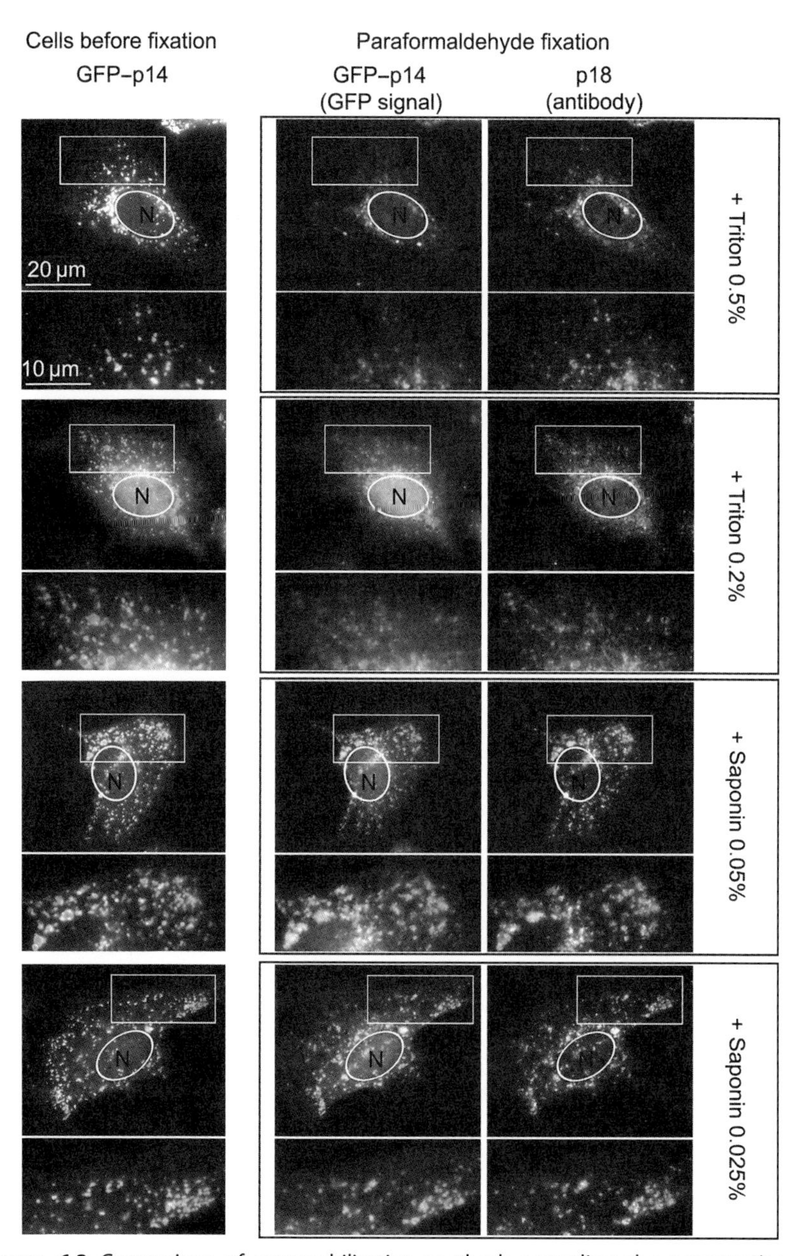

Figure 6.3 Comparison of permeabilization methods regarding the preservation of peripheral late endosomes. Pictures were first taken of live cells expressing GFP–p14. Then, cells were fixed and stained for p18, and pictures for GFP–p14 and p18 of the same cell as depicted in live were taken. On the upper two panels, cells were permeabilized with Triton X-100 with indicated concentrations. On the lower two panels, cells were permeabilized with saponin with indicated concentrations. N is indicating the position of cell nucleus.

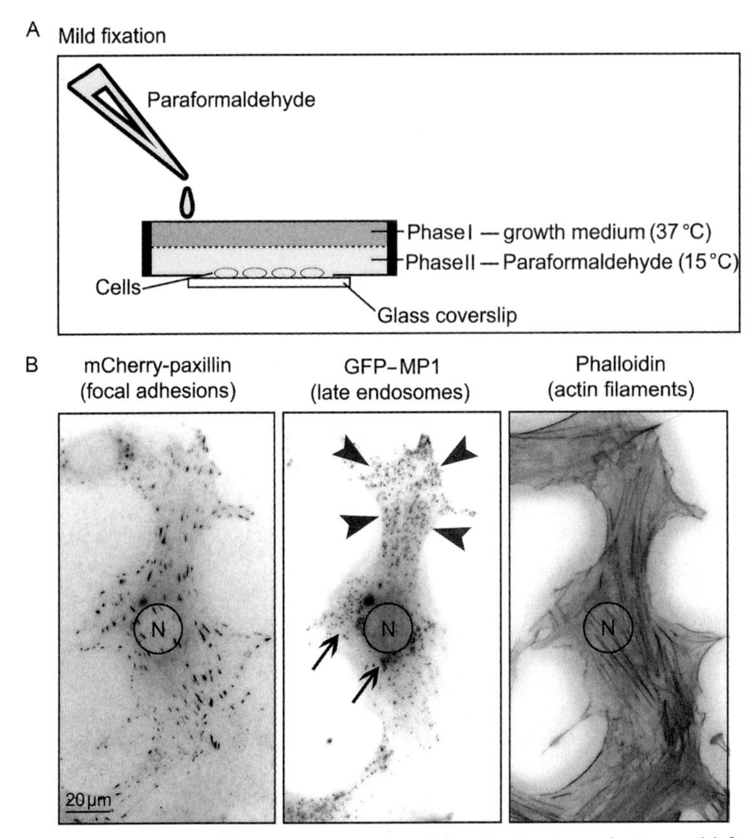

Figure 6.4 Preservation of late endosomes and focal adhesions during mild fixation. (A) Scheme of mild fixation. Paraformaldehyde was added dropwise directly on the cells with generous shaking. Due to the temperature difference, paraformaldehyde went to the dish bottom; it resulted in clear distinction between two phases: growth medium (phase I) and paraformaldehyde (phase II). (B) NIH3T3 cells were cotransfected with mCherry-paxillin and GFP–MP1, fixed using mild fixation as on (A) and further stained with phalloidin. Black arrows point to the late endosomes preserved in the perinuclear area; gray arrowheads point to late endosomes preserved in the cell periphery and in the leading edge. N is indicating the position of cell nucleus.

between two phases: medium (rose color) and PFA (light-yellow color) as it is depicted on Fig. 6.4.

7. Leave cells for 15 min at RT.
8. Wash cells three times with CB for 10 min.
9. Add solution containing 1:300 phalloidin (Alexa Fluor 350, Molecular Probes, Invitrogen), 0.05% saponin, and 50 m*M* NH_4Cl in CB on cells

and incubate for 30 min. Carefully wash cells three times with CB for 5 min. Leave cells in CB or PBS.

10. Perform microscopy using inverted microscope (Fig. 6.4). Pictures were taken on an Axiovert 200M microscope equipped with 1.4NA 63× oil objective (Carl Zeiss MicroImaging GmbH) and a CoolSnapHQ2 CCD camera (Photometrics). Acquisition was controlled by AxioVision software (Carl Zeiss MicroImaging GmbH).

5. SUMMARY

Permeabilization is a crucial step for the preservation of small peripheral organelles like endosomes. Comparing two commonly used permeabilization protocols revealed that the use of saponin in comparison to Triton X-100 yielded improved results (Fig. 6.3). These results were especially visible at the peripheral areas of the cell (magnifications in Fig. 6.3), where permeabilization with Triton X-100 disturbed organelle integrity when compared to the live-cell image. The use of saponin in comparison achieved a better conservation of these peripheral endosomes. In addition, the method of mild fixation results in an excellent preservation of focal adhesions (visualized by mCherry-paxillin) and late endosomes (visualized by GFP–MP1) as shown in Fig. 6.4 (the lower panel). Late endosomes are clearly preserved in the perinuclear area (Fig. 6.4, black arrows), in the cell periphery and in the leading edge (Fig. 6.4, gray arrowheads). Additionally, visualizing actin filaments after fixation is possible without disturbing the overall cell morphology.

ACKNOWLEDGMENT

This work was supported by the special research program *Cell proliferation and cell death in tumors*—SFB021, funded by the FWF, Austria.

REFERENCES

Bar-Peled, L., Schweitzer, L. D., Zoncu, R., & Sabatini, D. M. (2012). Ragulator is a GEF for the rag GTPases that signal amino acid levels to mTORC1. *Cell*, *150*(6), 1196–1208.

Brock, R., Hamelers, I. H., & Jovin, T. M. (1999). Comparison of fixation protocols for adherent cultured cells applied to a GFP fusion protein of the epidermal growth factor receptor. *Cytometry*, *35*(4), 353–362.

Domin, A., Lan, M. J., & Kaminski, C. (2004). *Laser applications to chemical and environmental analysis*. Annapolis, MD: Optical Society of America.

Efimov, A., Schiefermeier, N., Grigoriev, I., Ohi, R., Brown, M. C., Turner, C. E., et al. (2008). Paxillin-dependent stimulation of microtubule catastrophes at focal adhesion sites. *Journal of Cell Science*, *121*(Pt 2), 196–204.

Nada, S., Hondo, A., Kasai, A., Koike, M., Saito, K., Uchiyama, Y., et al. (2009). The novel lipid raft adaptor p18 controls endosome dynamics by anchoring the MEK-ERK pathway to late endosomes. *The EMBO Journal*, *28*(5), 477–489.

Obexer, P., Geiger, K., Ambros, P. F., Meister, B., & Ausserlechner, M. J. (2007). FKHRL1-mediated expression of Noxa and Bim induces apoptosis via the mitochondria in neuroblastoma cells. *Cell Death and Differentiation*, *14*(3), 534–547.

Piper, R. C., & Katzmann, D. J. (2007). Biogenesis and function of multivesicular bodies. *Annual Review of Cell and Developmental Biology*, *23*, 519–547.

Saftig, P., & Klumperman, J. (2009). Lysosome biogenesis and lysosomal membrane proteins: Trafficking meets function. *Nature Reviews Molecular Cell Biology*, *10*(9), 623–635.

Sancak, Y., Bar-Peled, L., Zoncu, R., Markhard, A. L., Nada, S., & Sabatini, D. M. (2010). Ragulator-Rag complex targets mTORC1 to the lysosomal surface and is necessary for its activation by amino acids. *Cell*, *141*(2), 290–303.

Schnell, U., Dijk, F., Sjollema, K. A., & Giepmans, B. N. (2012). Immunolabeling artifacts and the need for live-cell imaging. *Nature Methods*, *9*(2), 152–158.

Stasyk, T., Holzmann, J., Stumberger, S., Ebner, H. L., Hess, M. W., Bonn, G. K., et al. (2010). Proteomic analysis of endosomes from genetically modified p14/MP1 mouse embryonic fibroblasts. *Proteomics*, *10*(22), 4117–4127.

Teis, D., Wunderlich, W., & Huber, L. A. (2002). Localization of the MP1-MAPK scaffold complex to endosomes is mediated by p14 and required for signal transduction. *Developmental Cell*, *3*(6), 803–814.

Wang, D. S., Miller, R., Shaw, R., & Shaw, G. (1996). The pleckstrin homology domain of human beta I sigma II spectrin is targeted to the plasma membrane in vivo. *Biochemical and Biophysical Research Communications*, *225*(2), 420–426.

Zerial, M., & McBride, H. (2001). Rab proteins as membrane organizers. *Nature Reviews Molecular Cell Biology*, *2*(2), 107–117.

CHAPTER SEVEN

Characterizing and Measuring Endocytosis of Lipid-Binding Effectors in Mammalian Cells

Helen R. Clark[*,†], **Tristan A. Hayes**[*,‡], **Shiv D. Kale**[*,1]
[*]Virginia Bioinformatics Institute, Virginia Tech, Blacksburg, Virginia, USA
[†]Department of Biochemistry, Virginia Tech, Blacksburg, Virginia, USA
[‡]Department of Biological Sciences, Virginia Tech, Blacksburg, Virginia, USA
[1]Corresponding author: e-mail address: sdkale@vt.edu

Contents

Abstract

Pathogen–host interactions are mediated in part by secreted microbial proteins capable of exploiting host cells for their survival. Several of these manipulations involve, but are not limited to, suppression of defense responses, alterations in host vesicular trafficking, and manipulation of gene expression. The delivery of such molecules from microbe to host has been of intense interest in several microbe–host systems. Several well-studied bacterial effectors are delivered directly into host cells through a needle injection apparatus. Conversely, there have been several examples of secreted effectors and protein toxins from bacteria and eukaryotic microbes, such as fungi and oomycetes, being internalized into host cells by receptor-mediated endocytosis. In the following chapter, we discuss various techniques utilized to measure these endocytosed lipid-binding effectors that can be delivered in the absence of the pathogen.

Methods in Enzymology, Volume 535
ISSN 0076-6879
http://dx.doi.org/10.1016/B978-0-12-397925-4.00007-9

1. INTRODUCTION

Microbe–host interactions are considered an exciting yet challenging research area of the biological sciences. Many microbe–host systems parallel each other in areas such as basal defense recognition utilized by plants and animals to detect foreign nonself components and the ability of microbes to suppress such defense pathways (Diacovich & Gorvel, 2010; Schwessinger & Ronald, 2012). Moreover, microbes have evolved mechanisms to disrupt basal and specific defense signaling pathways through the use of small-secreted proteins (effectors), many of which are capable of translocating into cells (Kale, 2012; Kale & Tyler, 2011). The delivery and trafficking of these effectors is of intense interest in several microbe–host communities. A large subset of bacteria deliver effectors into the host cytoplasm via a specialized needle-like apparatus (type III secretion systems) capable of directly piercing the host cell membrane (Backert & Meyer, 2006; Hueck, 1998; Pukatzki, McAuley, & Miyata, 2009). Conversely, other secreted bacterial toxins, such as Shiga, Tetanus, and Botulinum toxins, enter cells by binding host cell surface glycolipids at nanomolar affinities, thereby triggering endocytosis (Jacewicz, Clausen, Nudelman, Donohue-Rolfe, & Keusch, 1986; MacKenzie, Hirama, Lee, Altman, & Young, 1997; Singh, Harrison, & Schoeniger, 2000). The effector CagA from *Helicobacter pylori* binds extracellular phosphatidylserine and is believed to enter through a membrane-flipping mechanism rather than endocytosis (Murata-Kamiya, Kikuchi, Hayashi, Higashi, & Hatakeyama, 2010). Bacterial effectors have adapted to exploit a number of cell surface receptors to gain entry into host cells.

Among the oomycetes, a group of economically important plant pathogens that parallel fungi in their method of pathogenesis, a large superfamily of secreted proteins share a conserved N-terminal RxLR–dEER motif. This motif has been shown to mediate effector entry into plant and animal cells in the absence of the pathogen by binding cell surface phosphatidylinositiol-3-phosphate at various nano- and micromolar affinities (PtdIns-3-P) (Dou et al., 2008; Gu et al., 2011; Kale et al., 2010; Plett et al., 2011; Sun et al., 2013). These intracellular fungal and oomycete effectors have the ability to target different defense components and basal machinery of the host plant. A subset of *Plasmodium falciparum* effectors harboring a similar PEXEL (RxLxE/D/Q) motif are targeted to the host cell by binding PtdIns-3-P in the endoplasmic reticulum (Bhattacharjee, Stahelin, Speicher, Speicher, &

Haldar, 2012). It is thought-provoking to note that these two motifs are interchangeable in both systems, since both clades are evolutionarily distinct (Bhattacharjee et al., 2006; Jiang & Tyler, 2012). The dynamic role of PtdIns-3-P has emerged recently to be relevant in several microbe–host pathosystems (reviewed in Jiang, Stahelin, Bhattacharjee, & Haldar, 2013). As several fungal plant pathogens utilize these RxLR-like domains to deliver effectors to plant cells via PtdIns-3-P receptor-mediated endocytosis, it would be interesting to determine if potential effectors from human fungal pathogens or certain fungal allergens utilize such mechanisms in mammals to facilitate infection or allergy.

The focus of this chapter is to describe several assays that characterize and measure endocytosis of lipid-binding effectors both qualitatively and quantitatively. These assays can be utilized to study cell entry with other cell-penetrating proteins and/or cell lines. Here, we describe the use of specific cell lines, instruments, and reagents for these experiments. The use of other cell lines, instruments, and reagents may require optimization to achieve comparable results.

2. CULTURING AND MAINTENANCE OF MAMMALIAN CELL LINES

Culturing of immortalized mammalian cells is a standard laboratory technique utilized extensively by most biomedically oriented laboratories. Here, we briefly describe our procedure to prepare healthy and robust human bronchial epithelium (BEAS-2B) cells for several endocytosis assays. All culturing techniques should be conducted in a biosafety level 2 hood. The following procedure may be applied to other cell lines with minimal modifications. Appropriate reagents and conditions, such as media, humidity, CO_2 levels, and temperature, should be modified for different cell lines.

1. Remove frozen cryogenic vials from liquid nitrogen storage and thaw at room temperature. Do not shake or invert.
2. Transfer 1 ml of cells in the cryosolution to a 75 cm^2 tissue culture flask containing 25 ml of RPMI (Roswell Park Memorial Institute medium) + 10% FBS (fetal bovine serum) + 1 × penicillin/streptomycin (referred to as complete media).
3. Incubate cells at 37 °C in 5% CO_2. Change the media every other day or more often if necessary. Flasks generally take 3–4 days to reach >80% confluence.

4. Once cells are adequately confluent, remove the media and add 5 ml of trypsin and incubate for 5 min at 37 °C in 5% CO_2. With certain cell lines, it is important to wash cells once with dPBS (Dulbecco's phosphate buffered saline) prior to trypsinization.
5. Transfer trypsinized cells to a 15 ml falcon tube and spin for 5 min at $500 \times g$. Remove trypsin solution and gently resuspend cells in 25 ml of dPBS.
6. Spin the cells for 5 min at $500 \times g$ and resuspend them once again in 25 ml of dPBS.
7. Spin the cells for 5 min at $500 \times g$ and resuspend them in 5 ml of complete media.
8. Determine cell density by hemocytometer measurement.
9. Depending on use, plate cells according to Table 7.1. Only use sterile glass-bottom multiwell plates for fluorescent microtiter plate reader and confocal microscopy-based measurements. Sterile tissue culture grade multiwell plates are appropriate for flow cytometry-based experiments.
10. Incubate cells for 2 days at 37 °C in 5% CO_2. Wash cells and replace the media every other day when passaging cells. As a rule of thumb, do not passage cells more than 10 times. The maximum number of passages varies depending on cell line used.
11. Cells are ready for various treatments once they have reached $\geq 80\%$ confluence. For internalization experiments, see Section 5. For transfection of cells, see Section 3.

3. TRANSFECTION OF MAMMALIAN CELLS

Mammalian cell transfection can be done utilizing a variety of techniques. We utilize lipofection (Lipofectamine 2000) for transient gene expression. For the following protocol, we specify volumes for a 12-well plate. See Table 7.1 for appropriate dilutions. Prior to use of lipofection on an alternative cell line, we strongly recommend doing several transfections using different concentrations of Lipofectamine 2000, as unintended cell line specific responses may occur.

1. Add 5 μl of Lipofectamine 2000 reagent and 1 μg plasmid DNA of interest to separate tubes containing RPMI (no FBS or penicillin/streptomycin). Gently mix the solution by finger tap, and let it sit at room temperature for 5 min. Combine volumes into a single 1.7 ml microcentrifuge tube and mix them by finger tap. Let the combined solution sit for

Table 7.1 Media and transfection volumes for mammalian cell culture line BEAS-2B

Plate type	No. of cells plated per well	Vol. of media passaging	Vol. of transfection reagent (μl)	Weight of plasmid DNA for transfection (ng)	Separate transfection volume[a] (μl)	Combined transfection volume (μl)	Final volume in well during transfection
75 cm^2	Frozen stock	25 ml					
6-Well	1,000,000	5 ml	10	2000	100	200	2 ml
12-Well	500,000	2 ml	5	1000	50	100	1 ml
24-Well	250,000	1 ml	2.5	500	25	50	0.5 ml
48-Well	100,000	0.5 ml					
96-Well	50,000	200 μl	1	250	5	10	100 μl

[a]Total volume for the separate transfection components. It is very important to keep the reagents separated as described in the Section 3. Failure to do this will result in extremely poor transfection rate.
Volumes work well for other cell lines such as HEK293, HepG2, and A549. Use appropriate media for each cell line.

20 min. Wash cells with 1 ml dPBS and then replace with 0.9 ml complete RPMI media.

2. Add the 100 μl Lipofectamine 2000/DNA solution to the plated cells and incubate for ~6 h. Wash cells with two volumes of dPBS and add fresh complete RPMI media.
3. It is critical to wash cells with one volume of dPBS and change media every day.
4. Two days posttransfection, cells are now ready for further treatment or confocal microscopy viewing.

4. PROTEIN PURIFICATION AND PREPARATION

Protein purification and preparation are critical to the success of endocytosis experiments. For the described experiments, we will utilize the fluorescent Avr1b(N)–GFP fusion protein as an example (Dou et al., 2008; Gu et al., 2011; Kale et al., 2010; Shan, Cao, Dan, & Tyler, 2004). This recombinant protein encodes the lipid-binding RxLR–dEER domain of Avr1b with a carboxy-terminus GFP and an amino-terminus hexahistidine tag.

Avr1b is a secreted protein from *Phytophthora sojae*, a severe plant pathogen of soybean accounting for hundreds of millions of dollars in lost agricultural revenue each year. The protein is known to localize inside soybean cells where it triggers a hypersensitive defense response on cultivars of soybean expressing the resistance protein Rps1b. It has repeatedly been shown that Avr1b can enter plant cells in the absence of pathogen-encoded machinery (Dou et al., 2008; Gu et al., 2011; Kale et al., 2010; Shan et al., 2004). Kale et al. (2010) found that this mechanism occurs in A549 cells, and subsequent follow-up work indicates that this process occurs in a number of mammalian cell lines (Clark, H.R., Drews, K.C., Hayes, T.A., Kale, S.D., unpublished communication):

1. Plasmids, in this case pGFP–Avr1b(N)WT, are transformed into BL21 (DE3), plated on LB Amp_{100} plates, and grown overnight at 37 °C to produce a lawn.
2. Inoculate two 2 l baffled flasks (800 ml LB broth) with a single loop full of cells. Allow cells to grow to an optical density of ~0.7 at 37 °C and 240 rpm. Induce cell with 1 m*M* IPTG (Isopropyl β-D-1-thiogalactopyranoside) final concentration and grow cells for an additional 4 h at 37 °C and 240 rpm. Different proteins require alternate conditions. We strongly recommend trying different temperatures, concentrations of IPTG, and induction times to optimize protein expression.

3. Pellet cells at 6000 × *g* for 10 min in a precooled (∼4 °C) rotor. Immediately resuspend cells in 10 ml of lysis buffer (50 mM sodium phosphate, 300 mM sodium chloride, 20 mM imidazole) and add 50 μl of PMSF (200 m*M* stock concentration in isopropanol and 1 m*M* final concentration).
4. Sonicate cells four times: 15 s "on" and 59 s "off" on ice in a 15 ml falcon tube. Transfer sonicated cells to a precooled SS34 tube and centrifuge for 30 min at 25,000 × *g*. Make sure that the rotor is also precooled to 4 °C.
5. Load supernatant into a 10 ml superloop attached to ÄKTAprime plus fast-phase liquid chromatography (FPLC) system (GE Healthcare, Inc.) with a 1 ml nickel NTA column.
6. Run the template program His-tag HisTrap. Wash buffer—50 m*M* sodium phosphate, 300 m*M* sodium chloride, 20 m*M* imidazole, pH 8.0. Elution buffer—50 m*M* sodium phosphate, 300 m*M* sodium chloride, 1 *M* imidazole, pH 8.0. Column wash phase can be increased as required.
7. Concentrate fractions using a 30 kDa cutoff Amicon centrifuge filter to 1.5 ml. Spin fractions for approximately 30 min at 4000 × *g* at 4 °C. Use alternate cutoff for different-sized proteins.
8. Run sample through 16/600 Superdex 75 column at 0.2–0.5 ml per minute using the ÄKTAprime plus FPLC system. Liquid phase: 50 m*M* sodium phosphate and 300 m*M* sodium chloride, pH 8.0. Collect 1 ml fractions (Fig. 7.1).

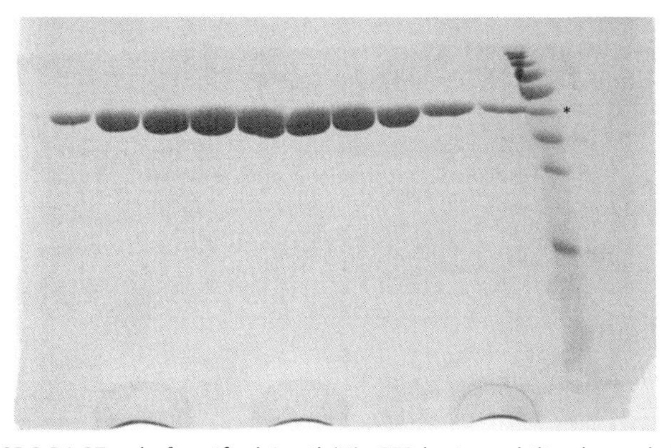

Figure 7.1 SDS-PAGE gel of purified Avr1b(N)–GFP by immobilized metal affinity chromatography followed by gel-filtration chromatography (16/600 Superdex 75) using the ÄKTAprime plus system. * Indicates lane marker of ∼40 kDa. Ten microliter is loaded per lane from 1 ml fractions.

9. Elution fractions containing the protein of interest can be further concentrated using an appropriate kDa cutoff Amicon centrifuge filter.
10. We prefer to buffer exchange with two rounds of ~12 h dialysis—1 l of PBS for every 1 ml of protein.

5. TREATMENT OF CELLS

1. On the morning of the experiment, remove complete media from microtiter plate wells and replace with serum-free media (no FBS). Incubate for 2 h at 37 °C in 5% CO_2. The purpose of this step is to provide synchronization.
2. Replace media again with serum-free media and use cells within 30 min. Keep cells at 37 °C in 5% CO_2.
3. Remove 10% of media volume (100 μl if cells are in 1 ml) from the growing cells. Add 10% volume of 2 mg/ml protein in PBS. Final concentration 0.2 mg/ml. For Avr1b(N)–GFP, this works out to ~5 μ*M* final concentration.
4. Incubate cells for 5–30 min, depending on protein being tested. In the case for Avr1b(N)–GFP, 20 min is sufficient.
5. Wash cells once with one volume of dPBS with various marker dyes, such as propidium iodide (0.2 μg/ml final concentration) for cell death. Let the cell sit for 5 min.
6. Wash cells twice, with one volume of dPBS.
7. Incubate cells in formalin solution to immediately stop the reaction from proceeding for microscopy (Section 6) and fluorescent microtiter plate assays (Section 7). For flow cytometry preparation, see Section 8.

6. TRACKING ENDOCYTOSIS BY CONFOCAL MICROSCOPY

1. Take glass-bottom plates to confocal microscope (in our case, Zeiss LSM 510 Meta Confocal Microscope) for imaging.
2. Table 7.2 highlights several settings associated with appropriate excitation and emission.
3. Negative control samples must be viewed first to establish a baseline well below autofluorescence. It is important to view negative samples randomly throughout your experiment whenever the detector or any other setting is adjusted. It is strongly advised to determine appropriate settings

Table 7.2 Confocal microscopy settings for viewing fluorescent protein and bead internalization
Zeiss LSM 510 Meta Confocal Microscopy

Lasers	Enterprise: 351 nm, 364 nm
	Argon: 458 nm, 488 nm, 514 nm
	HeNe1: 543 nm
	HeNe2: 633 nm
Multitrack feature	Yes, line by line, each signal gets own track
mCherry observation	Excitation: 543 nm
	Emission capture: 585–615 nm
GFP observation	Excitation: 488 nm
	Emission capture: 505–530 nm
Blue polystyrene beads observation[a]	Excitation: 364 nm
	Emission capture: 385–470 nm
Scan mode	Frame
Frame size	1024 × 1024
Pixel dwell	1.0–1.5 μs
Average number	8
Bit depth	8 bit
Pinhole	1 μm section
Digital offset[b]	−0.1 to 0.0
Gain[b]	500–750

[a]Settings to observe blue polystyrene beads can also be used for the nuclear stain DAPI. When using DAPI, a bead that fluoresces in an alternate channel must be used.
[b]Gain and amplifier offset can be set to maximize clarity of fluorescent images. We have found this range appropriate for viewing fluorescence. It is important to note that when imagining outside or inside this range, a control sample is imaged to show that there is no autofluorescence.

at the start of a microscopy session. Do not adjust excitation or emission settings once image acquisition has started.

4. All pictures should be captured as an average of at least eight scans with a pixel dwell intensity of at least 1.26 μs and an image size of 1024 × 1024. Though these images take 1–2 min to acquire, the quality of the image is well worth the time (Figs. 7.2 and 7.3).

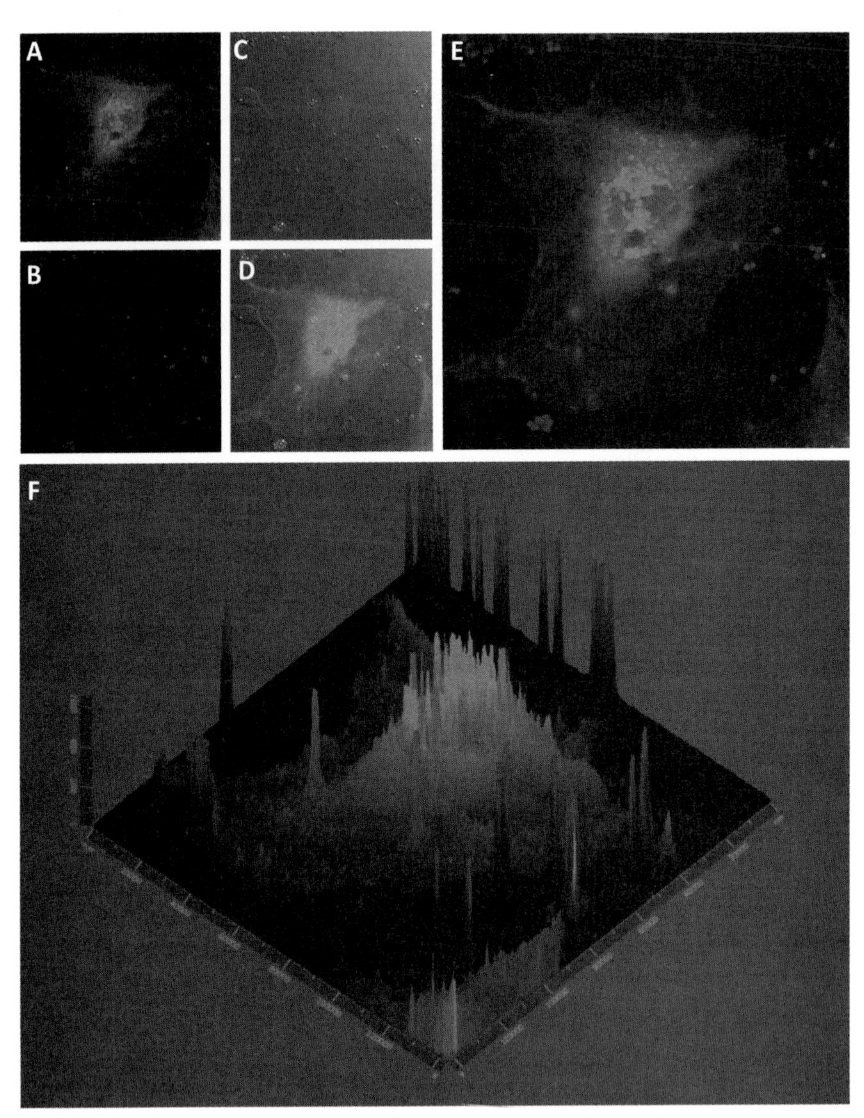

Figure 7.2 Transient expression of endocytic marker fused to mCherry in mammalian cell line BEAS-2B colocalizing with internalized fluorescent polystyrene beads. Transfected cells were incubated with polystyrene beads for 9 h, washed with dPBS, and stored in formalin until imaging by confocal microscopy. (A) mCherry fluorescent channel. (B) Polystyrene fluorescent channel. (C) Light image. (D) Total overlay. (E) Fluorescent overlay. (F) 2.5 D fluorescent channel, *x*- and *y*-axes indicate spatial position of fluorescence; *z*-axis indicates intensity at given position. An increase in intensity around an internalized fluorescent bead in comparison to other parts of the cells is indicative of colocalization. (For color version of this figure, the reader is referred to the online version of this chapter.)

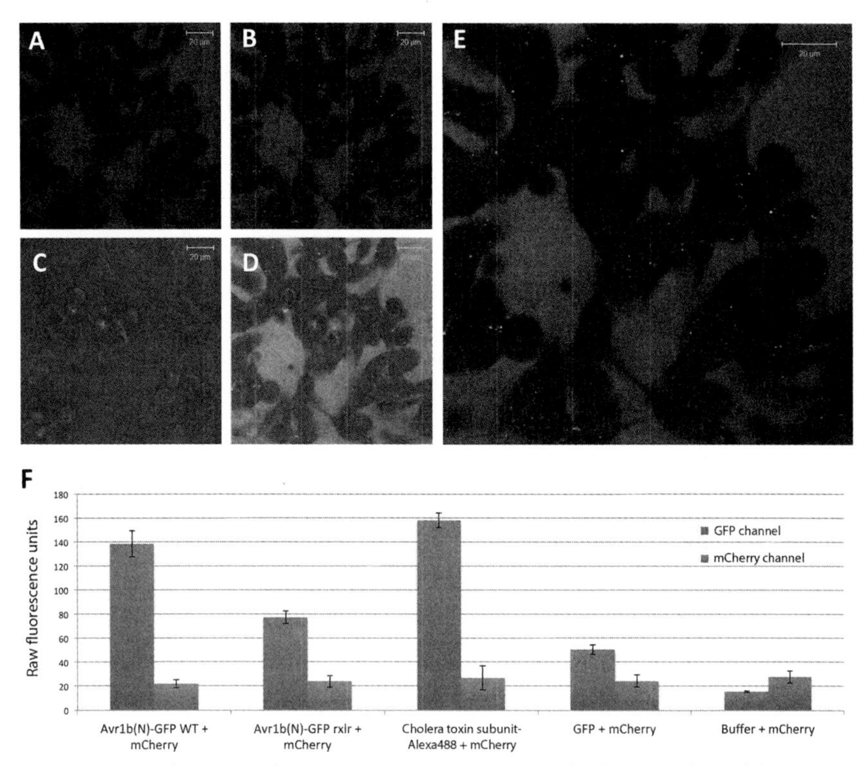

Figure 7.3 Qualitative and quantitative measurements of cell entry of Avr1b(N)–GFP in BEAS-2B mammalian cells. (A) Fluorescence from control mCherry samples. (B) Fluorescence from Avr1b(N)–GFP. (C) Light image. (D) Complete overlay. (E) Fluorescence overlay. (F) Bar graphs represent the mean raw signal of six replicates from fluorescent microplate reader assay from one experiment. In both cases, cells were incubated with construct of interest and control mCherry. Avr1b(N)–GFP WT refers to the wild-type protein purified in Section 4. Avr1b(N)–GFP RxLR refers to the same fusion protein except that the RxLR motif has been replaced with four alanines. Internalized fluorescent protein levels are compared using Duncan's multiple-range test. Control cholera toxin beta subunit was conjugated with Alexa Fluor 488 (1 m*M* of cholera toxin = 5 m*M* Alexa Fluor 488). (For color version of this figure, the reader is referred to the online version of this chapter.)

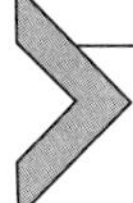

7. QUANTIFICATION OF CELL ENTRY BY FLUORESCENT MICROTITER PLATE READER

The sensitivity of spectrofluorometers has increased significantly over the past 20 years. Many systems, such as the SpectraMax M5, capable of analysis in a variety of multiwell formats, can accurately resolve the difference between different rates of effector entry (Kale et al., 2010; Sun et al.,

2013) in a very short period of experimental time allowing for many biological replicates per experiment. It is strongly suggested that these experiments are performed using glass-bottom microtiter plate dishes. In our hands, we find the glass-bottom plates strongly reduce the baseline background fluorescence up to 50-fold:

1. Prior to beginning protein uptake experiments, start SpectraMax M5. The machine takes several min to startup.
2. Open up SoftMax Pro software and set the microtiter plate temperature to room temperature, ~21 °C.
3. See Table 7.3 for microtiter plate settings.
4. Start and complete the uptake experiment (Section 4). We generally do six replicates per experiment, and pool the results from 2 to 3 experiments.
5. Prior to plate reading, use empty wells to set a standard curve. Concentrations from 0 to 2 μg of fluorescent GFP protein in 100 μl of formalin are sufficient.

Table 7.3 SpectraMax M5 reader settings for measurement of internalized GFP and mCherry proteins

SpectraMax M5 settings

Read mode	Fluorescence (RFUs) top read
Wavelengths (up to 4)	2
Wavelength 1:	Ex: 485; Em: 525; Cutoff: 515
Wavelength 2:	Ex: 584; Em: 612; Cutoff: 610
Sensitivity	
Readings	20 (precise)
PMT sensitivity	Medium
Automix	Off
Autocalibrate	On
Assay plate type	Dependent on brand[a]
Wells to read	Read only your wells of interest
Setting time	Off
Column/wavelength priority	Column priority
Carriage speed	Normal
Autoread	Off

[a]The SpectraMax M5 has precalibrated setting to maximize resolution of signal for a given brand of plate. We still strongly recommend using glass-bottom micro plate dishes.

6. Check plate for air bubbles in well. Remove the bubbles with a pipette.
7. Read the plate.
8. Average (mean) readings from the six independent treatments and calculate standard error.
9. Experiments must be repeated using a fresh protein preparation.
10. Perform an analysis of variance. In our experiences, Duncan's multiple-range test and Tukey's HSD have been appropriate for post hoc testing.

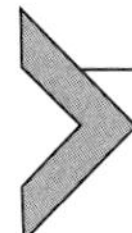

8. QUANTIFICATION OF CELL ENTRY BY FLOW CYTOMETRY

Flow cytometry is an incredibly sensitive and high-throughput technique for the analysis of multiple parameters of an individual cell from a hetero- or homogenous population. Flow cytometry is capable of analyzing thousands of cells per second allowing for rapid analysis of individual members of a population. This is in contrast to measurements using the microtiter plate assay, which measure the summation of signals from a population. In the past 10 years, the application range of flow cytometry has grown dramatically. We assume that the user is able to turn on and calibrate their flow cytometer. It is incredibly important to calibrate the flow cytometer using calibration beads. The following cytometry experiments were done with a FACS LSRII.

8.1. Prepping treated cells for analysis by flow cytometry

1. After washing with dPBS as described in Section 5 (step 4), cells are trypsinized for 5 min at 37 °C in 5% CO_2 and transferred to BD 5 ml polystyrene tubes.
2. These tubes are then spun for 5 min at $1000 \times g$.
3. The trypsin solution is removed and cells are then resuspended in formalin solution. It is very important to resuspend the cells very well as clumping will result in a poor singleton population during flow cytometry analysis.

8.2. Cytometer calibration and setup

1. It is essential to appropriately calibrate the flow cytometer prior to use using the manufacturer-specified protocols.
2. Prior to running samples, dot plots to view data should be set up as followed: FSC-A versus SSC-A, FSC-A versus FSC-W, FITC versus PI, and histograms of PI and FITC channels (Fig. 7.4).

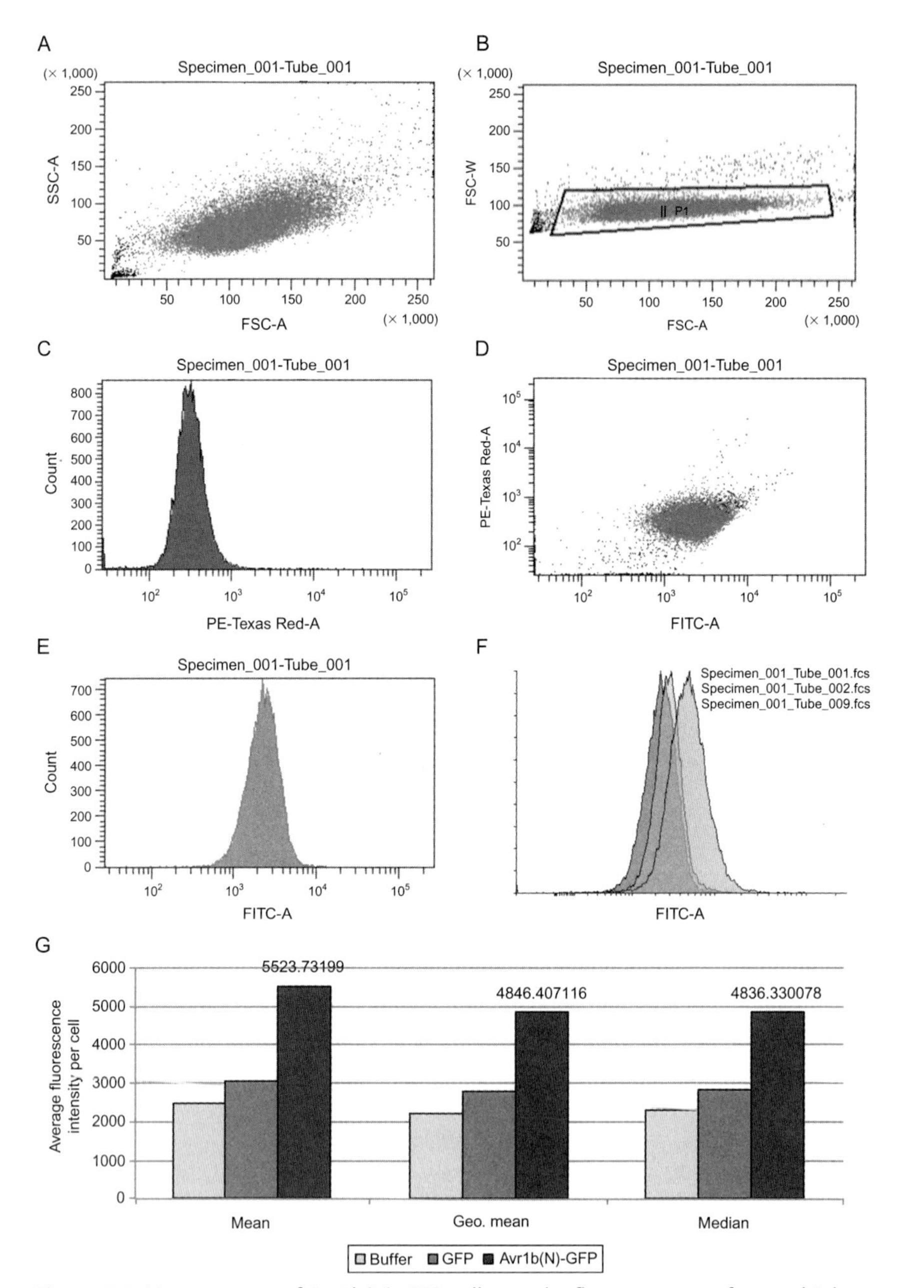

Figure 7.4 Measurement of Avr1b(N)–GFP cell entry by flow cytometry for one biological replicate. An example given is the gating of 20,000 cells positive for singleton status and PI-negative. (A–E) Gating a setup for analysis prior to beginning cytometry. (A) Forward scatter area (FSC-A) versus side scatter area (SSC-A). (B) FSC-A versus FSC width (W) to separate singletons from doublets. Singleton population is gated, shown in red.

3. Set PMT voltages and compensation gates to appropriate values. For BEAS-2B cells, we find the following voltage setting work well: FSC, 240 V; SSC, 280 V; FITC, 500 V; and PI, 635 V.
4. Begin running a negative control sample through the cytometer.
5. Set up corresponding gates in real time to filter out doublet cells and dead cells (PI-positive cells).
6. Gate 20,000 cells that fall under the categories of singleton and PI-negative. In cases where 20,000 cells cannot be gated in a reasonable period of time, analyze samples for 5 min and then move to the next sample. Generally, it should take 2 min at most to gate 20,000 cells. If cell density is low, resuspend cells in lower volume.

8.3. Data analysis and presentation

There are several commercially available programs, most notably the control software from the LSRII FACSDiva (recommended) and FlowJo (recommended) that may be utilized for data analysis and presentation. For the research scientist on a tight budget or doing large-scale analysis (100+ samples at once), we strongly recommend Flowing Software (http://www.flowingsoftware.com). This free software provides user-friendly tutorials and simple analysis tools that are easily utilized. We recommend the following resource to provide guidelines on appropriate experimental setup and data presentation (Alvarez, Helm, Degregori, Roederer, & Majka, 2010).

9. SUMMARY

Effectors play an integral role in mediating microbe–host interactions. While a number of these effectors are delivered directly into the host cell by the pathogen, others are capable of hijacking host cell trafficking machinery. This model provides the ability of the microbe to target an essential mechanism of cellular function thereby abrogating the intense selection pressure on the host

(C) Histogram display of propidium iodide fluorescence (log scale). (D) Dot plot of cellular fluorescence associated with propidium iodide and GFP (FITC). Histogram display of GFP (FITC channel) fluorescence. Logarithmic histogram overlay of FITC channel fluorescence for control (red, left), GFP (green, middle), and Avr1b(N)–GFP (right). (E) Presentation of different methods to determine average fluorescence among population. Utilization of mean results in artificial skewing. Populations of cells that respond homogeneously and have parametric distribution, such as these, have similar values for geometric means, medians, and modes (not shown). (For interpretation of the references to color in this figure legend, the reader is referred to the online version of this chapter.)

to lose the receptor. When performing cell entry experiments, it is important to include both positive and negative controls. Quality of protein preparation and health of cells play an important role in maximizing cell entry. We strongly recommend the use of intracellular markers to validate that the protein, organism, or bead is in fact entering the cells and the use of a quantitative measurement such as fluorescent microtiter plate assay or flow cytometry to have an understanding of the amount of protein entering cells.

ACKNOWLEDGMENTS

We would like to thank Professor Chris Lawrence for his reviews and comments. This work was funded through internal funds from the Virginia Bioinformatics Institute at Virginia Tech to SDK. TAH was supported in part by a NIH IMSD grant GM072767 to Dr. Edward J. Smith.

REFERENCES

Alvarez, D. F., Helm, K., Degregori, J., Roederer, M., & Majka, S. (2010). Publishing flow cytometry data. *American Journal of Physiology Lung Cellular and Molecular Physiology*, *298*, L127–L130.

Backert, S., & Meyer, T. F. (2006). Type IV secretion systems and their effectors in bacterial pathogenesis. *Current Opinion in Microbiology*, *9*, 207–217.

Bhattacharjee, S., Hiller, N. L., Liolios, K., Win, J., Kanneganti, T. D., Young, C., et al. (2006). The malarial host-targeting signal is conserved in the Irish potato famine pathogen. *PLoS Pathogens*, *2*, 453–465.

Bhattacharjee, S., Stahelin, R. V., Speicher, K. D., Speicher, D. W., & Haldar, K. (2012). Endoplasmic reticulum PI(3)P lipid binding targets malaria proteins to the host cell. *Cell*, *148*, 201–212.

Diacovich, L., & Gorvel, J. P. (2010). Bacterial manipulation of innate immunity to promote infection. *Nature Reviews Microbiology*, *8*, 117–128.

Dou, D. L., Kale, S. D., Wang, X., Jiang, R. H. Y., Bruce, N. A., Arredondo, F. D., et al. (2008). RXLR-mediated entry of Phytophthora sojae effector Avr1b into soybean cells does not require pathogen-encoded machinery. *Plant Cell*, *20*, 1930–1947.

Gu, B. A., Kale, S. D., Wang, Q. H., Wang, D. H., Pan, Q. N., Cao, H., et al. (2011). Rust secreted protein Ps87 is conserved in diverse fungal pathogens and contains a RXLR-like motif sufficient for translocation into plant cells. *PLoS One*, *6*(11): e27217.

Hueck, C. J. (1998). Type III protein secretion systems in bacterial pathogens of animals and plants. *Microbiology and Molecular Biology Reviews*, *62*, 379–433.

Jacewicz, M., Clausen, H., Nudelman, E., Donohue-Rolfe, A., & Keusch, G. T. (1986). Pathogenesis of shigella diarrhea. XI. Isolation of a shigella toxin-binding glycolipid from rabbit jejunum and HeLa cells and its identification as globotriaosylceramide. *The Journal of Experimental Medicine*, *163*, 1391–1404.

Jiang, R. H. Y., Stahelin, R. V., Bhattacharjee, S., & Haldar, K. (2013). Eukaryotic virulence determinants utilize phosphoinositides at the ER and host cell surface. *Trends in Microbiology*, *21*, 145–156.

Jiang, R. H., & Tyler, B. M. (2012). Mechanisms and evolution of virulence in oomycetes. *Annual Review of Phytopathology*, *50*, 295–318.

Kale, S. D. (2012). Oomycete and fungal effector entry, a microbial Trojan horse. *The New Phytologist*, *193*, 874–881.

Kale, S. D., Gu, B. A., Capelluto, D. G. S., Dou, D. L., Feldman, E., Rumore, A., et al. (2010). External lipid PI3P mediates entry of eukaryotic pathogen effectors into plant and animal host cells. *Cell*, *142*, 284–295.

Kale, S. D., & Tyler, B. M. (2011). Entry of oomycete and fungal effectors into plant and animal host cells. *Cellular Microbiology*, *13*, 1839–1848.

MacKenzie, C. R., Hirama, T., Lee, K. K., Altman, E., & Young, N. M. (1997). Quantitative analysis of bacterial toxin affinity and specificity for glycolipid receptors by surface plasmon resonance. *The Journal of Biological Chemistry*, *272*, 5533–5538.

Murata-Kamiya, N., Kikuchi, K., Hayashi, T., Higashi, H., & Hatakeyama, M. (2010). Helicobacter pylori exploits host membrane phosphatidylserine for delivery, localization, and pathophysiological action of the CagA oncoprotein. *Cell Host & Microbe*, 7, 399–411.

Plett, J. M., Kemppainen, M., Kale, S. D., Kohler, A., Legue, V., Brun, A., et al. (2011). A secreted effector protein of Laccaria bicolor is required for symbiosis development. *Current Biology*, *21*, 1197–1203.

Pukatzki, S., McAuley, S. B., & Miyata, S. T. (2009). The type VI secretion system: Translocation of effectors and effector-domains. *Current Opinion in Microbiology*, *12*, 11–17.

Schwessinger, B., & Ronald, P. C. (2012). Plant innate immunity: Perception of conserved microbial signatures. *Annual Review of Plant Biology*, *63*, 451–482.

Shan, W. X., Cao, M., Dan, L. U., & Tyler, B. M. (2004). The Avr1b locus of Phytophthora sojae encodes an elicitor and a regulator required for avirulence on soybean plants carrying resistance gene Rps1b. *Molecular Plant-Microbe Interactions*, *17*, 394–403.

Singh, A. K., Harrison, S. H., & Schoeniger, J. S. (2000). Gangliosides as receptors for biological toxins: Development of sensitive fluoroimmunoassays using ganglioside-bearing liposomes. *Analytical Chemistry*, *72*, 6019–6024.

Sun, F., Kale, S. D., Azurmendi, H. F., Li, D., Tyler, B. M., & Capelluto, D. G. (2013). Structural basis for interactions of the Phytophthora sojae RxLR effector Avh5 with phosphatidylinositol 3-phosphate and for host cell entry. *Molecular Plant-Microbe Interactions*, *26*, 330–344.

CHAPTER EIGHT

Measuring the Role for Met Endosomal Signaling in Tumorigenesis

Rachel Barrow[*], **Carine Joffre**[*,†], **Ludovic Ménard**[*], **Stéphanie Kermorgant**[*,1]

[*]Centre for Tumour Biology, Barts Cancer Institute – A Cancer Research UK Centre of Excellence, Queen Mary University of London, John Vane Science Centre, London, United Kingdom
[†]UNITE 830 INSERM, Institut Curie Centre de Recherche, Paris Cedex 05, France
[1]Corresponding author: e-mail address: s.kermorgant@qmul.ac.uk

Contents

Abstract

Met is a receptor tyrosine kinase, often overexpressed or mutated in human cancer. Upon activation by its ligand, the hepatocyte growth factor, Met controls several cell functions such as proliferation, migration, and survival through the activation of multiple pathways.

Upon ligand binding, Met rapidly internalizes and continues to signal from endosomal compartments prior to its degradation. Importantly, this "endosomal

Methods in Enzymology, Volume 535
ISSN 0076-6879
http://dx.doi.org/10.1016/B978-0-12-397925-4.00008-0

signaling" has recently been shown to be involved in tumorigenesis and experimental metastasis. Consequently, interfering with Met endosomal signaling may provide a novel therapeutic approach in cancer treatment. However, there is a need for additional studies in various experimental models to confirm this and find the most specific ways of achieving it. Thus, outlined in this review are the techniques and tools we have been using to study Met endocytosis and Met endosomal signaling.

1. INTRODUCTION

Met, the receptor of hepatocyte growth factor (HGF), is a receptor tyrosine kinase (RTK) overexpressed or mutated in various human cancers (Birchmeier, Birchmeier, Gherardi, & Vande Woude, 2003). HGF or Met triggers several cell functions such as proliferation, migration, and survival through the activation of multiple pathways, thanks to its multifunctional docking site (see Trusolino, Bertotti, and Comoglio (2010) and Gherardi, Birchmeier, Birchmeier, and Vande Woude (2012) for recent reviews).

Following HGF binding, Met rapidly internalizes via clathrin- and dynamin-mediated endocytosis (Hammond, Urbe, Vande Woude, & Clague, 2001; Kermorgant, Zicha, & Parker, 2003, 2004). At 15 min, HGF-bound Met is recruited to EEA1 (early endosome antigen 1)-positive peripheral endosomes and has trafficked to perinuclear endosomes by 120 min (Kermorgant & Parker, 2008; Kermorgant et al., 2003). The detailed mechanisms of Met endocytosis and trafficking can be found in our recent review (Joffre, Barrow, Ménard, & Kermorgant, 2013).

Met is able to signal from endosomes prior to its degradation (Kermorgant & Parker, 2008; Kermorgant et al., 2004). This "endosomal signaling" appears required for an optimum activation of Met downstream pathways such as ERK1/2, STAT3, and Rac1 and subsequent cell functions such as cell migration (Joffre et al., 2011, 2013; Kermorgant & Parker, 2008; Kermorgant et al., 2003, 2004; Trusolino et al., 2010). Interestingly, two constitutively active Met mutants, M1268T and D1246N (Jeffers et al., 1997), found in human cancer (Schmidt et al., 1997) were shown to be oncogenic not only because of their high activation status but also because of their increased endocytosis (Joffre et al., 2011).

In this review, the assays and tools we have been using to study Met endocytic trafficking and endosomal signaling are provided with a focus on tumorigenesis.

2. ANALYZING MET TRAFFICKING WITH FLUORESCENCE AND CONFOCAL MICROSCOPY

2.1. Immunostaining of Met

This technique allows visualization of Met at the plasma membrane followed by its intracellular trafficking upon stimulation with HGF for various amounts of time using a goat anti-Met affinity purified polyclonal antibody (AF276 (anti-human) or AF527 (anti-mouse) R&D Systems). DAPI (4′,6-diamidino-2-phenylindole) may be used to label the nucleus, especially if the study involves the investigation of the perinuclear location of Met:

1. Plate cells in 24 well plates onto 13 mm acid washed glass coverslips such that they will be at 50–60% confluence on the day of fixation. Poorly adherent cell types may need to be plated on coverslips coated with 0.01% poly-L-lysine.
2. After 24 h, or 48 h if 24 h of starvation is needed, pretreat cells with inhibitors where required and stimulate with HGF (typically 50–100 ng/ml) (R&D Systems) for the required time points as a count-down. At the end of the final time point (0 min), fix cells in 4% paraformaldehyde for 5 min.
3. Quench free aldehydes in NH_4Cl 50 m*M* for 10 min, followed by a wash in PBS.
4. Block specific antibody binding sites and permeabilize the cells in PBS containing 2% BSA and 0.1% Triton × 100 for 15 min.
5. Incubate coverslips in the primary antibodies (usual dilution 1:100 or 1:50 depending on cell systems/antibody) in PBS–2% BSA for 30 min, ensuring that if multiple primary antibodies are used, they are raised in different species. The final desired concentration for the goat anti-Met affinity purified polyclonal antibody (R&D Systems) is 1–2 μg/ml.
6. Wash coverslips three times in PBS prior to incubation in the fluorochrome-labeled secondary antibodies (dilution 1:500 to 1:1000 depending on the strength of the antibody) in PBS–2% BSA for 30 min, ensuring that the antibodies were raised in different species to which the primary antibodies were made.
7. After three washes in PBS and one wash in distilled H_2O to remove salts, mount coverslips on glass slides with ProLong Gold antifade reagent (Invitrogen, Life Technologies), containing DAPI.
8. After drying at 37 °C for 1 h, slides are stored at 4 °C prior to analysis using confocal microscopy.

In some cells and/or cell culture conditions, a substantial amount of Met staining may be observed in the Golgi. This staining corresponds in part to the precursor form of Met, which localizes to the Golgi prior to its trafficking to the plasma membrane. Indeed, to our knowledge, all commercial Met antibodies recognize both the mature and precursor forms of Met. Due to the fact that Met traffics from the plasma membrane to a perinuclear endosome and also partially to the Golgi (Kermorgant & Parker, 2008), it can be difficult to distinguish between the pool of Met internalized from the plasma membrane and the pool of Met present initially in the perinuclear/Golgi compartments. Pretreating the cells with 50 μg/ml cycloheximide (from 1 h to overnight), in order to inhibit protein synthesis, can clear the staining of Met precursor in the Golgi.

Another way of overcoming this problem is to use a fluorophore-labeled HGF that only reveals the Met that is internalized upon HGF treatment (see Section 2.2).

2.2. Fluorescence analysis with Alexa Fluor® 555-conjugated HGF

Alexa Fluor® 555-conjugated HGF (HGF*) allows visualization of HGF, which has been shown in several cell lines to traffic with Met (Kermorgant & Parker, 2008; Kermorgant et al., 2004). Figure 8.1 shows the overlap between Met (green) and HGF* (red) in endosomes after 60 min of stimulation, indicating that HGF* is a good tool to follow Met trafficking as an alternative to immunostaining of Met itself:

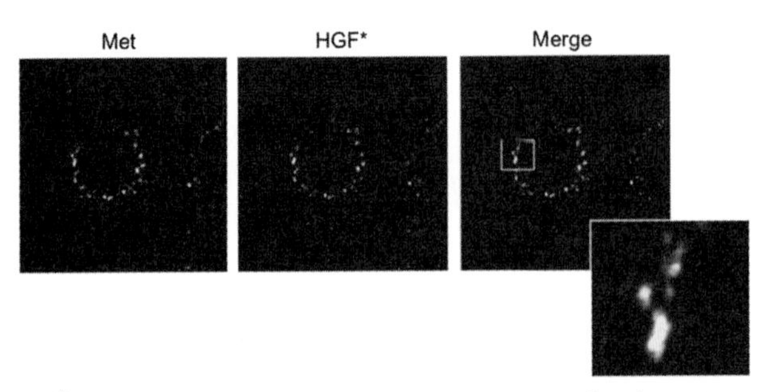

Figure 8.1 HGF* colocalizes with Met immunostaining. Confocal sections of cells stimulated with HGF* (red) for 60 min and then stained with an antibody against Met (green) and with DAPI (blue). (For interpretation of the references to color in this figure legend, the reader is referred to the online version of this chapter.)

1. 25 μg of recombinant HGF (R&D Systems) is labeled, according to the manufacturer protocol, using Alexa Fluor® 555 Microscale Protein Labeling Kit obtained from Life Technologies.
2. The concentration of the labeled HGF is obtained using a NanoDrop ND-1000 spectrophotometer.
3. Treat cells with the desired concentration of HGF* (same concentration as recombinant HGF) for the required time.
4. At the end point of HGF stimulation, fix the cells in 4% paraformaldehyde for 5 min. When costaining is required, continue from point 3 in part 2.1. Otherwise, wash the cells three times in PBS, once in distilled H_2O and mount the coverslips on glass slides with ProLong Gold antifade reagent containing DAPI. Once the slides are dry, use confocal microscopy to analyze HGF* uptake as described in Section 2.5.

One of the advantages of visualizing HGF* instead of Met is that only Met bound to HGF will be detected, meaning that only Met present at the plasma membrane at the time of the stimulation is observed, eliminating the problem often associated with Met staining (see Section 2.1). This is shown in Fig. 8.1 where the HGF* signal is very clean with no diffuse staining observed. One disadvantage of this tool is that HGF* is not normally able to be visualized at the plasma membrane, due to the fact that once HGF binds to Met, Met is quickly internalized. If necessary, this could be achieved by stimulating with HGF*, while the cells have been precooled and kept on ice at 4 °C. Of note, in some cell systems, Met-unbound HGF* sticks to the extracellular matrix, and thus, extracellular background can be observed. This can be avoided through decreasing the time between plating the cells and the start of the experiment or through plating cells on poly-L-lysine-coated coverslips.

2.3. Fluorescence analysis with Alexa Fluor® 546-labeled transferrin

Transferrin receptor is a non-RTK constitutively recycling receptor and thus can be used as a control to assess RTK specificity of Met trafficking. Cells can be treated with the ligand transferrin labeled with Alexa Fluor® 546.

1. Incubate cells with Alexa Fluor® 546-labeled transferrin (Molecular Probes) diluted 1:500 for 30 min prior to their fixation in 4% paraformaldehyde.
2. When costaining is required, continue from point 3 in Section 2.1; otherwise, wash the cells three times in PBS, once in distilled H_2O, and mount the coverslips on glass slides with ProLong Gold antifade reagent.

3. Once the slides are dry, use confocal microscopy to analyze Alexa Fluor® 546-labeled transferrin uptake as in points 1–3 (Section 2.5).

2.4. Colocalization with endosomal markers

Costaining of Met with endosomal markers can identify the subcellular compartments that Met traffics through. For example, EEA1 can be used to identify an early endosomal compartment, while Lamp1 is a marker of late endosomes. Triple staining can be carried out, and colocalization between various staining can be analyzed to give quantitative data. Table 8.1 provides a list of some commonly used endosomal markers and their localization.

Many of the markers can be observed using commercially available antibodies, and thus, immunofluorescence costaining between Met and the chosen marker can be performed as in Section 2.1. However, many antibodies against Rab GTPases do not give a satisfactory signal in several cell systems. To overcome this problem, cells can be transfected with a fluorescent tagged Rab construct usually 24 h prior to immunofluorescence staining. Low concentrations of DNA are used for this purpose (1 μg/1 × 10^6 cells if using electroporation). It is important, however, to control that the expressed Rab does not interfere with Met trafficking by comparing Met trafficking in transfected versus nontransfected cells (using methods from Sections 2.1 and 2.2).

2.5. Confocal microscopy and image analysis

1. Take single-section images of 0.7 μm thickness using a confocal-laser scanning microscope equipped with a 63×/1.4 Plan-Apochromat oil immersion objective.
2. For image quantifications, take a minimum of seven pictures per condition and choose picture fields arbitrarily on the basis of DAPI staining.

Table 8.1 The table provides a list of endosomal markers and their localization

Marker	Where found
EEA1	Early endosome
Lamp1	Lysosome
Rab4	Early recycling endosome
Rab5	Early endosome
Rab7	Late endosome/lysosome
Rab11	Late recycling endosome

Quantifications should be carried out ideally on 50–100 cells per condition, and experiments need to be repeated at least three times. When setting up the image acquisition for quantification, it is very important to ensure that signals are not saturated.

3. Image analysis can be done using various image analysis software (Zen, Photoshop, MetaMorph, ImageJ, etc.). Appropriate thresholds need to be set up and kept the same for all pictures within one given experiment. Unstimulated cells (e.g., no HGF*) or nonstained cells (or omission of the first antibody or by using an isotype control) should be used as controls to set up the threshold. There are many ways of analyzing confocal images. For quantification of HGF* uptake, one possibility is to measure the number of pixels per cell. For colocalization studies, there are usually two ways to analyze the percentage of colocalizations: (1) the percentage of total Met that is located in the given endosomal compartment (e.g., EEA1) or (2) the percentage of EEA1-positive compartments in which Met is present. Both results may be used as they give distinct information. A good negative control is the cells where Met endocytosis is impaired (see Section 5).

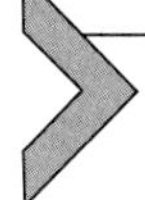

3. ANALYZING MET INTERNALIZATION USING FLOW CYTOMETRY

Rather than investigating the Met that gets internalized, another option is to analyze the level of Met expression at the plasma membrane upon HGF stimulation using a "flow cytometry internalization assay" (Abella et al., 2005; Joffre et al., 2011). A decrease in the level of Met expression at the plasma membrane indicates that Met is internalized.

Prior to performing this assay as part of an experiment, it is important to first perform a dose response test of the antibody being used to find the saturating concentration. The range of concentrations usually tested is between 1 and 20 μg/ml.

1. Plate 5×10^4 cells per well in 24 well plates for 24 h. Note that conditions should be performed in duplicate and two controls should be included: no primary antibody and an IgG isotype control for the primary antibody.
2. Stimulate the cells in complete medium with HGF (100 ng/ml) for various time points in order to induce internalization of Met. This should be performed as a countdown, starting with the longest time point, to

"0 min" so that the cells from all time points are stopped at the same time (0 min), to ensure consistency of the time spent on ice.

3. Place the cells immediately on ice at the end of the HGF stimulation to prevent further Met internalization and keep the cells on ice for the duration of the experiment.
4. Incubate the cells with cold acid wash medium at pH4 (DMEM and HCl + 1% BSA) for 7 min on a shaking platform at 4 °C, to dissociate any HGF bound to Met at the plasma membrane as this could prevent binding of the Met antibody. Wash the cells twice with cold flow cytometry buffer (PBS +2% FBS).
5. Incubate the cells at 4 °C with the primary antibody: goat anti-Met (R&D Systems) that recognizes an epitope located in the extracellular region of Met (or the control goat IgG isotype), diluted at 2 μg/ml in flow cytometry buffer, for 45 min on a shaking platform. The cells can be stained with multiple primary antibodies simultaneously. This allows Met co-internalization with other membrane proteins to be studied, for example.
6. Wash the cells three times in cold flow cytometry buffer and incubate with the secondary antibody Alexa Fluor® 488 Donkey Anti-Goat IgG 8 μg/ml diluted in flow cytometry buffer for 30 min on a shaking platform at 4 °C.
7. Wash the cells three times in cold flow cytometry buffer and incubate the cells in EDTA diluted in cold flow cytometry buffer at a concentration of 5 m*M* (300 μl/well) for 5 min in order to help the cells detach from the plastic and prevent cell clumping.
8. Detach the cells from the plate using a plunger from a 1 ml syringe as a scraper and harvest the cells in a flow cytometry tube.
9. Analyze on a flow cytometer. In flow cytometry experiments of this type, gate the population of healthy cells and ensure that a good peak is achieved using a histogram function showing that the majority of the cells within the population analyzed have the same level of fluorescence intensity and thus be homogenously stained and homogenously express Met.
10. Set the "no primary antibody" control close to the low end of the fluorescence intensity on the *x*-axis. The geometric mean is the value taken into account in this assay. Ideally, the "no primary antibody" control should display a low geometric mean. The geometric mean of the "isotype" control should not be much higher than that of the "no

primary antibody" control as this would indicate a high level of unspecific binding of the IgG. The Met-stained cells should display a much higher intensity of fluorescence than the isotype control, and the value of the corresponding isotype for each condition should be subtracted from the average value obtained from each time point. The raw geometric mean data for each time point with the corresponding isotype value subtracted are normalized to the geometric mean obtained at 0 min of HGF, which is placed at 100% (see Table 8.2).

While this assay measures the amount of Met at the cell surface, it is worth noting that it does not take into account any recycling of Met. Nevertheless, this could be revealed by pre-treatment of cells with inhibitors of recycling such as primaquine (van Weert, Geuze, Groothuis, & Stoorvogel, 2000), which requires 1 h pretreatment at 60 m*M* (Table 8.2). In addition, this assay does not distinguish between the Met that was present at the plasma membrane initially and the potential newly synthesized Met during the time of the assay. To overcome this, cycloheximide can be used as detailed in Section 2.1.

Table 8.2 An example of the raw geometric mean data obtained (duplicates and the average of duplicates) from a flow cytometry internalization assay in a breast cancer cell line, treated with DMSO or primaquine (60 μ*M*)

		No primary antibody	Isotype	0 min	15 min	30 min	60 min
DMSO	Geometric mean	5.17	7.25	101.27	75.07	65.61	56.35
		4.07	6.73	110.38	82.11	71.15	54.49
	Average	4.62	6.99	105.83	78.59	68.38	55.42
	Normalize 0 min to 100%			100.00	72.44	62.11	49.00
Primaquine	Geometric mean	5.82	7.93	89.55	51.55	50.58	48.85
		5.6	8.1	87.36	56.21	50.22	48.99
	Average	5.71	8.015	88.46	53.88	50.4	48.92
	Normalize 0 min to 100%			100.00	57.02	52.69	50.85

The data obtained for each time point of HGF stimulation have the isotype control values subtracted and are normalized to 0 min to give a percentage of Met at the cell surface. The data represent levels of Met at the cell surface upon HGF stimulation and demonstrate a proportion of Met that recycles as the level of Met at the cell surface following HGF stimulation is lower in cells treated with primaquine than those treated with DMSO.

4. ANALYZING MET INTERNALIZATION, RECYCLING, AND DEGRADATION WITH SURFACE MET BIOTINYLATION

4.1. Biotinylation surface removal

A modified method (from Gampel et al., 2006) was established to measure the proportion of intracellular over the total cellular Met. Cell surface proteins were labeled covalently using a membrane-impermeable biotinylation reagent (*N*-hydroxysulfosuccinimide (Sulfo-NHS-SS-Biotin, Pierce)) in order to pull down and thus remove the Met population present at the plasma membrane (Joffre et al., 2011).

The following steps were carried out in a room at 4 °C and on ice to prevent any trafficking events:

1. Incubate cells with 0.15 mg/ml biotin for 10 min.
2. Quench excess biotin by washing with 25 m*M* Tris at pH 8, 137 m*M* NaCl, 5 m*M* KCl, 2.3 m*M* $CaCl_2$, 0.5 m*M* $MgCl_2$, and 1 m*M* Na_2HPO_4.
3. Lyse cells in RIPA (radioimmunoprecipitation assay) buffer, containing 2 m*M* Na_3VO_4, 2 m*M* NaF, and a 1:100 dilution of protease inhibitor cocktail (Calbiochem), and centrifuge at 17,000 × *g*.
4. A fraction of the supernatant (total cellular Met = "total") can be collected. Incubate equal amounts of protein of the residual supernatant (the maximum possible, preferably greater than 200 μg) in an equal volume with streptavidin–agarose beads (Upstate) and agitate at 4 °C for 2 h.
5. Separate the beads by centrifugation (7000 × *g*) and collect the supernatant ("unbound" = internal pool of Met).
6. Wash the beads in lysis buffer at 4 °C and extract proteins ("bound" = surface pool of Met) by heating at 95 °C with sample buffer (Invitrogen, Life Technologies), containing DTT.
7. Analyze equivalent volumes in a Met Western blotting assay.
8. Densitometric analysis (ImageJ software) can be carried out to calculate the percentage of intracellular Met using the following formula:
   Intracellular Met receptor = (Met in unbound fraction)/(total Met) × 100.

4.2. Biotinylation internalization assay

This assay provides a method to measure the kinetics of Met internalization upon time (Roberts, Barry, Woods, van der Sluijs, & Norman, 2001). Cell surface proteins are biotinylated at 4 °C and then are allowed to internalize at 37 °C for the required time(s). We use 15 min here as an example. The remaining proteins on the cell surface are cleaved and a biotin–streptavidin pull down is performed, allowing the amount of internalized Met to be analyzed.

In this experiment, there are three conditions, which ideally should be performed in duplicate:

Total surface (TS): surface of cells undergoes biotinylation, but the biotin is not further cleaved, allowing the total amount of biotinylated Met on the cell surface to be measured.

"T0" minute: surface of cells undergoes biotinylation and biotin cleavage, allowing the efficiency of the cleavage and the inhibition of basal Met internalization on ice to be assessed.

"T15" minutes: following cell surface biotinylation, surface proteins are allowed to internalize for 15 min at 37 °C, and then, biotin is cleaved at 4 °C from remaining proteins on the cell surface. The period of incubation time can be adapted to the question being asked, and more time points can be added if required:

1. Pretreat cells with cycloheximide for 12 h.
2. The following steps should be carried out in a room at 4 °C and on ice to prevent any protein trafficking.
3. Label cell surface proteins with 0.2 mg/ml Sulfo-NHS-SS-Biotin in PBS for 45 min on a rocker.
4. Wash cells with cold PBS and place the T15 plates in an incubator at 37 °C in warm complete culture medium, with or without HGF, for 15 min to allow protein trafficking. After the indicated times, aspirate the medium and transfer the dishes onto ice and wash with cold PBS.
5. Remove biotin from proteins remaining at the cell surface in the T0 and T15 plates by reduction with 180 m*M* of the membrane-impermeant reducing agent MESNA (sodium 2-mercaptoethanesulfonate, Sigma) in 50 m*M* Tris and 100 m*M* NaCl (diluted in dH_2O) at pH 8.6. Cells should be incubated for 20 min on a rocker.
6. Quench the MESNA by adding 180 m*M* iodoacetamide (IAA, Sigma) for 10 min on a rocker.
7. Lyse cells in RIPA buffer, containing 2 m*M* Na_3VO_4, 2 m*M* NaF, and a 1:100 dilution of protease inhibitor cocktail (Calbiochem). Pass lysates three times through a 27-gauge needle; leave samples on a rotating wheel for 40 min and centrifuge (17,000 × *g* for 10 min).

8. Incubate equal amounts of protein (the maximum possible, preferably greater than 200 μg) with streptavidin–agarose beads and agitate at 4 °C for 2 h.
9. Collect the beads by centrifugation ($7000 \times g$) and discard the supernatant.
10. Wash the beads three times in lysis buffer and extract the proteins by heating at 95 °C with 25 μl of sample buffer.
11. Analyze equivalent volumes in a Met Western blotting assay.
12. Densitometric analysis (ImageJ software) can be carried out to calculate the percentage of Met internalization using the following formula:

$$(\text{Met level after first incubation at } 37\,^\circ\text{C}) - (\text{Met level at time 0 min})/(\text{total surface Met}) \times 100.$$

This assay reveals the rate of Met internalization from the cell surface over a set period of time. However, internalized Met that was then recycled to the cell surface within this incubation time would not be detected as the biotin would be cleaved during the cleavage step following this incubation (the aim of which is to clear the noninternalized biotinylated Met population). In order to see whether Met recycling is taking place, cells can be treated with primaquine to inhibit Met recycling or a biotinylation recycling assay would need to be performed (see Section 4.3).

One of the most common problems encountered with this assay is the presence of a band at T0, while it should give no band or a very faint one (see Fig. 8.2 as an example of an unsuccessful assay). It is important to ensure that the MESNA solution is at pH 8.6 to obtain complete cleavage of cell surface proteins. If the problem continues, it may be possible to perform several rounds of cleavage for shorter amounts of time rather than one round for 20 min. It is also very important to ensure that the cells are kept on ice in a cold room at 4 °C throughout the experiment to prevent Met

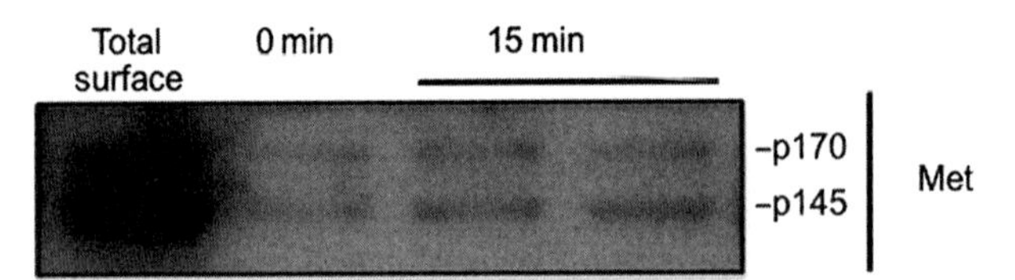

Figure 8.2 Biotinylation internalization assay (unsuccessful). Cells were surface biotinylated and then incubated for 15 min at 37 °C to allow internalization of surface proteins. The biotin was then cleaved from the surface. The remaining biotinylated proteins were pulled down using streptavidin beads and analyzed by Western blot for Met expression. The control T0 shows a too strong Met band indicating a poor biotin cleavage or a too high basal Met internalization, while cells are on ice.

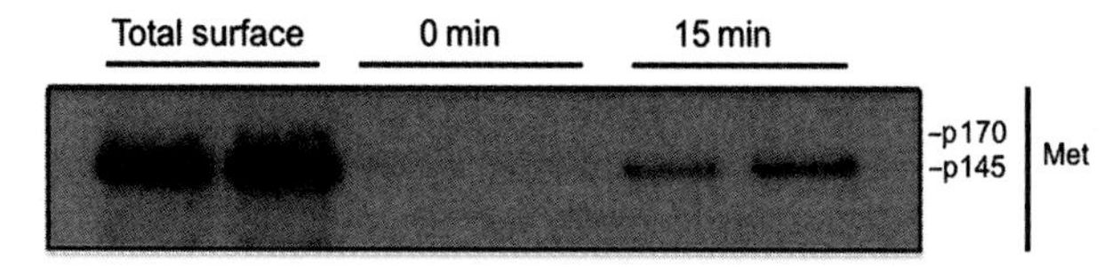

Figure 8.3 Biotinylation internalization assay (successful). Cells were surface biotinylated and then incubated for 15 min at 37 °C to allow internalization of surface proteins. The biotin was then cleaved from the surface. The remaining biotinylated proteins were pulled down using streptavidin beads and analyzed by Western blot for Met expression. All conditions were performed in duplicate, and a good biotin cleavage was achieved (T0) with an obvious internalization observed (T15).

trafficking. If this is not done, then it is possible that while full cleavage may be achieved, there may be some Met that internalizes. Figure 8.3 shows an example of a successful biotinylation internalization assay. Firstly, it was performed in duplicate, which is important, as there can be some variation between the duplicates such as that observed in the 15 min duplicates. Secondly, the T0 shows no band.

4.3. Biotinylation recycling assay

A biotinylation recycling assay measures the proportion of the internalized Met that recycles back to the cell surface. The assay starts as the biotinylation internalization assay (see Section 4.2). After the first biotin cleavage step, the internalized Met is allowed to recycle through reincubation of the cells at 37 °C for the required period(s) of time. We use 15 min here as an example. This is followed by a second round of biotin cleavage such that the amount of internalized Met detected after the second round of incubation at 37 °C can be compared to the amount after the initial round. If less internalized Met is observed after the second round, then the percentage of recycling can be calculated (Joffre et al., 2011).

In addition to the conditions used for a biotinylation internalization assay, a fourth condition needs to be added:

"*T15 recycle*" *minutes*: the cell surface proteins undergo two rounds of 15 min at 37 °C and of biotin cleavage:

1. Perform steps 1–6 of the biotinylation internalization assay at 4 °C (in a cold room).
2. Return the plates to 37 °C for 15 min.
3. Place the cells on ice and cleave the biotin from recycled proteins by a second reduction with 180 m*M* MESNA in 50 m*M* Tris and 100 m*M*

NaCl (diluted in dH_2O) at pH 8.6. Incubate for 20 min on a shaking platform at 4 °C.

4. Quench unreacted MESNA with 180 m*M* IAA (Sigma) for 10 min on a rocker.
5. Follow steps 7–11 of the biotinylation internalization assay.
6. Densitometric analysis (ImageJ software) can be carried out to calculate the percentage of recycled Met using the following formula:

   $100 - [(\text{Met level after 15 min of reincubation at } 37\,°\text{C}) - (\text{Met level at time 0 min})/(\text{Met level after first incubation at } 37\,°\text{C}) - (\text{Met level at time 0 min}) \times 100]$

Note: it is worth noting that although after only 15 min of internalization there is not usually any detectable Met degradation, if cells are incubated for longer time points, then the Met degradation would need to be taken into consideration.

4.4. Biotinylation degradation assay

A biotinylation degradation assay offers the advantage of following the degradation of the mature form of Met present at the plasma membrane at the beginning of the assay, compared to a Western blot, which shows the total amount of cellular Met (Kermorgant et al., 2003). To measure the degradation over a specified period of time, Met is allowed to internalize and degrade over much longer periods of time as compared to the biotinylation internalization assay, typically 45 min to 8 h. In this assay, the TS plate is considered to be time "0" and no cleavage of the Biotin is required (Joffre et al., 2011):

1. Perform steps 1–3 of the biotinylation internalization assay.
2. Wash cells with cold PBS and incubate the plates, starting with the longest time, at 37 °C in warm complete culture medium, with or without HGF, to allow Met internalization and subsequent degradation. After the indicated times, aspirate the medium, transfer the dishes on ice and wash with cold PBS.
3. No cleavage is required in this assay, so follow steps 7–12 of the biotinylation internalization assay. Here, following densitometric analysis of Met, Met is set at 100% at time 0 min and is then expressed as a percentage of Met at time 0 min for the other time points.

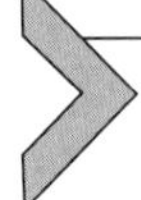

5. ANALYZING MET ENDOSOMAL SIGNALING *IN VITRO* AND *IN VIVO*

To demonstrate the existence of Met endosomal signaling, immunofluorescence analyses can be conducted to show that phosphorylated Met (using a specific anti-phospho-Met antibody) can be found on endosomes and the co-recruitment of activated signaling partners (Kermorgant & Parker, 2008; Kermorgant et al., 2004). Confocal live imaging using HGF* and GFP-tagged molecules (such as signaling molecules and/or endosomal markers) allows the direct observation of Met endosomal signaling/localization. It would be interesting in the future to complete these studies with a combination of subcellular fractionation coupled with phosphoproteomics, a method that allowed the discovery of organelle-specific EGFR signaling molecules (Stasyk et al., 2007).

To demonstrate the requirement of endosomal signaling for an optimal Met signaling and for correct cellular functions, one method in use is to interfere with Met endocytosis. For example, a reduction in cell migration upon HGF stimulation following a block in Met endocytosis suggests that Met endosomal signaling is required for optimal cell migration (Joffre et al., 2011; Kermorgant et al., 2004).

There are various means of interfering endocytosis, including pharmacological inhibition, dominant negative constructs, and knockdown of proteins required for endocytosis by either siRNA or shRNA. Obviously, it would be preferable that the tools used are specific to Met. However, presently, the majority of methods used lead to a broad inhibition of endocytosis (e.g., interference with dynamin or clathrin expression/activity). It is however possible to narrow this to inhibit RTK endocytosis, that is, knocking down the ubiquitin ligase c-Cbl (Joffre et al., 2011). There is a more specific way to inhibit Met endocytosis through mutating the Grb2 binding site within the Met (Joffre et al., 2011; Ponzetto et al., 1996). However, no method of targeting Met endocytosis is flawless, as even mutating the Grb2 binding site of Met itself may have side effects on cell signaling. Thus, it is useful to always use several methods in conjunction with each other to build a stronger conclusion. Furthermore, it is necessary to have a Met-dependent cell model when using broad endocytic inhibitors. This can be achieved either by comparing cells treated with HGF to those untreated (Hammond et al., 2001; Kermorgant et al., 2004) or by comparing a Wt Met to a constitutively active mutated Met (Joffre et al., 2011). A positive control of fluorescence analysis with HGF* can be used to check the inhibition of Met endocytosis.

5.1. Inhibiting Met endocytosis

The following endocytic blockers have been shown to significantly reduce Met endocytosis in cells in culture:

5.1.1 Pharmacological inhibition: Dynasore and Dynole 34-2

These are cell-permeable pharmacological inhibitors of the GTPase dynamin (Hill et al., 2009; Macia et al., 2006):

1. In the control, use DMSO (which is the diluent of these drugs).
2. If performing flow cytometry or immunofluorescence, be aware that Dynasore fluoresces in the green channel.
3. Starve cells in 0% serum for a minimum of 24 h.
4. Pretreat cells with Dynasore (80 μ*M*) or Dynole 34–2 (30 μ*M*) for 30 min prior to the start of the experiment.
5. Maintain the cells and any surrounding media in DMSO or Dynasore (80 μ*M*)/Dynole 34-2 (30 μ*M*) during the experiment.

5.1.2 Knockdown of clathrin, c-Cbl, or Grb2

These can be transient or permanent, using siRNA or shRNA (Joffre et al., 2011).

5.1.3 Expression of the dynamin-dominant-negative Dynamin2–K44A–GFP

The expression of this dynamin mutant leads to a reduction in Met internalisatation (Joffre et al., 2011; Hammond et al., 2001).

5.1.4 Expression of a Met mutant N1358H

The mutation N1358H, located on the +2 position of Y1356 in the Met docking site, abolishes the Grb2 consensus sequence without obstructing other molecules binding to Met (Ponzetto et al., 1996). As a consequence, this Met mutant is unable to bind to Grb2, which is required for Met endocytosis (Li, Lorinczi, Ireton, & Elferink, 2007). We have shown that introducing the mutation N1358H in the Met oncogenic mutant M1268T significantly reduces its endocytosis and subsequent oncogenicity (Joffre et al., 2011).

5.2. Analyzing the influence on Met signaling and tumorigenesis

5.2.1 Met signaling

The various "endocytic inhibitors" listed in Section 5.1 can be used to assess the influence on Met signaling. For example, Western blots can monitor pathway activation, such as phospho-ERK1/2 (Kermorgant et al., 2004) and phospho-STAT3 (Kermorgant & Parker, 2008); immunofluorescence can inform on signal location, such as ERK1/2 or STAT3 nuclear uptake

(Kermorgant & Parker, 2008); GTPase pull-down assays measure the activation of the Rho GTPase such as for Rac1 (Joffre et al., 2011).

5.2.2 *Cell transformation*

These tools can also be used to assess the influence of Met endosomal signaling on cellular functions/transformation. These include Transwell chemotactic migration assays, monitoring the organization of the actin cytoskeleton and focal adhesions by confocal microscopy and anchorage-independent soft agar cultures, which are usually maintained for 2 weeks (Joffre et al., 2011). Obviously, other assays could be performed.

5.2.3 In vivo *tumorigenesis*

The influence of Met endocytosis inhibition can be investigated in *in vivo* studies such as tumorigenesis and experimental metastasis assays. So far, two different means have been used successfully:

- Pharmacological inhibition (the painting method): Drugs that interfere with endocytosis are currently not bioavailable and so cannot be administered by oral gavage or intravenous injection. Thus, we have developed a protocol to use our two endocytic inhibitors cited in Section 5.1.1 in tumorigenesis experiments: the surface of the skin over the tumors (or where cells had been injected in the control mice) is painted daily with Dynasore or Dynole 34-2 or DMSO, as a control. DMSO is a solvent that is easily absorbed through the skin of nude mice bringing the drug with it.
- Stable knockdown with shRNA: we have so far used clathrin heavy chain knockdown for both *in vivo* tumorigenesis and metastasis (Joffre et al., 2011).

As previously mentioned, it is important to have a Met-dependent model. To achieve this, we have used NIH3T3 cells expressing a constitutively active, naturally occurring tumorigenic D1246N- or M1268T-mutated Met. These cells rapidly form tumors in nude mice compared to cells expressing Wt Met (Jeffers et al., 1997). In each experiment, cells expressing a Met mutant are compared to cells expressing Wt Met. Met dependence of tumor growth is verified using pharmacological Met inhibition (Joffre et al., 2011):

a. Tumor growth:

  1. 5×10^5 cells are inoculated subcutaneously in the flank region of nude mice.

  2. Once subcutaneous tumors have appeared, measure their size daily using calipers and calculate the volume using the following formula:

$$\text{Length} \times \text{width}^2 \times (\pi/6).$$

3. When the pharmacological inhibition is used, once the tumors have reached the set volume (in our case 50 mm^3), apply topically over the tumors Dynasore (80 μ*M*) or Dynole 34-2 (30 μ*M*), diluted in 100 μl DMSO.
4. When tumors have reached 1 cm in length, mice are killed humanely and tumors are resected, snap-frozen, or fixed in formal saline.
5. To observe Met on endosomes *in vivo* and check that the block in Met endocytosis has been achieved, fixed tumors were paraffin-embedded and sections of 4 μ*M* were processed for immunofluorescence (see the following point 6).
6. Heat the sections in citrate buffer for 20 min in a microwave for antigen retrieval.
7. Block sections in PBS containing 3% BSA and 0.5% donkey serum for 30 min, and then, permeabilize in 3% Triton.
8. Stain sections using a goat anti-mouse Met antibody (R&D Systems) (1/100) and an antibody against an endosomal marker such as a rabbit anti-EEA1 (Santa cruz Biothechnology, Inc) (1:100) overnight.
9. After washing, stain sections with appropriate fluorochrome-labeled secondary antibodies (1/500) for 2 h. Dehydrate sections in increasing concentrations of ethanol and mount coverslips with ProLong Gold antifade reagent. Analyze slides by confocal microscopy.

b. Experimental metastasis assay:
1. 5×10^5 cells are inoculated into the tail vein.
2. After the number of days required to observe tumors in the lung (10 in our study; Joffre et al., 2011), cull the mice.
3. Remove and weigh the lungs. Macroscopic tumors may be counted.
4. The paraffin-embedded lung sections, stained with hematoxylin/eosin, are further analyzed by microscopy.

6. SUMMARY

Endosomal signaling of Met may play a major role in tumor growth and metastasis. The techniques and tools presented here may help to improve our current understanding of Met endosomal signaling and can be adapted to study other RTKs. Such studies are particularly important, as it is vital that a wide range of tools are available and developed to investigate endosomal signaling across the field and to allow studies to be verified in multiple *in vitro* and *in vivo* tumor models. It is important to continue expanding the current understanding of the mechanisms involved, so as to develop inhibitors of Met/RTK endosomal signaling as potential novel therapeutic means. This may be of great therapeutic value to some cancer patients in the future.

REFERENCES

Abella, J. V., Peschard, P., Naujokas, M. A., Lin, T., Saucier, C., Urbe, S., et al. (2005). Met/Hepatocyte growth factor receptor ubiquitination suppresses transformation and is required for Hrs phosphorylation. *Molecular and Cellular Biology, 25*(21), 9632–9645. http://dx.doi.org/10.1128/MCB.25.21.9632-9645.2005.

Birchmeier, C., Birchmeier, W., Gherardi, E., & Vande Woude, G. F. (2003). Met, metastasis, motility and more [Review]. *Nature Reviews Molecular Cell Biology, 4*(12), 915–925. http://dx.doi.org/10.1038/nrm1261.

Gampel, A., Moss, L., Jones, M. C., Brunton, V., Norman, J. C., & Mellor, H. (2006). VEGF regulates the mobilization of VEGFR2/KDR from an intracellular endothelial storage compartment. *Blood, 108*(8), 2624–2631. http://dx.doi.org/10.1182/blood-2005-12-007484.

Gherardi, E., Birchmeier, W., Birchmeier, C., & Vande Woude, G. (2012). Targeting MET in cancer: Rationale and progress. *Nature Reviews Cancer, 12*(2), 89–103. http://dx.doi.org/10.1038/nrc3205.

Hammond, D. E., Urbe, S., Vande Woude, G. F., & Clague, M. J. (2001). Down-regulation of MET, the receptor for hepatocyte growth factor. *Oncogene, 20*(22), 2761–2770. http://dx.doi.org/10.1038/sj.onc.1204475.

Hill, T. A., Gordon, C. P., McGeachie, A. B., Venn-Brown, B., Odell, L. R., Chau, N., et al. (2009). Inhibition of dynamin mediated endocytosis by the dynoles–synthesis and functional activity of a family of indoles. *Journal of Medicinal Chemistry, 52*(12), 3762–3773. http://dx.doi.org/10.1021/jm900036m.

Jeffers, M., Schmidt, L., Nakaigawa, N., Webb, C. P., Weirich, G., Kishida, T., et al. (1997). Activating mutations for the met tyrosine kinase receptor in human cancer. *Proceedings of the National Academy of Sciences of the United States of America, 94*(21), 11445–11450.

Joffre, C., Barrow, R., Menard, L., Calleja, V., Hart, I. R., & Kermorgant, S. (2011). A direct role for Met endocytosis in tumorigenesis. *Nature Cell Biology, 13*(7), 827–837. http://dx.doi.org/10.1038/ncb2257.

Joffre, C., Barrow, R., Menard, L., & Kermorgant, S. (2013). RTKs as models for trafficking regulation—c-Met/HGF-receptor: c-Met signalling in cancer: Location counts. In Y. Yarden (Ed.), *Vesicle trafficking in cancer*. Pondicherry, India: Springer.

Kermorgant, S., & Parker, P. J. (2008). Receptor trafficking controls weak signal delivery: A strategy used by c-Met for STAT3 nuclear accumulation. *The Journal of Cell Biology, 182*(5), 855–863. http://dx.doi.org/10.1083/jcb.200806076.

Kermorgant, S., Zicha, D., & Parker, P. J. (2003). Protein kinase C controls microtubule-based traffic but not proteasomal degradation of c-Met. *The Journal of Biological Chemistry, 278*(31), 28921–28929. http://dx.doi.org/10.1074/jbc.M302116200.

Kermorgant, S., Zicha, D., & Parker, P. J. (2004). PKC controls HGF-dependent c-Met traffic, signalling and cell migration. *The EMBO Journal, 23*(19), 3721–3734. http://dx.doi.org/10.1038/sj.emboj.7600396.

Li, N., Lorinczi, M., Ireton, K., & Elferink, L. A. (2007). Specific Grb2-mediated interactions regulate clathrin-dependent endocytosis of the cMet-tyrosine kinase. *The Journal of Biological Chemistry, 282*(23), 16764–16775. http://dx.doi.org/10.1074/jbc.M610835200.

Macia, E., Ehrlich, M., Massol, R., Boucrot, E., Brunner, C., & Kirchhausen, T. (2006). Dynasore, a cell-permeable inhibitor of dynamin. *Developmental Cell*, *10*(6), 839–850. http://dx.doi.org/10.1016/j.devcel.2006.04.002.

Ponzetto, C., Zhen, Z., Audero, E., Maina, F., Bardelli, A., Basile, M. L., et al. (1996). Specific uncoupling of GRB2 from the Met receptor. Differential effects on transformation and motility. *The Journal of Biological Chemistry*, *271*(24), 14119–14123.

Roberts, M., Barry, S., Woods, A., van der Sluijs, P., & Norman, J. (2001). PDGF-regulated rab4-dependent recycling of alphavbeta3 integrin from early endosomes is necessary for cell adhesion and spreading. *Current Biology*, *11*(18), 1392–1402.

Schmidt, L., Duh, F. M., Chen, F., Kishida, T., Glenn, G., Choyke, P., et al. (1997). Germline and somatic mutations in the tyrosine kinase domain of the MET proto-oncogene in papillary renal carcinomas. *Nature Genetics*, *16*(1), 68–73. http://dx.doi.org/10.1038/ng0597-68.

Stasyk, T., Schiefermeier, N., Skvortsov, S., Zwierzina, H., Peranen, J., Bonn, G. K., et al. (2007). Identification of endosomal epidermal growth factor receptor signaling targets by functional organelle proteomics. *Molecular & Cellular Proteomics*, *6*(5), 908–922. http://dx.doi.org/10.1074/mcp.M600463-MCP200.

Trusolino, L., Bertotti, A., & Comoglio, P. M. (2010). MET signalling: Principles and functions in development, organ regeneration and cancer. *Nature Reviews Molecular Cell Biology*, *11*(12), 834–848. http://dx.doi.org/10.1038/nrm3012.

van Weert, A. W., Geuze, H. J., Groothuis, B., & Stoorvogel, W. (2000). Primaquine interferes with membrane recycling from endosomes to the plasma membrane through a direct interaction with endosomes which does not involve neutralisation of endosomal pH nor osmotic swelling of endosomes. *European Journal of Cell Biology*, *79*(6), 394–399. http://dx.doi.org/10.1078/0171-9335-00062.

CHAPTER NINE

Intracellular Toll-Like Receptor Recruitment and Cleavage in Endosomal/Lysosomal Organelles

Mira Tohmé[*,†,‡], **Bénédicte Manoury**[*,†,1]

[*]INSERM, Unité 1013, Paris, France
[†]Université Paris Descartes, Sorbonne Paris Cité, Faculté de médecine, Paris, France
[‡]INSERM, Unité 932, Institut Curie, Paris, France
[1]Corresponding author: e-mail address: benedicte.manoury@inserm.fr

Contents

Abstract

Microbial pathogens are recognized through multiple, distinct receptors such as intracellular Toll-like receptors (TLRs 3, 7, 8, 9, and 13) which reside in the endosomes and lysosomes. TLRs are sensitive to chloroquine, a lysomotropic agent that neutralizes acidic compartments indicating a role for endo/lysosomal proteases for their signaling. Indeed, upon stimulation, full-length TLR7 and 9 are cleaved into a C-terminal fragment and this processing is highly dependent on a cysteine protease named asparagine endopeptidase (AEP) in dendritic cells. A recruitment and a boost in AEP activity, which was induced shortly after TLR7 and 9 stimulation, are shown to promote TLR7 and 9 cleavage and correlate with an increased acidification in endosomes and lysosomes. Moreover, mutating a putative AEP cleavage site in TLR7 or 9 strongly decreases their signaling in DCs, suggesting perhaps a direct cleavage of TLR7 and 9 by AEP. These results demonstrate that TLR7 and 9 require a proteolytic cleavage for their signaling and identified a key endocytic protease playing a critical role in this process.

Methods in Enzymology, Volume 535
ISSN 0076-6879
http://dx.doi.org/10.1016/B978-0-12-397925-4.00009-2

1. INTRODUCTION

Toll-like receptors (TLRs) are proteins, which recognize conserved molecules from microorganisms, and in dendritic cells (DCs), they are crucial in linking innate to adaptive immunity. TLRs contain several leucine-rich repeats in an extracellular loop, a transmembrane domain, and a cytosolic domain and are expressed either at the plasma membrane or in the endosomal/lysosomal organelles. TLR stimulation is linked to MyD88 or TRIF-dependent signaling pathways that regulate the activation of different transcription factors, such as NF-kB (Janeway & Medzhitov, 2002). Specific interaction between TLRs and their ligands activates NF-kB, resulting in enhanced inflammatory cytokine responses, induction of DC maturation, and expression of chemokine receptors. TLRs expressed at the plasma membrane recognized Gram-negative bacteria, and endosomal TLRs sense viral and bacterial nucleic acids such as double/single-stranded RNA or DNA. Endogenous ligands called DAMPs (for damage-associated molecular patterns) may also activate TLRs during self-tissues or cell damage. Recent findings have described the importance of proteolysis for endosomal TLR function (Ewald et al., 2008; Garcia-Cattaneo et al., 2012; Park et al., 2008) and TLR activation has been shown to boost MHCI cross-presentation in DCs (de Brito et al., 2011; Sathe et al., 2011). Upon stimulation, full-length (FL) TLR9 is cleaved into a C-terminal fragment sufficient for signaling. This cleavage is realized by several cathepsins in different cells, including macrophages, while in DCs this cleavage is performed mainly by cathepsin K (CatK) and asparagine endopeptidase (AEP). In CatK-deficient DCs, TLR9 signaling is abrogated, and in DCs lacking AEP, TLR9 cleavage in phagosomal compartments as well as CD4+ antigen-specific T cell proliferation was greatly reduced upon CpG stimulation (Asagiri et al., 2008; Sepulveda et al., 2009). TLR7 is also subjected to similar proteolytic maturation and requires AEP for proper signaling (Maschalidi et al., 2012). Altogether these results indicate that endosomal proteases, which are key players in generating peptides for the MHC class II pathway (Manoury et al., 1998; Moss, Villadangos, & Watts, 2005; Nakagawa et al., 1998; Riese et al., 1996), play also an important role in intracellular TLR activation.

2. PURIFICATION OF ENDOSOMES AND LYSOSOMES

2.1. Sucrose gradient fractionation

All procedures, unless otherwise indicated, are carried out at 4 °C.

1. 10^8 dendritic cells (DCs) are generated from mouse bone marrow by culturing precursors for 7–10 days in Iscove's modified Dulbecco's medium (IMDM, Sigma, I3390) supplemented with 10 FBS, 1% PS, 1% GLN, and 10 ng/mL GM-CSF (Peprotech, 315-03).
2. Detach BMDCs (bone marrow-derived DCs) with PBS–EDTA 5 m*M* (10 min at 37 °C).
3. After two washes in PBS and one wash in homogenization buffer (HB: 3 m*M* imidazole, 8% sucrose, 1 m*M* DTT, 1 m*M* EDTA supplemented with one tablet of proteases inhibitor cocktail (Roche), pH 7.4), cells are homogenized with a cell cracker in HB through a ball of 0.006 mm of clearance (ball number 6) in order to have 80% of mortality.
4. Postnuclear supernatant (PNS) is prepared by centrifugation (1000 × *g* for 10 min).
5. PNS is adjusted with 62% of sucrose solution in order to have a final concentration of 40.6% of sucrose.
6. A gradient of sucrose is prepared following this order from the bottom to the top:
   2 mL of PNS containing endosomes and lysosomes
   3 mL of 35% of sucrose
   2 mL of 25% of sucrose
   4 mL of HB
7. After ultracentrifugation for 90 min with no brake using the SW41 rotor, fractions are collected. The interface/ring between the 35% and 25% sucrose corresponds to the early endosomes while the one between the 25% of sucrose and the HB corresponds to the late endosomes.
8. The fractions are concentrated by ultracentrifugation using the TLA110 rotor for 30 min at 35,000 rpm.
9. Pellets are resuspend in 50 μL of PBS and can be either frozen or used immediately for a Western blot.

2.2. Percoll gradient fractionation (lysosome Purification)

2.2.1 Lysosomes purification

1. DCs are prepared and homogenized as described in steps 2.1.1 and 2.1.2.
2. The homogenate is centrifuged at 2000 rpm for 10 min and the PNS is recovered and underlaid with 23 mL of 27% of percoll.
3. After centrifugation for 60 min at 23,000 rpm using the Ti45 rotor, fractions were collected by upward displacement using a fraction collector.
4. Fractions are collected and assayed for β-hexosaminidase activity (check Section 2.2.2).

5. Positive fractions corresponding to lysosomes are pooled and centrifuged for 60 min at 40,000 rpm using the Ti70.2 rotor.
6. The ring obtained is recovered with a Pasteur pipette and is subjected to one more centrifugation using the TLS-55 rotor at 50,000 rpm for 60 min.
7. Pellet containing concentrated lysosomes is collected and frozen.

2.2.2 β-Hexosaminidase assay

1. Add 20 μL of the sample (fractions collected in step 2.2.1.3) or the blank (27% percoll) to 80 μL of HB, prewarmed at 37 °C.
2. Prepare the standard curve using the 4-methylumbelliferone, starting concentration at 1 nmol.
3. Add 100 μL of substrate (4-methylumbelliferyl *N*-acetyl β-D-glucosamide).
4. Incubate 10 min at 37 °C.
5. Stop the reaction by adding 100 μL of stop buffer (0.1 *M* glycine, pH 10.3), store in the dark.
6. Read fluorescence on a fluorimeter: excitation 360 nm, emission 450 nm.

2.3. Endosomes and lysosome purification using magnetic beads

1. Detach BMDCs with PBS–EDTA 5 m*M* (10 min at 37 °C).
2. Wash the cells with PBS 2 ×.
3. Resuspend the cells in 15-mL Falcon tube (10^8 cells/1 mL) in IMDM alone.
4. Add 60 μL of magnetic nanoparticles (TurboBeads of 20 nm of diameter, 10 mg/mL) for 30 min at 4 ° C.
5. The cells are incubated for 20 min (pulse) in the water bath at 37 °C.
6. Stop the reaction by adding 10 mL cold PBS–BSA 0.1% and centrifuge for 10 min at 1300 rpm at 4 °C.
7. Repeat step 6 three times.
8. Resuspend the cells in 2 mL of complete IMDM and split the cells in two Falcon tubes of 15 mL (1 mL of cells each).
9. Add 10 mL cold PBS–BSA 0.1% for $t=0$ (pulse) and leave on ice, put the other tube at 37 °C for 100 min (chase).
10. $t=0$ corresponds to early endosomes (20 min of pulse), $t=120$ min corresponds to lysosomes.
11. At the end of the chase, centrifuge the cells.
12. Now, perform the end of the experiment on ice.

13. Wash the cells in HB.
14. Resuspend the cells in 1 mL of HB.
15. Break the cells using a 1-mL syringe and a 22-g needle. You need about 30 flushes to break the cells. Check the breaking of the cells under a microscope (about 80% of the cells should be dead).
16. Transfer the breaking cells into cold Eppendorf tubes and place the Eppendorf tubes on the magnetic stand on ice.
17. Leave 5 min; aspirate the supernatant with a thin tip. Keep the SNT aside.
18. Wash the beads carefully with 1 mL PBS–BSA 0.1% and repeat step 18 eight times.
19. Resuspend the beads in 50 μL of lysis buffer (50 m*M* Tris, pH 7.4, 150 m*M* NaCl, 0.5% NP40, 2 m*M* $MgCl_2$, and a tablet of a cocktail of proteases inhibitors) and leave on ice for 15 min.
20. Centrifuge at 12,000 rpm for 15 min and freeze the supernatant.

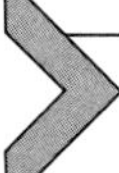

3. PROTEASES ASSAYS

1. Buffer: 50 m*M* citrate buffer, pH 5.5, 1 m*M* DTT.
2. Protease activity assays are performed on a Mithras LB940 (Berthold Technologies) by measuring the release of fluorescent *N*-acetyl-methyl-coumarin (NHMec) in citrate buffer (pH 5.5) at 37 °C. Specific substrates are the following: Z-Ala-Ala-Asn-NHMec for AEP, Z-Arg-Arg-NHMec for CatB, Z-Phe-Arg-NHMec for CatB/L, Z-Val-Val-Arg-NHMec for CatS, and Z-Gly-Pro-Arg-NHMec for CatK.
3. Activity of the proteases is assessed using specific fluorometric substrates. One microgram of early or late endosomes is incubated with the different specific substrates (see below) for different times. To calculate the specific activity of each protease (μmol of substrate release per minute), a standard curve is obtained using different concentrations of NHMec (from 0.2 to 2 μ*M*).
4. Same assays are performed with purified endosomes or lysosomes from BMDCs stimulated with intracellular TLR ligands.

4. INTRACELLULAR TLR PROCESSING

To check the purity of the endosomes and lysosomes obtained and to study TLR9 cleavage, proteins expressed in the endosomes and lysosomes are subjected to a Western blot (Burnette, 1981).

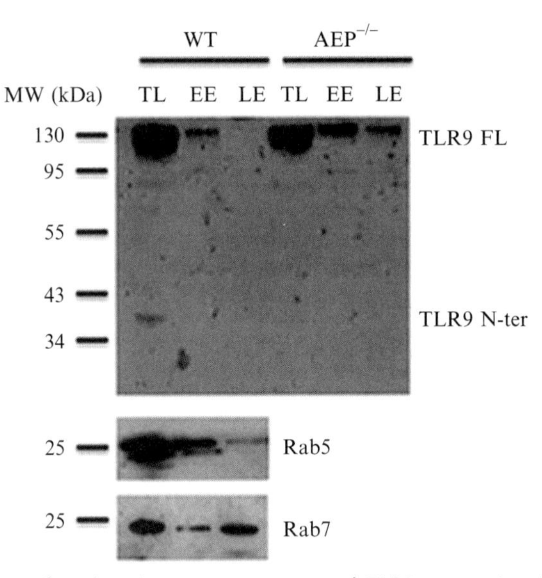

Figure 9.1 Purity of endocytic compartments and TLR9 expression in endosomes. Endosomes from wt (wild-type) and AEP-deficient BMDCs (AEP$^{-/-}$) were magnetically purified after 20 or 120 min. Protein expressed in total lysate (TL) or in endosomes (5 microg) was resolved by SDS-PAGE. TLR9 proteins in early (EE) and late (LE) endosomes from WT and AEP-deficient BMDCs were detected by immunoblot. Immunodetection of early (Rab5) and late (Rab7) markers in endosomes. FL TLR9, full-length TLR9; TLR9 N-ter, fragment corresponding to the N-terminal part of TLR9.

1. Protein concentration is measured with a colorimetric assay kit (Bio-Rad).
2. Samples (5 μg of endosomes or lysosomes) are heated at 80 °C for 10 min in the reducing buffer (10 μL of 4× Laemmli buffer supplemented with 8% of β-mercaptoethanol).
3. Following electrophoresis, proteins are transferred on a PVDF membrane, pore size 0.45 μm, and they are stained with antibodies specific for the target proteins: Rab5 for early endosomes and cathepsin D or rab7 for late endosomes. TLR9 cleavage is also monitored (see Fig. 9.1).

REFERENCES

Asagiri, M., Hirai, T., Kunigami, T., Kamano, S., Gober, H. J., Okamoto, K., et al. (2008). Cathepsin K-dependent Toll-like receptor 9 signaling revealed in experimental arthritis. *Science*, *319*, 624–627.

Burnette, W. N. (1981). Western blotting: Electrophoretic transfer of proteins from sodium dodecyl sulfate—Polyacrylamide gels to unmodified nitrocellulose and radiographic detection with antibody and radioiodinated protein A. *Analytical Biochemistry*, *112*, 195–203.

de Brito, C., Tomkowiak, M., Ghittoni, R., Caux, C., Leverrier, Y., & Marvel, J. (2011). CpG promotes cross-presentation of dead cell-associated antigens by pre-CD8+alpha dendritic cells. *Journal of Immunology*, *186*, 1503–1511.

Ewald, S. E., Lee, B. L., Lau, L., Wickliffe, K. E., Shi, G. P., Chapman, H. A., et al. (2008). The ectodomain of Toll-like receptor 9 is cleaved to generate a functional receptor. *Nature*, *456*, 658–662.

Garcia-Cattaneo, A., Gobert, F. X., Muller, M., Toscano, F., Flores, M., Lescure, A., et al. (2012). Cleavage of Toll-like receptor 3 by cathepsins B and H is essential for signaling. *Proceedings of the National Academy of Sciences of the United States of America*, *109*, 9053–9058.

Janeway, C. A., & Medzhitov, R. (2002). Innate immune recognition. *Annual Review of Immunology*, *20*, 197–216.

Manoury, B., Hewitt, E. W., Morrice, N., Dando, P. M., Barrett, A. J., & Watts, C. (1998). An asparaginyl endopeptidase processes a microbial antigen for class II MHC presentation. *Nature*, *396*, 695–699.

Maschalidi, S., Hässler, S., Blanc, F., Sepulveda, F., Tohme, M., Chignard, M., et al. (2012). Asparagine endopeptidase controls anti-influenza virus immune responses through TLR7 activation. *PLoS Pathogens*, *8*(8), e1002841.

Moss, C. X., Villadangos, J. A., & Watts, C. (2005). Destructive potential of the aspartyl protease cathepsin D in MHC class-restricted antigen processing. *European Journal of Immunology*, *35*, 3442–3451.

Nakagawa, T., Roth, W., Wong, P., Nelson, A., Farr, A., Deussing, J., et al. (1998). Cathepsin L: Critical role in Ii degradation and CD4 T cell selection in the thymus. *Science*, *280*, 450–453.

Park, B., Brinkmann, M. M., Spooner, E., Lee, C. C., Kim, Y. M., & Ploegh, H. H. (2008). Proteolytic cleavage in an endolysosomal compartment is required for activation of Toll-like receptor 9. *Nature Immunology*, *9*, 1407–1414.

Riese, R. J., Wolf, P. R., Bromme, D., Natkin, L. R., Villadangos, J. A., Ploegh, H. L., et al. (1996). Essential role for cathepsin S in MHC class II-associated invariant chain processing and peptide loading. *Immunity*, *4*, 357–366.

Sathe, P., Pooley, J., Vremec, D., Mintern, J., Jin, J. O., Wu, L., et al. (2011). The acquisition of antigen cross-presentation function by newly formed dendritic cell. *Journal of Immunology*, *186*, 5184–5192.

Sepulveda, F., Maschalidi, S., Colisson, R., Heslop, L., Sakka, E., Ghirelli, C., et al. (2009). Critical role for asparagine endopeptidase in endocytic TLR signalling in dendritic cells. *Immunity*, *31*, 731–748.

CHAPTER TEN

Assessment of the Toll-Like Receptor 3 Pathway in Endosomal Signaling

Misako Matsumoto[1], Kenji Funami, Megumi Tatematsu, Masahiro Azuma, Tsukasa Seya

Department of Microbiology and Immunology, Hokkaido University Graduate School of Medicine, Kita-ku, Sapporo, Japan

[1]Corresponding author: e-mail address: matumoto@pop.med.hokudai.ac.jp

Contents

Abstract

The innate immune system plays key roles in antimicrobial responses by developing the pattern-recognition receptors that recognize microbial components. The endosomal Toll-like receptors (TLRs) and cytosolic RIG-I-like receptors (RLRs) both recognize viral nucleic acids and are essential for antiviral immunity. Recent evidence suggests that compartmentalization of the receptors, and also their adaptor molecule, is important for discrimination between self and nonself and for distinct innate immune signals. TLR3 is a type I transmembrane protein that localizes in the endosomal membrane in myeloid dendritic cells (DCs) and fibroblasts/epithelial cells. TLR3 recognizes

Methods in Enzymology, Volume 535
ISSN 0076-6879
http://dx.doi.org/10.1016/B978-0-12-397925-4.00010-9

extracellular viral double-stranded RNA (dsRNA) and the synthetic dsRNA, poly(I:C). On recognition of dsRNA in the endosomes, TLR3 oligomerizes and induces type I interferon and proinflammatory cytokine production via an adaptor molecule, TICAM-1 (also known as TRIF). Additionally, the TLR3 signal in DCs triggers gene transcription required for DC maturation and the activation of natural killer cells and cytotoxic T lymphocytes. Remarkably, it has been reported that extracellular dsRNA is also recognized by cytosolic RLR. Making a distinction between TLR3-mediated endosomal signaling and RLR-mediated signaling is key to understanding the role of these receptors in innate immunity.

1. INTRODUCTION

The innate immune system senses microbial infection using pattern-recognition receptors and signals to activate innate and adaptive antimicrobial immunity (Akira, Uematsu, & Takeuchi, 2006; Janeway & Medzhitov, 2002). The endosomal Toll-like receptors (TLRs 3, 7, 8, and 9) and cytoplasmic RIG-I-like receptors (RLRs) both recognize viral nucleic acids, playing essential roles in protection against viral infection (Diebold, 2008; Yoneyama & Fujita, 2010). TLR3 has been functionally identified as a sensor for viral double-stranded RNA (dsRNA) and the synthetic dsRNA analog polyriboinosinic–polyribocytidylic acid (poly(I:C)) using TLR3-deficient mice or an anti-human TLR3 blocking antibody (Alexopoulou, Holt, Medzhitov, & Flavell, 2001; Matsumoto, Kikkawa, Kohase, Miyake, & Seya, 2002). On recognition of dsRNA in the endosomes, TLR3 oligomerizes and induces type I interferon (IFN-α/β) and proinflammatory cytokine production from host cells via Toll–IL-1 receptor (TIR)-domain-containing adaptor molecule-I (TICAM-1, also known as TRIF) (Oshiumi, Matsumoto, Funami, Akazawa, & Seya, 2003; Yamamoto et al., 2003). Additionally, activation of TLR3 in myeloid dendritic cells (DCs) leads to the maturation of DCs, which activates natural killer (NK) cells and cytotoxic T lymphocytes (CTLs). Because TLR3 signals effectively induce cellular immunity, TLR3 ligands, such as poly(I:C), are considered to be promising adjuvants for cancer and infectious disease vaccines (Seya & Matsumoto, 2009). However, when dsRNA is added to cells extracellularly *in vitro* or administered to mice, it also activates cytosolic dsRNA sensors, including RIG-I and MDA5, in addition to TLR3, inducing type I IFNs and proinflammatory cytokines via adaptor protein MAVS (also known as IPS-1, Cardif, and VISA) (Kato et al., 2006; Kawai et al., 2005; Meylan et al., 2005; Seth, Sun, Ea, & Chen, 2005; Xu et al., 2005). Thus, it is important to assess which pathway(s) contributes to dsRNA-induced cellular responses in various cells or situations. In this chapter, we focus on the TLR3–TICAM-1 pathway and describe methods used for studying TLR3-mediated signaling.

2. ANALYSES OF TLR3 EXPRESSION AND LOCALIZATION

TLR3 is a type I transmembrane protein and is expressed in fibroblasts, epithelial cells of various tissues, and neural cells, including neurons, astrocytes, and microglia. Among immune cells, only myeloid DCs and macrophages express TLR3 (Muzio et al., 2000; Visintin et al., 2001). Among myeloid DC subsets, TLR3 is highly expressed in professional antigen-presenting DCs, including mouse $CD8\alpha^{+}$ DCs and human CD141 $(BDCA3)^{+}$ DCs (Jelinek et al., 2011; Jongbloed et al., 2010). Unlike other nucleic acid-sensing TLRs, the subcellular localization of TLR3 depends on the cell type; human fibroblasts, macrophages, and some epithelial cell lines express TLR3 in both the cell surface and the endosomal membrane, while myeloid DCs express it in early endosomes (Matsumoto et al., 2003, 2002). Cell-surface TLR3 appears to recognize extracellular dsRNA, because an anti-human TLR3 mAb (TLR3.7) partially inhibited poly(I:C)-induced IFN-β production by fibroblasts (Matsumoto et al., 2002). However, TLR3-mediated signaling is initiated from endosomal compartments in either type of cell, requiring endosomal maturation. The expression and subcellular localization of TLR3 are assessed by flow cytometric and immunofluorescent analyses with an anti-TLR3 mAb (Funami et al., 2004, 2007) (Fig. 10.1).

2.1. Flow cytometric analysis

2.1.1 Cell-surface TLR3 (human cells)

1. All procedures are carried out at 4 °C.
2. Harvest and wash cells ($\sim10^{5}$–10^{6} cells/sample) three times with FACS buffer (PBS, 0.5% BSA, 0.1% NaN_3).
3. Add 50 μL of anti-human TLR3 mAb (e.g., TLR3.7) or isotype control Ab (10 μg/mL in FACS buffer) together with human IgG (for blocking Fc receptors, final conc 0.1 mg/mL) and incubate for 0.5–1.0 h.
4. Wash cells three times with FACS buffer.

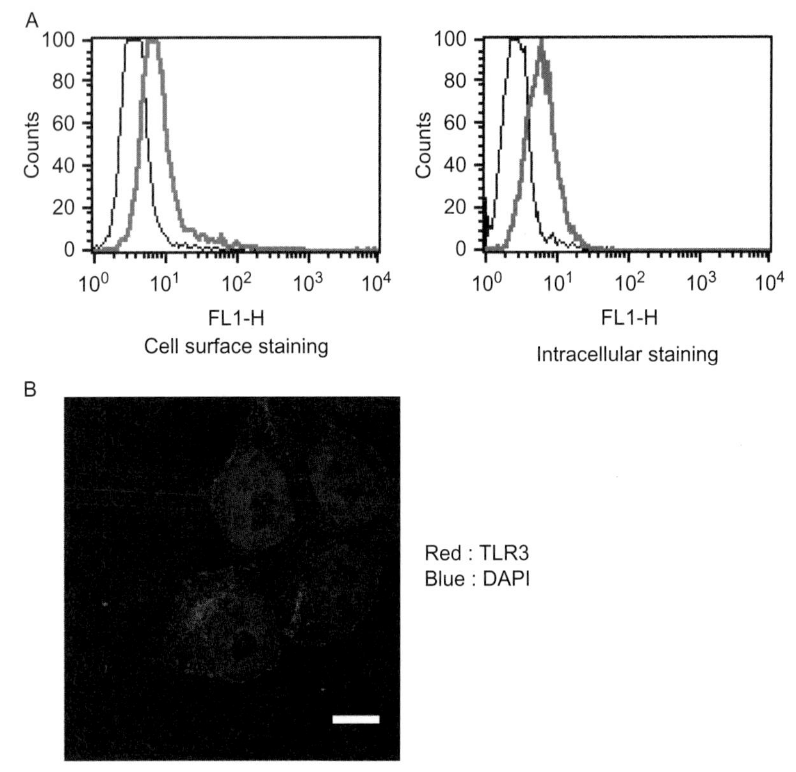

Figure 10.1 Expression of human TLR3 in HeLa cells. (A) TLR3 is expressed on the cell surface and inside the cells in HeLa cells. Cell-surface (left) staining and intracellular (right) staining were performed using the TLR3.7 mAb and analyzed by flow cytometry. The black line indicates control mouse IgG staining. The red line indicates TLR3.7 staining. (B) TLR3 is localized to intracellular vesicles in HeLa cells. Cells were fixed and permeabilized, and endogenous TLR3 was stained with the TLR3.7 mAb. The red signal indicates endogenous TLR3 and the blue signal indicates DAPI staining. Scale bar, 10 μm. (For interpretation of the references to color in this figure legend, the reader is referred to the online version of this chapter.)

5. Add 50 μL of fluorescent-conjugated goat anti-mouse IgG (highly absorbed and appropriately diluted with FACS buffer), mix thoroughly, and incubate for 30 min in the dark.
6. Wash cells three times with FACS buffer.
7. Resuspend cells in 0.5 mL of FACS buffer and analyze fluorescence signals using a flow cytometer. If samples will not be analyzed immediately, add 0.5 mL of 1% paraformaldehyde in PBS, mix thoroughly, and keep them in the dark at 4 °C and analyze within 1 week.

2.1.2 Intracellular TLR3

1. Wash cells (~10^5–10^6 cells/sample) three times with DPBS at 4 °C.
2. Add 1 mL of BD FACS Permeabilizing Solution 2 (Becton Dickinson); incubate for 10 min at room temperature.
3. Wash cells three times with FACS buffer.
4. Add 50 μL of anti-human TLR3 mAb (e.g., TLR3.7) or isotype control Ab (10 μg/mL in FACS buffer) together with 1/10 volume of goat serum and incubate for 30 min at room temperature.
5. Wash cells three times with FACS buffer.
6. Add 50 μL of fluorescent-labeled goat anti-mouse IgG (highly absorbed and appropriately diluted with FACS buffer) together with 1/10 volume of goat serum, mix thoroughly, and incubate for 30 min at room temperature in the dark.
7. Wash cells three times with FACS buffer.
8. Resuspend cells in 0.5 mL of FACS buffer and analyze fluorescence signals using a flow cytometer (Fig. 10.1A).

2.2. Immunofluorescent analysis

1. Plate cells (5.0×10^4 cells/well) on micro cover glasses (Matsunami, Tokyo, Japan) in a 24-well plate.
2. On the next day, wash the cells twice with PBS.
3. In the case of cell-surface staining, fix cells with PBS containing 4% paraformaldehyde for 15 min. In the case of intracellular staining, fix cells with PBS containing 4% paraformaldehyde for 30 min and permeabilize with PBS containing 0.2% Triton X-100 for 15 min at room temperature.
4. Wash cells four times with PBS.
5. Incubate cells in blocking buffer (PBS containing 1% BSA and 10% goat serum) for at least 10 min.
6. Add 20 μg/mL anti-TLR3 mAb (TLR3.7) or an isotype control Ab in blocking buffer and incubate for 1 h.
7. Wash cells four times with PBS.
8. Incubate in Alexa-conjugated secondary Ab (Molecular Probes, highly absorbed, 1:1000 dilution) in blocking buffer for 30 min.
9. Wash cells four times with PBS.
10. Mount cover glasses onto slide glass using ProLong Gold antifade reagent with DAPI (Molecular Probes).
11. Analyze fluorescence signals by confocal microscopy (Fig. 10.1B).

In the case of DCs,

1. stain TLR3 as described for FACS sample preparation (intracellular staining 1-5);
2. add Alexa-conjugated goat anti-mouse IgG (Molecular Probes, highly absorbed, 1:1000 dilution) together with 1/10 volume of goat serum, mix thoroughly, and incubate for 30 min at room temperature;
3. wash cells four times with PBS;
4. plate cells onto slide glass by centrifugation using a cytospin;
5. mount cover glasses onto slide glass using ProLong Gold antifade reagent with DAPI (Molecular Probes);
6. analyze fluorescence signals by confocal microscopy.

3. ASSAY FOR TLR3-MEDIATED SIGNALING

TLR3 consists of an extracellular domain containing 23 leucine-rich repeats and N- and C-terminal flanking regions, a transmembrane domain, and an intracellular TIR domain (Bell et al., 2003). Based on structural analyses of the TLR3 ectodomain–dsRNA complex, it has been proposed that 40–50 bp dsRNA is the minimum signaling unit with two TLR3 molecules (Liu et al., 2008). On a cellular basis, *in vitro*-transcribed dsRNAs, >90 bp in length, trigger TLR3 oligomerization, and effectively induce IFN-β and proinflammatory cytokine production in murine myeloid DCs (Jelinek et al., 2011; Leonard et al., 2008).

After oligomerization in the endosomes, TLR3 recruits adaptor molecule TICAM-1 into the cytoplasmic TIR domains (Funami et al., 2007). Once TICAM-1 is oligomerized, the transcription factors IRF-3, NF-κB, and AP-1 are activated, which then induce IFN-β and proinflammatory cytokine production. The N-terminal region of TICAM-1, where TRAF2, TRAF6, and TBK1 binding sites exist, is crucial for IRF-3 activation (Sasai et al., 2010; Tatematsu et al., 2010). TRAF3 and NAP-1 participate in the recruitment and activation of the IRF-3 kinase, TBK1 (Fitzgerald et al., 2003; Oganesyan et al., 2006; Sasai et al., 2005). On the other hand, RIP1 associates with TICAM-1 via the RHIM domain in the C-terminal region and, together with TRAF6, mediates NF-κB activation (Meylan et al., 2004; Sato et al., 2003) (Fig. 10.2).

Because TLR3-mediated signaling is initiated from endosomal compartments, uptake and delivery of TLR3 ligands to endosomes is critical for TLR3 activation. Indeed, poly(I:C) is internalized via clathrin-mediated endocytosis (Itoh, Watanabe, Funami, Seya, & Matsumoto, 2008).

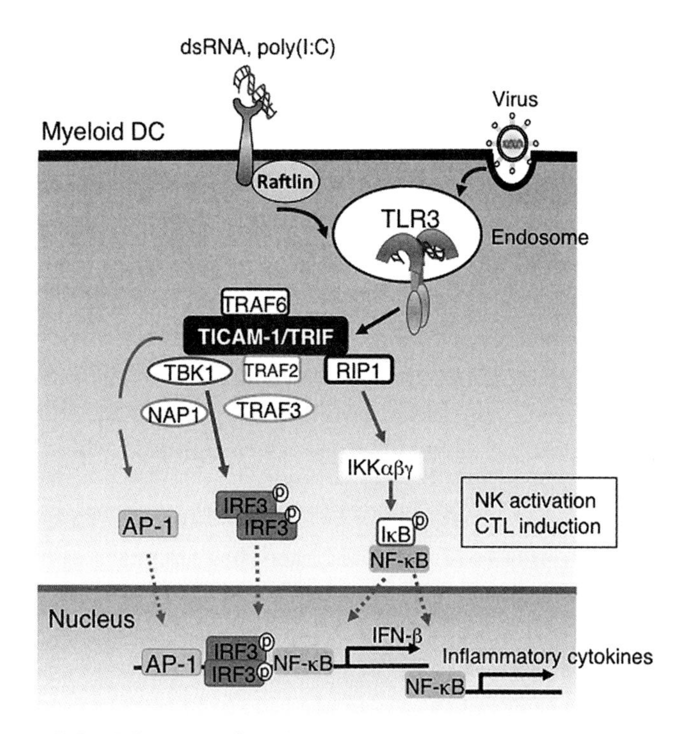

Figure 10.2 TLR3–TICAM-1-mediated signaling in myeloid DCs. An extracellular dsRNA is delivered to early endosomes via clathrin–Raftlin-dependent endocytosis. Once TLR3 is oligomerized with internalized dsRNA, it recruits TICAM-1. After the transient association of TLR3 with TICAM-1 through the TIR domains, TICAM-1 dissociates from TLR3 to form a speckle-like structure containing downstream signaling molecules where TICAM-1-mediated signaling is initiated. The TICAM-1–TBK-1–IRF-3 axis is essential for TLR3-mediated IFN-β production and DC-mediated activation of NK cells and CTLs. (For color version of this figure, the reader is referred to the online version of this chapter.)

Watanabe et al. reported that the cytoplasmic protein Raftlin induced poly (I:C) internalization through an interaction with the clathrin–AP-2 complex in human myeloid DCs and fibroblasts/epithelial cells (Watanabe et al., 2011). CD14 and the scavenger receptor class A were reported to act as the poly(I:C) uptake receptor in mouse macrophages and human bronchial epithelial cells, respectively, but this is apparently not the case in human fibroblasts/DCs. These results suggest that the uptake machinery for TLR3 ligands may differ between mouse and human cells and also by cell type. To assess TLR3-activating ability, direct delivery of ligands into endosomes with cationic liposome, such as DOTAP, is required in addition to extracellular stimulation of TLR3-negative or TLR3-positive cells.

3.1. Reporter assay for IFN-β promoter activation and NF-κB activation

1. Plate HEK293 cells in a 96-well plate and grow them to approximately 80% confluence.
2. Transfect the cells with the expression vector of TLR3 or empty vector (25 ng), together with IFN-β promoter plasmid (25 ng) (Fujita et al., 1998) or NF-κB reporter plasmid (15 ng), and internal control vector (1.25 ng) using Lipofectamine 2000. Total DNA is kept constant by adding an empty vector.
3. At 24 h after transfection, wash the cells once with fresh culture medium and stimulate with poly(I:C) (5–50 μg/mL) for 6 h. For endosomal delivery, poly(I:C) is preincubated with DOTAP (Roche; 0.5 μL for 1.0 μg poly(I:C)).
4. Lyse the cells in Passive Lysis Buffer (50 μL/well).
5. Measure dual luciferase activities according to the manufacturer's instructions (Promega, Dual-Luciferase Reporter Assay System) (Fig. 10.3).

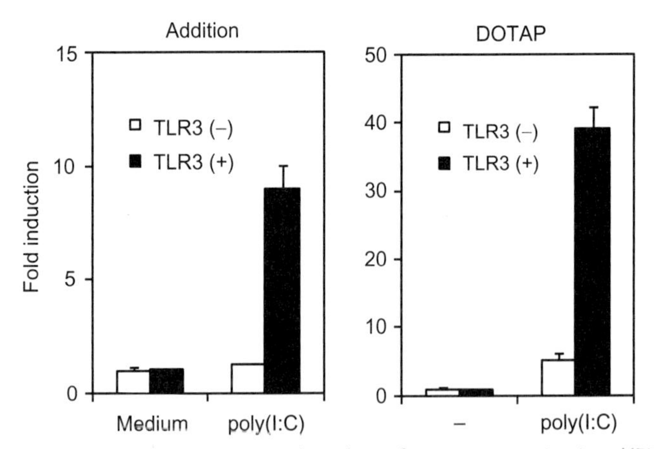

Figure 10.3 Poly(I:C)-induced TLR3-mediated IFN-β promoter activation. HEK293 cells in 96-well plates were transfected with an expression vector for human TLR3 (pEFBOS/hTLR3: filled column) or empty vector (open column) together with the IFN-β reporter plasmid (p125-luc). Then, 24 h after transfection, cells were stimulated with 10 μg/mL poly(I:C) or medium alone (left panel) or stimulated with 1 μg poly(I:C) complexed with DOTAP or DOTAP alone (–) (right panel). After 6 h, luciferase reporter activities were measured and expressed as the fold induction relative to the activity of unstimulated vector-transfected cells.

3.2. Assay for IRF-3 activation

IRF-3 has an essential role in the TLR3-mediated IFN-β gene transcription. IRF-3 is expressed ubiquitously as an inactive monomer in the cytosol. When cells are stimulated with poly(I:C) or virus infection, serine residues in the C-terminal region of IRF-3 are phosphorylated by the serine/threonine kinases, TBK1 and IKKε, and form homodimers (Sato et al., 2000). Homodimerized IRF-3 translocates from the cytosol into the nucleus and binds to responsive elements for IFN-β gene transcription. Fujita et al. developed a sensitive assay for activated IRF-3 using native PAGE, which clearly detects the inactive and active forms of IRF-3 (Iwamura et al., 2001). Compared with the mobility change on SDS–PAGE used to assess the phosphorylation status of the IRF-3 molecules, the native PAGE assay simply and sensitively detects the functional IRF-3 homodimer.

3.3. Cytokine assay

Cytokine production in response to dsRNA has been examined using human or mouse TLR3-expressing cells. In the case of human cells, the normal embryonic lung fibroblast MRC-5, the cervical epithelial cell line HeLa, or monocyte-derived DCs and macrophages are often used for poly(I:C)-induced cytokine assays. MRC-5 cells produce considerable amounts of IFN-β in response to poly(I:C), which is detectable by ELISA (Matsumoto et al., 2002). In HeLa cells, IFN-β mRNA expression can be detected at 3 h after poly(I:C) stimulation. Human monocyte-derived DCs produce large amounts of IFN-α/β, IL-12p70, and proinflammatory cytokines, TNF-α and IL-6, in response to poly(I:C). The TLR3.7 mAb does not inhibit poly(I:C)-induced cytokine production by DCs, because DCs do not express TLR3 at the cell surface (Matsumoto et al., 2003). In the case of mouse cells, MEFs, bone marrow-derived DCs (BMDCs), or splenic DCs from wild-type or gene-disrupted mice ($TLR3^{-/-}$, $TICAM\text{-}1^{-/-}$, $MDA5^{-/-}$, or $MAVS^{-/-}$) are used. Studies using knockout mice have demonstrated that the TLR3 pathway is primarily involved in IL-12p40 production, while the RLR pathway is important for type I IFN production both *in vitro* and *in vivo*. On the other hand, IL-6 production depends on RLR and TLR3 pathways (Kato et al., 2006).

4. ASSAY FOR DC-MEDIATED NK ACTIVATION

NK cells do not express TLR3, and therefore, dsRNA such as poly (I:C) does not activate NK cells directly *in vitro*. However, when mouse NK cells are cocultured with BMDCs or splenic DCs in the presence of poly(I:C), NK cells are activated through cell–cell contact between DC and NK cells, leading to IFN-γ production. Additionally, activated NK cells are able to kill MHC class I-negative mouse tumor cell lines, including YAC-I and B16 cells (Akazawa et al., 2007). Among splenic DC subsets, CD8α^+ DCs, in which the TLR3–TICAM-1 pathway is predominant in response to dsRNA, are mainly involved in NK activation. The TLR3–TICAM-1 pathway in DCs plays important roles in NK activation (Ebihara et al., 2010), while RLR-mediated type I IFNs from nonimmune cells also contribute to NK activation *in vivo* (McCartney et al., 2009).

4.1. Preparation of bone marrow-derived DCs

BMDCs are prepared from the bone marrow of wild-type (C57BL/6) or various gene-disrupted mice by a reported method (Inaba et al., 1992) with minor modifications.

Day 0:

1. Sacrifice a mouse and remove the legs.
2. Remove muscle tissues from the femurs and tibias and place the bones in a 60 mm dish with 70% ethanol for 2 min.
3. Wash the bones twice with PBS.
4. Transfer the bones into a fresh dish with RPMI 1640 and cut out both ends of the bones.
5. Flush out bone marrow cells using 5 mL RPMI 1640 with a syringe and 26G needle until the bones turn white.
6. Transfer the cell suspension into a 15 mL tube through a stainless mesh.
7. Wash the dish with 5 mL of medium described in the succeeding text and transfer the cell suspension into the tube through a stainless mesh.
8. Centrifuge at 750 × *g* for 5 min.
9. Suspend the cell pellets in ACK buffer and incubate for 1 min at room temperature to lyse red cells.
10. Add 9 mL of medium and centrifuge cells at 750 × *g* for 5 min.
11. Count the cells and resuspend in medium containing 10 ng/mL rmGM-CSF at 1×10^6/mL.
12. Seed cells in a 24-well plate (1 mL/well) and incubate at 37 °C in 5% CO_2.

Day 2:

1. Aspirate the medium and add prewarmed fresh medium containing 10 ng/mL GM-CSF.

Day 4:

1. Wash the adherent cells with culture medium gently and aspirate the supernatant.
2. Add prewarmed medium containing 10 ng/mL GM-CSF.

Day 6 or Day 7:

1. Collect cells into a 50 mL tube after pipetting gently.
2. Centrifuge the cells at 750 × *g* for 5 min.
3. Aspirate the supernatant and count the cells.
4. Resuspend the cells and seed with appropriate conditions for respective experiments.

Medium: RPMI 1640 supplemented with 10% FCS, 10 m*M* HEPES, 55 μ*M* 2-ME, and penicillin/streptomycin.

ACK buffer: 150 m*M* NH_4Cl, 10 m*M* $KHCO_3$, 0.1 m*M* Na_2EDTA (pH 7.4).

4.2. Preparation of NK cells

1. Harvest spleens from C57BL/6 mice and prepare a single-cell suspension by mashing with slide glasses.
2. Pass the cells through a 70 μm cell strainer (BD Falcon) to remove debris.
3. Centrifuge at 750 × *g* for 5 min.
4. Resuspend cell pellet in 1 mL of ACK buffer per 10^8 splenocytes and incubate for 1 min at room temperature.
5. Wash cells twice with 10 mL of medium, and centrifuge at 750 × *g* for 5 min.
6. Isolate NK cells using CD49b MicroBeads (Miltenyi Biotec, 130-052-501) according to manufacturer's protocol.

4.3. Assay for NK activation (IFN-γ production, killing assay)

1. Coculture 5×10^5 NK cells and 2.5×10^5 BMDC in 500 μL of medium per well in a 24-well plate.
2. Stimulate with 10 μg/mL poly(I:C).
3. After 24 h, collect supernatants for evaluating IFN-γ production by ELISA and collect cells for ^{51}Cr release assay.
4. Wash the cells twice with medium and centrifuge at 750×g for 5 min.
5. Count only NK cells.

6. Add 100 μL of DC–NK mixture to 100 μL of target cells at an E/T ratio of 5:1 to 100:1 in a 96-well round-bottom plate. As controls, add 100 μL of medium or 10% NP-40 to the target cells for measuring spontaneous release or total release of chromium, respectively. Target cell preparation is described in the succeeding text.
7. After 4 h, centrifuge plates at 1300 g for 2 min.
8. Collect 150 μL of the supernatant and measure radiation with a gamma counter.
9. Specific cytotoxicity is determined by the following formula: specific cytotoxicity (%) = [(experimental release − spontaneous release)/(total release − spontaneous release)] × 100.

4.3.1 Target cell preparation

1. B16D8, YAC-1, and RMA-S cells are generally used as NK target cells.
2. Suspend 2×10^5 target cells in 180 μL medium and add 50 μCi of $Na_2^{51}CrO_4$.
3. Incubate for 1 h at 37 °C.
4. Wash cells twice with 1 mL of medium, and centrifuge at $1300 \times g$ for 2 min.
5. Add 1 mL of medium and keep at 4 °C for 15 min.
6. Centrifuge at $1300 \times g$ for 2 min.
7. Suspend cells in medium and plate $0.2–1.0 \times 10^4$/100 μL target cells in a 96-well round-bottom plate.

5. ASSAY FOR DC-MEDIATED CTL ACTIVATION

Myeloid DCs, especially mouse $CD8\alpha^+$ DCs and human $CD141^+$ DCs, are the best professional antigen-presenting cells that can cross present exogenous antigens to $CD8^+$ T lymphocytes (Shen & Lock, 2006). Using TLR3-deficient mice, Schultz et al. showed that TLR3 played an important role in cross priming (Schulz et al., 2005). Immunization with virally infected cells, which contain dsRNA or cells containing poly(I:C), both carrying ovalbumin antigen, induced ovalbumin-specific $CD8^+$ T lymphocyte responses, which were largely dependent on TLR3-expressing DCs. Additionally, Jongbloed et al. reported that $CD141^+$ DCs were able to cross present viral antigens from human cytomegalovirus-infected necrotic fibroblasts (Jongbloed et al., 2010). Physiologically, endosomal TLR3 in a DC subset specialized for antigen presentation encounters viral dsRNAs when apoptotic or necrotic virus-infected cells are phagocytosed and signals for cross presentation of viral antigens.

Cross priming is also important for induction of CTLs against tumor cells. Azuma et al. recently demonstrated that antitumor CTL induced by a tumor antigen and poly(I:C) depended on the TLR3–TICAM-1 pathway in mouse splenic CD8α^+ DCs (Azuma, Ebihara, Oshiumi, Matsumoto, & Seya, 2012). They showed that IRF-3/7 were essential but MAVS and type I IFNs were minimally involved in poly(I:C)-mediated CTL proliferation. Here, we describe the *in vitro* assay for DC-mediated CTL induction (Datta et al., 2003):

1. Plate 5×10^5/500 μL medium BMDC in 24-well plate.
2. Stimulate cells with 10 μg/mL poly(I:C) and incubate for 18 h at 37 °C.
3. Add soluble ovalbumin (Sigma) to a final concentration of 100 ng/mL.
4. After 4 h, collect cells and wash twice with medium. Centrifuge at $750 \times g$ for 5 min.
5. Coculture 1×10^5 BMDC and 1×10^5 CFSE-labeled OT-1 T lymphocytes in 200 μL/well in a 96-well round-bottom plate. The CFSE labeling protocol is described in the succeeding text.
6. After 60 h, collect cells and stain with anti-CD8α and anti-TCR-Vα2 Abs.
7. CFSE diminution is evaluated by flow cytometry. It is recommended to count 20,000 cells gated on $CD8^+$ TCR-$V\alpha2^+$ cells.

5.1. CFSE labeling for OT-1 T lymphocytes

1. OT-1 $CD8^+$ T lymphocytes are positively isolated by CD8(Ly-2) microbeads (Miltenyi Biotech, 130-049-401) according to the manufacturer's protocol.
2. Suspend cells in PBS to a concentration of 4×10^7 cells/mL.
3. Mix the cells with same volume of 2 μ*M* CFSE solution and keep it for 10 min at room temperature.
4. Add an equal volume of FCS and keep it for 1 min at room temperature.
5. Wash three times with medium. Centrifuge at $750 \times g$ for 5 min.
6. Plate 1×10^5/100 μL CFSE-labeled OT-1 T lymphocytes in a 96-well round-bottom plate.

Medium: RPMI 1640 supplemented with 10% FCS, 10 m*M* HEPES, 55 μ*M*, 2-ME, and penicillin/streptomycin.

6. SUMMARY

TLR3-mediated endosomal signaling induces a wide range of cellular responses (Matsumoto & Seya, 2008). Activation of the TICAM-1–TBK1–IRF-3 axis is critical for IFN-β gene expression and DC-mediated NK and CTL induction. Interestingly, recent reports have further demonstrated that TLR3 signaling participates in chromatin remodeling and nuclear reprogramming (Lee et al., 2012) and also in the control of endogenous retroviruses in mice (Yu et al., 2012). TLR3 recognizes viral dsRNA and poly(I:C), but it is still unknown what RNA molecules activate TLR3 in viral infection and sterile inflammatory states. Recently, it has been reported that TLR3 recognizes virus-derived ssRNA with mismatched stems, host cell mRNA, and RNA from necrotic cells, indicating that structures other than dsRNA can activate TLR3 (Bernard et al., 2012; Cavassani et al., 2008; Karikó, Ni, Capodici, Lamphier, & Weissman, 2004; Tatematsu, Nishikawa, Seya, & Matsumoto, 2013). Thus, identification of endogenous/exogenous TLR3 ligands and assessment of their signaling are important for a full understanding of the role of the TLR3 pathway in innate and adaptive immunity. Additionally, the development of the TLR3-specific ligands would be useful to analyze human cells and also to discriminate between endosomal and cytoplasmic signaling.

ACKNOWLEDGMENTS

We thank Drs. H. Oshiumi, T. Akazawa, T. Ebihara, H. Shime, H. Takaki, and J. Kasamatsu for invaluable discussions. This work was supported, in part, by Grants-in-Aid from the Ministry of Education, Science, and Culture; by the Ministry of Health, Labour, and Welfare of Japan; by the Naito Foundation; by the Uehara Memorial Foundation; and by the Akiyama Life Science Foundation.

REFERENCES

Akazawa, T., Ebihara, T., Okuno, M., Okuda, Y., Shingai, M., Tsujimura, K., et al. (2007). Antitumor NK activation induced by the Toll-like receptor 3-TICAM-1 (TRIF) pathway in myeloid dendritic cells. In: *Proceedings of the National Academy of Sciences of the United States of America, 104*, 252–257.

Akira, S., Uematsu, S., & Takeuchi, O. (2006). Pathogen recognition and innate immunity. *Cell, 124*, 783–801.

Alexopoulou, L., Holt, A. C., Medzhitov, R., & Flavell, R. A. (2001). Recognition of double-stranded RNA and activation of NF-kappaB by Toll-like receptor 3. *Nature, 413*, 732–738.

Azuma, M., Ebihara, T., Oshiumi, H., Matsumoto, M., & Seya, T. (2012). Cross-priming for antitumor CTL induced by soluble Ag+polyI:C depends on the TICAM-1 pathway in mouse CD11c(+)/CD8α(+) dendritic cells. *Oncoimmunology, 1*, 581–592.

Bell, J. K., Mullen, G. E. D., Leifer, C. A., Mazzoni, A., Davies, D. R., & Segal, D. M. (2003). Leucine-rich repeats and pathogen recognition in Toll-like receptors. *Trends in Immunology, 24*, 528–533.

Bernard, J. J., Cowing-Zitron, C., Nakatsuji, T., Muehleisen, B., Muto, J., Borkowski, A. W., et al. (2012). Ultraviolet radiation damages self noncoding RNA and is detected by TLR3. *Nature Medicine, 18*(8), 1286–1290. http://dx.doi.org/10.1038/nm.2861.

Cavassani, K. A., Ishii, M., Wen, H., Schaller, M. A., Lincoln, P. M., Lukacs, N. W., et al. (2008). TLR3 is an endogenous sensor of tissue necrosis during acute inflammatory events. *The Journal of Experimental Medicine, 205*, 2609–2621.

Datta, S. K., Redecke, V., Prilliman, K. R., Takabayashi, K., Corr, M., Tallant, T., et al. (2003). A subset of Toll-like receptor ligands induces cross-presentation by bone marrow-derived dendritic cells. *The Journal of Immunology, 170*, 4102–4110.

Diebold, S. S. (2008). Recognition of viral single-stranded RNA by Toll-like receptors. *Advanced Drug Delivery Reviews, 60*, 813–823.

Ebihara, T., Azuma, M., Oshiumi, H., Kasamatsu, J., Iwabuchi, K., Matsumoto, K., et al. (2010). Identification of a polyI:C-inducible membrane protein that participates in dendritic cell-mediated natural killer cell activation. *The Journal of Experimental Medicine, 207*, 2675–2687.

Fitzgerald, K. A., McWhirter, S. M., Faia, K. L., Rowe, D. C., Latz, E., Golenbock, D. T., et al. (2003). IKKε and TBK1 are essential components of the IRF3 signaling pathway. *Nature Immunology, 4*, 491–496.

Fujita, T., Sakakibara, J., Sudo, Y., Miyamoto, M., Kimura, Y., & Taniguchi, T. (1998). Evidence for a nuclear factor(s), IRF-1, mediating induction and silencing properties to human IFN-β gene regulatory elements. *The EMBO Journal, 7*, 3397–3405.

Funami, K., Matsumoto, M., Oshiumi, H., Akazawa, T., Yamamoto, A., & Seya, T. (2004). The cytoplasmic 'linker region' in Toll-like receptor 3 controls receptor localization and signaling. *International Immunology, 16*, 1143–1154.

Funami, K., Sasai, M., Ohba, Y., Oshiumi, H., Seya, T., & Matsumoto, M. (2007). Spatiotemporal mobilization of TICAM-1 in response to dsRNA. *The Journal of Immunology, 179*, 6867–6872.

Inaba, K., Inaba, M., Romani, N., Aya, H., Deguchi, M., Ikehara, S., et al. (1992). Generation of large numbers of DC from mouse bone marrow cultures supplemented with GM-CSF. *The Journal of Experimental Medicine, 176*, 1693–1702.

Itoh, K., Watanabe, A., Funami, K., Seya, T., & Matsumoto, M. (2008). The clathrin-mediated endocytic pathway participates in dsRNA-induced IFN-beta production. *The Journal of Immunology, 181*, 5522–5529.

Iwamura, T., Yoneyama, M., Yamaguchi, K., Suhara, W., Mori, W., Shiota, K., et al. (2001). Induction of IRF-3/-7 kinase and NF-kappaB in response to double-stranded RNA and virus infection: Common and unique pathways. *Genes to Cells, 6*, 375–388.

Janeway, C. A., Jr., & Medzhitov, R. (2002). Innate immune recognition. *Annual Review of Immunology, 20*, 197–216.

Jelinek, I., Leonard, J. N., Price, G. E., Brown, K. N., Meyer-Manlapat, A., Goldsmith, P. K., et al. (2011). TLR3-specific double-stranded RNA oligonucleotide adjuvants induce dendritic cell cross-presentation, CTL responses, and antiviral protection. *The Journal of Immunology, 186*, 2422–2429.

Jongbloed, S. L., Kassianos, A. J., McDonald, K. J., Clark, G. J., Ju, X., Angel, C. E., et al. (2010). Human CD141+ (BDCA-3)+dendritic cells (DCs) represent a unique myeloid

DC subset that cross-presents necrotic cell antigens. *The Journal of Experimental Medicine*, *207*, 1247–1260.
Karikó, K., Ni, H., Capodici, J., Lamphier, M., & Weissman, D. (2004). mRNA is an endogenous ligand for Toll-like receptor 3. *The Journal of Biological Chemistry*, *279*, 12542–12550.
Kato, H., Takeuchi, O., Sato, S., Yoneyama, M., Yamamoto, M., Matsui, K., et al. (2006). Differential roles of MDA5 and RIG-I helicases in the recognition of RNA viruses. *Nature*, *441*, 101–105.
Kawai, T., Takahashi, K., Sato, S., Coban, C., Kumar, H., Kato, H., et al. (2005). IPS-1, an adaptor triggering RIG-I- and Mda5-mediated type I interferon induction. *Nature Immunology*, *6*, 981–988.
Lee, J., Sayed, N., Hunter, A., Au, K. F., Wong, W. H., Mocarski, E. S., et al. (2012). Activation of innate immunity is required for efficient nuclear reprogramming. *Cell*, *151*, 547–558.
Leonard, J. N., Ghirlando, R., Askins, J., Bell, J. K., Margulies, D. H., Davies, D. R., et al. (2008). The TLR3 signaling complex forms by cooperative receptor dimerization. In: *Proceedings of the National Academy of Sciences of the United States of America*, *105*, 258–263.
Liu, L., Botos, I., Wang, Y., Leonard, J. N., Shiloach, J., Segal, D. M., et al. (2008). Structural basis of Toll-like receptor 3 signaling with double-stranded RNA. *Science*, *320*, 379–381.
Matsumoto, M., Funami, K., Tanabe, M., Oshiumi, H., Shingai, M., Seto, Y., et al. (2003). Subcellular localization of Toll-like receptor 3 in human dendritic cells. *The Journal of Immunology*, *171*, 3154–3162.
Matsumoto, M., Kikkawa, S., Kohase, M., Miyake, K., & Seya, T. (2002). Establishment of a monoclonal antibody against human Toll-like receptor 3 that blocks double-stranded RNA-mediated signaling. *Biochemical and Biophysical Research Communications*, *293*, 1364–1369.
Matsumoto, M., & Seya, T. (2008). TLR3: Interferon induction by double-stranded RNA including poly(I:C). *Advanced Drug Delivery Reviews*, *60*, 805–812.
McCartney, S., Vermi, W., Gilfillan, S., Cella, M., Murphy, T. L., Schreiber, R. D., et al. (2009). Distinct and complementary functions of MDA5 and TLR3 in poly(I:C)-mediated activation of mouse NK cells. *The Journal of Experimental Medicine*, *206*, 2967–2976.
Meylan, E., Burns, K., Hofmann, K., Blancheteau, V., Martinon, F., Kelliher, M., et al. (2004). RIP1 is an essential mediator of Toll-like receptor 3-induced NF-kappa B activation. *Nature Immunology*, *5*, 503–507.
Meylan, E., Curran, J., Hofmann, K., Moradpour, D., Binder, M., Bartenschlager, R., et al. (2005). Cardif is an adaptor protein in the RIG-I antiviral pathway and is targeted by hepatitis C virus. *Nature*, *437*, 1167–1172.
Muzio, M., Bosisio, D., Polentarutti, N., D'amico, G., Stoppacciaro, A., Mancinelli, R., et al. (2000). Differential expression and regulation of Toll-like receptors (TLR) in human leukocytes: Selective expression of TLR3 in dendritic cells. *The Journal of Immunology*, *64*, 5998–6004.
Oganesyan, G., Saha, S. K., Guo, B., He, J. Q., Shahangian, A., Zarnegar, B., et al. (2006). Critical role of TRAF3 in the Toll-like receptor-dependent and -independent antiviral response. *Nature*, *439*, 208–211.
Oshiumi, H., Matsumoto, M., Funami, K., Akazawa, T., & Seya, T. (2003). TICAM-1, an adaptor molecule that participates in Toll-like receptor 3-mediated interferon-beta induction. *Nature Immunology*, *4*, 161–167.
Sasai, M., Oshiumi, H., Matsumoto, M., Inoue, N., Fujita, F., Nakanishi, M., et al. (2005). Cutting edge: NF-kappaB-activating kinase-associated protein 1 participates in TLR3/Toll-IL-1 homology domain-containing adapter molecule-1-mediated IFN regulatory factor 3 activation. *The Journal of Immunology*, *174*, 27–30.

Sasai, M., Tatematsu, M., Oshiumi, H., Funami, K., Matsumoto, M., Hatakeyama, S., et al. (2010). Direct binding of TRAF2 and TRAF6 to TICAM-1/TRIF adaptor participates in activation of the Toll-like receptor 3/4 pathway. *Molecular Immunology*, *47*, 1283–1291.

Sato, M., Suemori, H., Hata, N., Asagiri, M., Ogasawara, K., Nakao, K., et al. (2000). Distinct and essential roles of transcription factors IRF-3 and IRF-7 in response to viruses for IFN-α/β gene induction. *Immunity*, *13*, 539–548.

Sato, S., Sugiyama, M., Yamamoto, M., Watanabe, Y., Kawai, T., Takeda, K., et al. (2003). Toll/IL-1 receptor domain-containing adaptor inducing IFN-β (TRIF) associates with TNF receptor-associated factor 6 and TANK-binding kinase 1, and activates two distinct transcription factors, NF-κB and IFN-regulatory factor-3, in the Toll-like receptor signaling. *The Journal of Immunology*, *171*, 4304–4310.

Schulz, O., Diebold, S. S., Chen, M., Näslund, T. I., Nolte, M. A., Alexopoulou, L., et al. (2005). Toll-like receptor 3 promotes cross-priming to virus-infected cells. *Nature*, *433*, 887–892.

Seth, R. B., Sun, L., Ea, C. K., & Chen, Z. J. (2005). Identification and characterization of MAVS, a mitochondrial antiviral signaling protein that activates NF-kappaB and IRF 3. *Cell*, *122*, 669–682.

Seya, T., & Matsumoto, M. (2009). The extrinsic RNA-sensing pathway for adjuvant immunotherapy of cancer. *Cancer Immunology, Immunotherapy*, *58*, 1175–1184.

Shen, L., & Lock, K. L. (2006). Priming of T cells by exogenous antigen cross-presented on MHC class I molecules. *Current Opinion in Immunology*, *18*, 85–91.

Tatematsu, M., Ishii, A., Oshiumi, H., Horiuchi, M., Inagaki, F., Seya, T., et al. (2010). A molecular mechanism for Toll/IL-1 receptor domain-containing adaptor molecule-1-mediated IRF-3 activation. *The Journal of Biological Chemistry*, *285*, 20128–20136.

Tatematsu, M., Nishikawa, F., Seya, T., & Matsumoto, M. (2013). Toll-like receptor 3 recognizes incomplete stem structures in single-stranded viral RNA. *Nature Communications*, *4*, 1833.

Visintin, A., Mazzoni, A., Spitzer, J. H., Wyllie, D. H., Dower, S. K., & Segal, D. M. (2001). Regulation of Toll-like receptors in human monocytes and dendritic cells. *The Journal of Immunology*, *166*, 249–254.

Watanabe, A., Tatematsu, M., Saeki, K., Shibata, S., Shime, H., Yoshimura, A., et al. (2011). Raftlin is involved in the nucleocapture complex to induce poly(I:C)-mediated TLR3 activation. *The Journal of Biological Chemistry*, *286*, 10702–10711.

Xu, L. G., Wang, Y. Y., Han, K. J., Li, L. Y., Zhai, Z., & Shu, H. B. (2005). VISA is an adapter protein required for virus-triggered IFN-beta signaling. *Molecular Cell*, *19*, 727–740.

Yamamoto, M., Sato, S., Hemmi, H., Hoshino, K., Kaisho, T., Sanjo, H., et al. (2003). Role of adaptor TRIF in the MyD88-independent Toll-like receptor signaling pathway. *Science*, *301*, 640–643.

Yoneyama, M., & Fujita, T. (2010). Recognition of viral nucleic acids in innate immunity. *Reviews in Medical Virology*, *20*, 4–22.

Yu, P., Lübben, W., Slomka, H., Gebler, J., Konert, M., Cai, C., et al. (2012). Nucleic acid-sensing Toll-like receptors are essential for the control of endogenous retrovirus viremia and ERV-induced tumors. *Immunity*, *37*, 867–879.

CHAPTER ELEVEN

Labeling of Platelet-Derived Growth Factor by Reversible Biotinylation to Visualize Its Endocytosis by Microscopy

Łukasz Sadowski*, Kamil Jastrzębski*, Elżbieta Purta†, Carina Hellberg‡, Marta Miaczynska[1,*]

*Laboratory of Cell Biology, International Institute of Molecular and Cell Biology, Warsaw, Poland
†Laboratory of Bioinformatics and Protein Engineering, International Institute of Molecular and Cell Biology, Warsaw, Poland
‡University of Birmingham, School of Biosciences, Birmingham, United Kingdom
[1]Corresponding author: e-mail address: miaczynska@iimcb.gov.pl

Contents

Abstract

Microscopical analyses of endocytic trafficking require tools for efficient detection of internalized cargo. Due to the lack of suitable reagents and limitations related to its biological properties, visualization of platelet-derived growth factor (PDGF) by microscopy remained a challenge. To overcome these restrictions, we generated a biologically active PDGF labeled with up to five biotins on cleavable linkers. Subsequently, we stimulated cells with such ligand followed by removal of extracellular biotins. PDGF captured in endocytic vesicles was successfully detected with antibiotin antibodies with parallel detection of PDGF receptor, as well as other markers of endocytic compartments. Labeled PDGF was successfully validated and can be utilized in various microscopical techniques.

Methods in Enzymology, Volume 535
ISSN 0076-6879
http://dx.doi.org/10.1016/B978-0-12-397925-4.00011-0

1. INTRODUCTION

Platelet-derived growth factors (PDGFs) are a family of polypeptides exhibiting mitogenic activity towards multiple types of cells, such as fibroblasts, vascular endothelial cells, or glial cells (Heldin et al., 1981; Kaplan et al., 1979; Zetter and Antoniades, 1979). PDGFs are secreted as disulfide-bonded homodimers consisting of two A-, B-, C-, or D-polypeptide chains (PDGF-AA, PDGF-BB, PDGF-CC, and PDGF-DD) and one heterodimer formed by the A- and B-chains (PDGF-AB) (Bergsten et al., 2001; Heldin et al., 1988; Li et al., 2000). PDGFs act through binding to the extracellular domains of PDGF receptors (PDGFRs) that possess tyrosine kinase activity. Ligand binding triggers receptor autophosphorylation what ultimately leads to the activation of downstream signaling and evokes physiological responses like proliferation or migration (Bishayee et al., 1989; Eriksson et al., 1992). Following ligand binding, the ligand–receptor complexes are internalized into endocytic vesicles and targeted to lysosomes for degradation and termination of the signal propagation (Rosenfeld et al., 1984) or are recycled back to the cell membrane (Karlsson et al., 2006; Schmees et al., 2012).

It is widely accepted that ligand-induced downstream signaling is not only limited to the plasma membrane, but also internalized receptors are still capable of signal propagation from distinct endosomal compartments (Sadowski et al., 2009). It has been shown that following ligand-induced internalization, PDGFRs retain their activity within intracellular organelles (Sorkin et al., 1993). Moreover, modulation of either internalization routes (Schmees et al., 2012) or endocytic trafficking of PDGFRs (Hellberg et al., 2009) affects the subcellular localization of signaling molecules (Huang et al., 2007; Wang et al., 2004) and in consequence cell fate. Such findings have spurred a growing interest in this field of research, highlighting the need for development of new tools that enable investigations of the links between tyrosine kinase receptor endocytosis and signaling.

Up to date, there were no commercially available reagents, like antibodies or fluorescently labeled ligands, allowing detection and analysis of PDGF trafficking by microscopy. At physiological pH, PDGFs are very sticky and exhibit strong nonspecific binding to most surfaces. For this reason, attempts to directly or indirectly detect internalized PDGFs result in a strong extracellular background that prevents analysis of endocytic transport of the ligand. The stickiness of PDGF is mainly due to the C-terminal retention motif, which contains basic amino acids, that mediates binding of PDGF to the extracellular matrix (Andersson et al., 1994). This interaction restricts the bioavailability of PDGF, which is important for autocrine and paracrine signaling during development (Erlandsson et al., 2006; Nystrom et al., 2006).

In order to visualize endocytosis of PDGF by microscopy and overcome the aforementioned issues, we developed a method utilizing reversible labeling of PDGF-BB (thereafter referred to as PDGF) with biotin. The complete procedure includes stimulation of cells expressing PDGFR with biotinylated PDGF-BB (bt-PDGF), followed by removal of extracellular biotins using a cell-impermeable reducing agent and detection of internalized biotins attached to PDGF with antibiotin antibodies (Fig. 11.1).

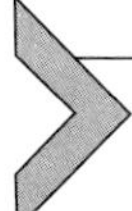

2. PDGF LABELING WITH SULFO-*N*-HYDROXYSUCCINIMIDE-SS-BIOTIN

For PDGF labeling, we chose sulfo-*N*-hydroxysuccinimide-SS-biotin (Sulfo-NHS-SS-Biotin, Pierce). The presence of a disulfide bond in the spacer arm allows removal of biotins from labeled proteins with cell-impermeable reducing agents such as glutathione (GSH), sodium 2-mercaptoethanesulfonate, tris(2-carboxyethyl)phosphine hydrochloride, or dithiothreitol.

Sulfo-NHS-SS-Biotin is an amine-reactive reagent, that is, it will bind to NH_2 groups present on lysine residues and to the *N*-terminal NH_2 group of a target protein. Some of the lysines present in both PDGF-BB and PDGF-AA mediate ligand binding to the receptor and are thus important for mitogenic activities of the ligand (Fenstermaker et al., 1993; Schilling et al., 1998). Therefore, it is crucial to control the extent of biotinylation in order to avoid inactivation of PDGF:

1. Dissolve 1 mg of PDGF (PeproTech) in 1 ml ultrapure water. The final solution will have acidic pH, as PDGF was lyophilized from 10 m*M* acetic acid. The solution can be stored at −20 °C in aliquots to avoid repeated freeze–thaw cycles.
2. For a single labeling reaction, transfer 50 μl of PDGF solution to a Slide-A-Lyzer MINI Dialysis Device (Pierce) with molecular weight cutoff (MWCO) 3500 Da (Pierce). Dialyze PDGF solution at 4 °C twice for 30 min against 500 ml PBS with mild stirring. This is required to exchange buffer with neutral pH that is optimal for the coupling reaction.

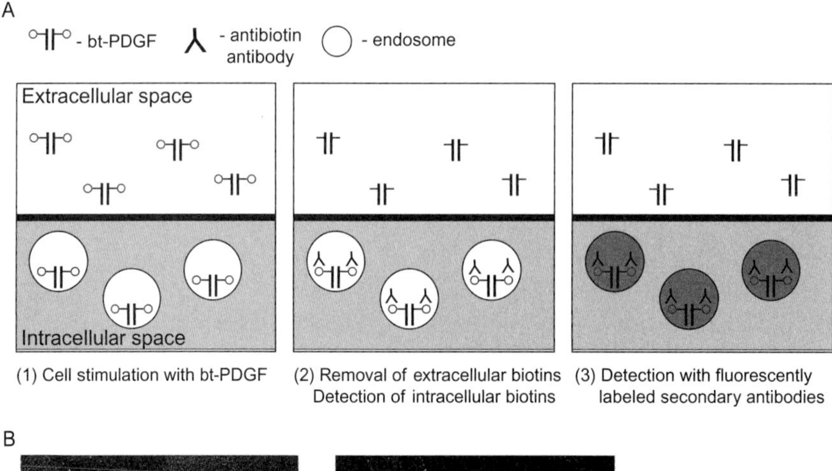

B

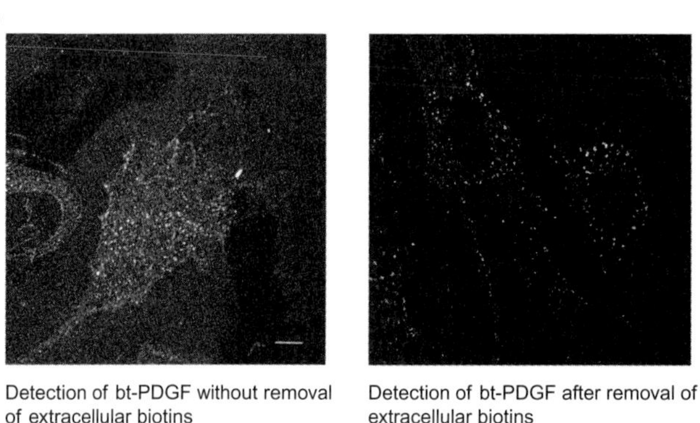

Figure 11.1 *A method to detect bt-PDGF by microscopy.* (A) The principle of intracellular bt-PDGF detection. (1) Cells are stimulated with bt-PDGF. (2) Subsequently, extracellular biotins are removed with a reducing reagent and intracellular bt-PDGF is detected with antibiotin antibodies. (3) Internalized bt-PDGF is visualized by microscopy using fluorescently labeled secondary antibodies. (B) Detection of bt-PDGF without removal (left image) and following removal (right image) of extracellular biotins. Cells were stimulated with 100 ng/ml of bt-PDGF for 30 min (left) and 40 min (right). Scale bar 10 μm. (For color version of this figure, the reader is referred to the online version of this chapter.)

3. Recover the PDGF solution from the dialysis unit.
4. Determine concentration of the recovered PDGF by measuring absorbance at 280 nm using $E_{1\%}^{280\mathrm{nm}}$ of 5. Calculate the molar concentration of PDGF that is required to determine the volume of 10 m*M* Sulfo-NHS-SS-Biotin to obtain a desired molar excess of a label against a target. Based on our empirical observations, the recommended molar ratio

for the coupling reaction is 12:1 Sulfo-NHS-SS-Biotin to PDGF-BB. For details of calculations, refer to the manufacturer's protocol (Pierce).

5. Immediately before use dissolve 6 mg Sulfo-NHS-SS-Biotin in 1 ml of ultrapure water in order to obtain 10 m*M* solution.
6. Add the required volume of Sulfo-NHS-SS-Biotin solution to the PDGF solution and incubate at room temperature for 30 min. During incubation, the mixture will become turbid.
7. Add 2 μl of 2 *M* glycine and incubate 5 min at room temperature. Glycine will quench the excess Sulfo-NHS-SS-Biotin and terminate the coupling reaction.
8. Transfer bt-PDGF solution to a Slide-A-Lyzer MINI Dialysis Device with molecular weight cutoff 3500 Da (Pierce). Dialyze bt-PDGF solution twice in 4 °C for 30 min against 500 ml 0.01 *M* sodium acetate pH 4.5 with mild stirring.
9. Recover bt-PDGF solution from the Slide-A-Lyzer MINI Dialysis Device. Bt-PDGF solution can be stored in aliquots at −20 °C.

3. DETERMINATION OF THE EXTENT OF PDGF BIOTINYLATION WITH MASS SPECTROMETRY

Since attachment of too many biotin molecules may result in loss of biological activity, it is critical to verify the level of PDGF biotinylation. The extent of labeling can be measured by MS analysis:

1. Precipitate 1 μg of bt-PDGF with trichloroacetic acid.
2. Dry and dissolve in 5% formic acid (FA).
3. Apply the sample onto StageTips™ C18 (Proxeon).
4. Wash with 5% FA.
5. Elute with 70% acetonitrile and 0.1% trifluoroacetic acid (TFA).
6. Evaporate to dryness and resuspend in 2 μl of 0.1% TFA.
7. Acidify sample with 2 μl of 2% TFA.
8. Add 2 μl of freshly prepared 2,5-dihydroxyacetophenone matrix solution and mix by pipetting.
9. Spot 1 μl of the mixture onto the ground steel target plate and dry at room temperature.

MS analysis was performed using Bruker Daltonics ultrafleXtreme™. Mass spectra were acquired by MALDI time of flight in the linear positive mode using ~500 laser shots, and the masses were assigned using flexAnalysis™ software (Bruker Daltonics). The reference protein standards (protein calibration standard II (Bruker Daltonics)—trypsinogen, protein $A^{1+/2+}$, and serum albumin-bovine $^{1+/2+}$) were used for external calibration.

For the analysis of bt-PDGF, addition of an active ester group of Sulfo-NHS-SS-Biotin derivatives, which reacts with the primary amines of proteins and/or the amino group of lysine residues, results in an increase in mass of 390 Da. For example, if the mass of labeled protein increased by ~2000 Da, this corresponds to the attachment of approximately five biotin residues per PDGF-BB molecule, that is, about 2.5 biotins per polypeptide chain. In our experience, attachment of approximately 3–5 biotin residues per PDGF-BB molecule allows for efficient detection of PDGF and does not affect its biological activity.

4. DETERMINATION OF A BIOLOGICAL ACTIVITY OF bt-PDGF

It is essential to ensure that PDGF labeling does not affect its biological activity. Since PDGF induces autophosphorylation of the PDGFR (Heldin et al., 1989), the most direct assay to determine the activity of bt-PDGF is to detect tyrosine phosphorylation of PDGFR by Western blotting (Fig. 11.2):

1. Plate CCD-1070Sk human fibroblasts (or another cell line expressing PDGFR) in a 24-well dish (5×10^4 cells per well), in Minimum Essential Medium Eagle (MEM) supplemented with 10% fetal bovine serum (FBS), 2 mM L-glutamine, 100 U/ml penicillin, and 100 μg/ml streptomycin.

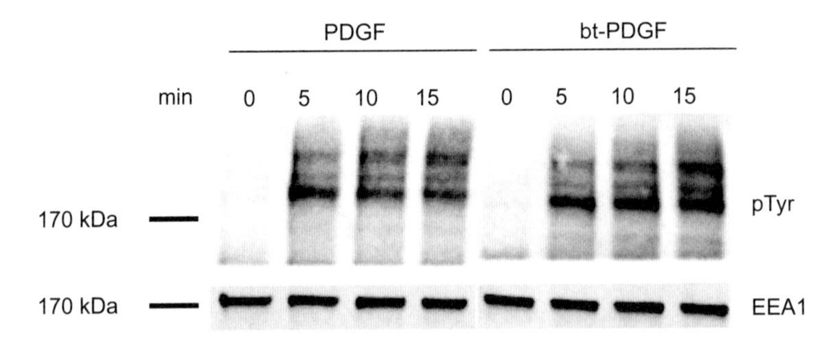

Figure 11.2 *Bt-PDGF exhibits similar biological activity as unlabeled PDGF.* Lysates of CCD-1070Sk human foreskin fibroblasts stimulated with 100 ng/ml of unlabeled PDGF or bt-PDGF for up to 15 min were subjected to Western blot analysis. Activation of PDGFR was detected by immunoblotting with antiphosphotyrosine (pTyr) antibodies (the visualized bands represent different modified forms of the receptor). Anti-EEA1 antibodies were used as a loading control.

2. Twenty-four hours before stimulation, replace the growth medium with the starvation medium (MEM supplemented with 0.2% bovine serum albumin (BSA), 2 m*M* L-glutamine, 100 U/ml penicillin, and 100 μg/ml streptomycin) in order to deprive cells of serum.
3. On the day of stimulation, prepare 100 ng/ml bt-PDGF in the starvation medium.
4. Stimulate cells with medium containing bt-PDGF for the required periods of time.
5. Transfer cells on ice and wash twice with ice-cold PBS.
6. Lyse cells in RIPA lysis buffer (150 m*M* NaCl, 1% NP-40, 0.1% SDS, 0.5% sodium deoxycholate, and 50 m*M* Tris–HCl, pH 8.0) in the presence of protease inhibitors (6 μg/ml chymostatin, 0.5 μg/ml leupeptin, 10 μg/ml antipain, 2 μg/ml aprotinin, 0.7 μg/ml pepstatin A, and 10 μg/ml 4-amidinophenylmethanesulfonyl fluoride hydrochloride) and phosphatase inhibitors (phosphatase inhibitor cocktails 2 and 3 from Sigma-Aldrich). Centrifuge lysates for 10 min at 20,000 × *g*.
7. Measure protein concentration and boil lysates in Laemmli sample buffer for 10 min.
8. Apply 30 μg of protein per lane of an 8% polyacrylamide gel. Detect activation of PDGFR on Western blot using antiphosphotyrosine (anti-pTyr) antibodies (e.g., clone 4G10®, Millipore).

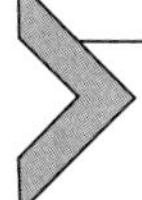

5. STIMULATION OF CELLS WITH bt-PDGF AND REMOVAL OF EXTRACELLULAR BIOTINS

Following cell stimulation, bt-PDGF is internalized in a complex with PDGFR into endocytic structures. However, due to its adhesive properties, PDGF also binds unspecifically to surfaces such as plastic or glass, on which the cells are plated (Fig. 11.1B). To specifically detect internalized PDGF, biotins present on the outside of the cells have to be removed by applying a cell-impermeable reducing agent. Such reagent will reduce the disulfide bond present in the linker between PDGF and biotin, allowing for the free biotin to be washed off the glass coverslip. Out of several reagents tested, the most efficient and reproducible results were obtained with a slightly alkaline solution of glutathione (GSH):

1. Plate CCD-1070Sk human fibroblasts (or another cell line expressing PDGFR) on 12 mm glass coverslips in a 24-well dish (5×10^4 cells per well) in MEM supplemented with 10% FBS, 2 m*M* L-glutamine, 100 U/ml penicillin, and 100 μg/ml streptomycin.

2. Follow Steps 2–5 from paragraph 4.
3. Remove extracellular biotins: incubate cells on ice for 5 min with a freshly prepared glutathione stripping solution (50 m*M* reduced glutathione, 150 m*M* NaCl, 70 m*M* NaOH, 1.25 m*M* $MgSO_4$, 1.25 m*M* $CaCl_2$, and 1 m*M* EDTA, pH 8.5).
4. Wash twice with ice-cold PBS.
5. Incubate for 5 min with 30 m*M* iodoacetamide in order to terminate the reducing reaction.
6. Wash twice with ice-cold PBS.
7. Fix cells for 12 min with 3% paraformaldehyde in PBS.
8. Wash twice for 5 min in PBS.

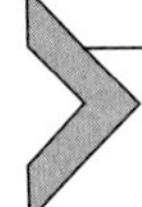

6. VALIDATION AND DETECTION OF bt-PDGF IN MICROSCOPICAL ASSAYS

Upon PDGF binding to PDGFR, the ligand–receptor complex is internalized and targeted to lysosomes for degradation (Heldin et al., 1982). Therefore, the activity of the ligand and its specific binding to the receptor can be confirmed using microscopy by detecting PDGF and PDGFR in the same endosomes. In parallel, PDGF can be visualized in specific endocytic compartments, such as early endosomes positive for EEA1 (early endosome antigen 1) or multivesicular bodies positive for CD63 (Fig. 11.3):

1. Incubate fixed cells for 10 min in solution I (0.1% [w/v] saponin, 0.2% [w/v] gelatin, and 5 mg/ml BSA in PBS) in order to permeabilize cell membranes and block unspecific binding sites.
2. Dilute primary antibodies in solution II (0.2% gelatin and 0.01% saponin in PBS). To detect biotinylated PDGF, we recommend antibiotin antibodies (#B3640 from Sigma-Aldrich, used at dilution of 1:200). Fluorescently labeled streptavidin conjugates are not suitable, as they generate extensive intracellular background effectively masking endocytic structures.
3. Apply 40 μl of diluted primary antibody on each coverslip. Incubate in a humid chamber for 30 min.
4. Wash coverslips twice for 5 min in solution II.
5. Dilute secondary antibodies in solution II.
6. Apply 40 μl of diluted secondary antibody on each coverslip. Incubate in a humid chamber for 30 min.
7. Wash coverslips twice for 5 min in PBS.
8. Mount coverslips on ethanol-cleaned glass slides and view under the microscope.

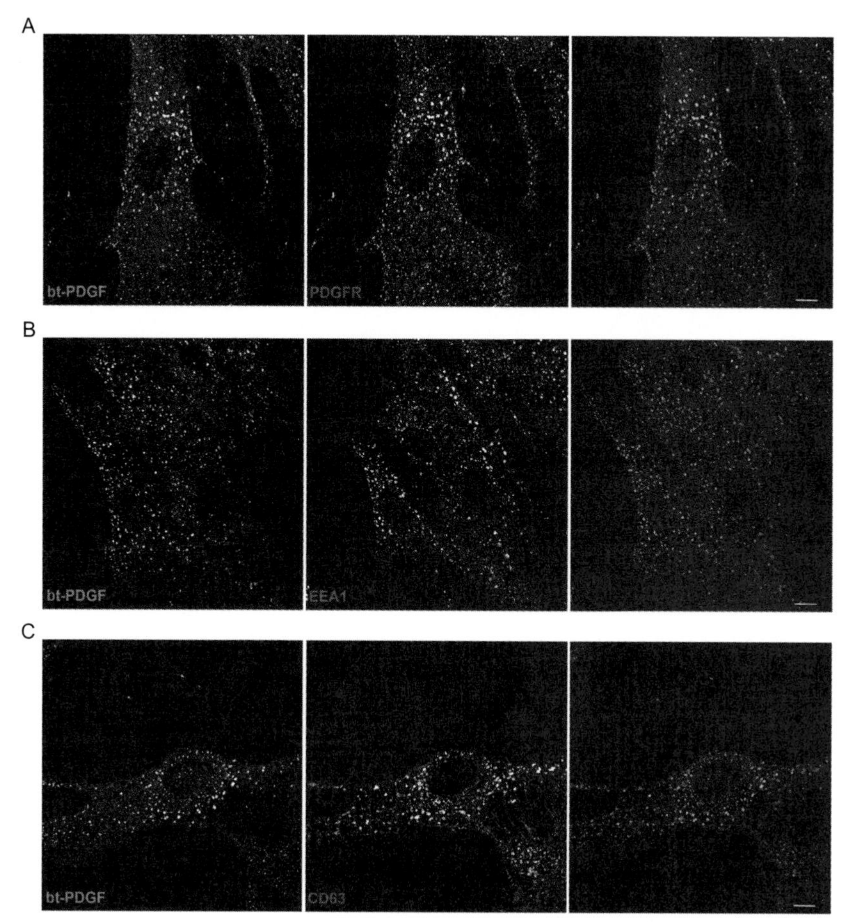

Figure 11.3 *Colocalization analysis of internalized bt-PDGF with PDGFR and endocytic markers.* CCD-1070Sk human foreskin fibroblasts were stimulated with bt-PDGF for 40 min (A), 30 min (B), or 60 min (C). Cells were immunostained for biotin (in order to detect bt-PDGF) and either PDGFRβ (A), EEA1 (marker of early endosomes; B), or CD63 (marker of multivesicular bodies; C). Scale bar 10 μm. (For color version of this figure, the reader is referred to the online version of this chapter.)

7. CONCLUSION

Reversible biotinylation is a valuable tool for tracking internalized PDGF in the endosomal compartments by microscopy. A carefully controlled labeling reaction generates a fully active bt-PDGF that is readily detectable with antibiotin antibodies and secondary antibodies coupled to fluorescent dyes. This strategy of ligand labeling combined with removal of extracellular background allowed us recently to perform quantitative analysis of endocytic trafficking of PDGF-BB (Sadowski et al., 2013). This novel tool can be applied in microscopical analysis of PDGF colocalization with various signaling molecules, in parallel to markers of endosomal compartments. In addition to PDGF-BB described in this chapter, we succeeded to label PDGF-AA in the same way (data not shown), indicating that the method may be suitable for other proteins that cannot be detected with conventional reagents or techniques.

ACKNOWLEDGMENTS

We thank Dr. Carl-Henrik Heldin for advice on the project and the critical reading of the manuscript. This work was supported by EU Grant LSHG-CT-2006-019050 (EndoTrack) to M. M. and C. H. and by EU Grant 229676 (HEALTH-PROT) to M. M. K. J. was supported by Foundation for Polish Science within International PhD Project "Studies of nucleic acids and proteins—from basic to applied research," cofinanced from European Union—Regional Development Fund.

REFERENCES

Andersson, M., Ostman, A., Westermark, B., & Heldin, C. H. (1994). Characterization of the retention motif in the C-terminal part of the long splice form of platelet-derived growth factor A-chain. *The Journal of Biological Chemistry*, *269*, 926–930.

Bergsten, E., Uutela, M., Li, X., Pietras, K., Ostman, A., Heldin, C. H., et al. (2001). PDGF-D is a specific, protease-activated ligand for the PDGF beta-receptor. *Nature Cell Biology*, *3*, 512–516.

Bishayee, S., Majumdar, S., Khire, J., & Das, M. (1989). Ligand-induced dimerization of the platelet-derived growth factor receptor. Monomer-dimer interconversion occurs independent of receptor phosphorylation. *The Journal of Biological Chemistry*, *264*, 11699–11705.

Eriksson, A., Rorsman, C., Ernlund, A., Claesson-Welsh, L., & Heldin, C. H. (1992). Ligand-induced homo- and hetero-dimerization of platelet-derived growth factor alpha- and beta-receptors in intact cells. *Growth Factors*, *6*, 1–14.

Erlandsson, A., Brannvall, K., Gustafsdottir, S., Westermark, B., & Forsberg-Nilsson, K. (2006). Autocrine/paracrine platelet-derived growth factor regulates proliferation of neural progenitor cells. *Cancer Research*, *66*, 8042–8048.

Fenstermaker, R. A., Poptic, E., Bonfield, T. L., Knauss, T. C., Corsillo, L., Piskurich, J. F., et al. (1993). A cationic region of the platelet-derived growth factor (PDGF) A-chain (Arg159-Lys160-Lys161) is required for receptor binding and mitogenic activity of the PDGF-AA homodimer. *The Journal of Biological Chemistry*, *268*, 10482–10489.

Heldin, C. H., Backstrom, G., Ostman, A., Hammacher, A., Ronnstrand, L., Rubin, K., et al. (1988). Binding of different dimeric forms of PDGF to human fibroblasts: Evidence for two separate receptor types. *The EMBO Journal*, 7, 1387–1393.

Heldin, C. H., Ernlund, A., Rorsman, C., & Ronnstrand, L. (1989). Dimerization of B-type platelet-derived growth factor receptors occurs after ligand binding and is closely associated with receptor kinase activation. *The Journal of Biological Chemistry*, *264*, 8905–8912.

Heldin, C. H., Wasteson, A., & Westermark, B. (1982). Interaction of platelet-derived growth factor with its fibroblast receptor. Demonstration of ligand degradation and receptor modulation. *The Journal of Biological Chemistry*, *257*, 4216–4221.

Heldin, C. H., Westermark, B., & Wasteson, A. (1981). Specific receptors for platelet-derived growth factor on cells derived from connective tissue and glia. In: *Proceedings of the National Academy of Sciences of the United States of America*, *78*, 3664–3668.

Hellberg, C., Schmees, C., Karlsson, S., Ahgren, A., & Heldin, C. H. (2009). Activation of protein kinase C alpha is necessary for sorting the PDGF beta-receptor to Rab4a-dependent recycling. *Molecular Biology of the Cell*, *20*, 2856–2863.

Huang, M., Duhadaway, J. B., Prendergast, G. C., & Laury-Kleintop, L. D. (2007). RhoB regulates PDGFR-beta trafficking and signaling in vascular smooth muscle cells. *Arteriosclerosis, Thrombosis, and Vascular Biology*, *27*, 2597–2605.

Kaplan, D. R., Chao, F. C., Stiles, C. D., Antoniades, H. N., & Scher, C. D. (1979). Platelet alpha granules contain a growth factor for fibroblasts. *Blood*, *53*, 1043–1052.

Karlsson, S., Kowanetz, K., Sandin, A., Persson, C., Ostman, A., Heldin, C. H., et al. (2006). Loss of T-cell protein tyrosine phosphatase induces recycling of the platelet-derived growth factor (PDGF) beta-receptor but not the PDGF alpha-receptor. *Molecular Biology of the Cell*, *17*, 4846–4855.

Li, X., Ponten, A., Aase, K., Karlsson, L., Abramsson, A., Uutela, M., et al. (2000). PDGF-C is a new protease-activated ligand for the PDGF alpha-receptor. *Nature Cell Biology*, *2*, 302–309.

Nystrom, H. C., Lindblom, P., Wickman, A., Andersson, I., Norlin, J., Faldt, J., et al. (2006). Platelet-derived growth factor B retention is essential for development of normal structure and function of conduit vessels and capillaries. *Cardiovascular Research*, *71*, 557–565.

Rosenfeld, M. E., Bowen-Pope, D. F., & Ross, R. (1984). Platelet-derived growth factor: Morphologic and biochemical studies of binding, internalization, and degradation. *Journal of Cellular Physiology*, *121*, 263–274.

Sadowski, L., Jastrzebski, K., Kalaidzidis, Y., Heldin, C. H., Hellberg, C., & Miaczynska, M. (2013). Dynamin inhibitors impair endocytosis and mitogenic signaling of PDGF. *Traffic*, *14*, 725–736.

Sadowski, L., Pilecka, I., & Miaczynska, M. (2009). Signaling from endosomes: Location makes a difference. *Experimental Cell Research*, *315*, 1601–1609.

Schilling, D., Reid, I. J., Hujer, A., Morgan, D., Demoll, E., Bummer, P., et al. (1998). Loop III region of platelet-derived growth factor (PDGF) B-chain mediates binding to PDGF receptors and heparin. *The Biochemical Journal*, *333*(Pt. 3), 637–644.

Schmees, C., Villasenor, R., Zheng, W., Ma, H., Zerial, M., Heldin, C. H., et al. (2012). Macropinocytosis of the PDGF beta-receptor promotes fibroblast transformation by H-RasG12V. *Molecular Biology of the Cell*, *23*, 2571–2582.

Sorkin, A., Eriksson, A., Heldin, C. H., Westermark, B., & Claesson-Welsh, L. (1993). Pool of ligand-bound platelet-derived growth factor beta-receptors remain activated and tyrosine phosphorylated after internalization. *Journal of Cellular Physiology*, *156*, 373–382.

Wang, Y., Pennock, S. D., Chen, X., Kazlauskas, A., & Wang, Z. (2004). Platelet-derived growth factor receptor-mediated signal transduction from endosomes. *The Journal of Biological Chemistry*, *279*, 8038–8046.

Zetter, B. R., & Antoniades, H. N. (1979). Stimulation of human vascular endothelial cell growth by a platelet-derived growth factor and thrombin. *Journal of Supramolecular Structure*, *11*, 361–370.

CHAPTER TWELVE

Endosomal Signaling and Oncogenesis

Nikolai Engedal*, Ian G. Mills*,†,‡,§,1

*Prostate Cancer Research Group, Centre for Molecular Medicine Norway, Nordic EMBL Partnership, University of Oslo and Oslo University Hospital, Oslo, Norway
†Department of Cancer Prevention, Institute of Cancer Research, Oslo University Hospital, Oslo, Norway
‡Department of Urology, Oslo University Hospital, Oslo, Norway
§Uro-Oncology Research Group, Cambridge Research Institute, University of Cambridge, Cambridge, United Kingdom
[1]Corresponding author: e-mail address: ian.mills@ncmm.uio.no; ian.mills@cruk.cam.ac.uk

Contents

Abstract

The endosomal system provides a route whereby nutrients, viruses, and receptors are internalized. During the course of endocytosis, activated receptors can accumulate within endosomal structures and certain signal-transducing molecules can be recruited to endosomal membranes. In the context of signaling and cancer, they provide platforms within the cell from which signals can be potentiated or attenuated. Regulation of the duration of receptor signaling is a pivotal means of refining growth responses in cells. In cancers, this is often considered in terms of mutations that affect receptor tyrosine kinases and maintain them in hyperactivated states of dimerization and/or phosphorylation. However, disruption to the regulatory control exerted by the assembly of protein complexes within the endosomal network can also contribute to disease among which oncogenesis is characterized in part by dysregulated growth, enhanced cell survival, and changes in the expression of markers of differentiation.

Methods in Enzymology, Volume 535
ISSN 0076-6879
http://dx.doi.org/10.1016/B978-0-12-397925-4.00012-2

In this chapter, we will discuss the role of proteins that regulate in endocytosis as tumor suppressors or oncogenes and how changing the fate of internalized receptors and concomitant endosomal signaling can contribute to cancer.

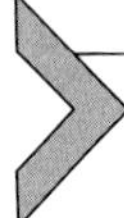

1. INTRODUCTION

1.1. Endocytosis

This is the process by which receptors, nutrients, viruses, and other extracellular factors are internalized by the cell. It is beyond the scope of this chapter to describe the routes of entry in great detail and they are well covered by other reviews (Rao, Ruckert, Saenger, & Haucke, 2012; Sandvig, Pust, Skotland, & van Deurs, 2011). Broadly, they are defined as clathrin-dependent and clathrin-independent. Clathrin is composed of heavy and light chains, which assemble into triskelia and polymerize on the surface of vesicles assembling at the plasma membrane (Pearse, 1976; Pearse & Bretscher, 1981; Pfeffer & Kelly, 1981). A range of adaptor proteins contribute to cargo recruitment, curvature induction, clathrin polymerization, and vesicle release. These include AP2 adaptors, heterotetramers with subunits that contain binding sites for peptide motifs in receptor tails, clathrin, and phosphoinositide—4,5-bisphosphate lipid (McMahon & Mills, 2004). Other adaptors have been classified as curvature inducers and curvature sensors based on their ability to deform membranes *in vitro* through helical insertion (Epsin1) (Ford et al., 2002) or to bind with affinities that are proportionate to the reduced diameter and increased curvature of artificial liposomes (e.g., amphiphysin) (Peter et al., 2004). Debate rages in the field over which adaptor is the initiator of curvature and which mechanism for curvature induction is dominant (Cocucci, Aguet, Boulant, & Kirchhausen, 2012; Stachowiak et al., 2012). Recently, studies have shown that if recombinant proteins are tagged with histidine residues to pack them at high density onto charged artificial membranes, then deformation can also be induced (Stachowiak, Hayden, & Sasaki, 2010; Stachowiak et al., 2012). This molecular crowding model is difficult to validate in cells since it would require quantitation of both lipid and protein levels at single-molecule resolution. The extensive repertoire of adaptors that have been identified make clathrin-mediated endocytosis the most thoroughly explored internalization route for cargo. However, the large number of molecules identified now compels the field to try to define hierarchies for recruitment (Schmid &

McMahon, 2007). Given the large number of adaptors with overlapping properties as lipid, clathrin, and receptor binders, it remains possible that multiple recruitment sequences will elicit vesicle formation and that the process is somewhat stochastic. Resolving the question of hierarchy and committed vesicle formation will require new thinking and approaches. By contrast, clathrin-independent endocytosis encompasses a number of molecularly distinct internalization routes including pathways associated with caveolae/lipid rafts and flotillin. The recruitment of cargo into these structures and the mechanisms for membrane deformation and internalization not only have been less extensively investigated but also provide routes of entry for signaling receptors.

Once internalized, receptors are trafficked to early endosomal structures that are marked by early endosomal autoantigen 1 (EEA1) and a small GTPase (Rab5) (Mu et al., 1995; Stenmark, Aasland, Toh, & D'Arrigo, 1996). They are referred to as early endosomes because they were originally defined as isolatable structures into which fluid-phase markers were internalized "early" following addition to cell culture media, typically within a few minutes of addition (Griffiths, Back, & Marsh, 1989). At this juncture, receptor tails remain exposed to the cytoplasm and the phosphopeptides and peptide motifs that can act as docking platforms for signaling molecules are therefore available for the assembly of signal-transduction complexes (Fig. 12.1). From this compartment, receptors either are trafficked back to the cell surface via recycling endosomes or are further internalized into multivesicular bodies (MVBs) at which point their tails are sequestered from the cytoplasm altering their signaling properties. In the course of recycling, signaling may also be altered upon ligand dissociation, a process that is promoted at least in part by the acidification of the lumen of vesicles. MVBs are delivered to lysosomes by membrane fusion, and postfusion, the content is degraded proteolytically.

2. RECEPTORS

Receptors form the principal initiators of signaling from endosomal structures and it is important to consider them as distinct classes of signal inducer. Owing to space constraints, we will focus predominantly on two classes, receptor tyrosine kinases (RTKs) and G protein-coupled receptors (GPCRs) (Table 12.1).

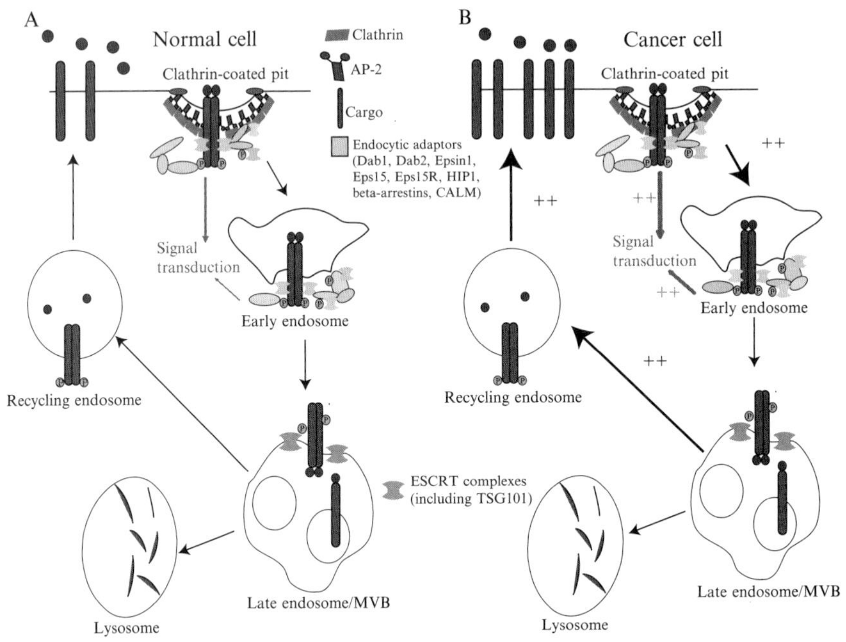

Figure 12.1 *Putative roles of membrane trafficking proteins in potentiating receptor signaling.* In the normal cell (A), binding of ligand to its receptor results in endocytosis of the ligand–receptor complex and triggers the activation of signaling cascades. Endocytic adaptor proteins including HIP1, Dab2, beta-arrestins, Eps15/15R, and Epsin1 facilitate the clustering of the ligand–receptor cargo into coated pits and the delivery of this cargo to early endosomes. Receptor-mediated signal transduction then occurs both from the plasma membrane and from early endosomes and the coated pits themselves owing to sustained ligand binding at a neutral pH. Ligand is dissociated from the receptor as the complex is trafficked through compartments of increasing acidity via late endosomes and multivesicular bodies to lysosomes for degradation. The incorporation of the cargo into multivesicular bodies requires the Endosomal Sorting Complexes Required for Transport (ESCRT) complexes, which include TSG101, a putative tumor suppressor. In the cancer cell (B), changes in the expression or mutagenic status of endocytic adaptors and ESCRT proteins alter the rate-limiting steps of vesicle trafficking and endocytosis and may preferentially increase recycling of the receptors to the cell surface or reduce the rate of internalization thus amplifying signal transduction from growth factor receptors at the expense of receptor degradation. (For color version of this figure, the reader is referred to the online version of this chapter.)

2.1. Receptor tyrosine kinases

Fractionation and coimmunoprecipitation have played a critical role in redefining our view of RTK signaling in cells. Early work showed that epidermal growth factor (EGF) induced accumulation of EGFR and downstream signaling molecules such as SOS, Grb2, and SHC in liver

Table 12.1 Endosomal signaling mechanisms used by selected RTKs and GPCRs

Receptor class	Receptor	Endosomal signaling transducers	References
Receptor tyrosine kinase	EGFR	ERK1/2 PI 3-kinase/Akt	Vieira, Lamaze, and Schmid (1996) and Wang, Pennock, Chen, and Wang (2002b)
	Insulin receptor	ERK1/2 PI 3-kinase	Kelly and Ruderman (1993), Kelly, Ruderman, and Chen (1992), and Li, Stolz, and Romero (2005)
	Platelet-derived growth factor receptor	ERK1/2, PI 3-kinase/Akt	Wang, Pennock, Chen, Kazlauskas, and Wang (2004)
	Vascular endothelial growth factor receptor-2	ERK1/2	Lampugnani, Orsenigo, Gagliani, Tacchetti, and Dejana (2006)
G protein-coupled receptor	Beta2-adrenergic receptor	ERK1/2	Daaka et al. (1998)
	NK1R	ERK1/2	Jafri et al. (2006)
	PAR2	ERK1/2, PI 3-kinase	Ge, Ly, Hollenberg, and DeFea (2003) and Wang and DeFea (2006)
	AT1AR	ERK1/2, JNK3	Ahn, Shenoy, Wei, and Lefkowitz (2004) and McDonald et al. (2000)
	V2R	ERK1/2	Tohgo et al. (2003)
	D2R	PP2A/Akt	Beaulieu et al. (2005)
	CXCR4	p38	Sun, Cheng, Ma, and Pei (2002)

parenchymal cells (Di Guglielmo, Baass, Ou, Posner, & Bergeron, 1994). Similarly, in insulin-treated adipocytes, insulin receptor was found to be more highly phosphorylated upon internalization than at the plasma membrane and insulin receptor substrate 1 (IRS-1) associated with internal membranes where the phosphorylation level paralleled that of the receptor (Kublaoui, Lee, & Pilch, 1995). Activation of PI 3-kinase by the receptor was found to occur at internal membranes rather than at the plasma membrane (Kelly & Ruderman, 1993) and could promote recruitment of mitogen-activated protein kinases (MAPKs) to endosomes (Li et al., 2005). Similarly, nerve growth factor treatment of pheochromocytoma PC12 cells and nociceptive neurons caused accumulation of components of PI 3-kinase and MAPK signaling pathways and activated TrkA receptor and phospholipase C-γ1 (PLC-γ1) (Delcroix et al., 2003; Yu et al., 2011).

While this collectively showed that signaling proteins associated with endosomal membranes, it did not provide insights into whether new patterns of signaling could be initiated from these compartments. The focus of initial studies to address this question was EGF-induced signal transduction and the activation of MAPK and PI 3-kinase/Akt signaling pathways, which regulate cell proliferation and survival. Expression of a mutant of an endocytic protein called dynamin, which is required for GTP-dependent vesicle release from the plasma membrane, suppressed ERK1/2 and PI 3-kinase signaling in response to EGF treatment (Vieira et al., 1996). This confirmed that internalization was necessary to promote the activation of a full spectrum of signaling responses. However, because the mutant dynamin suppresses the internalization of a broad array of proteins, it was difficult to attribute this directly to EGFR internalization. Indeed, subsequently, it was reported that trafficking of the downstream kinase MEK from the plasma membrane was the key step in the process (Kranenburg, Verlaan, & Moolenaar, 1999). An alternative approach was also used that did not involve treatment with endocytic inhibitors, but rather a pulse-chase treatment with tyrosine kinase inhibitors (Pennock & Wang, 2003; Wang, Pennock, Chen, & Wang, 2002a). Treatment of cells with the EGFR tyrosine kinase inhibitor AG-1478 blocked activation of EGFR at the plasma membrane and permitted endocytosis. Withdrawal of AG-1478 activated endosomal EGFR and induced recruitment of signaling factors (SHC, Grb2, and p85 subunit of PI 3-kinase) to endosomes and leading to the activation of ERK1/2 and Akt. This endosome-specific signaling of activated EGFR was sufficient to promote cell survival by the PI 3-kinase/Akt pathway. Consequently, signaling can originate from activated EGFR within endosomes. However, AG-1478 is not an entirely optimal drug for the study of endosome-specific signaling since it can also perturb receptor internalization. Endocytosis is also required for the full biological activity of platelet-derived growth factor receptor (Wang et al., 2004) and vascular endothelial growth factor receptor-2 (Lampugnani et al., 2006).

2.2. G protein-coupled receptors

While best characterized as seven-transmembrane span receptors signaling at the cell surface through heterotrimeric G proteins, these receptors can also signal from endosomes through G protein-independent routes. Arrestins are vitally important for desensitization, internalization, and G protein-independent signaling of GPCRs from endosomes (Shenoy & Lefkowitz, 2011). Arrestins were first identified as inhibitors of GPCR signaling, and beta-arrestins 1 and 2 were originally classified as inhibitors of the beta2-adrenergic receptor and subsequently found to regulate many other GPCRs. The interaction with beta-arrestins is marked by the uncoupling of the receptors from G proteins and their association with clathrin and AP2 adaptors. This switch is mediated by binding to agonist-occupied G protein-coupled receptor kinase (GRK) phosphorylated GPCRs. There are however many additional signaling molecules that are recruited by this vital scaffolding factor both at the plasma membrane and on endosomes. Although beyond the scope of this chapter, it is also worth noting that the scaffolding functions of beta-arrestins and indeed other adaptor proteins may not be restricted to the plasma membrane and endosomes. In the case of beta-arrestin, a number of studies have reported associations with transcription factors such as beta-catenin in the nucleus and also with chromatin-modifying enzymes such as the p300 acetyltransferase (Kang et al., 2005; Rosano et al., 2012). Dissecting the relative contributions of these functions to cellular phenotypes represents a significant challenge not only for beta-arrestins but also for all multifunctional adaptors that act as scaffolding factors for signaling complexes.

In the context of receptor signaling, the most thoroughly studied contribution of beta-arrestins is to MAPK cascades and principally ERK and c-Jun N-terminal kinases (JNKs, p38) (DeWire, Ahn, Lefkowitz, & Shenoy, 2007). The first indication that beta-arrestins are active contributors to signaling dates back to the studies on beta2-adrenergic receptor, in which expression of dominant-negative mutants of beta-arrestins were found to inhibit adrenergic receptor-induced activation of ERK1/2 (Daaka et al., 1998). Subsequently, beta-arrestins were found to couple beta2-adrenergic receptor to c-Src and mediate ERK1/2 activation (Luttrell et al., 1999). Beta-arrestins similarly participate in ERK1/2 signaling by other GPCRs, including neurokinin-1 receptor (NK1R) and vasopressin V2 receptor (V2R) (Table 12.1). These observations led to the view that beta-arrestins are scaffolds that couple activated GPCRs with MAPK signaling complexes or "signalsomes" (Luttrell, 2005). Beta-arrestins thereby mediate a second wave of GPCR signaling that is distinct from G protein-dependent signaling at the plasma membrane. The importance of this mechanism depends on the affinity with which GPCRs interact with beta-arrestins, which varies depending on the extent of GPCR phosphorylation by GRKs and has allowed GPCRs to be divided into distinct classes on this basis.

"Class A" GPCRs (e.g., beta1- and beta2-adrenergic receptors) have few phosphorylation sites and transiently interact with beta-arrestins 1 and 2, mostly at the plasma membrane, with a higher affinity for beta-arrestin 2. "Class B" receptors are phosphorylated at multiple sites (e.g., $AT_{1A}R$ and PAR_2) and interact with high affinity with both ARRB1 and ARRB2 for extended periods at both endosomal membranes and the plasma membrane. "Class C" receptors (e.g., bradykinin B_2 receptor) internalize with beta-arrestin into endosomes followed by rapid dissociation of beta-arrestin upon agonist removal (Simaan, Bedard-Goulet, Fessart, Gratton, & Laporte, 2005). Across these classes of receptor, the duration of beta-arrestin-regulated MAPK signaling is affected by the affinity of beta-arrestins for receptors, the structure of the receptors, and which one of the seven GRKs is responsible for phosphorylating the receptors (Tohgo et al., 2003).

By recruiting receptors and MAPK to endosomes, beta-arrestins can determine the subcellular location and function of activated ERKs. As is the case for RTKs, receptors in endosomes may activate signals that differ from those originating from G proteins at the plasma membrane, resulting in distinct physiological responses. These distinct mechanisms of signaling have been evaluated by disrupting beta-arrestin or G proteins, by studying mutant receptors that are unable to interact with beta-arrestins, or by using agonists that selectively activate particular pathways. When beta-arrestin 1 is fused to the neurokinin-1 receptor C-terminus, the receptor is constitutively associated with a c-Raf/MEK1/2/ERK1/2 complex in endosomes, leading to robust activation of cytosolic but not nuclear ERK1/2 (Jafri et al., 2006). Beta-arrestins similarly participate in activation of the JNK–MAPK cascade, a regulator of stress-induced apoptosis, cell survival, and morphogenesis. Stimulation of angiotensin II receptor type 1A promotes assembly of a signaling complex in endosomes comprising beta-arrestin 2, the upstream kinases MAP kinase kinase (MKK4), apoptosis signaling kinase (ASK1), and active JNK3 (McDonald et al., 2000).

In contrast to the role of beta-arrestins as activators of JNK–MAPK–ERK, beta-arrestins exert an inhibitory influence over PI 3-kinase and Akt. PI 3-kinase is a regulator of cell growth, movement, and apoptosis. Activation of protease-activated receptor-2 (PAR2) promotes interaction of beta-arrestin and PI 3-kinase, which inhibits PI 3-kinase catalytic activity (Wang & DeFea, 2006; Wang, Kumar, Wang, & Defea, 2007). This mechanism opposes PAR2-induced stimulation of PI 3-kinase, which is mediated by $G\alpha q$. The result of these opposing mechanisms depends on the level of beta-arrestin expression, with PI 3-kinase inhibition predominating in cells that highly express beta-arrestin. Akt is a downstream target of PI 3-kinase that controls transcription, apoptosis, and cell cycle. Dopamine 2 receptor (D2R) stimulation induces formation of a beta-arrestin 2/Akt/PP2A complex, identified in striatal extracts by pull-down assays (Beaulieu et al., 2005). Whether this complex forms at the plasma membrane or endosomes is unknown. Sustained stimulation of D2R in the mouse striatum inactivates Akt by a beta-arrestin 2-dependent mechanism (Beaulieu et al., 2007). This mechanism is another example, along with regulation of JNK3, of beta-arrestin recruiting both activators and inhibitors (PP2A, MKP7) to signaling complexes.

2.3. Proteases regulate receptor signaling

This too is an entire field in its right and extensively reviewed elsewhere (Beckett, Nalivaeva, Belyaev, & Turner, 2012; Blobel, 2005). In brief, although many signal-transduction cascades are propagated by phosphorylation, signaling can also require proteolysis, which may activate a substrate or allow a product to translocate to a different cellular location to exert its effect. Proteases are essential for Notch signaling, a regulator of development (Schweisguth, 2004). Notch receptors exist in the plasma membrane as heterodimers composed of the Notch extracellular domain and the membrane-anchored intracellular domain. Ligand binding results in cleavage of the membrane-anchored intracellular domain at an extracellular site by metalloproteases. Notch then undergoes intramembrane cleavage of the membrane-anchored intracellular domain by γ-secretase to liberate the Notch intracellular domain, which translocates to the nucleus to regulate gene transcription (Tagami et al., 2008). Although the role of endocytosis in Notch cleavage and signaling is poorly understood, observations of *Drosophila melanogaster* mutants with defects in the endocytic pathway indicate that entry of Notch into early endosomes is required for efficient γ-secretase-mediated cleavage of Notch and Notch signaling (Giebel & Wodarz, 2006; Lu & Bilder, 2005). Alterations in Notch trafficking in endosomes may underlie developmental abnormalities that are related to defects in Notch signaling. Cleavage of the intracellular domain of ErbB4 by gamma secretase has also been reported but is less well characterized (Lee et al., 2002). Proteolytic processing of ErbB4 includes a basal level, which can be increased by TPA in all cells or by the addition of neuregulin (heregulin) to some but not all cells (Zhou & Carpenter, 2000). This cleavage results in the formation of two receptor fragments: a 120 kDa ectodomain fragment that is released into the media and an 80 kDa membrane-bound fragment, termed m80. Cleavage requires ADAM 17 (TACE) and it is likely that this is the enzyme that also executes cleavage of ErbB4 between His651 and Ser652 within the extracellular stalk or ecto-juxtamembrane region (Cheng, Tikhomirov, Zhou, & Carpenter, 2003). Various functions have been ascribed to the intracellular domain of ErbB4 but the most ubiquitous can be termed chaperoning. This means in general terms a role for ErbB4 as a scaffolding factor for the assembly of transcription factor and coregulator complexes and examples include STAT5 (Williams et al., 2004), ER (Zhu et al., 2006), and WWOX (Aqeilan et al., 2005). Gamma-secretase cleavage has been proposed as a mechanism to release intracellular fragment of a number of other receptors to facilitate their role as transcriptional coregulators in the nucleus; however, detailed discussion of this lies beyond the scope of the chapter.

3. ADAPTOR PROTEINS AND ONCOGENESIS

As discussed briefly in the preceding text, adaptor proteins such as beta-arrestins play critical roles in regulating the subcellular distribution and signaling pathway that are activated by receptors. The roles of adaptor proteins in clathrin-mediated endocytosis have been described comprehensively in other reviews. In the context of cancer, changes in the expression of adaptor proteins can contribute to cellular transformation (Table 12.2). In this section, I will briefly provide some examples of this. Due to space constraints I will now describe a subset of these proteins which represent some of the more extensively studied adaptor proteins in cancers.

3.1. Huntingtin-interacting protein 1

Huntingtin-interacting protein 1 (HIP1) was initially described as an adaptor protein associated with the clathrin light chain and AP2 adaptor complex with a role in clathrin-mediated endocytosis having been first identified in a yeast two-hybrid screen for interacting partners of huntingtin, which is encoded by the huntingtin chorea gene, a gene which when mutated causes Huntington's disease. (Mishra et al., 2001; Waelter et al., 2001). It was shown additionally able to bind to phosphoinositides through an AP180 N-terminal homology (ANTH) domain and thus to act as an adaptor between membranes and the clathrin coat involved in endocytosis (Hyun et al., 2004). HIP1 is predominantly expressed in neurons and the brain. HIP1 is however also overexpressed in prostate, brain, breast, and colon cancers (Bradley, Holland, et al., 2007; Rao et al., 2002). In prostate cancer, HIP1 expression correlates with histotype and grade at diagnosis and is predictive of a poor clinical outcome

Table 12.2 Selected regulators of endocytosis with dysregulated expression and functions in human tumors

Protein	Function in endocytosis	Aberrations in cancer	Impact on cancer phenotypes	References
HIP1 and HIP1R	Coordinates actin remodeling and clathrin and AP2 adaptor recruitment during endocytosis	Overexpressed in primary epithelial tumors and gliomas, also reported as a gene fusion (HIP1–PDGFBR) in leukemia	Transforms mouse fibroblasts and induces tumors in xenograft models. Also promotes cytokine-independent growth	Rao et al. (2003) and Ross, Bernard, Berger, and Gilliland (1998)
CBL	Regulates ubiquitylation of RTKs	Point-mutated (e.g., R420Q in the RING domain) and also subject to deletions and insertions in acute myeloid leukemia (AML)	Inhibits ubiquitylation of RTKs such as FLT3 and their endocytosis	Abbas, Rotmans, Lowenberg, and Valk (2008), Caligiuri et al. (2007), and Langdon, Hartley, Klinken, Ruscetti, and Morse (1989)
Clathrin heavy chain	Principal coat protein that polymerizes to form a molecular brace on the membrane	Found in gene fusions in large B-cell lymphoma and in pediatric renal cell carcinoma	Constitutive activation of fusion partners and aberrant transcription control	De Paepe et al. (2003), Shinmura et al. (2008), and Wang et al. (2011)
NDRG1	RAB4A effector involved in the recycling of E-cadherin	Downregulated in prostate, breast, and pancreatic carcinomas. Fused to ERG in prostate cancer	Suppression of metastasis of colon and prostate cells in mouse models	Pflueger et al. (2009) and Tu, Yan, Hood, and Lin (2007)

Continued

Table 12.2 Selected regulators of endocytosis with dysregulated expression and functions in human tumors—cont'd

Protein	Function in endocytosis	Aberrations in cancer	Impact on cancer phenotypes	References
Eps15	Scaffolding protein originally identified as an RTK substrate, which enhanced clathrin-mediated endocytosis	Expressed as a fusion with MLL in acute myeloid leukemia	Coiled-coil domain of Eps15 mediates the oligomerization of MLL, a histone methyltransferase resulting in aberrant chromatin modifications and transcription	Fazioli, Minichiello, Matoskova, Wong, and Di Fiore (1993) and Kotecha et al. (2012)
Caveolin 1	Essential for the biogenesis of caveolae and endocytosis	Downregulation and sporadic mutation in breast cancer, correlating with estrogen receptor expression. Upregulation in multiple cancer types. Amplification in aggressive breast carcinomas	Expression can correlate inversely with cell cycle progression and transformation. Ectopic expression however can confer resistance to oncogene-induced apoptosis	Sotgia et al. (2006)

(Rao et al., 2002). Overexpression of HIP1 was later shown to induce the transformation of a fibroblast cell line when stably overexpressed and confer the ability to form tumors in a xenograft model through growth factor hypersensitization and increased steady-state levels of EGFR and FGFR in the fibroblast lines (Rao et al., 2003). Overexpression of HIP1 reduces the intracellular levels of the clathrin adaptor AP2 and relocalizes clathrin to the perinuclear region, implying that removal of clathrin from the plasma membrane may explain the defect in EGFR internalization. This mechanism is yet to be proven categorically since HIP1 knockout mice do not generate fibroblasts with detectable trafficking defects (Bradley, Hyun, et al., 2007). A similar transformation phenotype can be induced in NIH3T3 cells through the overexpression of another endocytic adaptor, Eps15 (Fazioli et al., 1993).

3.2. Endosomal Sorting Complexes Required for Transport (ESCRT) machinery-mediated sorting on to intraluminal vesicle

The ESCRT machinery is a series of protein complexes that are thought to act sequentially to irreversibly concentrate ubiquitinated cargo into domains on the perimeter membrane of the MVB (multivesicular body)/endosome. The later components of the machinery then initiate the formation of the inwardly budding intraluminal vesicle (ILV). EGFR ubiquitination and the ESCRT machinery have been shown to be necessary for sorting of EGFRs on to ILVs and for EGF-stimulated ILV formation. The ESCRT components are soluble proteins/complexes that can be recruited from the cytoplasm. Clathrin-containing coats were identified on flattened domains of endosomes more than 10 years ago and these domains were shown to contain Hrs (Hepatocyte growth factor-regulated tyrosine kinase substrate), a critical ubiquitinated protein-binding component of ESCRT-0. These coats are readily identified by conventional EM (electron microscopy) and are frequently observed on MVBs that have a clearly identifiable contact site with the ER. Sometimes, the coat is observed immediately adjacent to the contact site.

Tumor-suppressor gene 101 (TSG101) is a well-studied endocytic adaptor for its potential role in oncogenesis. TSG101 is one of the three proteins forming the ESCRT-I complex, which is involved in the recruitment of ubiquitinated cargo and receptors, following their internalization from the cell surface into MVBs and ultimately for degradation following their internalization from the cell surface (Winter & Hauser, 2006). Its important role in this process is highlighted by the finding that inactivation with a neutralizing antibody specific to TSG101 promotes EGFR recycling and blocks its degradation (Bishop, Horman, & Woodman, 2002). It was therefore proposed that an inhibition of TSG101 function could lead to a block in RTK sorting and potentiated and aberrant endosomal signaling contributing to tumor formation (Bishop et al., 2002). ESCRT-I is itself recruited to endosomal membranes through an interaction with hepatocyte growth factor receptor substrate (Hrs), a protein also implicated in both endosomal trafficking and nuclear hormone receptor transactivation (Lu, Hope, Brasch, Reinhard, & Cohen, 2003).

Overexpression and depletion of TSG101 can result in aberrant HIV budding and receptor sorting and have been implicated in cellular transformation through the potentiation of cell signaling. Inactivation of TSG101 using an antisense approach transforms a fibroblast cell line allowing the formation of colonies on soft agar and driving the development of metastatic tumors in nude mice (Li & Cohen, 1996). Partial deletions and aberrant splicing of *TSG101* have been identified in human cancers, although variant transcripts can also be found in normal tissue (Gayther et al., 1997; Lee & Feinberg, 1997). Its role in cancer has been hard to validate however since a complete knockout in transgenic mice is embryonic lethal (Wagner et al., 2003).

It was first identified in a random insertional mutagenesis experiment to identify tumor suppressor genes in NIH3T3 cells (Li & Cohen, 1996). Subcutaneous injection of NIH3T3 fibroblasts in which TSG101 has been knocked out results in *in vitro* transformation and metastatic tumors. *TSG101* is found at 11p15, which is a frequent site of loss of heterozygosity in breast, ovarian, testicular, and Wilm's cancers, and this further implies that TSG101 may be a bona fide tumor suppressor (Zhong, Chen, Chen, Chen, & Lee, 1997). It is also known to interact with a regulator of cell growth and differentiation known as oncoprotein 18/stathmin, which is overexpressed in neuroblastoma and acute leukemias (Li & Cohen, 1996).

The hypothesis that *TSG101* is a bona fide tumor suppressor owing to its role in regulating receptor signaling is yet to be definitively confirmed owing to a number of conflicting pieces of data. A knockout of *Tsg101* in the mouse mammary gland fails to elicit tumors (Wagner et al., 2003). A complete *Tsg101* knockout is embryonic lethal and heterozygous mice or mammary gland-specific knockout mice show impaired mammogenesis in the late stages of pregnancy (Wagner et al., 2003). However, by defining *Tsg101* as a tumor susceptibility gene, the implication is that overexpression could also contribute to tumor formation. Tsg101 has been shown to be upregulated in thyroid tumors, a subset of breast tumors, and other human cancers (Liu et al., 2002; Oh, Stanton, West, Todd, & Wagner, 2007). There is evidence that the upregulation of Tsg101 inhibits the degradation of the E3 ubiquitin ligase MDM2, enhancing polyubiquitylation and proteasomal degradation of the tumor suppressor p53 (Li, Liao, Ruland, Mak, & Cohen, 2001). More recently, overexpression of Tsg101 in a transgenic model of breast cancer was shown to contribute to tumor progression rather than cancer initiation (Oh et al., 2007). In this model, phosphorylation of EGFR and MAPK was increased but only contributed to tumor formation in aging females implying weak oncogenic properties for Tsg101 in keeping with a molecule with a regulatory role but no intrinsic enzymatic activity. Until these apparently conflicting data are definitively reconciled, the implication is that Tsg101 functions as a rheostat to fine tune tumor development dependent on tissue and genetic context.

3.3. Beta-arrestins

Chemotaxis, migration, and metastasis are essential functions to many cellular events including development, wound healing, immune responses, and cancer metastasis. Recent reports have indicated a role for beta-arrestin in these processes. Beta-arrestin 2 is a critical component in CXCR4-mediated chemotaxis in HeLa and HEK-293 cells (Sun et al., 2002). The forced expression of beta-arrestin 2 augmented stromal cell-derived factor1a-induced chemotaxis in this system while the suppression of beta-arrestin 2 severely inhibited migration. Also, a role for beta-arrestin 2 in AT1AR-stimulated chemotaxis has been observed (Hunton et al., 2005). The use of small interfering RNA directed at beta-arrestin 2 almost abolished AT1AR-mediated chemotaxis. Interestingly, the beta-arrestin 2-meditated migration in both of these systems was mediated by p38 mitogen-activated protein kinase. As mentioned earlier, the regulation of p38 activity has been associated with beta-arrestin function, implying a direct connection between these observations. Beta-arrestins 1 and 2 also play a pivotal role in the protease-activated receptor-2-mediated migration of MDA-MB-231 breast cancer cells (Ge, Shenoy, Lefkowitz, & DeFea, 2004). Furthermore, the importance of beta-arrestin 2 in cellular migration and chemotaxis has been observed in beta-arrestin 2-deficient mouse models (Fong et al., 2002). Chemokine-meditated CD4+ T-cell migration was impaired in these mice and suggests that beta-arrestin 2 may play a major role in the development of allergic asthma (Walker et al., 2003). Also, splenocytes derived from beta-arrestin 2 knockout mice showed decreased CXCL12-induced migration (Fong et al., 2002).

These observations are now being extended to cancer models. A prostaglandin E2-induced signaling complex consisting of a PGE2 receptor, beta-arrestin 1, and c-Src was recently shown to be critical for cell migration *in vitro* and metastatic spread of colorectal carcinoma to the liver *in vivo* (Buchanan et al., 2006). Owing to the involvement of beta-arrestins both in the internalization and downregulation of activated receptors and in the promotion of cell migration, the functional contribution of these proteins to cancer progression through the modulation of signaling will need to be explored across a panel of tumor models.

3.4. Future perspectives

The rate-limiting steps in exploring and exploiting the impact of endocytosis on signaling and oncogenesis are fundamental in defining the interplay between competing proteins and signals at a network level. Since most of the proteins touched upon in this chapter lack intrinsic enzymatic activity, they are not viewed classically as drug targets. However, there is an increasing number of therapeutics under development that utilize antisense approaches or seek to target protein–protein interactions. The means to impact on trafficking pathways will therefore surely exist in the near future. The cell is however inherently maintained in a signaling and membrane flux equilibrium by endocytic pathways, and relative to other drug targets, endocytic proteins at least so far are not commonly somatically mutated in cancers. Consequently, overcoming the bottleneck in developing a regulatory hierarchy for the assembly of trafficking and signaling complexes is a key to focusing on these proteins in disease progression. This will require a cross-community effort during in industry and academia and spanning mathematical modeling, synthetic biology, imaging, structural biology, and biochemistry. It remains to be seen whether the will exists to step beyond traditional boundaries to achieve this in this area.

Although long regarded as a conduit for the degradation or recycling of cell surface receptors, the endosomal system is also an essential site of signal transduction. Activated receptors accumulate in endosomes, and certain signaling components are exclusively localized to endosomes. Receptors can continue to transmit signals from endosomes that are different from those that arise from the plasma membrane, resulting in distinct physiological responses. Endosomal signaling is widespread in metazoans and plants, where it transmits signals for diverse receptor families that regulate essential processes including growth, differentiation, and survival. Receptor signaling at endosomal membranes is tightly regulated by mechanisms that control agonist availability, receptor coupling to signaling machinery, and the subcellular localization of signaling components. Drugs that target mechanisms that initiate and terminate receptor signaling at the plasma membrane are widespread and effective treatments for disease. Selective disruption of receptor signaling in endosomes, which can be accomplished by targeting endosome-specific signaling pathways or by selective delivery of drugs to the endosomal network, may provide novel therapies for disease.

REFERENCES

Abbas, S., Rotmans, G., Lowenberg, B., & Valk, P. J. (2008). Exon 8 splice site mutations in the gene encoding the E3-ligase CBL are associated with core binding factor acute myeloid leukemias. *Haematologica, 93*, 1595–1597.

Ahn, S., Shenoy, S. K., Wei, H., & Lefkowitz, R. J. (2004). Differential kinetic and spatial patterns of beta-arrestin and G protein-mediated ERK activation by the angiotensin II receptor. *The Journal of Biological Chemistry, 279*, 35518–35525.

Aqeilan, R. I., Donati, V., Palamarchuk, A., Trapasso, F., Kaou, M., Pekarsky, Y., et al. (2005). WW domain-containing proteins, WWOX and YAP, compete for interaction with ErbB-4 and modulate its transcriptional function. *Cancer Research, 65*, 6764–6772.

Beaulieu, J. M., Sotnikova, T. D., Marion, S., Lefkowitz, R. J., Gainetdinov, R. R., & Caron, M. G. (2005). An Akt/beta-arrestin 2/PP2A signaling complex mediates dopaminergic neurotransmission and behavior. *Cell, 122*, 261–273.

Beaulieu, J. M., Tirotta, E., Sotnikova, T. D., Masri, B., Salahpour, A., Gainetdinov, R. R., et al. (2007). Regulation of Akt signaling by D2 and D3 dopamine receptors in vivo. *The Journal of Neuroscience: The Official Journal of the Society for Neuroscience, 27*, 881–885.

Beckett, C., Nalivaeva, N. N., Belyaev, N. D., & Turner, A. J. (2012). Nuclear signalling by membrane protein intracellular domains: The AICD enigma. *Cellular Signalling, 24*, 402–409.

Bishop, N., Horman, A., & Woodman, P. (2002). Mammalian class E vps proteins recognize ubiquitin and act in the removal of endosomal protein-ubiquitin conjugates. *The Journal of Cell Biology, 157*, 91–101.

Blobel, C. P. (2005). ADAMs: Key components in EGFR signalling and development. *Nature Reviews Molecular Cell Biology, 6*, 32–43.

Bradley, S. V., Holland, E. C., Liu, G. Y., Thomas, D., Hyun, T. S., & Ross, T. S. (2007). Huntingtin interacting protein 1 is a novel brain tumor marker that associates with epidermal growth factor receptor. *Cancer Research, 67*, 3609–3615.

Bradley, S. V., Hyun, T. S., Oravecz-Wilson, K. I., Li, L., Waldorff, E. I., Ermilov, A. N., et al. (2007). Degenerative phenotypes caused by the combined deficiency of murine HIP1 and HIP1r are rescued by human HIP1. *Human Molecular Genetics, 16*, 1279–1292.

Buchanan, F. G., Gorden, D. L., Matta, P., Shi, Q., Matrisian, L. M., & DuBois, R. N. (2006). Role of beta-arrestin 1 in the metastatic progression of colorectal cancer. In: *Proceedings of the National Academy of Sciences of the United States of America, 103*, 1492–1497.

Caligiuri, M. A., Briesewitz, R., Yu, J., Wang, L., Wei, M., Arnoczky, K. J., et al. (2007). Novel c-CBL and CBL-b ubiquitin ligase mutations in human acute myeloid leukemia. *Blood, 110*, 1022–1024.

Cheng, Q. C., Tikhomirov, O., Zhou, W., & Carpenter, G. (2003). Ectodomain cleavage of ErbB-4: Characterization of the cleavage site and m80 fragment. *The Journal of Biological Chemistry, 278*, 38421–38427.

Cocucci, E., Aguet, F., Boulant, S., & Kirchhausen, T. (2012). The first five seconds in the life of a clathrin-coated pit. *Cell, 150*, 495–507.

Daaka, Y., Luttrell, L. M., Ahn, S., Della Rocca, G. J., Ferguson, S. S., Caron, M. G., et al. (1998). Essential role for G protein-coupled receptor endocytosis in the activation of mitogen-activated protein kinase. *The Journal of Biological Chemistry, 273*, 685–688.

Delcroix, J. D., Valletta, J. S., Wu, C., Hunt, S. J., Kowal, A. S., & Mobley, W. C. (2003). NGF signaling in sensory neurons: Evidence that early endosomes carry NGF retrograde signals. *Neuron, 39*, 69–84.

De Paepe, P., Baens, M., van Krieken, H., Verhasselt, B., Stul, M., Simons, A., et al. (2003). ALK activation by the CLTC-ALK fusion is a recurrent event in large B-cell lymphoma. *Blood, 102*, 2638–2641.

DeWire, S. M., Ahn, S., Lefkowitz, R. J., & Shenoy, S. K. (2007). Beta-arrestins and cell signaling. *Annual Review of Physiology, 69*, 483–510.

Di Guglielmo, G. M., Baass, P. C., Ou, W. J., Posner, B. I., & Bergeron, J. J. (1994). Compartmentalization of SHC, GRB2 and mSOS, and hyperphosphorylation of Raf-1 by EGF but not insulin in liver parenchyma. *The EMBO Journal, 13*, 4269–4277.

Fazioli, F., Minichiello, L., Matoskova, B., Wong, W. T., & Di Fiore, P. P. (1993). eps15, a novel tyrosine kinase substrate, exhibits transforming activity. *Molecular and Cellular Biology, 13*, 5814–5828.

Fong, A. M., Premont, R. T., Richardson, R. M., Yu, Y. R., Lefkowitz, R. J., & Patel, D. D. (2002). Defective lymphocyte chemotaxis in beta-arrestin2- and GRK6-deficient mice. In: *Proceedings of the National Academy of Sciences of the United States of America, 99*, 7478–7483.

Ford, M. G., Mills, I. G., Peter, B. J., Vallis, Y., Praefcke, G. J., Evans, P. R., et al. (2002). Curvature of clathrin-coated pits driven by epsin. *Nature, 419*, 361–366.

Gayther, S. A., Barski, P., Batley, S. J., Li, L., de Foy, K. A., Cohen, S. N., et al. (1997). Aberrant splicing of the TSG101 and FHIT genes occurs frequently in multiple malignancies and in normal tissues and mimics alterations previously described in tumours. *Oncogene, 15*, 2119–2126.

Ge, L., Ly, Y., Hollenberg, M., & DeFea, K. (2003). A beta-arrestin-dependent scaffold is associated with prolonged MAPK activation in pseudopodia during protease-activated receptor-2-induced chemotaxis. *The Journal of Biological Chemistry, 278*, 34418–34426.

Ge, L., Shenoy, S. K., Lefkowitz, R. J., & DeFea, K. (2004). Constitutive protease-activated receptor-2-mediated migration of MDA MB-231 breast cancer cells requires both beta-arrestin-1 and -2. *The Journal of Biological Chemistry, 279*, 55419–55424.

Giebel, B., & Wodarz, A. (2006). Tumor suppressors: Control of signaling by endocytosis. *Current Biology, 16*, R91–R92.

Griffiths, G., Back, R., & Marsh, M. (1989). A quantitative analysis of the endocytic pathway in baby hamster kidney cells. *The Journal of Cell Biology, 109*, 2703–2720.

Hunton, D. L., Barnes, W. G., Kim, J., Ren, X. R., Violin, J. D., Reiter, E., et al. (2005). Beta-arrestin 2-dependent angiotensin II type 1A receptor-mediated pathway of chemotaxis. *Molecular Pharmacology, 67*, 1229–1236.

Hyun, T. S., Rao, D. S., Saint-Dic, D., Michael, L. E., Kumar, P. D., Bradley, S. V., et al. (2004). HIP1 and HIP1r stabilize receptor tyrosine kinases and bind 3-phosphoinositides via epsin N-terminal homology domains. *The Journal of Biological Chemistry, 279*, 14294–14306.

Jafri, F., El-Shewy, H. M., Lee, M. H., Kelly, M., Luttrell, D. K., & Luttrell, L. M. (2006). Constitutive ERK1/2 activation by a chimeric neurokinin 1 receptor-beta-arrestin1 fusion protein. Probing the composition and function of the G protein-coupled receptor "signalsome" *The Journal of Biological Chemistry, 281*, 19346–19357.

Kang, J., Shi, Y., Xiang, B., Qu, B., Su, W., Zhu, M., et al. (2005). A nuclear function of beta-arrestin1 in GPCR signaling: Regulation of histone acetylation and gene transcription. *Cell, 123*, 833–847.

Kelly, K. L., & Ruderman, N. B. (1993). Insulin-stimulated phosphatidylinositol 3-kinase. Association with a 185-kDa tyrosine-phosphorylated protein (IRS-1) and localization in a low density membrane vesicle. *The Journal of Biological Chemistry, 268*, 4391–4398.

Kelly, K. L., Ruderman, N. B., & Chen, K. S. (1992). Phosphatidylinositol-3-kinase in isolated rat adipocytes. Activation by insulin and subcellular distribution. *The Journal of Biological Chemistry, 267*, 3423–3428.

Kotecha, R. S., Ford, J., Beesley, A. H., Anderson, D., Cole, C. H., & Kees, U. R. (2012). Molecular characterization of identical, novel MLL-EPS15 translocation and individual genomic copy number alterations in monozygotic infant twins with acute lymphoblastic leukemia. *Haematologica, 97*, 1447–1450.

Kranenburg, O., Verlaan, I., & Moolenaar, W. H. (1999). Dynamin is required for the activation of mitogen-activated protein (MAP) kinase by MAP kinase kinase. *The Journal of Biological Chemistry, 274*, 35301–35304.

Kublaoui, B., Lee, J., & Pilch, P. F. (1995). Dynamics of signaling during insulin-stimulated endocytosis of its receptor in adipocytes. *The Journal of Biological Chemistry, 270*, 59–65.

Lampugnani, M. G., Orsenigo, F., Gagliani, M. C., Tacchetti, C., & Dejana, E. (2006). Vascular endothelial cadherin controls VEGFR-2 internalization and signaling from intracellular compartments. *The Journal of Cell Biology, 174*, 593–604.

Langdon, W. Y., Hartley, J. W., Klinken, S. P., Ruscetti, S. K., & Morse, H. C., 3rd. (1989). v-cbl, an oncogene from a dual-recombinant murine retrovirus that induces early B-lineage lymphomas. In: *Proceedings of the National Academy of Sciences of the United States of America, 86*, 1168–1172.

Lee, M. P., & Feinberg, A. P. (1997). Aberrant splicing but not mutations of TSG101 in human breast cancer. *Cancer Research, 57*, 3131–3134.

Lee, H. J., Jung, K. M., Huang, Y. Z., Bennett, L. B., Lee, J. S., Mei, L., et al. (2002). Presenilin-dependent gamma-secretase-like intramembrane cleavage of ErbB4. *The Journal of Biological Chemistry, 277*, 6318–6323.

Li, L., & Cohen, S. N. (1996). Tsg101: A novel tumor susceptibility gene isolated by controlled homozygous functional knockout of allelic loci in mammalian cells. *Cell, 85*, 319–329.

Li, L., Liao, J., Ruland, J., Mak, T. W., & Cohen, S. N. (2001). A TSG101/MDM2 regulatory loop modulates MDM2 degradation and MDM2/p53 feedback control. In: *Proceedings of the National Academy of Sciences of the United States of America, 98*, 1619–1624.

Li, H. S., Stolz, D. B., & Romero, G. (2005). Characterization of endocytic vesicles using magnetic microbeads coated with signalling ligands. *Traffic, 6*, 324–334.

Liu, R. T., Huang, C. C., You, H. L., Chou, F. F., Hu, C. C., Chao, F. P., et al. (2002). Overexpression of tumor susceptibility gene TSG101 in human papillary thyroid carcinomas. *Oncogene, 21*, 4830–4837.

Lu, H., & Bilder, D. (2005). Endocytic control of epithelial polarity and proliferation in Drosophila. *Nature Cell Biology, 7*, 1232–1239.

Lu, Q., Hope, L. W., Brasch, M., Reinhard, C., & Cohen, S. N. (2003). TSG101 interaction with HRS mediates endosomal trafficking and receptor down-regulation. In: *Proceedings of the National Academy of Sciences of the United States of America, 100*, 7626–7631.

Luttrell, L. M. (2005). Composition and function of g protein-coupled receptor signalsomes controlling mitogen-activated protein kinase activity. *Journal of Molecular Neuroscience, 26*, 253–264.

Luttrell, L. M., Ferguson, S. S., Daaka, Y., Miller, W. E., Maudsley, S., Della Rocca, G. J., et al. (1999). Beta-arrestin-dependent formation of beta2 adrenergic receptor-Src protein kinase complexes. *Science, 283*, 655–661.

McDonald, P. H., Chow, C. W., Miller, W. E., Laporte, S. A., Field, M. E., Lin, F. T., et al. (2000). Beta-arrestin 2: A receptor-regulated MAPK scaffold for the activation of JNK3. *Science, 290*, 1574–1577.

McMahon, H. T., & Mills, I. G. (2004). COP and clathrin-coated vesicle budding: Different pathways, common approaches. *Current Opinion in Cell Biology, 16*, 379–391.

Mishra, S. K., Agostinelli, N. R., Brett, T. J., Mizukami, I., Ross, T. S., & Traub, L. M. (2001). Clathrin- and AP-2-binding sites in HIP1 uncover a general assembly role for endocytic accessory proteins. *The Journal of Biological Chemistry, 276*, 46230–46236.

Mu, F. T., Callaghan, J. M., Steele-Mortimer, O., Stenmark, H., Parton, R. G., Campbell, P. L., et al. (1995). EEA1, an early endosome-associated protein. EEA1 is a conserved alpha-helical peripheral membrane protein flanked by cysteine "fingers" and contains a calmodulin-binding IQ motif. *The Journal of Biological Chemistry, 270*, 13503–13511.

Oh, K. B., Stanton, M. J., West, W. W., Todd, G. L., & Wagner, K. U. (2007). Tsg101 is upregulated in a subset of invasive human breast cancers and its targeted overexpression in transgenic mice reveals weak oncogenic properties for mammary cancer initiation. *Oncogene, 26*, 5950–5959.

Pearse, B. M. (1976). Clathrin: A unique protein associated with intracellular transfer of membrane by coated vesicles. In: *Proceedings of the National Academy of Sciences of the United States of America, 73*, 1255–1259.

Pearse, B. M., & Bretscher, M. S. (1981). Membrane recycling by coated vesicles. *Annual Review of Biochemistry, 50*, 85–101.

Pennock, S., & Wang, Z. (2003). Stimulation of cell proliferation by endosomal epidermal growth factor receptor as revealed through two distinct phases of signaling. *Molecular and Cellular Biology, 23*, 5803–5815.

Peter, B. J., Kent, H. M., Mills, I. G., Vallis, Y., Butler, P. J., Evans, P. R., et al. (2004). BAR domains as sensors of membrane curvature: The amphiphysin BAR structure. *Science, 303*, 495–499.

Pfeffer, S. R., & Kelly, R. B. (1981). Identification of minor components of coated vesicles by use of permeation chromatography. *The Journal of Cell Biology, 91*, 385–391.

Pflueger, D., Rickman, D. S., Sboner, A., Perner, S., LaFargue, C. J., Svensson, M. A., et al. (2009). N-myc downstream regulated gene 1 (NDRG1) is fused to ERG in prostate cancer. *Neoplasia, 11*, 804–811.

Rao, D. S., Bradley, S. V., Kumar, P. D., Hyun, T. S., Saint-Dic, D., Oravecz-Wilson, K., et al. (2003). Altered receptor trafficking in Huntingtin Interacting Protein 1-transformed cells. *Cancer Cell, 3*, 471–482.

Rao, D. S., Hyun, T. S., Kumar, P. D., Mizukami, I. F., Rubin, M. A., Lucas, P. C., et al. (2002). Huntingtin-interacting protein 1 is overexpressed in prostate and colon cancer and is critical for cellular survival. *The Journal of Clinical Investigation, 110*, 351–360.

Rao, Y., Ruckert, C., Saenger, W., & Haucke, V. (2012). The early steps of endocytosis: From cargo selection to membrane deformation. *European Journal of Cell Biology, 91*, 226–233.

Rosano, L., Cianfrocca, R., Tocci, P., Spinella, F., Di Castro, V., Spadaro, F., et al. (2012). Beta-arrestin-1 is a nuclear transcriptional regulator of endothelin-1-induced beta-catenin signaling. *Oncogene*.

Ross, T. S., Bernard, O. A., Berger, R., & Gilliland, D. G. (1998). Fusion of Huntingtin interacting protein 1 to platelet-derived growth factor beta receptor (PDGFbetaR) in chronic myelomonocytic leukemia with t(5;7)(q33;q11.2). *Blood, 91*, 4419–4426.

Sandvig, K., Pust, S., Skotland, T., & van Deurs, B. (2011). Clathrin-independent endocytosis: Mechanisms and function. *Current Opinion in Cell Biology, 23*, 413–420.

Schmid, E. M., & McMahon, H. T. (2007). Integrating molecular and network biology to decode endocytosis. *Nature, 448*, 883–888.

Schweisguth, F. (2004). Regulation of notch signaling activity. *Current Biology, 14*, R129–R138.

Shenoy, S. K., & Lefkowitz, R. J. (2011). Beta-Arrestin-mediated receptor trafficking and signal transduction. *Trends in Pharmacological Sciences, 32*, 521–533.

Shinmura, K., Kageyama, S., Tao, H., Bunai, T., Suzuki, M., Kamo, T., et al. (2008). EML4-ALK fusion transcripts, but no NPM-, TPM3-, CLTC-, ATIC-, or TFG-ALK fusion transcripts, in non-small cell lung carcinomas. *Lung Cancer, 61*, 163–169.

Simaan, M., Bedard-Goulet, S., Fessart, D., Gratton, J. P., & Laporte, S. A. (2005). Dissociation of beta-arrestin from internalized bradykinin B2 receptor is necessary for receptor recycling and resensitization. *Cellular Signalling, 17*, 1074–1083.

Sotgia, F., Rui, H., Bonuccelli, G., Mercier, I., Pestell, R. G., & Lisanti, M. P. (2006). Caveolin-1, mammary stem cells, and estrogen-dependent breast cancers. *Cancer Research, 66*, 10647–10651.

Stachowiak, J. C., Hayden, C. C., & Sasaki, D. Y. (2010). Steric confinement of proteins on lipid membranes can drive curvature and tubulation. In: *Proceedings of the National Academy of Sciences of the United States of America, 107*, 7781–7786.

Stachowiak, J. C., Schmid, E. M., Ryan, C. J., Ann, H. S., Sasaki, D. Y., Sherman, M. B., et al. (2012). Membrane bending by protein-protein crowding. *Nature Cell Biology, 14*, 944–949.

Stenmark, H., Aasland, R., Toh, B. H., & D'Arrigo, A. (1996). Endosomal localization of the autoantigen EEA1 is mediated by a zinc-binding FYVE finger. *The Journal of Biological Chemistry, 271*, 24048–24054.

Sun, Y., Cheng, Z., Ma, L., & Pei, G. (2002). Beta-arrestin2 is critically involved in CXCR4-mediated chemotaxis, and this is mediated by its enhancement of p38 MAPK activation. *The Journal of Biological Chemistry, 277*, 49212–49219.

Tagami, S., Okochi, M., Yanagida, K., Ikuta, A., Fukumori, A., Matsumoto, N., et al. (2008). Regulation of Notch signaling by dynamic changes in the precision of S3 cleavage of Notch-1. *Molecular and Cellular Biology, 28*, 165–176.

Tohgo, A., Choy, E. W., Gesty-Palmer, D., Pierce, K. L., Laporte, S., Oakley, R. H., et al. (2003). The stability of the G protein-coupled receptor-beta-arrestin interaction determines the mechanism and functional consequence of ERK activation. *The Journal of Biological Chemistry, 278*, 6258–6267.

Tu, L. C., Yan, X., Hood, L., & Lin, B. (2007). Proteomics analysis of the interactome of N-myc downstream regulated gene 1 and its interactions with the androgen response program in prostate cancer cells. *Molecular & Cellular Proteomics, 6*, 575–588.

Vieira, A. V., Lamaze, C., & Schmid, S. L. (1996). Control of EGF receptor signaling by clathrin-mediated endocytosis. *Science, 274*, 2086–2089.

Waelter, S., Scherzinger, E., Hasenbank, R., Nordhoff, E., Lurz, R., Goehler, H., et al. (2001). The Huntingtin interacting protein HIP1 is a clathrin and alpha-adaptin-binding protein involved in receptor-mediated endocytosis. *Human Molecular Genetics, 10*, 1807–1817.

Wagner, K. U., Krempler, A., Qi, Y., Park, K., Henry, M. D., Triplett, A. A., et al. (2003). Tsg101 is essential for cell growth, proliferation, and cell survival of embryonic and adult tissues. *Molecular and Cellular Biology, 23*, 150–162.

Walker, J. K., Fong, A. M., Lawson, B. L., Savov, J. D., Patel, D. D., Schwartz, D. A., et al. (2003). Beta-arrestin-2 regulates the development of allergic asthma. *The Journal of Clinical Investigation, 112*, 566–574.

Wang, P., & DeFea, K. A. (2006). Protease-activated receptor-2 simultaneously directs beta-arrestin-1-dependent inhibition and Galphaq-dependent activation of phosphatidylinositol 3-kinase. *Biochemistry, 45*, 9374–9385.

Wang, W. Y., Gu, L., Liu, W. P., Li, G. D., Liu, H. J., & Ma, Z. G. (2011). ALK-positive extramedullary plasmacytoma with expression of the CLTC-ALK fusion transcript. *Pathology, Research and Practice, 207*, 587–591.

Wang, P., Kumar, P., Wang, C., & Defea, K. A. (2007). Differential regulation of class IA phosphoinositide 3-kinase catalytic subunits p110 alpha and beta by protease-activated receptor 2 and beta-arrestins. *The Biochemical Journal, 408*, 221–230.

Wang, Y., Pennock, S. D., Chen, X., Kazlauskas, A., & Wang, Z. (2004). Platelet-derived growth factor receptor-mediated signal transduction from endosomes. *The Journal of Biological Chemistry, 279*, 8038–8046.

Wang, Y., Pennock, S., Chen, X., & Wang, Z. (2002a). Internalization of inactive EGF receptor into endosomes and the subsequent activation of endosome-associated EGF receptors. Epidermal growth factor. *Science's STKE*, *2002*, pl17.

Wang, Y., Pennock, S., Chen, X., & Wang, Z. (2002b). Endosomal signaling of epidermal growth factor receptor stimulates signal transduction pathways leading to cell survival. *Molecular and Cellular Biology*, *22*, 7279–7290.

Williams, C. C., Allison, J. G., Vidal, G. A., Burow, M. E., Beckman, B. S., Marrero, L., et al. (2004). The ERBB4/HER4 receptor tyrosine kinase regulates gene expression by functioning as a STAT5A nuclear chaperone. *The Journal of Cell Biology*, *167*, 469–478.

Winter, V., & Hauser, M. T. (2006). Exploring the ESCRTing machinery in eukaryotes. *Trends in Plant Science*, *11*, 115–123.

Yu, T., Calvo, L., Anta, B., Lopez-Benito, S., Southon, E., Chao, M. V., et al. (2011). Regulation of trafficking of activated TrkA is critical for NGF-mediated functions. *Traffic*, *12*, 521–534.

Zhong, Q., Chen, C. F., Chen, Y., Chen, P. L., & Lee, W. H. (1997). Identification of cellular TSG101 protein in multiple human breast cancer cell lines. *Cancer Research*, *57*, 4225–4228.

Zhou, W., & Carpenter, G. (2000). Heregulin-dependent trafficking and cleavage of ErbB-4. *The Journal of Biological Chemistry*, *275*, 34737–34743.

Zhu, Y., Sullivan, L. L., Nair, S. S., Williams, C. C., Pandey, A. K., Marrero, L., et al. (2006). Coregulation of estrogen receptor by ERBB4/HER4 establishes a growth-promoting autocrine signal in breast tumor cells. *Cancer Research*, *66*, 7991–7998.

CHAPTER THIRTEEN

ROS-Containing Endosomal Compartments: Implications for Signaling

A. Paige Davis Volk, Jessica G. Moreland[1]
Division of Critical Care, Department of Pediatrics and the Inflammation Program,
The University of Iowa, Iowa City, Iowa, USA
[1]Corresponding author: e-mail address: jessica-moreland@uiowa.edu

Contents

Abstract

The endogenous generation of reactive oxygen species (ROS), previously perceived as a detrimental by-product of cellular processes, is now recognized as a critical component of intracellular signaling. Exploration of these biological signaling functions requires understanding the complex redox biochemistry and recognizing the compartment-specific elements of ROS generation. The endosomal compartment is increasingly recognized as a source for NADPH oxidase (NOX)-generated signaling ROS. Despite this

Methods in Enzymology, Volume 535
ISSN 0076-6879
http://dx.doi.org/10.1016/B978-0-12-397925-4.00013-4

growing understanding, there are significant limitations to the available detection and measurement systems for endogenous ROS. This chapter provides information about specific methodologies and redox-sensitive probes to guide the investigator and define the critical limitations for many of the available approaches. Although measurement continues to be challenging, the rapid growth and development of new detection systems suggests that our capacity to assign specific signaling roles to endosomal ROS will expand markedly in the next several years.

1. INTRODUCTION

ROS generated by the NOX in phagocytes are critical for innate immune defense against certain microbial pathogens. ROS for bacterial killing are generated into phagosomes, but there is accumulating evidence for ROS generation into endosomal compartments in both phagocytic and nonphagocytic cells. These ROS are generated in response to a wide variety of stimuli and function as intracellular signaling molecules. Regulation of phosphatase activity by oxidation of cysteines is the best-described ROS-based signaling modification (Janssen-Heininger et al., 2008), although a number of other mechanisms have been elucidated. In addition, ROS are implicated as metabolites initiating host cell damage under conditions leading to oxidative stress. Scientific investigation of the biological importance of these ROS signals requires sensitive and specific tools to allow spatial and quantitative analysis. However, by definition, these reactive species present a number of challenges to those who are investigating their effects. ROS have relatively short lifetimes and a number of antioxidants exist *in vivo* that may impair detection and measurement. These two facts suggest that signaling ROS are not likely to migrate a great distance from where they are generated. The spatial and temporal specificity of ROS generation is likely to provide highly relevant information about what are the physiological roles for those ROS.

As we select the best tools or probes to utilize in our investigations of endosomal ROS, there are several critical questions to be answered. What oxidant molecule(s) do we seek to measure? How stable is that molecule? For this chapter, we focus on ROS generated into endosomal or intracellular membrane-bound compartments; thus, the distribution of the probe is critical to what will be measured. Equally important questions include the following: What catalysts or cofactors are needed for the probe to detect the ROS under study? Which cellular antioxidants (enzymatic or non-enzymatic) will be in competition with the probe for reactivity with the

ROS? What is the stability of the reaction product and can intermediates formed by the probe generate ROS themselves? Importantly, the available technology for the design and generation of probes for various ROS is advancing rapidly. Thus, the potential to analyze specific ROS in living cell systems continues to improve.

2. OVERVIEW OF ENDOCYTOSIS

Cellular endocytic mechanisms range from simple ways to obtain nutrients and sample environmental conditions to highly complex systems of host defense. Moreover, endosomes provide an intracellular compartment for spatiotemporal control of highly reactive (and potentially damaging) signaling molecules, such as ROS (Lamb et al., 2012; Oakley, Abbott, Li, & Engelhardt, 2009). A summary of endocytic processes, with a focus on size and composition, follows.

2.1. Pinocytosis and macropinocytosis

Several diverse mechanisms meet the definition of "cell drinking" or the cellular internalization of solutes and extracellular fluid (ECF). Pinocytosis is a constitutive process in virtually every cell type and, though size varies, pinosomes are <150 nm in diameter. In contrast, macropinocytosis is an active process driven not by external forces but by actin-dependent localized membrane ruffling to form large intracellular vacuoles for the nonselective uptake of ECF and membrane (Mercer & Helenius, 2009). Under certain conditions, macropinocytosis may drive the uptake of viral and bacterial particles and may be useful for the uptake of bioprobes such as fluorescent proteins and nanoparticles (NPs), discussed in the succeeding text.

2.2. Clathrin-dependent and clathrin-independent endocytic pathways

In recent years, the complexity of endocytic pathways has become clear, as has the difficulty in investigating subclasses of endosomes. This difficulty is due in large part to the lack of specific endocytosis inhibitors (Doherty & McMahon, 2009). The best-studied endocytic pathway, clathrin-dependent endocytosis, occurs through a series of intermediate steps, from adaptor protein-dependent cargo recruitment and invagination of clathrin-coated pits to dynamin-dependent scission events releasing clathrin-coated vesicles (CCVs) from the plasma membrane. EM studies indicate that cargo-enriched CCVs may range in size from <100 to >300 nm in diameter.

There is evidence to support endocytosis of nearly every receptor family via the clathrin-dependent mechanism (Sorkin & von Zastrow, 2009). However, many cargo types also undergo endocytosis via diverse clathrin-independent pathways. Ultimately, endocytic cargo proceeds to lysosomal degradation or is recycled (Sorkin & von Zastrow, 2009). Maturing endolysosomes (250–500 nm in diameter) undergo significant composition changes, with drops in pH and activation of degradative enzymes. These changes may significantly affect the function of probes intended to study endosomal signaling events.

2.3. Phagocytosis

A mechanism restricted to professional phagocytes, phagocytosis proceeds through close interactions between cell surface receptors (especially Fcγ and complement receptors) and opsonins on the surface of invading microbes or foreign particles (Sansonetti, 2001). As a result, relatively little ECF is included in the composition of the phagosome (500 nm to >10 μm in diameter) compared to endosomes or macropinosomes. Neutrophil phagosomes mature through mobilization of intracellular granules (Nordenfelt & Tapper, 2011). In contrast, maturation of the macrophage phagolysosome is similar to that of endolysosomes (Allen, 2003).

3. SOURCES OF ENDOSOMAL ROS

The primary sources of endosomal ROS are members of the NOX family. Assembly of phagocyte NOX2 was initially described to occur only on the plasma or phagosomal membranes, but accumulating evidence supports assembly and activation of NOX2 on endosomes within neutrophils (Lamb et al., 2012). Moreover, there is extensive evidence supporting endosomal assembly of a number of other NOX isoforms (Lamb, Moreland, & Miller, 2009; Ushio-Fukai, 2009). Specifically, both cytokine and other ligand-receptor interactions trigger endosomal ROS generation (Miller, Filali, et al., 2007; Oakley et al., 2009). For some NOX isoforms/cell types, the complex is constitutively assembled and active on membranes within unstimulated cells. In other cell types, full assembly of the functional NOX complex on the plasma or intracellular membranes occurs in response to cell stimulation (Ushio-Fukai, 2009).

Under all of the circumstances described in the preceding text, NADPH binding occurs on the cytoplasmic face of the endosome with electron

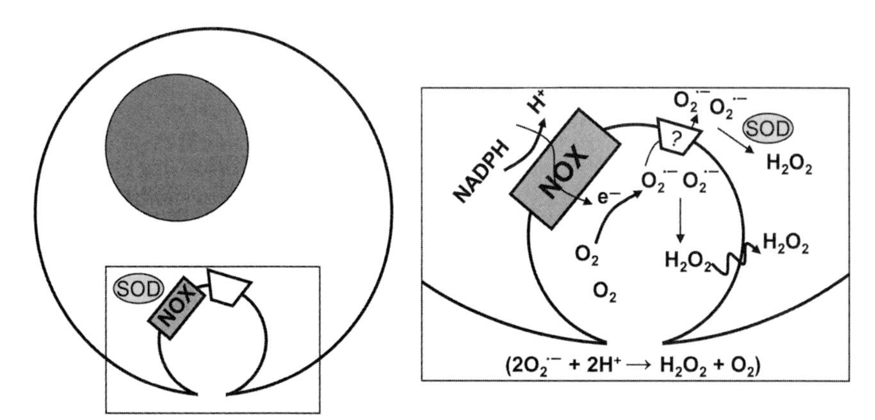

Figure 13.1 Endosomal topology and ROS generation. Model of ROS-signaling endosome with NOX-dependent superoxide generation into the endosomal compartment. Spontaneous dismutation of superoxide occurs within the endosome, with H_2O_2 freely diffusing to the cytosol. Alternatively, superoxide may traverse a putative anion channel (Oakley et al. 2009) with subsequent SOD-catalyzed dismutation to H_2O_2 in the cytosolic space.

passage from the cytoplasm to the interior of the endosome and thus intra-endosomal generation of superoxide. This topology is consistent regardless of the NOX enzyme complex involved, cell type under study, or agonist that elicited production (Fig. 13.1). The subsequent conversion of superoxide to H_2O_2, via either spontaneous dismutation or catalysis by SOD, and the complex redox chemistry that ensues are likely to be cell type- and stimulus-specific. Thus, in order to assign a specific signaling role to a distinct reactive molecule, measurement of endosomal ROS must be approached with a clear understanding of each of the probes.

4. MEASUREMENT OF ENDOSOMAL ROS

Investigators seeking to measure endosomal ROS should first give careful consideration to their goals. The specific probe and methodological approach selected depends on whether quantitative and/or kinetic information about ROS generation is desired or necessary to answer the question being studied. Information regarding spatial specificity of the ROS signal may be preferred. General guidelines regarding methodology for consideration prior to selection of a specific probe follow. See Table 13.1 for a summary of probe characteristics.

Table 13.1 Characteristics of ROS detecting probes

Probe	ROS detected	Delivery/cell distribution	Pros	Cons	Ex/Em
DCFH	Nonspecific	Membrane permeant	Easy to use	Prone to auto-oxidation and oxidized by cytochrome *c*	492–495/517–527
DHR	Nonspecific	Membrane permeant	Easy to use	Lipophilic: Sticks to membranes and very reactive with $ONOO^-$	500/536
HE	Specific for $O_2^{\cdot -}$	Membrane permeant	Specific for single ROS	Oxidized by cytochrome *c* and DNA intercalation → fluorescence	518/606
OxyBURST-BSA	Nonspecific	Membrane impermeant/ endocytosed	Signal can survive fixation and useful for flow or microscopy	Photosensitivity	488–490/520–530
Amplex Red	Specific for H_2O_2	Membrane impermeant/ endocytosed	Sensitive, red fluorescence, and quantitative for H_2O_2	pH sensitive below 7.0 and requires peroxidase	571/585
Boronate Probes	Specific for H_2O_2	Probe-dependent	High specificity	Limited fluorescent output	Variable
CellROX	Nonspecific	Membrane permeant	Near IR signal and survives fixation		640/655
GFPs/YFPs	Specific for H_2O_2	Transfection	Reversible and can be organelle specific	Requires transfection	Variable
HyPer (cpYFP)	Detects H_2O_2, may detect $O_2^{\cdot -}$	Targeted transfection/ cytoplasm or organelle specific	Ratiometric detection	Very sensitive to pH below 7.0 and requires transfection	420 and 500/516

Lucigenin	Nonspecific	Membrane permeant	Easy to use	Auto-oxidation at ↑ concentration and may generate $O_2^{\cdot -}$	
Luminol	Nonspecific	Membrane permeant	Minimal background signal and not cumulative ROS measurement	Requires peroxidase, weak signal, and no optical microscopy	
Isoluminol	Nonspecific	Membrane impermeant/ endocytosed?	Minimal background signal and not cumulative ROS measurement	Requires peroxidase, weak signal, and no optical microscopy	
Cytochrome *c*	Specific for $O_2^{\cdot -}$	Membrane impermeant/no uptake by endocytosis	Calibration to quantify $O_2^{\cdot -}$ and quantitative	Limited sensitivity	
NBT	Specific for $O_2^{\cdot -}$	Membrane impermeant/ endocytosed	Easy to use and stable colorimetric product	Limited quantitation	
Ormosil DCFH	Specific for H_2O_2	Membrane permeant	Specific for single ROS	Emerging method and limited validation	505/523

4.1. ROS measurement in intact cells

Investigation of endosomal ROS generation is most often undertaken in intact cells. Three distinct methodologies are routinely used: microplate measurement using luminometer, spectrophotometer, or fluorometer; assessment by flow cytometry; and direct observation by microscopy. Use of flow cytometry for analysis is possible with a number of fluorescent probes and provides some degree of quantitative information about intracellular ROS only. Depending on the distribution of the probe selected, this method may or may not provide specific information about endosomal generation/spatial distribution, as many of the probes are membrane-permeable and freely diffusible in the cell. Concurrently, it is possible to obtain kinetic and semiquantitative information about the process of endocytosis in the cell of interest and use this as complementary information to the ROS assessment.

4.1.1 Measurement of endocytosis of fluorescent dextran by flow cytometry

Cells are suspended in phenol-free media of choice at 37 °C in the presence of Texas Red dextran (MW: 10,000) (125–250 μg/ml). At predetermined time points, aliquots are removed, placed on ice, and then washed and centrifuged twice to remove nonendocytosed dextran. After washing and resuspension in buffer, each aliquot must be kept on ice and in the dark until analysis by flow cytometry. The dynamin-II inhibitor Dynasore (100 μ*M*) can be used to demonstrate inhibition of dynamin-dependent endocytosis using this assay. We routinely perform simultaneous evaluation of endosomal ROS generation using fluorescent probes (Lamb et al., 2012).

4.1.2 Microplate analysis of ROS generation

Microplate assays can similarly be used for intact cell ROS generation assays. Depending on the probe selected, both intracellular and extracellular ROS may be assayed. The primary advantage of this approach over flow cytometry is ease of continuous kinetic readout. For filter set-based (often multipurpose) microplate readers, potential excitation and emission will be limited by the filter sets available. The Gemini Fluorescence Microplate Reader (Molecular Devices, CA, United States) is a fluorescence-only microplate reader that allows for variable wavelength selection (in 1 nm increments) through the use of dual monochromators, thus allowing optimization of signal for each fluorescence probe. Although, for many probes, filter-based technology is adequate, some of the ratiometric probe choices will deliver much enhanced ratio-based quantitation with dual monochromator technology.

4.1.3 Confocal microscopy

Confocal assessment provides the greatest opportunity to obtain spatially specific information about ROS generation into endosomes. Used in combination with fluorescent dextran loading of endosomes, this is a powerful tool. These analyses can be performed on living cells or postfixation. Specific fixation methods must be optimized for each probe tested as there is frequent interference with the intensity of the probe. In addition, based on photo-oxidation of many probes, microscopic analysis must be performed expeditiously.

4.2. ROS measurement in isolated endosomes

Although most investigations will focus on analysis of endosomal ROS in the intact cell, ROS measurement in isolated endosomes can be performed in a highly quantitative fashion. We isolate the neutrophil light membrane fraction from granule fractions by nitrogen cavitation followed by three-layer Percoll density gradient centrifugation. The light membrane fraction is then subjected to high-voltage free-flow electrophoresis to separate endosomes from the plasma membrane (Sengelov, Nielsen, & Borregaard, 1992). Endosomes can also be isolated from nonphagocytic cells using gradient separation of postnuclear supernatants (Oakley et al., 2009). Following endosome isolation, cell-free superoxide generation is measured after addition of NADPH using colorimetric (cytochrome *c*) or luminescence methods. The relative superoxide-generating capacity of endosomes isolated from cells under unstimulated versus agonist-stimulated conditions provides quantitative information (Lamb et al., 2012).

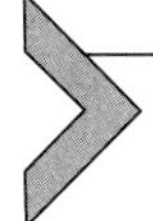

5. FLUORESCENT PROBES

5.1. General considerations

Fluorescent probes can be very sophisticated ROS sensors based on spatial resolution and high sensitivity, but the diffusion properties and specificity of each probe varies significantly. A common property of many of the probes currently in use is that they are stable (and nonfluorescent) in their reduced form but subject to oxidation by one or more ROS, resulting in a change in bonding pattern and subsequent fluorescence. Many probes are sensitive to pH (particularly DCFH and OxyBURST), such that fluorescence decreases with acidification of the endosome. Interestingly, the two most commonly employed subgroups of fluorescent probes (fluorescein- and rhodamine-based dyes) do not react directly with either superoxide or H_2O_2 (Kalyanaraman et al., 2012). A greater understanding of the limitations of this redox chemistry has led to the more recent development of new technology including small molecule sensors and protein-based redox probes. These emerging technologies allow for not only greater specificity and localization of specific ROS signaling molecules but also much greater potential to study live cells as they react to specific stimuli.

5.2. Fluorescent dyes

5.2.1 General experimental setup

For use of the fluorescent dyes described in the succeeding text, the same general experimental protocol can be applied. Cells are incubated with a stimulus, if desired, and with the probe at the concentrations described later. During stimulation, cell/probe mixtures should be protected from light as many probes are subject to photo-oxidation. For microscopy experiments, excess probe will need to be washed extensively from the media overlying cells prior to fixation and mounting. For flow cytometry, the need for washing is probe-specific and may diminish signal intensity.

5.2.2 DCFH

DCFH (2,7-dichlorodihydrofluorescein) is the most widely used fluorescent probe for the measurement of intracellular ROS, but is not specific for detection of endosomal ROS. For use in cellular assays, the diacetate form (DCFH-DA) is freely diffusible across cell membranes and subsequently hydrolyzed by intracellular esterases to the polar and nonfluorescent DCFH form that is retained intracellularly. Free diffusion limits the potential for this probe to measure endosomal ROS specifically; however, the chloromethyl derivative (CM-H_2DCFCA) has improved cell retention and may be a better choice. In addition, this probe is minimally reactive with superoxide and does not react directly with H_2O_2, although DCFH can detect H_2O_2 in the presence of cellular peroxidases. DCFH is not specific for the detection of H_2O_2 but also is oxidized by other ROS including $HO^{\cdot}$ and $ROO^{\cdot}$. The oxidized form is quite prone to photo-oxidation and thus light exposure has to be very carefully managed, especially in microscopy experiments. An additional caveat in the use of this probe is the recognition that the DCF intermediate radical reacts independently with O_2 thus forming additional $O_2^{\cdot-}$, an artifactual amplification of the ROS signal (Wardman, 2007).

DCFH-DA should be loaded into the cells at a concentration of 2–5 μ*M* (final), with optimization of concentration for particular cell type/condition. High concentrations of the probe must be used to compete with endogenous antioxidants, and even under optimal circumstances, assays are likely performed under nonsaturating conditions of the probe. The implication of use under nonsaturating conditions is that quantitative analysis is not possible, and interexperiment variability may be very high. Generation of fluorescent product can be measured by microplate reader or followed by microscopy, with $\lambda_{\text{excitation}}$ 498 nm and $\lambda_{\text{emission}}$ 522 nm. Based on the described limitations, DCFH can be used as a general measure of cellular redox state, but cannot provide specific or quantitative analysis of endosomal superoxide or H_2O_2 generation.

5.2.3 DHR

DHR123 is also a nonfluorescent molecule that diffuses across cell membranes based on its lipophilicity. The molecule is oxidized to the fluorescent rhodamine123 ($\lambda_{\text{excitation}}$ 505 nm and $\lambda_{\text{emission}}$ 529 nm) and is trapped within cells in this form. Like DCFH, DHR is not oxidized by superoxide or H_2O_2 alone, but can detect H_2O_2 in the presence of endogenous peroxidases. DHR has low specificity for H_2O_2 and readily detects the reactive nitrogen species (RNS), $ONOO^-$. Although discussion of RNS is beyond the scope of this chapter, it is critical to recognize that many of these probes lack specificity for ROS over RNS. This probe is also reactive toward chloramines, with taurine chloramine as the most likely product based on the availability of taurine in the cytoplasm (Dypbukt et al., 2005). DHR123 is routinely used to measure intracellular ROS generation by flow cytometry and has been used as a diagnostic test for chronic granulomatous disease (Roesler & Emmendorffer, 1991). Concerns regarding the optimal concentration to generate adequate intracellular levels are similar to those with DCFH, although, once loaded, leakage from the cell is less of a concern. Also similar to the DCFH radical, the DHR radical can react directly with oxygen, generating a falsely elevated signal. Cells are loaded with DHR123 in the dark at a concentration range of 50–150 m*M*. Saturating conditions for the probe should also be tested.

5.2.4 HE

Cell-permeable dihydroethidium (HE) is the most specific of the fluorescent probes for the detection of superoxide. Due to the ability of this probe to cross all membranes, the spatial specificity for endosomal ROS measurement is very low. Until quite recently, the assumption in using this probe was that ethidium (E^+) was generated in the reaction between HE and superoxide, which subsequently bound to DNA creating red fluorescence. However, it has been elegantly demonstrated that the fluorescent reaction product of HE and superoxide is 2-hydroxyethidium (2-OH-E^+) ($\lambda_{excitation}$ 520 nm and $\lambda_{emission}$ 567 nm) (Wardman, 2007; Zhao et al., 2003). This distinction is relevant as the fluorescence properties of E^+ ($\lambda_{excitation}$ 520 nm and $\lambda_{emission}$ 610 nm) are different from 2-hydroxyethidium. Importantly, a number of investigators report that 2-hydroxyethidium cannot be accurately measured using standard fluorescence approaches, either microplate or microscopy. These authors contend that HPLC analysis is the only appropriate assay for quantitative superoxide generation (Zielonka & Kalyanaraman, 2010). Other considerations in the use of HE include the ability of cytochrome *c* (or other mitochondrial cytochromes) to oxidize the probe, thus making it difficult to distinguish endosomal superoxide generation from other ROS sources. This is of particular concern in the setting of mitochondrial cytochrome *c* release, especially in cell populations that may have ongoing apoptosis. Starting probe concentrations for HE range from 2 to 5 μ*M* (stock at 10 m*M*).

5.2.5 OxyBURST

The earlier-mentioned fluorescent probes share the property of relatively free diffusion, although the chemistry differs among them. Conjugation of fluorescent probes to particles for measurement of intraphagosomal ROS in neutrophils has been possible for some time, with the more recent addition of bovine serum albumin (BSA) conjugates. OxyBURST Green H_2HFF BSA, for example, although initially marketed to measure extracellular ROS, is taken up into endosomes and retained there. The probe is used at a stock concentration of 1 mg/ml and diluted to 100 μg/ml (final). The stock concentration can be used for up to 4 weeks if refrigerated and protected from light. Cells are incubated with the probe during agonist stimulation in suspension or under adherent conditions for microscopy. Measurement of the fluorescent product ($\lambda_{excitation}$ 488 nm and $\lambda_{emission}$ 530 nm) can be either accomplished using flow cytometry or directly observed by confocal microscopy (Fig. 13.2; Lamb et al., 2012). OxyBURST detection of endosomal ROS can be used in combination with quantitative analysis of endocytosis using Texas Red dextran. By flow cytometry, simultaneous kinetic measurement of relative rates of endocytosis and endosomal ROS generation can be obtained. Although limitations include significant

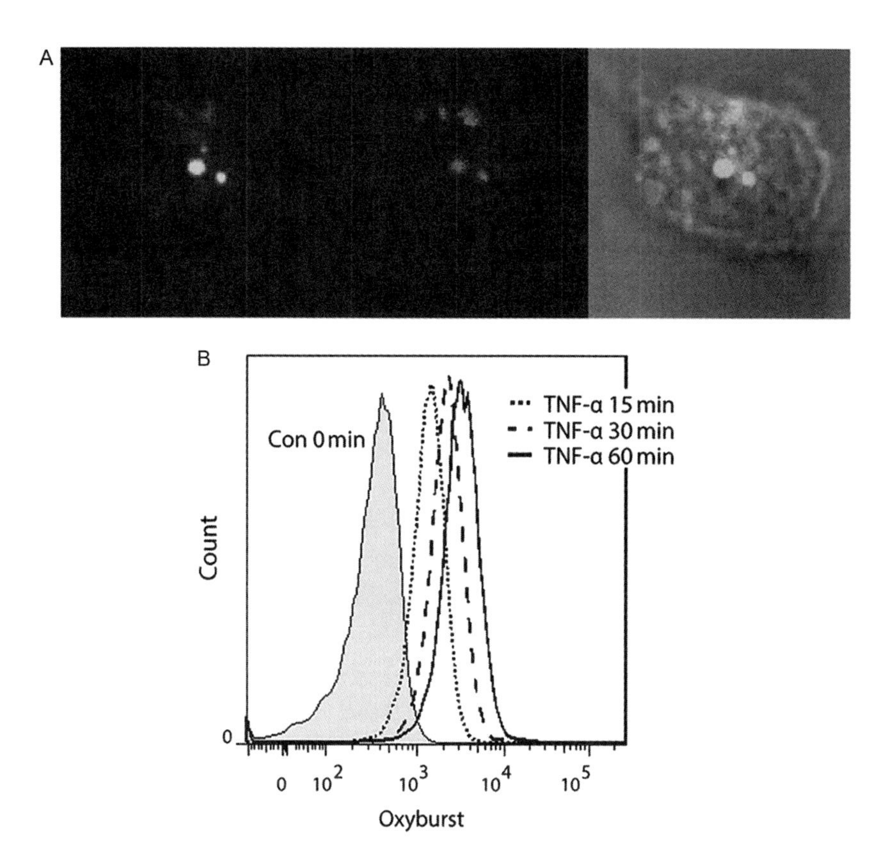

Figure 13.2 Endosomal ROS evaluated by confocal microscopy and flow cytometry. (A) Representative confocal image of an endotoxin-stimulated neutrophil. OxyBURST positive, ROS-containing vesicles are seen in green (left panel), which colocalize with TR-dextran containing vesicles in red (middle), and merged image (right). (B) Representative histogram of intracellular ROS measured by flow cytometry using the OxyBURST fluorescent probe. TNF-α (1 ng/ml) elicited a time-dependent increase in intracellular ROS as compared to control. *Panel A originally published in* The Journal of Biological Chemistry *(Lamb et al., 2012)*. (For interpretation of the references to color in this figure legend, the reader is referred to the online version of this chapter.)

photosensitivity and unclear longevity of the signal, this probe provides the greatest spatial specificity for detection of endosomal ROS.

5.2.6 Amplex Red

N-Acetyl-3,7-dihydroxyphenoxazine is a nonfluorescent molecule that reacts with H_2O_2 in the presence of HRP to generate the highly fluorescent molecule resorufin ($\lambda_{excitation}$ 568 nm and $\lambda_{emission}$ 581 nm). The probe can detect 50 n*M* H_2O_2 and has predictable (linear) stoichiometry with H_2O_2

within the biological range measured. Amplex Red does not cross cell membranes and has minimal background fluorescence, whereas the oxidation product resorufin is intensely fluorescent and quite stable (Freitas, Lima, & Fernandes, 2009). This probe has primarily been utilized for the detection of extracellular H_2O_2 levels, but uptake of Amplex Red into endosomes is possible, depending on the mode of endocytosis. Thus, this probe can measure H_2O_2 in the endosome, although the free diffusion of H_2O_2 will limit the concentration of this ROS in an endosomal location. Although the fluorescence of Amplex Red-generated resorufin was noted to be highly pH-sensitive below pH of 7, Amplex UltraRed has markedly reduced pH sensitivity and greater ROS sensitivity. Amplex UltraRed is reconstituted in high-quality anhydrous DMSO with minimal exposure to light and air and used at final concentration of 50 μM. An excess of horseradish peroxidase should be available in the range of 0.1 U/ml final concentration. To utilize this probe in a quantitative fashion, a standard curve of fluorescence can be generated by adding H_2O_2 at known concentrations to duplicate wells without any cells, with Amplex Red and HRP as previously mentioned. Cell-permeant analogs of Amplex Red are less useful than the previously described fluorescent probes based on direct oxidation of resorufin by other molecules.

5.2.7 Boronate Probes

This new generation of probes is undergoing rapid development with the potential for detection of a number of distinct ROS, but initial probes have focused specifically on detection of H_2O_2. First-generation boronate probes, including Peroxyfluor 1, involve boronate-masked fluorescein. The probe has no fluorescent properties in the absence of ROS but, upon reaction with H_2O_2, is transformed into a phenol with 1000-fold increase in fluorescence. Subsequent probes have been generated on blue xanthone and red resorufin scaffolds (PX1 and PR1) to emit in different fluorescence ranges. They are reported to detect H_2O_2 in the high μM range but in practice have sensitivity too low for detection of endogenous H_2O_2 production (Lippert, Van de Bittner, & Chang, 2011). Building on this concept, a second generation of boronate probes has been constructed with enhanced sensitivity, including Peroxy Green 1 and Peroxy Crimson 1. These probes are highly specific for H_2O_2 detection and have shown the capacity to detect this species in living cells stimulated with growth factors (Miller, Tulyathan, Isacoff, & Chang, 2007). Dual-wavelength boronate probes are now reported that may provide ratiometric peroxide imaging. Ratio Peroxyfluor 1 operates by fluorescence resonance energy transfer on reaction with H_2O_2 (Lippert et al., 2011) and has been used to visualize H_2O_2 in macrophages.

5.2.8 CellROX

A newer fluorescent reagent, CellROX, is freely diffusible and non-fluorescent in the reduced state. Following oxidation, a near-infrared fluorescence is generated ($\lambda_{excitation}$ 640 nm and $\lambda_{emission}$ 655 nm) that is easily combined with other fluorescent probes without concern for overlap based on the peaks. This signal is reported to be retained following formaldehyde fixation, enhancing its potential for microscopy studies (Eshraghi et al., 2013). The fluorescence can also be utilized in flow cytometry-based assays and for microplate fluorometry. Based on its membrane-permeant properties, CellROX cannot be utilized specifically for endosomal ROS.

5.3. Redox-sensitive fluorescent proteins

5.3.1 GFPs/YFPs

Redox-sensitive fluorescent proteins are genetically encoded biosensors that have been developed to overcome the lack of specificity of the fluorescent dyes for individual ROS. The first probes were genetically modified fluorescent proteins with artificial disulfide bonds placed on the scaffold, including redox-sensitive YFP (rxYFP) and redox-sensitive GFP (roGFP) (Meyer & Dick, 2010). rxYFP ($\lambda_{excitation}$ 512 nm and $\lambda_{emission}$ 527 nm) is engineered with cysteine pairs in the β-strands such that disulfide bonds can be generated in the β-barrel by oxidation. Formation of the disulfide bond results in slightly greater than twofold change in fluorescence, but analysis of the fluorescence is complicated by autofluorescence. roGFPs were then generated to overcome some of these limitations. Similarly, cysteine residues were engineered into the β-strands; however, ratiometric sensing was possible. These "first-generation" probes roGFP1 and roGFP2 display an increased dynamic range compared with rxYFP when utilized as ratiometric sensors. For example, using roGFP2, two distinct excitation points of 390 and 480 nm can be utilized if using a plate reader with the capability to excite simultaneously at both wavelengths ($\lambda_{emission}$ 510 nm). Using both specific excitation points, a dynamic range of 12 is possible. It is important to note that the dynamic range is much more limited under conditions of "off-peak" excitation as might be present when using confocal microscopy. Although the majority of initial studies utilized exogenously added oxidants, the roGFPs have now been used to measure endogenously generated ROS in kinetic assays (Meyer & Dick, 2010).

5.3.2 HyPer

HyPer, as the name indicates, was generated as a specific sensor for H_2O_2 by inserting circularly permuted yellow fluorescent protein (cpYFP) into the prokaryotic regulatory domain of an H_2O_2-sensing protein, OxyR. This probe has a single emission maximum of 516 nm but with two distinct excitation points, 420 and 500 nm. Generation of the disulfide bond by oxidation elicits a shift in the relative intensity of the two peaks with decreased intensity at 420 nm and increased at 500 nm. These probes display a relatively low dynamic range (three to fourfold at best *in vitro*) but can still be detected using flow cytometry with optimization of technical parameters. The most significant disadvantage of this and some of the other fluorescent proteins is the strong sensitivity to pH, making use in the fluctuating environment of the endosome difficult (Belousov et al., 2006). Using various targeting sequences, HyPer has been specifically targeted to subcellular compartments (Meyer & Dick, 2010). For the measurement of endosomal ROS, HyPer targeted to the cytosol would likely be the most sensitive, measuring H_2O_2 as it diffuses locally across endosomal membranes. In addition, cpYFP was reported to respond to superoxide directly using a protein expressed in the mitochondria (Wang et al., 2008). Although the chemistry remains unclear and more extensive analyses are required, the potential for direct measurement of NOX-generated superoxide is exciting.

General protocol for redox-sensitive fluorescent proteins includes standard transfection reagents and techniques according to manufacturer's instruction for specific cell type. At 24 h posttransfection, cells are washed and placed in phenol-free media specific to cell type. Cells are treated with stimulus if desired prior to measuring output fluorescence. Positive controls for maximal reduction of HyPer are additional wells/tubes treated with 1 mM DTT for 5–15 min or with direct addition of H_2O_2 (50–100 μM). Cells can be visualized by confocal microscopy or studied by ratiometric detection using the dual excitation peaks described earlier in a microplate reader.

6. CHEMILUMINESCENT PROBES

6.1. General considerations

Chemiluminescent assays have been used for decades to study ROS generated by professional phagocytes (Dahlgren & Karlsson, 1999) and more recently for detection of superoxide in vascular biology (Miller & Griendling, 2002). Despite their high sensitivity and ubiquitous presence in the field of NOX research, complicated redox chemistry underlies the release of detectable photons from luminol and lucigenin intermediates and these interactions are not specific to select ROS. To react with ROS, luminol must undergo univalent oxidation (a step that requires catalysis by a peroxidase) while lucigenin must be univalently reduced. For an excellent review of this redox chemistry, see Faulkner and Fridovich (1993). Whenever possible, complimentary assays are advised in conjunction with the use of chemiluminescent probes. All chemiluminescent probes are affected by changes in pH with decreased luminescence at lower pH (Miller & Griendling, 2002; Oosthuizen, Engelbrecht, Lambrechts, Greyling, & Levy, 1997). Chemiluminescent assays are performed in a cuvette-based spectrophotometer or plate-based luminometer with cell-free background luminescence subtracted.

6.1.1 Lucigenin

Lucigenin (bis-*N*-methylacridinium nitrate) is perhaps the most commonly used chemiluminescent probe for the detection of superoxide in the cells and tissue. It is important to note that lucigenin is subject to redox cycling in the presence of endogenous cellular reductases, which can result in direct superoxide generation by the probe itself (Liochev & Fridovich, 1997). Despite its common use, there is controversy in the field of NOX biology over whether lucigenin is an appropriate probe to detect ROS generation and caution is advised. However, validation studies using a wide range of techniques for superoxide generation show that auto-oxidation does not occur at lucigenin concentrations of 1–5 μ*M* (Li et al., 1998). For most cell types, a 5 μ*M* concentration is adequate for superoxide detection and higher concentrations should not be used (Miller & Griendling, 2002). However, for neutrophil assays, where ROS generation is at its highest, we have used higher concentrations of lucigenin to ensure that the probe is not limiting. Lucigenin is made fresh from powder and is maintained in the dark.

6.1.2 Isoluminol/luminol/L-012

Freely membrane diffusible use of luminol (5-amino-2,3-dihydro-1,4-phthalazinedione) to study intravesicular ROS is problematic because its activation is peroxidase-dependent. Isoluminol differs from luminol only in the location of an amino group on its phthalate ring, which makes isoluminol membrane-impermeant, but does not change its chemiluminescent properties compared to luminol (Dahlgren & Karlsson, 1999). Isoluminol in the presence of a peroxidase (typically HRP) could be endocytosed, but at such low levels, the sensitivity of the compound is unlikely to detect signaling concentrations of ROS. Also of note, luminol reacts with H_2O_2, not superoxide, so detection of endosomally generated H_2O_2 would occur immediately outside the endosomal membrane as H_2O_2 freely diffuses (Faulkner & Fridovich, 1993). Isoluminol and luminol are both used in final concentrations of 50–100 μM (with 4–20 U/ml HRP added to isoluminol assays) (Dahlgren & Karlsson, 1999). Recently, the probe L-012, a modification of luminol, has been touted as being specific for superoxide (Nishinaka et al., 1993). Other reports indicate that L-012 also detects peroxynitrite and likely other ROS (Daiber et al., 2004). Its sensitivity and signal/noise ratio are reported to be greater than lucigenin, luminol, or MCLA, but beyond this, L-012 may not provide any clear advantage to other chemiluminescent probes (Dikalov, Griendling, & Harrison, 2007; Nishinaka et al., 1993). No issues with redox cycling have been reported for L-012, but comparison to the structure of luminol suggests it has the potential to do so.

6.1.3 Coelentrazine/MCLA

Two derivatives from bioluminescent marine organisms in the genus *Cypridina*, coelenterazine and MCLA, have been developed recently for the study of ROS. Of these, MCLA has a better signal/noise ratio and emits >100-fold more light than coelenterazine under similar conditions. MCLA not only has been employed to detect superoxide production in an array of vascular tissues but also has been shown to measure hydroxyl radical and other ROS (Oosthuizen & Greyling, 2001). Coelenterazine and MCLA are promising probes for detection of superoxide, but their use has been limited and not yet validated under a variety of conditions (Dikalov et al., 2007).

6.2. Colorimetric probes

6.2.1 Cytochrome c

Among optical methods for ROS detection, the SOD-inhibitable reduction of ferricytochrome *c* is still widely used to detect superoxide, primarily in the investigation of neutrophils. The electron acceptor cytochrome *c* is one of the very few highly specific probes, detecting only superoxide. Cytochrome *c* is not cell-permeable and thus detects primarily extracellular superoxide. The caveat to this statement is that cytochrome *c* can be taken into the phagosome to some extent and thus detects intraphagosomal ROS. For the purposes of this chapter, this leads to the question of whether cytochrome *c* can be taken up into the endosome and measure endosomal ROS. Our experience using TNF-α stimulation of human neutrophils would suggest that this does not occur as there is significant generation of endosomal ROS by flow cytometry methods, but no change in cytochrome *c* signal in response to stimulation with increasing concentrations of TNF-α. Superoxide generation is measured with a spectrophotometer as a change in absorbance measured at OD 550 nm. Despite the potential to quantify $O_2^{\cdot -}$ production using this method, the sensitivity for low levels of superoxide is poor. In addition, cytochrome *c* can be reoxidized, thus leading to reductions in absorbance and artificially low levels of detection.

Absorbance can be measured as an endpoint reading or as continuous kinetic assay. The latter is preferable to perform duplicate samples with one of each pair of samples pretreated with SOD 50 μg/ml (150 U/ml), cytochrome *c* is then added at a final concentration of 50–100 μ*M*, and the cells are stimulated. If there is only low-level superoxide generation, assay can be measured continuously for 60 min. Superoxide is quantified using the following equation:

$$\frac{\Delta \mathrm{O.D.} - \Delta \mathrm{O.D.\ for\ SOD\ containing\ wells}}{21.1\ \mathrm{m}M^{-1}\ \mathrm{cm}^{-1}(\mathrm{path\ length})} \times 1000 = \mu M\, \mathrm{O_2}^{\cdot -}$$

$$= \mathrm{nmol\ O_2}^{\cdot -}/\mathrm{ml}$$

21.1 is the extinction coefficient for ferricytochrome *c*. The path length = 1 cm for a 1 ml cuvette and should be 0.3 cm for a 100 μl sample or 0.6 cm for a 200 μl sample.

6.2.2 Nitroblue tetrazolium

The tetrazolium salts have been used extensively for the detection of superoxide in phagocytes. Nitroblue tetrazolium (NBT) is pale yellow in color prior to reduction by superoxide. The tetrazoinyl radical is generated by interaction with superoxide and disruption of the tetrazole ring, which subsequently dismutates to generate the bright blue, stable formazan product. Although primarily used to date for phagocytic superoxide generation, NBT is taken up into endosomes and detects superoxide in that compartment. The final concentration for NBT should be in the range of 0.025–0.5 mg/ml. The reduction product can be measured by spectrophotometer or by microscopy but is likely to be more useful for providing spatial and qualitative superoxide detection rather than quantitative analysis.

7. EMERGING ROS PROBES

7.1. General considerations

An emerging technology with a great deal of promise for the study of biological processes, including measurement of endocytic ROS, NP probes are also fraught with difficulties for the researcher to consider. NP size, shape, concentration, uptake kinetics, temperature, and surface charge all affect the interaction between NPs and a cell. In addition, different cell types handle NP uptake differently, so NP methods established in one cell type are not necessarily reproducible in other cells. Of particular importance for the study of endosomal ROS, many NPs have been shown to induce cytotoxicity and NOX-dependent intracellular ROS generation, which may confound the use of a given probe. For further reviews on NP technology, see Iversen, Skotland, and Sandvig (2011), Lee and Kopelman (2012), and Zhang et al. (2012).

7.1.1 NP-based fluorescent sensors (aka PEBBLEs)

ROS sensing "photonic explorers for biomedical use with biologically localized embedding" (PEBBLEs) have been designed to detect specific ROS molecules, especially H_2O_2. Perhaps the most specific NP for H_2O_2 detection, the hydrophobic ormosil-2,7-dichlorofluorescin NP's "*or*ganically *mo*dified *sil*icate" matrix provides time, size, and hydrophobic energy barriers that prevent interaction of ROS other than H_2O_2 with the probe (Kim, Lee, Xu, Philbert, & Kopelman, 2010). The DCFDA dye is maintained in its lipophilic/inactive form until interaction with H_2O_2 leads to fluorescence detectable by standard fluorospectrophotometer techniques. The ormosil-2,7-dichlorofluorescin NP is used at a concentration of 0.1 mg/ml and detects H_2O_2 in a concentration range of 10–100 μM (Kim et al., 2010). Specific NP fluorescent probes for detection of the hydroxyl radical and singlet oxygen are also available. A silica-based NP, with covalently linked 5(6)-carboxyfluorescein diacetate (Hammond et al., 2008), detects most ROS, including superoxide anion, H_2O_2, hydroxyl radical, and some RNS, all in the range of 1–30 nM. Of note, the fluorescent NPs are subject to the same pH-dependence associated with their fluorescein derivatives, with optimal fluorescence at physiological pH.

7.1.2 Other NP probes

Polymer-coated quantum dots (QDots) have the advantage of inherent fluorescence with narrow emission spectra, photostability, and exceptional brightness. Depending on the application, QDots may be an appropriate technology for study of endosomal ROS, but concentration and polymer coating plays a crucial role in the tendency of the QDot to cause cytotoxicity and/or generation of ROS. Nanodiamonds (NDs) have exciting potential as intracellular probes due to their greater cell biocompatibility, their synthesis advantages (high purity/high yield/autoclave-friendly sterilizing properties), stable fluorescence, and lack of induction of ROS generation (Schrand, Lin, Hens, & Hussain, 2011). To our knowledge, no ND has been generated for specific detection of intracellular ROS. A multiparametric protocol should be employed to test the effects of all NP probes on a researcher's cell type of interest, with particular attention paid to cell viability, NP location and concentration (as many NPs tend to form cytoplasmic aggregates when released from the endolysosomal pathway), NP-induced ROS generation, and cell morphology and functionality (Soenen, Demeester, De Smedt, & Braeckmans, 2012).

8. SUMMARY

The study of signaling ROS requires highly sensitive and specific tools to permit accurate spatial, kinetic, and quantitative analysis. As we have outlined, endosomal ROS present a number of challenges to investigators. Although techniques continue to improve for the study of intracellular ROS, there are still significant limitations to endosomal ROS detection (Kalyanaraman et al., 2012). At present, it remains difficult to achieve good spatial resolution and it is very hard to measure superoxide alone. Therefore, in general, we recommend the use of more than one approach to verify the species and location of the signaling ROS under study. Herein, we have outlined multiple considerations and limitations investigators should ponder in the measurement of endosomal ROS. We look forward to the continued development of probes that will optimize the detection of the most challenging/evanescent signaling ROS.

ACKNOWLEDGMENT

The authors would like to thank Dr. Francis Miller for his critical review of the manuscript.

REFERENCES

Allen, L. A. (2003). Mechanisms of pathogenesis: Evasion of killing by polymorphonuclear leukocytes. *Microbes and Infection/Institut Pasteur, 5*, 1329–1335.

Belousov, V. V., Fradkov, A. F., Lukyanov, K. A., Staroverov, D. B., Shakhbazov, K. S., Terskikh, A. V., et al. (2006). Genetically encoded fluorescent indicator for intracellular hydrogen peroxide. *Nature Methods, 3*, 281–286.

Dahlgren, C., & Karlsson, A. (1999). Respiratory burst in human neutrophils. *Journal of Immunological Methods, 232*, 3–14.

Daiber, A., Oelze, M., August, M., Wendt, M., Sydow, K., Wieboldt, H., et al. (2004). Detection of superoxide and peroxynitrite in model systems and mitochondria by the luminol analogue L-012. *Free Radical Research, 38*, 259–269.

Dikalov, S., Griendling, K. K., & Harrison, D. G. (2007). Measurement of reactive oxygen species in cardiovascular studies. *Hypertension, 49*, 717–727.

Doherty, G. J., & McMahon, H. T. (2009). Mechanisms of endocytosis. *Annual Review of Biochemistry, 78*, 857–902.

Dypbukt, J. M., Bishop, C., Brooks, W. M., Thong, B., Eriksson, H., & Kettle, A. J. (2005). A sensitive and selective assay for chloramine production by myeloperoxidase. *Free Radical Biology & Medicine, 39*, 1468–1477.

Eshraghi, A. A., Gupta, C., Van De Water, T. R., Bohorquez, J. E., Garnham, C., Bas, E., et al. (2013). Molecular mechanisms involved in cochlear implantation trauma and the protection of hearing and auditory sensory cells by inhibition of c-Jun-N-terminal kinase signaling. *Laryngoscope, 123* (Suppl. 1): S1–S14.

Faulkner, K., & Fridovich, I. (1993). Luminol and lucigenin as detectors for O2. *Free Radical Biology & Medicine, 15*, 447–451.

Freitas, M., Lima, J. L., & Fernandes, E. (2009). Optical probes for detection and quantification of neutrophils' oxidative burst. A review. *Analytica Chimica Acta, 649*, 8–23.

Hammond, V. J., Aylott, J. W., Greenway, G. M., Watts, P., Webster, A., & Wiles, C. (2008). An optical sensor for reactive oxygen species: Encapsulation of functionalised silica nanoparticles into silicate nanoprobes to reduce fluorophore leaching. *Analyst, 133*, 71–75.

Iversen, T. G., Skotland, T., & Sandvig, K. (2011). Endocytosis and intracellular transport of nanoparticles: Present knowledge and need for future studies. *Nano Today, 6*, 176–185.

Janssen-Heininger, Y. M., Mossman, B. T., Heintz, N. H., Forman, H. J., Kalyanaraman, B., Finkel, T., et al. (2008). Redox-based regulation of signal transduction: Principles, pitfalls, and promises. *Free Radical Biology & Medicine, 45*, 1–17.

Kalyanaraman, B., Darley-Usmar, V., Davies, K. J., Dennery, P. A., Forman, H. J., Grisham, M. B., et al. (2012). Measuring reactive oxygen and nitrogen species with fluorescent probes: Challenges and limitations. *Free Radical Biology & Medicine, 52*, 1–6.

Kim, G., Lee, Y. E., Xu, H., Philbert, M. A., & Kopelman, R. (2010). Nanoencapsulation method for high selectivity sensing of hydrogen peroxide inside live cells. *Analytical Chemistry, 82*, 2165–2169.

Lamb, F. S., Hook, J. S., Hilkin, B. M., Huber, J. N., Volk, A. P., & Moreland, J. G. (2012). Endotoxin priming of neutrophils requires endocytosis and NADPH oxidase-dependent endosomal reactive oxygen species. *The Journal of Biological Chemistry, 287*, 12395–12404.

Lamb, F. S., Moreland, J. G., & Miller, F. J., Jr. (2009). Electrophysiology of reactive oxygen production in signaling endosomes. *Antioxidants & Redox Signaling, 11*, 1335–1347.

Lee, Y. E., & Kopelman, R. (2012). Nanoparticle PEBBLE sensors in live cells. *Methods in Enzymology, 504*, 419–470.

Li, Y., Zhu, H., Kuppusamy, P., Roubaud, V., Zweier, J. L., & Trush, M. A. (1998). Validation of lucigenin (bis-N-methylacridinium) as a chemilumigenic probe for detecting

superoxide anion radical production by enzymatic and cellular systems. *The Journal of Biological Chemistry, 273*, 2015–2023.
Liochev, S. I., & Fridovich, I. (1997). Lucigenin (bis-N-methylacridinium) as a mediator of superoxide anion production. *Archives of Biochemistry and Biophysics, 337*, 115–120.
Lippert, A. R., Van de Bittner, G. C., & Chang, C. J. (2011). Boronate oxidation as a bioorthogonal reaction approach for studying the chemistry of hydrogen peroxide in living systems. *Accounts of Chemical Research, 44*, 793–804.
Mercer, J., & Helenius, A. (2009). Virus entry by macropinocytosis. *Nature Cell Biology, 11*, 510–520.
Meyer, A. J., & Dick, T. P. (2010). Fluorescent protein-based redox probes. *Antioxidants & Redox Signaling, 13*, 621–650.
Miller, F. J., Jr., Filali, M., Huss, G. J., Stanic, B., Chamseddine, A., Barna, T. J., et al. (2007). Cytokine activation of nuclear factor {kappa}B in vascular smooth muscle cells requires signaling endosomes containing Nox1 and ClC-3. *Circulation Research, 101*, 663–671.
Miller, F. J., Jr., & Griendling, K. K. (2002). Functional evaluation of nonphagocytic NAD(P)H oxidases. *Methods in Enzymology, 353*, 220–233.
Miller, E. W., Tulyathan, O., Isacoff, E. Y., & Chang, C. J. (2007). Molecular imaging of hydrogen peroxide produced for cell signaling. *Nature Chemical Biology, 3*, 263–267.
Nishinaka, Y., Aramaki, Y., Yoshida, H., Masuya, H., Sugawara, T., & Ichimori, Y. (1993). A new sensitive chemiluminescence probe, L-012, for measuring the production of superoxide anion by cells. *Biochemical and Biophysical Research Communications, 193*, 554–559.
Nordenfelt, P., & Tapper, H. (2011). Phagosome dynamics during phagocytosis by neutrophils. *Journal of Leukocyte Biology, 90*, 271–284.
Oakley, F. D., Abbott, D., Li, Q., & Engelhardt, J. F. (2009). Signaling components of redox active endosomes: The redoxosomes. *Antioxidants & Redox Signaling, 11*, 1313–1333.
Oosthuizen, M. M., Engelbrecht, M. E., Lambrechts, H., Greyling, D., & Levy, R. D. (1997). The effect of pH on chemiluminescence of different probes exposed to superoxide and singlet oxygen generators. *Journal of Bioluminescence and Chemiluminescence, 12*, 277–284.
Oosthuizen, M. M., & Greyling, D. (2001). Hydroxyl radical generation: The effect of bicarbonate, dioxygen and buffer concentration on pH-dependent chemiluminescence. *Redox Report, 6*, 105–116.
Roesler, J., & Emmendorffer, A. (1991). Diagnosis of chronic granulomatous disease. *Blood, 78*, 1387–1389.
Sansonetti, P. (2001). Phagocytosis of bacterial pathogens: Implications in the host response. *Seminars in Immunology, 13*, 381–390.
Schrand, A. M., Lin, J. B., Hens, S. C., & Hussain, S. M. (2011). Temporal and mechanistic tracking of cellular uptake dynamics with novel surface fluorophore-bound nanodiamonds. *Nanoscale, 3*, 435–445.
Sengelov, H., Nielsen, M. H., & Borregaard, N. (1992). Separation of human neutrophil plasma membrane from intracellular vesicles containing alkaline phosphatase and NADPH oxidase activity by free flow electrophoresis. *The Journal of Biological Chemistry, 267*, 14912–14917.
Soenen, S. J., Demeester, J., De Smedt, S. C., & Braeckmans, K. (2012). The cytotoxic effects of polymer-coated quantum dots and restrictions for live cell applications. *Biomaterials, 33*, 4882–4888.
Sorkin, A., & von Zastrow, M. (2009). Endocytosis and signalling: Intertwining molecular networks. *Nature Reviews Molecular Cell Biology, 10*, 609–622.
Ushio-Fukai, M. (2009). Compartmentalization of redox signaling through NADPH oxidase-derived ROS. *Antioxidants & Redox Signaling, 11*, 1289–1299.

Wang, W., Fang, H., Groom, L., Cheng, A., Zhang, W., Liu, J., et al. (2008). Superoxide flashes in single mitochondria. *Cell, 134*, 279–290.
Wardman, P. (2007). Fluorescent and luminescent probes for measurement of oxidative and nitrosative species in cells and tissues: Progress, pitfalls, and prospects. *Free Radical Biology & Medicine, 43*, 995–1022.
Zhang, Y., Tekobo, S., Tu, Y., Zhou, Q., Jin, X., Dergunov, S. A., et al. (2012). Permission to enter cell by shape: Nanodisk vs nanosphere. *ACS Applied Materials & Interfaces, 4*, 4099–4105.
Zhao, H., Kalivendi, S., Zhang, H., Joseph, J., Nithipatikom, K., Vasquez-Vivar, J., et al. (2003). Superoxide reacts with hydroethidine but forms a fluorescent product that is distinctly different from ethidium: Potential implications in intracellular fluorescence detection of superoxide. *Free Radical Biology & Medicine, 34*, 1359–1368.
Zielonka, J., & Kalyanaraman, B. (2010). Hydroethidine- and MitoSOX-derived red fluorescence is not a reliable indicator of intracellular superoxide formation: Another inconvenient truth. *Free Radical Biology & Medicine, 48*, 983–1001.

CHAPTER FOURTEEN

SARA and RNF11 at the Crossroads of EGFR Signaling and Trafficking

Eleftherios Kostaras[*,†], **Nina Marie Pedersen**[‡], **Harald Stenmark**[‡], **Theodore Fotsis**[*,†], **Carol Murphy**[†,1]

[*]Laboratory of Biological Chemistry, Medical School, University of Ioannina, Ioannina, Greece
[†]Department of Biomedical Research, Foundation for Research & Technology – Hellas, Institute of Molecular Biology & Biotechnology, University Campus of Ioannina, Ioannina, Greece
[‡]Centre for Cancer Biomedicine, Faculty of Medicine, University of Oslo, Oslo, Norway
[1]Corresponding author: e-mail address: cmurphy@cc.uoi.gr

Contents

Abstract

The classical view that endocytosis serves only for growth factor receptor degradation and signaling termination has recently been challenged by an increasing number of reports showing that various growth factor receptors such as epidermal growth factor receptor (EGFR) continue to activate downstream signaling molecules en route to lysosomes prior to their degradation. Moreover, the trafficking route that the ligand–receptor complexes follow to enter the cell is mutually interconnected with the final signaling output.

Endosomal resident effector proteins are compartmentalized and regulate the signaling and trafficking of the ligand-bound receptor complexes. Smad anchor for receptor activation (SARA) is an early endosomal protein facilitating TGF-β signaling cascade. Even though SARA was identified as an adaptor protein that regulates SMAD2 activation and TGF-β signal propagation, an increasing number of reports in various systems

Methods in Enzymology, Volume 535
ISSN 0076-6879
http://dx.doi.org/10.1016/B978-0-12-397925-4.00014-6

describe SARA as a trafficking regulator. Recently, SARA has been shown to interact with the E3 ubiquitin ligase RNF11 (RING finger protein 11) and members of the ESCRT-0 (endosomal sorting complex required for transport) complex functionally participating in the degradation of EGFR.

1. INTRODUCTION

Smad anchor for receptor activation (SARA) is a FYVE (Fab1p–YOTB–Vps27p–EEA1) domain containing protein that resides on early endosomal (EE) membranes (Panopoulou et al., 2002) originally identified as a protein that recruits nonphosphorylated SMAD2/3 proteins to the activated transmembrane TβRI (TGF-β type I receptor) for their phosphorylation and subsequent signal propagation (Tsukazaki, Chiang, Davison, Attisano, & Wrana, 1998). The predominant localization of SARA on EE raised the hypothesis that Smad2/3 phosphorylation is facilitated from the EE compartment with clathrin-mediated endocytosis serving as internalization route that propagates TGF-β signaling (Di Guglielmo, Le Roy, Goodfellow, & Wrana, 2003; Hayes, Chawla, & Corvera, 2002; Penheiter et al., 2002).

Following the studies mentioned earlier, in which SARA was characterized as a regulator of TGF-β signaling, various reports have suggested that SARA may be a general regulator of trafficking. Overexpression of SARA causes early endosome enlargement in a Rab5-dependent manner and impairs transferrin recycling (Hu, Chuang, Xu, McGraw, & Sung, 2002). In addition, SARA has been found to interact with syntaxin-3 and other members of the SNARE family facilitating targeting of the rhodopsin-bearing axonemal vesicles in mammalian rods' organelle (Chuang, Zhao, & Sung, 2007). More recently, SARA has been shown to interact with the RING-H2 E3 ubiquitin ligase RING finger protein 11 (RNF11) and has been proposed to mimic HRS function by assembling an alternative endosomal sorting complex required for transport (ESCRT-0) complex. Endosomal membranes serve as intracellular platforms where RNF11 and SARA associate with each other and the ESCRT-0 core proteins regulating epidermal growth factor (EGF) activated EGFR degradation and signaling (Kostaras et al., 2012).

2. RNF11 INTRACELLULAR COMPARTMENTALIZATION

The information acquired by intracellular localization studies is important, especially when investigating the role of protein–protein interactions. The presence of a protein in a cellular compartment indicates the protein's site of action and can lead to functional correlations with other proteins localized at the same sites. Various proteins are located to specific intracellular compartments and can be used as markers (Table 14.1). We used each of the markers listed in the succeeding text to characterize the intracellular localization of RNF11 (Fig. 14.1 and Table 14.2).

Before undertaking localization experiments, there are a number of things to consider:

1. Which antibody to use: The best option is to use well-characterized and specific antibodies that will recognize the endogenous protein. The lack of a specific antibody detecting endogenous levels of RNF11 led us to the overexpression of fluorescent protein fusions (GFP, mCherry) or epitope-tagged (HA, myc) versions of the protein. Before deciding on which tag to use, one should be certain that tagging the protein will not alter its intracellular localization. In the case of RNF11, tagging at the N-terminus mislocalizes the protein due to interference with the myristoylation signal, and therefore, tags should be added to the C-terminus. One should first overexpress the untagged protein if possible—this is normally recognized even using a polyclonal antibody that fails to recognize the endogenous protein. Then, we check that the tagged protein has the same localization as the untagged. This is especially important when using large tags such as GFP or mCherry.
2. How to detect the compartment markers: In some cases, antibodies against the endogenous proteins are available. However, when antibodies are not available, the tagged marker (e.g., GFP-Rabs) can be overexpressed and multilabeling immunofluorescence performed. Overexpression should mimic the endogenous levels so that the localization, and/or the compartment to which the protein localizes, is not altered due to overexpression. Overexpression of Rab proteins will increase their cytosolic fraction, and the removal of the cytoplasmic nonmembrane-associated Rab proteins can be carried out by pre-extraction prior to fixation, with a mild detergent (0.05% saponin in phosphate-buffered (saline) PBS for 5 min).
3. Which secondary antibodies to use: When performing multilabeling immunofluorescence, always use secondary antibodies that are specifically designed for this purpose and are preabsorbed. A good source of these antibodies is Jackson ImmunoResearch.
4. Which cells to use: Select cells that have a high endogenous level of your protein, are easily transfected, have good morphology, and can withstand the washing steps easily without detaching. We use primary

Table 14.1 Membrane-associated proteins used to label the various endocytic compartments of the cell

Marker	Compartment	References
Ap2	Cell membrane and CCV	Robinson (2004)
Clathrin	Cell membrane, CCV, and EE	McMahon and Boucrot (2011) and Raiborg, Bache, Mehlum, Stang, and Stenmark (2001)
Rab5a,b,c	CCV and EE	Bucci et al. (1992)
EEA1	EE	Mu et al. (1995)
Rabankyrin5, 70 kD dextran	Macropinosomes	Schnatwinkel et al. (2004)
SARA	EE, outer membrane of MVB	Panopoulou et al. (2002) and Bokel, Schwabedissen, Entchev, Renaud, and Gonzalez-Gaitan (2006)
HRS	EE, outer membrane of MVB	Raiborg, Bremnes, et al. (2001) and Bache, Brech, Mehlum, and Stenmark (2003)
Rab4a	EE and early recycling endosomes	van der Sluijs et al. (1992)
Rabenosyn-5	EE	Nielsen et al. (2000)
Transferrin	EE, early and late recycling endosomes	Sonnichsen, De Renzis, Nielsen, Rietdorf, and Zerial (2000)
Rab11a,b	Golgi, perinuclear recycling endosomes and microtubule organizing center	Ullrich, Reinsch, Urbe, Zerial, and Parton (1996)
Rab7	Late endosomes, lysosomes	Chavrier, Parton, Hauri, Simons, and Zerial (1990)
Rab9a,b,c	Late endosomes	Lombardi et al. (1993)
Lamp1	Late endosomes, lysosomes	Rohrer, Schweizer, Russell, and Kornfeld (1996)
CD63	Late endosomes, lysosomes	Metzelaar et al. (1991)
Caveolin	Caveolae	Rothberg et al. (1992)
APPL1, APPL2	APPL-positive early endosomal compartment	Miaczynska et al. (2004)

CCV, clathrin-coated vesicles; MVB, multivesicular bodies.

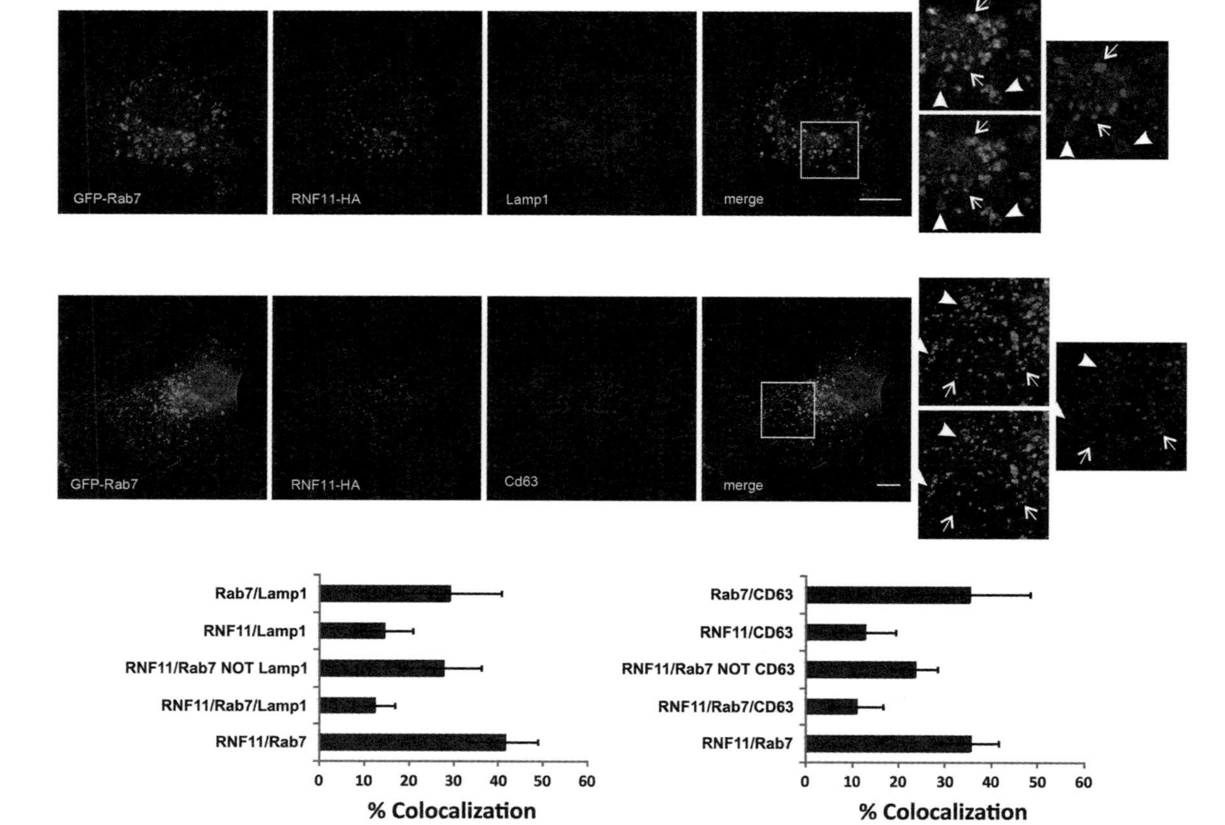

Figure 14.1 Colocalization of RNF11 with late endosomal and not lysosomal markers. Arrowheads show GFP-Rab7 vesicles positive for Lamp1 or CD63 but RNF11-HA-negative. Arrows indicate that the RNF11-positive Rab7 endosomes are Lamp1- or CD63-negative. Scale bars, 10 μm. Quantitation of the triple colocalization was calculated using Kalaimoscope motionTracker and error bars depict the standard deviation of the mean. *From Kostaras et al. (2012).* (For color version of this figure, the reader is referred to the online version of this chapter.)

Table 14.2 Quantitation of RNF11 localization with early and late endosomal markers using Kalaimoscope motionTracker

Marker	Colocalization
EEA1	33 ± 5.7
GFP-SARA	42 ± 5
GFP-Rab5	46 ± 8
myc-Rabenosyn-5	34 ± 7.5
GFP-Rab7	42 ± 8
Lamp1	14.46 ± 6
CD63	12.75 ± 7

± Represents the standard deviation of the mean.

endothelial cells and HeLa cells for this purpose. Cells can be plated on coverslips with the appropriate thickness and quality for confocal microscopy (170 μm ± 10). The mountant must have an additive to prevent photobleaching, and ProLong antifade (Invitrogen) is a good choice.

5. During the acquisition of the images on the confocal microscope, check that the signal is not saturated by activating the Glow (OU) LUT (Leica SP5) and adjust appropriately the offset, the PMTs (Photomultiplier tube), and the laser power. When performing multichannel simultaneous scan, always check for cross talk between channels. If cross talk is present, use sequential scanning.
6. For statistical analysis of the colocalization, at least 10 images (with optimized sections of planes at various depths within the sample) with each marker should be acquired and quantitated. The percentage of colocalization was assessed using Kalaimoscope motionTracker (Kalaidzidis Y, MPI-CBG, Dresden, Germany) (http://www.kalaimoscope.com/). This program assigns an x–y position to objects in your bioimages, such as intracellular vesicles, and quantifies their intensity profiles. Figure 14.1 shows triple immunofluorescence staining of RNF11-HA, GFP-Rab7, and the lysosomal markers Lamp1 or CD63. Extensive analysis of the images with Kalaimoscope motionTracker, with >50% overlap considered as colocalized, allowed us to quantitatively define RNF11 localization between the late endosomal–lysosomal compartments. RNF11 mainly colocalizes with Rab7-positive vesicles that are Lamp1- or CD63-negative.

3. Biochemical Analysis of Interacting Proteins

Many methods have been developed for studying protein interactions. Microscopy-based methods such as FRET (Förster resonance energy transfer) or BiFC (bimolecular fluorescence complementation) are extensively used to verify protein–protein interactions in live or fixed cells. Yeast two-hybrid screening is a versatile technique that allows the identification of interacting proteins. Other more classical biochemical approaches, such as copurification, affinity purification, pull down, or coimmunoprecipitation of protein complexes, require *in vitro* handling of protein extracts and are useful for verifying the interaction of two or more proteins.

3.1. Mapping the SARA interaction domains of RNF11

Y2H analysis of SARA (SARAΔ1–664) revealed RNF11 (aa 62–154) as a putative interacting protein. Full-length RNF11 or various deletion mutants were expressed as GST (glutathione S-transferase) fusion proteins, and the interaction domains were characterized biochemically by pull-down experiments (Fig. 14.2). Alternatively, proteins can be expressed in *Escherichia coli* as maltose-binding protein fusion or His-tagged proteins and purified into amylose/dextrin Sepharose or with Ni-NTA agarose columns, respectively.

3.1.1 Bacterial protein expression

1. Insert cDNA in frame into the appropriate GST bacterial expression vector (the pGEX vectors, GE Healthcare, are convenient) and transform the DNA plasmid into BL21 (DE3) competent *E. coli* cells, a strain defective in OmpT and Lon protease production and harboring the T7 RNA polymerase gene.
2. Inoculate a single bacterial colony into 2 ml sterile LB broth plus ampicillin (50 μg/ml) and incubate overnight at 37 °C at 180–200 rpm.

   Note: It is wise to test protein expression on a small scale (1 ml culture ± IPTG) before proceeding to large volume cultures.
3. Take 1 ml of the culture and inoculate 1 l of sterile LB broth (RT) plus ampicillin (50 μg/ml). Incubate at 37 °C at 180–200 rpm under continuous aeration till OD_{600} reaches 0.6.

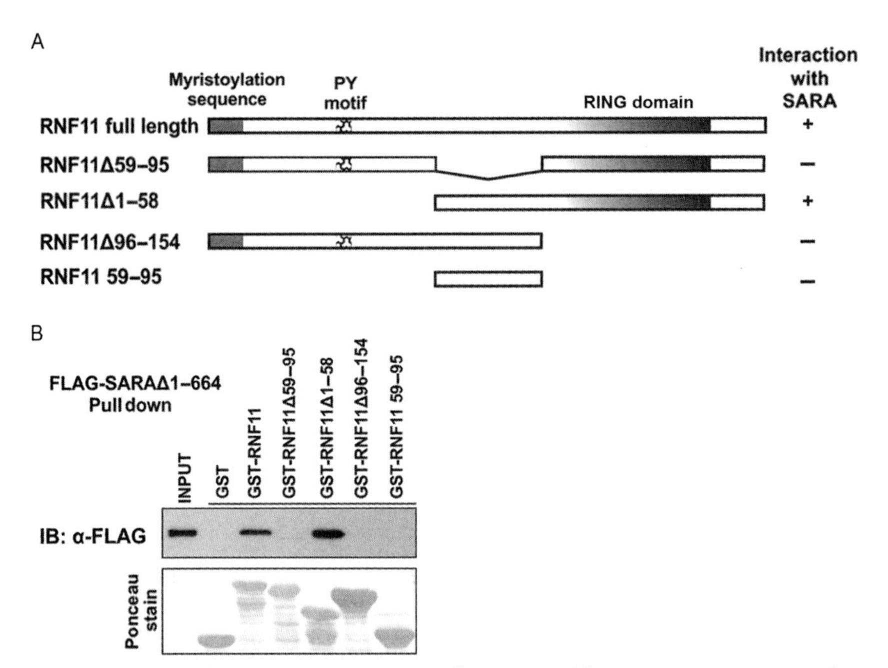

Figure 14.2 (A) Schematic representation of RNF11 and fusion proteins generated to test its interaction with SARA. (B) Pull-down experiment of GST-RNF11 fusion proteins with HEK293 cell lysates expressing FLAG-SARAΔ1–664. The experiment revealed that both aa 59–95 and the RING domain are important for the interaction. *From Kostaras et al. (2012).*

4. Transfer 1 ml of culture into an Eppendorf, spin at 5000 rpm for 5 min, and resuspend pellet in 1× Laemmli buffer (this is your −IPTG sample).
5. Induce the rest of the culture with 0.3 m*M* IPTG (Fermentas) for 2 h.
6. Transfer 1 ml of culture into an Eppendorf, spin at 5000 rpm for 5 min, and resuspend pellet in 1× Laemmli buffer (this is your +IPTG sample).
7. Collect cells from Step 5 by spinning bacteria at 6000 rpm, at 4 °C in SLA-3000 rotor (Sorvall).
8. Resuspend bacterial pellet on ice in precooled lysis buffer containing 20 m*M* Tris–HCl pH 7.8, 100 m*M* NaCl, and complete protease inhibitors (Roche).
9. Pass the bacteria through a French press at 1000 psi, keeping the bacteria on ice.
10. Spin the bacterial lysate at 14,000 rpm 4 °C for 30 min in SS-34 rotor (Sorvall).

11. Transfer supernatant into a new falcon tube and also keep a small amount of the pellet resuspended in 100 μl 1 × Laemmli buffer. This will indicate the solubilization of the protein in the final gel analysis.
12. Run all samples on a 12% SDS-PAGE using appropriate volume of the pellets prior to and post-IPTG induction (Steps 4 and 6) and 10 μl of the pellet resuspended in Step 11.
13. Stain gel with 0.1% Coomassie Brilliant Blue R 250 (Fluka) in fixative (40% methanol/10% acetic acid) for 1 h.
14. Destain with 20% methanol/10% acetic acid for 1 h. Refresh buffer every 15 min.
15. A clear band at the desired molecular weight (26 kDa plus that of the protein of interest) should be visible in the +IPTG lane, whereas no protein at that size should be visible in −IPTG and in the resuspended pellet from Step 11.

    Note: In some cases, especially when expressing high-molecular-weight proteins in *E. coli*, GST fusion proteins are poorly folded and rapidly degraded. Overnight induction at 16 °C with 0.05 m*M* IPTG in Step 5 can sometimes solve this problem.

3.1.2 Protein purification with affinity chromatography

1. Transfer 300–400 μl glutathione Sepharose 4B beads (GE Healthcare) into a Poly-Prep chromatography column (Bio-Rad), allow beads to settle, and wash 3 × the column with 10 bed volumes of PBS to remove residual ethanol.
2. Allow PBS to flow through (without drying out) and equilibrate the beads with 10 bed volumes of equilibration buffer containing 20 m*M* Tris–HCl pH 7.8 and 100 m*M* NaCl. Allow to flow through.
3. Load the extracted lysate (Section 3.1.1, Step 11) of the bacterially produced GST fusion protein to the beads and allow to flow through.

    Note: In order to decrease the protein degradation, you can mix the equilibrated beads with lysate in a falcon tube, rotate for 1 h at 4 °C, and reload the beads on the column. The disadvantage is that the protein does not elute concentrated into one or two fractions but rather appears in all of the eluted fractions.
4. Wash the column 2 × with 10 ml precooled lysis buffer (20 m*M* Tris–HCl pH 7.8, 100 m*M* NaCl, and protease inhibitors). Allow to flow through.
5. Prepare elution buffer containing 10 m*M* reduced glutathione (Sigma), 20 m*M* Tris–HCl pH 7.8, 100 m*M* NaCl, and protease inhibitors. Adjust pH with NaOH to 7.8.

6. Elute GST-bound protein by adding 0.5 ml of the elution buffer to the column and collect the flow through into an Eppendorf tube. Repeat Step 6 and collect eight fractions of 0.5 ml each.

   Note: The beads can be regenerated as follows: perform three cycles of extensive wash with 0.1 *M* Tris, pH 8.5, and 0.5 *M* NaCl followed by 0.1 *M* sodium acetate, pH 4.5, and 0.5 *M* NaCl. Wash once with deionized water and 20% ethanol and store column in 20% ethanol at 4 °C. Reuse up to four times.
7. Run 10 μl of each fraction on a 12% SDS-PAGE. Keep remainder of the fractions at −80 °C.
8. Stain gel with Coomassie as described in Section 3.1.1, Step 13.
9. Pool relevant fractions containing the eluted protein in dialysis tubing cellulose membrane (Sigma) and dialyze against equilibration or desired buffer overnight at 4 °C.
10. Measure protein concentration with Bradford assay (Bio-Rad), aliquot, and store protein at −80 °C.

3.1.3 In vitro *GST pull-down assay*

1. Transfer 15–20 μl of packed glutathione Sepharose 4B beads (GE Healthcare) in an Eppendorf tube and wash 3 × with 10 bed volumes of PBS. Spin down the beads at 2000 rpm for 2 min.
2. Equilibrate beads with 10 bed volumes of buffer containing 150 m*M* NaCl and 20 m*M* Tris–HCl pH 7.5. Spin down the beads at 2000 rpm for 2 min and remove supernatant.
3. Incubate beads with 100 μl of the buffer mentioned earlier plus 1% BSA (bovine serum albumin) for 1 h at 4 °C under continuous rotation. This step will decrease the nonspecific binding to the beads.
4. Spin beads 2 min at 2000 rpm and discard the supernatant.
5. Incubate beads with 15–20 μg purified GST protein alone or GST fusion protein at final volume of 200 μl or more assay buffer for 1 h at 4 °C under continuous rotation.
6. Spin beads for 2 min 2000 rpm (4 °C), remove supernatant, and wash the beads with 0.5 ml of assay buffer. Repeat washing step for two more times.
7. Add 100–250 μg of cell lysate (0.1% NP-40, 150 m*M* NaCl, 20 m*M* Tris–HCl pH 7.5, and complete protease inhibitors (Roche)) overexpressing the protein of interest, for example, FLAG-SARAΔ1–664 to the beads in the presence of 0.1% BSA, and incubate for 3 h at 4 °C under continuous rotation.

8. Spin beads for 2 min at 2000 rpm (4 °C), remove supernatant, and wash the beads with 1 ml of lysis buffer. Repeat wash step four more times.
9. Elute GST–protein complexes with 20 μl 2 × Laemmli buffer, boil for 5 min at 100 °C, centrifuge full speed for 1 min, and transfer all supernatant into a new Eppendorf.
10. Run complexes on an SDS-PAGE and transfer proteins onto nitrocellulose membranes. Stain with Ponceau dye and ensure equimolar amounts of the GST fusion proteins in case multiple conditions or proteins have been used. Destain the membranes with deionized water and probe with the appropriate antibodies detecting the overexpressed or the endogenous protein.

3.2. RNF11 and SARA interaction with members of the ESCRT-0

Among the SARA interacting proteins are the SMAD2/3 proteins and the TGF-β receptors (Tsukazaki et al., 1998), the LAP family member ERBIN (Sflomos et al., 2011), and the catalytic subunit of type 1 serine/threonine protein phosphatase (PP1c) (Bennett & Alphey, 2002). Moreover, SARA interacts with RNF11 and both proteins associate with members of the ESCRT-0 core proteins such as STAM2 and Eps15b (Fig. 14.3). RNF11 also associates with HRS, whereas SARA has been found in complex with the ESCRT-I protein TSG101 and clathrin (Kostaras et al., 2012).

Immunoprecipitating endogenous proteins is the best option when studying protein–protein interactions, but this is sometimes impossible due to the lack of specific antibodies detecting the endogenous protein of interest and low abundance in specific cell lines. In order to overcome this

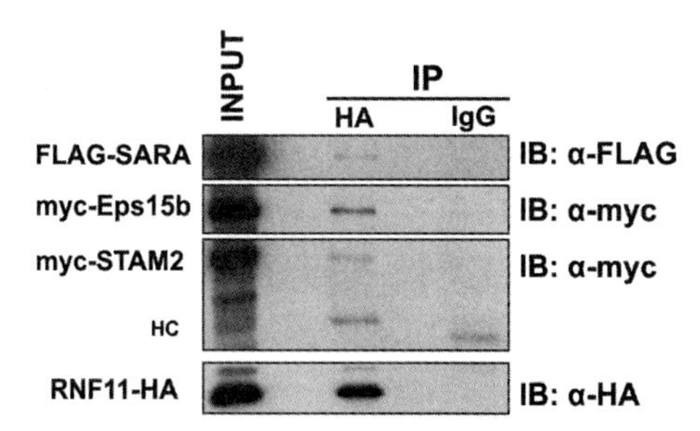

Figure 14.3 RNF11-HA, myc-Eps15b, myc-STAM2, and FLAG-SARA were overexpressed in BHK cells and cell lysates were subjected to immunoprecipitation with α-HA or rat IgG control antibody. Immunocomplexes revealed that RNF11 coimmunoprecipitates SARA, STAM2, and EPS15 proteins. *From Kostaras et al. (2012).*

drawback, we performed immunoprecipitations with overexpressed proteins. Overexpression of the proteins was performed in HEK293 cells by transient transfection using ExtremeGene9 (Roche), Lipofectamine (Invitrogen) or other lipids, PEI (polyethylenimine), or calcium phosphate methods. In some cases, proteins were overexpressed with recombinant adenoviruses or the vaccinia T7 promoter system (Stenmark, Bucci, & Zerial, 1995):

1. Overexpress the proteins of interest in HEK293 cells or any other cell line. For each experimental condition, one 60 mm dish is enough when immunoprecipitating overexpressed proteins. In the case of endogenous protein, immunoprecipitations from cell lysates especially for low-abundance proteins scale up to 2×100 mm per condition.
2. Wash cells once with 4 ml PBS. Be careful to remove all PBS from the cells to avoid altering the volume and salt concentration of the lysis buffer.
3. Add 0.5 ml of precooled lysis buffer containing 0.1–1% NP-40, 100 m*M* NaCl, 20 m*M* Tris–HCl pH 7.5, and complete protease inhibitors (Roche) directly to the dish and transfer dish on ice. Scrape cells and transfer cell lysate into an Eppendorf tube. Incubate on ice for 15 min.

   Note: To break the cells, several methods can be used. A cell cracker or passing the cell lysate five times through an insulin syringe is the most common. Choose the minimum concentration of detergent that allows maximum solubilization of the proteins, as increased detergent concentrations might disrupt the protein–protein interactions. In case phosphorylation-sensitive interactions are assessed, include the appropriate phosphatase inhibitors (NaF and β-glycerol phosphate for Ser/Thr phosphatases and sodium orthovanadate for Tyr phosphatases) in the lysis buffer. Moreover, when performing immunoprecipitations with proteins that contain modular domains that depend on trace elements (such as the Zn in the RING domain of RNF11) do not use chelating reagents such as EDTA in the lysis buffer. In the case of RNF11, 1 m*M* of $ZnCl_2$ was added to the lysis buffer.
4. Centrifuge full speed (13,200 rpm) for 20 min at 4 °C.
5. Collect supernatant and centrifuge at $100{,}000 \times g$ in a benchtop ultracentrifuge for 60 min at 4 °C using Beckman MLA-130 rotor (optional).

6. Collect supernatant and measure cell lysate protein concentration using Bradford assay kit (Bio-Rad) or BCA (Pierce), depending on the content of the lysis buffer. Keep 20–50 μg of lysate for the input (5–10% of the total).
7. In the meantime, wash 10–15 μl/reaction protein A or protein G beads with 10 bed volumes of PBS. Repeat washing and equilibrate beads with 10 bed volumes of lysis buffer.
8. Preclear cell lysate with beads for 1 h at 4 °C under continuous rotation.
9. Centrifuge 2000 rpm at 4 °C, transfer supernatant in a new Eppendorf tube, and throw away the beads.
10. Incubate about 300 μg precleared cell lysate with 1 μg of purified antibody at 4 °C overnight under continuous rotation.

    Note: A negative control for the co-IP must be included and can be either an isotype control IgG antibody or cell lysates that do not overexpress one or more of the proteins.
11. Transfer the lysate with the antibody into 10–15 μl of clean and equilibrated beads and incubate for 4 h at 4 °C under continuous rotation.
12. Spin 2 min at 2000 rpm, and discard the supernatant.
13. Wash the beads with 0.5–1 ml of lysis buffer. Spin at 2000 rpm for 2 min at 4 °C. Discard the supernatant. Repeat four times the washing step.
14. Elute protein complexes with 20 μl 2× Laemmli buffer and boil for 5 min at 100 °C.
15. Full spin 1 min and transfer all supernatant in new Eppendorf.

    Note: Sometimes, the protein of interest comigrates at the same molecular weight as the heavy or the light chain of the antibody. Preabsorbed secondary antibodies recognizing only the light or the heavy chain of the primary antibody can be used to overcome the issue mentioned earlier (from Jackson ImmunoResearch). Alternatively, this can be overcome using antibody-conjugated beads and elution of the immunocomplexes by peptide competition or reduced pH as described in the succeeding text.

3.2.1 FLAG-peptide elution of the immunocomplexes

1. Immunoprecipitate FLAG-tagged proteins in the lysates using α-FLAG M2 affinity resin following the steps outlined earlier. Wash beads well.
2. Incubate beads with affinity immunocomplexes with 1.2 μg/μl of 3× FLAG-peptide (Sigma) in 50 μl TBS solution containing 150 m*M* NaCl and 50 m*M* Tris–HCl pH 7.5, 1 h at 4 °C using end-over-end rotation.

3. Spin down the beads at 13,200 rpm for 2 min at 4 °C and transfer supernatant in a new Eppendorf tube very carefully avoiding the beads.
4. Incubate supernatant with 20 μl washed protein G beads for 1 h at 4 °C to remove antibody that might have leaked from the beads during the elution procedure.
5. Spin down the beads as mentioned in the preceding text and transfer supernatant in a new Eppendorf tube very carefully avoiding the beads.
6. Add the appropriate volume of sample buffer and analyze immunocomplexes by SDS-PAGE.

3.2.2 Acidic elution of immunocomplexes from the beads

1. Immunocomplexes can be very easily eluted by incubating the antibody-conjugated beads with 50 μl of glycine 0.1 *M* pH 2.5 for 5 min on ice. Flick the tube from time to time to achieve complete elution.
2. Spin down the beads at 13,200 rpm for 2 min at 4 °C and transfer supernatant in a new Eppendorf tube very carefully avoiding the transfer of the antibody-conjugated beads.
3. Add the appropriate amount of 4 × Laemmli buffer (it will be yellow) and adjust the pH by exposing the eluate to ammonia.

   Note: The disadvantage of the glycine versus the FLAG-peptide elution method is that the former will elute also the nonspecific proteins that bind to the beads leading to increased background. A plus is that you achieve complete removal of the bound proteins from the beads that sometimes is hard to achieve with FLAG-peptide competition.

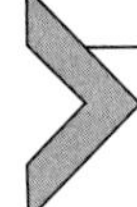

4. TRACKING EGF LIGAND–RECEPTOR COMPLEX DYNAMICS

Various methods have been used for following the internalization and intracellular trafficking of either the ligands or their binding receptors. Fluorescent-conjugated, biotin-tagged, or radiolabeled ligands are important tools for the study of internalized ligands in live, fixed cells or cell lysates and have been extensively used for transferrin and EGF growth factors.

EGF and its receptor (EGFR) complexes are internalized through clathrin-dependent and clathrin-independent mechanisms and sorted in early endosomes. The main part of EGFR is rapidly degraded through the lysosomal compartment, whereas a small portion recycles back to the cell surface. The internalization, trafficking, and degradation fate of EGF–EGFR were assessed in HeLa cells after RNF11 or SARA downregulation or overexpression using the methodology described in the succeeding text.

Depletion of the proteins was performed by siRNA transfection (20 n*M*) using RNAiMAX (Invitrogen) according to the manufacturer's protocol, and knockdown was verified with either pRT-PCR (RNF11) or Western probing with specific antibodies (SARA). For the overexpression experiments, lipids achieved only 20–50% transfection efficiency in HeLa cells so infection with recombinant adenoviruses was the best choice achieving overexpression in 100% of the cells. Both depletion and overexpression of SARA led to a decreased degradation of EGF, or EGFR and RNF11 downregulation caused increased EGF recycling (Fig. 14.4 and Kostaras et al., 2012).

4.1. Monitoring Rh-EGF internalization and degradation

1. Seed cells on coverslips and incubate overnight. At the time of the experiment, the cells should be no more than 80% confluent. Serum starve cells in 0.1% FCS for 5 h to decrease the basal phosphorylation status of downstream signaling molecules.
2. Wash once with PBS and add 100 ng/ml rhodamine-EGF (Rh-EGF) (Molecular Probes) in prewarmed serum-starved medium supplemented with 0.1% BSA for 5 min.
3. Remove ligand, wash two times with prewarmed PBS, and chase for 5 min, 30 min, 1 h, and 2 h with prewarmed medium containing 0.1% FCS and 25 μg/ml cycloheximide.
4. Wash two times with PBS and fix with 3.7% paraformaldehyde for 15 min.
5. Wash once with PBS and quench with 50 m*M* NH_4Cl_2 for 10 min. In case that the colocalization of EGF with an endosomal marker such as EEA1 has to be assessed, proceed to immunofluorescence with a specific antibody against the protein of interest.
6. Mount coverslips on slides with ProLong antifade reagent (Invitrogen) and acquire images using standard confocal microscope with the same parameters at all time points and in all experimental conditions.

   Note: The 2 h time point should be used so as to set the laser intensities for the background remaining fluorescence after Rh-EGF has been degraded, whereas the 10 min time point should specify the total internalized Rh-EGF and lasers should be set in nonsaturated linear conditions.
7. Quantitate total fluorescence/cell using ImageJ program after the same color balance adjustment in all images (http://rsbweb.nih.gov/ij/index.html) and plot values as a percentage of the remaining EGF compared to the total internalized (5 min time point).

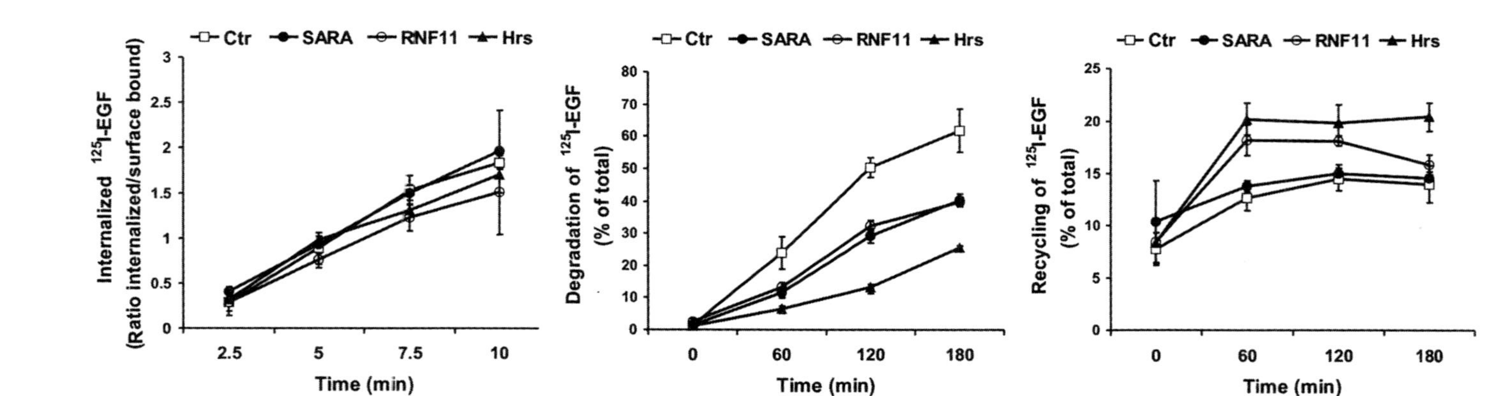

Figure 14.4 HeLa cells were transfected with siRNAs against SARA, RNF11, or siRNA control, and the internalization, degradation, or recycling of iodinated EGF was assessed. Depletion of HRS was used as a positive control. No difference in the internalization rate of ^{125}I-EGF is observed upon depletion of RNF11 or SARA, whereas the degradation rate seems to be altered. RNF11 and HRS depletion enhances ^{125}I-EGF recycling towards the cell membrane, whereas the depletion of SARA is without effect. *From Kostaras et al. (2012).*

4.2. ^{125}I-EGF trafficking kinetics

Assays using radiolabeled EGF (^{125}I-EGF) remain the gold standard for quantitative trafficking assays, due to the high sensitivity of radioactivity detection and linearity within a large range of ^{125}I-EGF concentrations. Here, we describe methods to study two different steps of the EGF–EGFR trafficking using ^{125}I-EGF.

4.2.1 Internalization of ^{125}I-EGF

The assay is based on separation of surface-bound and intercellularly localized ^{125}I-EGF at different time points, and this value will indicate the internalization rate. The assay discussed is based on Johannessen, Pedersen, Pedersen, Madshus, and Stang (2006), Sorkin and Duex (2010), and personal experience:

1. Seed cells in 12- or 24-well plates and let them grow for 2 or 3 days to confluence; this will reduce unspecific binding of ^{125}I-EGF to the plastic of the plates.
2. Wash cells twice with preheated complete medium containing 0.1% BSA and incubate with preheated medium containing 0.1% BSA and 1–2 ng/ml ^{125}I-EGF at the selected time points at 37 °C.
3. Upon incubation, place the plates on ice, remove the radioactive medium, and wash the cells three times with ice-cold PBS.
4. To make a control (time point 0), incubate with cold medium containing 0.1% BSA and 1–2 ng ^{125}I-EGF/ml on ice for 1 min.
5. For all wells, remove surface-bound ^{125}I-EGF by incubating the cells twice with a low-pH buffer (0.2 *M* acetate buffer, pH 2.8) for 5 min on ice.
6. These fractions are collected, counted in a γ-counter, and represent the surface-bound fraction.
7. The remaining ^{125}I-EGF (acid wash-resistant) represents ^{125}I-EGF internalized, and the cells are detached from the wells by adding 1% SDS in PBS for 10 min at room temperature and then collected and counted in a γ-counter.
8. The ratio of internalized to surface ^{125}I-EGF is plotted against time and represents a measurement of the internalization rate.

   Notes: The assay is based on brief incubations with ^{125}I-EGF and carried out at 37 °C since the endocytosis process is strongly temperature-dependent. A rapid handling with temperate solutions in all steps is important because of the short incubation times. The specific internalization rate depends on the concentration of EGFRs at the plasma membrane, and it is important that the EGFR concentration remains constant during the time course. It is also important to remember that the stripping buffer approximately strips off 90% of the ^{125}I-EGF and different acid buffers can be used. A concern is to use the "right" EGF concentration, by using a low ^{125}I-EGF concentration (1–2 ng/ml), and mostly, clathrin-dependent endocytosis is studied and this is important to have in mind. Rapid recycling of ^{125}I-EGF might underestimate the internalization rate, but by using a short time course (max. 15 min), ^{125}I-EGF will not have time to return to the cell surface. A good internalization experiment will include at least five time points, with two to four wells at each time point, and repeat at least twice to get significant statistics between the different conditions used (different mutants, siRNA, etc.). This assay represents a good way to monitor the initial kinetics of EGF–EGFR trafficking, but keep the pitfalls in mind.

4.2.2 Degradation and recycling of ^{125}I-EGF

Two important facts with ^{125}I-EGF have made it possible to monitor both recycling and degradation of the ^{125}I-EGF–EGFR complex in one assay. First, ^{125}I-EGF sorted to early endosomes is bound to EGFR all the way to the lysosome, and second, degradation of ^{125}I-EGF in lysosomes results in a proteolytic degradation product (mono- and di-^{125}I iodotyrosins) that can be separated from intact ^{125}I-EGF depending on precipitation with trichloroacetic acid (TCA) or not. The trick of the assay is to load early endosomes with ^{125}I-EGF, and then, monitor where the ^{125}I-EGF–EGFR complex traffics. Description and discussion is based on Sorkin, Teslenko, and Nikolsky (1988), Skarpen et al. (1998), Sorkin and Duex (2010) in addition to personal experience:

1. Seed cells in 12-well dishes and let them grow till confluent. Wash cells in complete medium containing 0.1% BSA and incubate cells for 5–15 min at 37 °C with complete medium containing 0.1% BSA and 1–200 ng/ml ^{125}I-EGF. This step will load early endosomes with ^{125}I-EGF–EGFR complexes.
2. Upon incubation at 37 °C, place the dishes on ice and rapidly wash three times with ice-cold medium.
3. Incubate the dishes with washing buffer (0.15 *M* NaCl and 0.1 *M* glycine, pH 3) for 5 min twice on ice to remove ^{125}I-EGF that have not

been internalized during the incubation at 37 °C. All the ^{125}I-EGF is now localized to early endosomes.

4. Wash once with preheated medium before adding preheated complete medium to the cells and place at 37 °C to allow trafficking of the ^{125}I-EGF–EGFR complexes. It is important to make a zero time point for the time course, so add medium to some dishes and keep the dishes on ice.
5. Chase from 15 to 240 min depending on the cell line and concentration of EGFR at the plasma membrane.
6. Upon chase incubation, place dishes on ice and transfer the chasing medium to tubes. Wash once with washing solution and transfer this to the same tubes as for the chasing medium.
7. Add 50% TCA–10% phosphotungstic acid (PTA) to the tubes for precipitation for 1 h or longer at 4 °C, and then, centrifuge for 10 min at 5000 × *g*, 4 °C.
8. Both the supernatant (TCA–PTA-soluble fraction, representing degraded ^{125}I-EGF) and the pellet (TCA–PTA-precipitable, representing recycled ^{125}I-EGF) are transferred to tubes and analyzed in γ-counter. The pellet is washed once in 5% TCA–PTA before being counted.
9. Wash the cells remaining in the wells with ice-cold PBS, add 1% SDS in PBS cells, and collect lysate in tubes for γ-counter analysis. This fraction represents internalized and cell-associated ^{125}I-EGF.
10. After all the counting is complete, the total amount of ^{125}I-EGF is calculated by summing the values from the three fractions; then, calculate the percent of each ^{125}I-EGF fraction (intracellular, recycled, and degraded) relative to the total cell-associated ^{125}I-EGF for each time point and plot it against time.

    Note: It is important to have in mind that it is difficult to calculate the specific rate constants, due to heterogeneity of endosomes and that there is a decrease in ^{125}I-EGF concentration in endosomes during chase time. It is important to load cells with sufficient amount of ^{125}I-EGF in early endosomes in the first step, and therefore, a higher ^{125}I-EGF concentration is often used. The length of the first step is based on cellular level of EGFR and internalization rates for each cell type, so this is important to investigate before starting the experiment. In most cases, 5–15 min is enough to find most ^{125}I-EGF on the early endosomes and to prevent interference of recycling. It is critical to have the same loading time between all the samples within one experiment since the experiment is based on the endosomal pool of ^{125}I-EGF. To get trustworthy results, the experiment should include at least four time points, with two to four wells at each time point, and repeat at least twice to get significant statistics, and time of chase incubations might depend on the main focus of the experiment. For recycling rate, shorter chase incubations are recommended, whereas longer chase incubations are used for a good detection of the degradation rate.

4.3. EGFR degradation monitoring

Several antibodies for EGFR are available that detect the endogenous preceptor levels, allowing to monitor the receptors degradation kinetics in Western blot. HeLa cells contain considerable levels of EGFR, and upon genetic manipulation of the cells, its degradation was followed over time:

1. Wash cells once with prewarmed PBS and serum starve cells in medium containing 0.1% FCS and 1% pen./strep. for 5 h.
2. Induce with 100 ng/ml EGF (ImmunoTools) for 30 min at 37 °C in medium containing 0.1% FCS. Do not induce in one dish. This will represent the total amount of EGFR.
3. Remove ligand and chase for 0, 30 min, 1 h, and 2 h in medium containing 0.1% FCS in the presence of the protein synthesis inhibitor cycloheximide at the concentration of 25 μg/ml. Alternatively, do not remove ligand nor chase with cycloheximide but leave EGF till the end of induction time point.
4. Run an SDS-PAGE and probe with α-EGFR and α-actin antibodies, quantitate with quantity one program after background subtraction, and correct α-EGFR values with the values measured by α-actin quantitation.
5. Plot the percentage of EGFR that has not been degraded compared to the total.

5. SUMMARY

Early endosomes function as intracellular platforms where multiprotein complexes assemble and regulate ligand–receptor trafficking fate. SARA on early endosomes is found in complexes with RNF11, STAM2, Eps15b, TSG101, and clathrin (Fig. 14.5b). RNF11 apart from STAM2 and Eps15b associates also with the HRS (Fig. 14.5a). Both overexpression and depletion of SARA or RNF11 delay the EGF-induced degradation of EGFR. By monitoring EGF intracellular trafficking with methodologies described earlier, we showed that depletion of RNF11 increased EGF

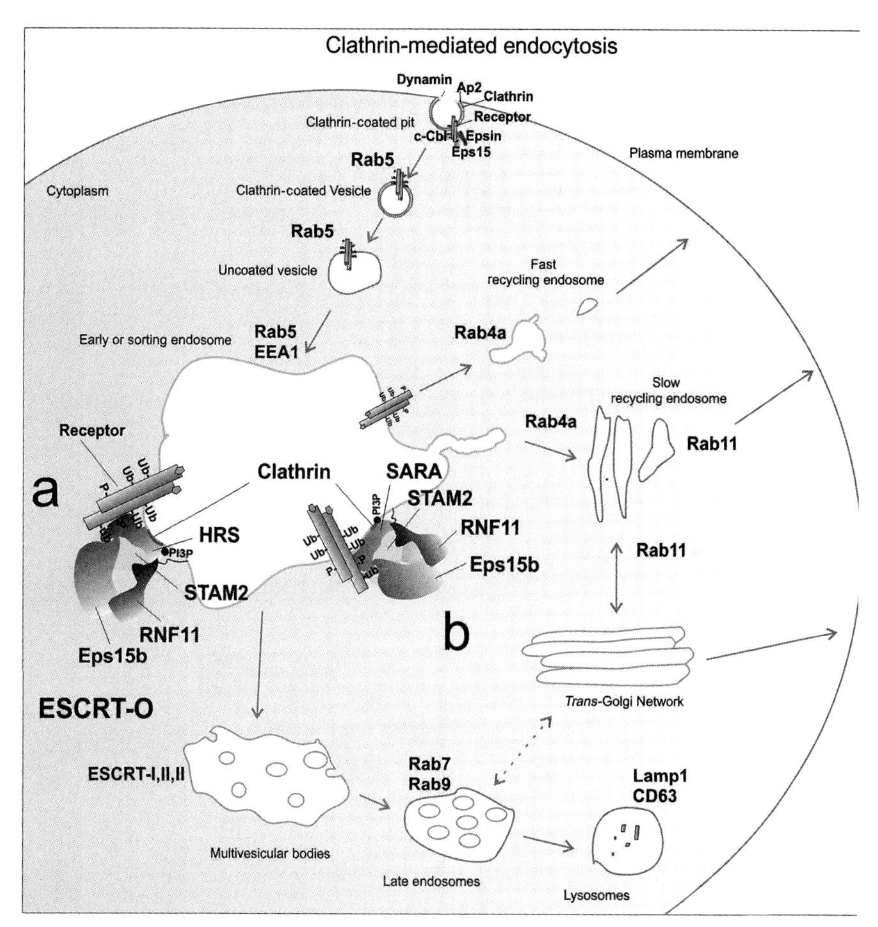

Figure 14.5 Clathrin-mediated endocytosis, trafficking along the endocytic compartments, and SARA/RNF11/ESCRT interactions on early endosomes. (For color version of this figure, the reader is referred to the online version of this chapter.)

recycling and enhanced EGF-induced ERK1/2 phosphorylation. SARA loss of function trapped EGF–EGFR in early endosomes, and the delayed degradation of EGF–EGFR upon SARA depletion results in increased ERK1/2 signaling. Our results extend the knowledge on SARA function and offer new insights into how the modulation of EGF–EGFR trafficking routes affects their subsequent signal propagation.

ACKNOWLEDGMENTS

This work was granted by the European Union Integrated Project ENDOTRACK (EU FP6, LSH-2004-1.1.5-2) and the Hellenic Ministry of Health and Social Solidarity, Central Health Council—KESY.

REFERENCES

Bache, K. G., Brech, A., Mehlum, A., & Stenmark, H. (2003). Hrs regulates multivesicular body formation via ESCRT recruitment to endosomes. *The Journal of Cell Biology, 162*, 435–442.

Bennett, D., & Alphey, L. (2002). PP1 binds Sara and negatively regulates Dpp signaling in Drosophila melanogaster. *Nature Genetics, 31*, 419–423.

Bokel, C., Schwabedissen, A., Entchev, E., Renaud, O., & Gonzalez-Gaitan, M. (2006). Sara endosomes and the maintenance of Dpp signaling levels across mitosis. *Science, 314*, 1135–1139.

Bucci, C., Parton, R. G., Mather, I. H., Stunnenberg, H., Simons, K., Hoflack, B., et al. (1992). The small GTPase rab5 functions as a regulatory factor in the early endocytic pathway. *Cell, 70*, 715–728.

Chavrier, P., Parton, R. G., Hauri, H. P., Simons, K., & Zerial, M. (1990). Localization of low molecular weight GTP binding proteins to exocytic and endocytic compartments. *Cell, 62*, 317–329.

Chuang, J. Z., Zhao, Y., & Sung, C. H. (2007). SARA-regulated vesicular targeting underlies formation of the light-sensing organelle in mammalian rods. *Cell, 130*, 535–547.

Di Guglielmo, G. M., Le Roy, C., Goodfellow, A. F., & Wrana, J. L. (2003). Distinct endocytic pathways regulate TGF-beta receptor signalling and turnover. *Nature Cell Biology, 5*, 410–421.

Hayes, S., Chawla, A., & Corvera, S. (2002). TGF beta receptor internalization into EEA1-enriched early endosomes: Role in signaling to Smad2. *The Journal of Cell Biology, 158*, 1239–1249.

Hu, Y., Chuang, J. Z., Xu, K., McGraw, T. G., & Sung, C. H. (2002). SARA, a FYVE domain protein, affects Rab5-mediated endocytosis. *Journal of Cell Science, 115*, 4755–4763.

Johannessen, L. E., Pedersen, N. M., Pedersen, K. W., Madshus, I. H., & Stang, E. (2006). Activation of the epidermal growth factor (EGF) receptor induces formation of EGF receptor- and Grb2-containing clathrin-coated pits. *Molecular and Cellular Biology, 26*, 389–401.

Kostaras, E., Sflomos, G., Pedersen, N. M., Stenmark, H., Fotsis, T., & Murphy, C. (2012). SARA and RNF11 interact with each other and ESCRT-0 core proteins and regulate degradative EGFR trafficking. *Oncogene*.

Lombardi, D., Soldati, T., Riederer, M. A., Goda, Y., Zerial, M., & Pfeffer, S. R. (1993). Rab9 functions in transport between late endosomes and the trans Golgi network. *The EMBO Journal, 12*, 677–682.

McMahon, H. T., & Boucrot, E. (2011). Molecular mechanism and physiological functions of clathrin-mediated endocytosis. *Nature Reviews Molecular Cell Biology, 12*, 517–533.

Metzelaar, M. J., Wijngaard, P. L., Peters, P. J., Sixma, J. J., Nieuwenhuis, H. K., & Clevers, H. C. (1991). CD63 antigen. A novel lysosomal membrane glycoprotein, cloned by a screening procedure for intracellular antigens in eukaryotic cells. *The Journal of Biological Chemistry, 266*, 3239–3245.

Miaczynska, M., Christoforidis, S., Giner, A., Shevchenko, A., Uttenweiler-Joseph, S., Habermann, B., et al. (2004). APPL proteins link Rab5 to nuclear signal transduction via an endosomal compartment. *Cell, 116*, 445–456.

Mu, F. T., Callaghan, J. M., Steele-Mortimer, O., Stenmark, H., Parton, R. G., Campbell, P. L., et al. (1995). EEA1, an early endosome-associated protein. EEA1 is a conserved alpha-helical peripheral membrane protein flanked by cysteine "fingers" and contains a calmodulin-binding IQ motif. *The Journal of Biological Chemistry, 270*, 13503–13511.

Nielsen, E., Christoforidis, S., Uttenweiler-Joseph, S., Miaczynska, M., Dewitte, F., Wilm, M., et al. (2000). Rabenosyn-5, a novel Rab5 effector, is complexed with hVPS45 and recruited to endosomes through a FYVE finger domain. *The Journal of Cell Biology, 151*, 601–612.

Panopoulou, E., Gillooly, D. J., Wrana, J. L., Zerial, M., Stenmark, H., Murphy, C., et al. (2002). Early endosomal regulation of Smad-dependent signaling in endothelial cells. *The Journal of Biological Chemistry, 277*, 18046–18052.

Penheiter, S. G., Mitchell, H., Garamszegi, N., Edens, M., Dore, J. J., Jr., & Leof, E. B. (2002). Internalization-dependent and -independent requirements for transforming growth factor beta receptor signaling via the Smad pathway. *Molecular and Cellular Biology, 22*, 4750–4759.

Raiborg, C., Bache, K. G., Mehlum, A., Stang, E., & Stenmark, H. (2001). Hrs recruits clathrin to early endosomes. *The EMBO Journal, 20*, 5008–5021.

Raiborg, C., Bremnes, B., Mehlum, A., Gillooly, D. J., D'Arrigo, A., Stang, E., et al. (2001). FYVE and coiled-coil domains determine the specific localisation of Hrs to early endosomes. *Journal of Cell Science, 114*, 2255–2263.

Robinson, M. S. (2004). Adaptable adaptors for coated vesicles. *Trends in Cell Biology, 14*, 167–174.

Rohrer, J., Schweizer, A., Russell, D., & Kornfeld, S. (1996). The targeting of Lamp1 to lysosomes is dependent on the spacing of its cytoplasmic tail tyrosine sorting motif relative to the membrane. *The Journal of Cell Biology, 132*, 565–576.

Rothberg, K. G., Heuser, J. E., Donzell, W. C., Ying, Y. S., Glenney, J. R., & Anderson, R. G. (1992). Caveolin, a protein component of caveolae membrane coats. *Cell, 68*, 673–682.

Schnatwinkel, C., Christoforidis, S., Lindsay, M. R., Uttenweiler-Joseph, S., Wilm, M., Parton, R. G., et al. (2004). The Rab5 effector Rabankyrin-5 regulates and coordinates different endocytic mechanisms. *PLoS Biology, 2*, E261.

Sflomos, G., Kostaras, E., Panopoulou, E., Pappas, N., Kyrkou, A., Politou, A. S., et al. (2011). ERBIN is a new SARA-interacting protein: Competition between SARA and SMAD2 and SMAD3 for binding to ERBIN. *Journal of Cell Science, 124*, 3209–3222.

Skarpen, E., Johannessen, L. E., Bjerk, K., Fasteng, H., Guren, T. K., Lindeman, B., et al. (1998). Endocytosed epidermal growth factor (EGF) receptors contribute to the EGF-mediated growth arrest in A431 cells by inducing a sustained increase in p21/CIP1. *Experimental Cell Research, 243*, 161–172.

Sonnichsen, B., De Renzis, S., Nielsen, E., Rietdorf, J., & Zerial, M. (2000). Distinct membrane domains on endosomes in the recycling pathway visualized by multicolor imaging of Rab4, Rab5, and Rab11. *The Journal of Cell Biology, 149*, 901–914.

Sorkin, A., & Duex, J. E. (2010). Quantitative analysis of endocytosis and turnover of epidermal growth factor (EGF) and EGF receptor. *Current Protocols in Cell Biology, 46*, 15.14.1–15.14.20.

Sorkin, A. D., Teslenko, L. V., & Nikolsky, N. N. (1988). The endocytosis of epidermal growth factor in A431 cells: A pH of microenvironment and the dynamics of receptor complex dissociation. *Experimental Cell Research, 175*, 192–205.

Stenmark, H., Bucci, C., & Zerial, M. (1995). Expression of Rab GTPases using recombinant vaccinia viruses. *Methods in Enzymology, 257*, 155–164.

Tsukazaki, T., Chiang, T. A., Davison, A. F., Attisano, L., & Wrana, J. L. (1998). SARA, a FYVE domain protein that recruits Smad2 to the TGFbeta receptor. *Cell, 95*, 779–791.

Ullrich, O., Reinsch, S., Urbe, S., Zerial, M., & Parton, R. G. (1996). Rab11 regulates recycling through the pericentriolar recycling endosome. *The Journal of Cell Biology, 135*, 913–924.

van der Sluijs, P., Hull, M., Webster, P., Male, P., Goud, B., & Mellman, I. (1992). The small GTP-binding protein rab4 controls an early sorting event on the endocytic pathway. *Cell, 70*, 729–740.

CHAPTER FIFTEEN

p18/LAMTOR1: A Late Endosome/Lysosome-Specific Anchor Protein for the mTORC1/MAPK Signaling Pathway

Shigeyuki Nada, Shunsuke Mori, Yusuke Takahashi, Masato Okada[1]

Department of Oncogene Research, Research Institute for Microbial Diseases, Osaka University, Suita, Osaka, Japan

[1]Corresponding author: e-mail address: okadam@biken.osaka-u.ac.jp

Contents

Abstract

p18/LAMTOR1 is a membrane protein specifically localized to the surface of late endosomes/lysosomes that serves as an anchor for the "Ragulator" complex, which contains p14/LAMTOR2, MP1/LAMTOR3, HBXIP, and C7orf59. The Ragulator interacts with RagAB/CD GTPases and V-ATPase and plays crucial roles for activation of mammalian target of rapamycin complex 1 (mTORC1) on the lysosomal surface. Activated mTORC1 orchestrates various cellular functions, for example, macromolecule biosynthesis, energy metabolism, autophagy, cell growth, responses to growth factors, and the trafficking and maturation of lysosomes. The Ragulator can also regulate a branch of the MAPK pathway by recruiting MEK1 to MP1/LAMTOR3. These findings suggest that p18/LAMTOR1 creates a core platform for intracellular signaling pathways that function via late endosomes/lysosomes.

Methods in Enzymology, Volume 535
ISSN 0076-6879
http://dx.doi.org/10.1016/B978-0-12-397925-4.00015-8

1. INTRODUCTION

The endosome system not only is a sorting compartment for incorporated and/or newly synthesized biomaterials but also acts as a regulatory platform for certain intracellular signaling pathways. It has been proposed that a branch of the MAPK pathway is regulated via the p14–MP1 complex, a MEK1-specific scaffold localized to late endosomes/lysosomes (LE/Lys) (Teis, Wunderlich, & Huber, 2002). Because loss of p14 results in severe defects in cell homeostasis, for example, growth retardation and aberrant trafficking of LE/Lys (Bohn et al., 2007; Teis et al., 2006), the LE/Lys-mediated signaling pathway may be crucial for control of cellular functions.

p18 (also LAMTOR1) was identified as an essential membrane anchor of the p14–MP1 complex on LE/Lys (Nada et al., 2009). p18 undergoes myristate and palmitate modifications, and it contains a highly specific LE/Lys targeting motif (diLeu motif) at its amino terminus, whereby it is exclusively localized to membrane microdomains (Rafts) on LE/Lys (Fig. 15.1A). Recently, the p18–p14–MP1 complex was demonstrated to serve as a scaffold ("Ragulator") for the Rag GTPase complex (RagAB/CD), which is required for amino acid-dependent activation of the mammalian target of rapamycin complex 1 (mTORC1) on LE/Lys (Sancak et al., 2010; Zoncu, Efeyan, & Sabatini, 2010). This finding demonstrates that the p18–p14–MP1 complex on LE/Lys is essential for controlling the functions of mTORC1, for example, macromolecule biosynthesis, energy metabolism, autophagy, cell growth, responses to growth factors, and trafficking of lysosomes. Furthermore, our research group showed that p18–mTORC1 on LE/Lys is involved in the regulation of lysosomal function by controlling lysosome maturation (Takahashi et al., 2012). Other groups have proposed potential roles for p18 in cellular cholesterol homeostasis (Guillaumot et al., 2010), cancer metastasis via activation of Rho GTPases (Hoshino, Koshikawa, & Seiki, 2010), and p53-dependent apoptosis via aberrant lysosomal activation (Malek et al., 2012). These findings underscore a pivotal role of p18 in controlling various cellular signaling pathways involving the endosome system, particularly LE/Lys (Fig. 15.1B).

2. IDENTIFICATION OF p18

p18 has been identified as a phosphoprotein in detergent-resistant membrane fractions (DRMs) of cultured cell lines (e.g., PC12 rat

A

```
1-MGCCYSSENE DSDQDREERK LLLDPSSPPT KALNGAEPNY HSLPSARTDE
  QALLSSILAK TASNIIDVSA ADSQGMEQHE YMDRARQYST RLAVLSSSLT
  HWKKLPPLPS LTSQPHQVLA SEPIPFSDLQ QVSRIAAYAY SALSQIRVDA
  KEELVVQFGI P-161
```

B

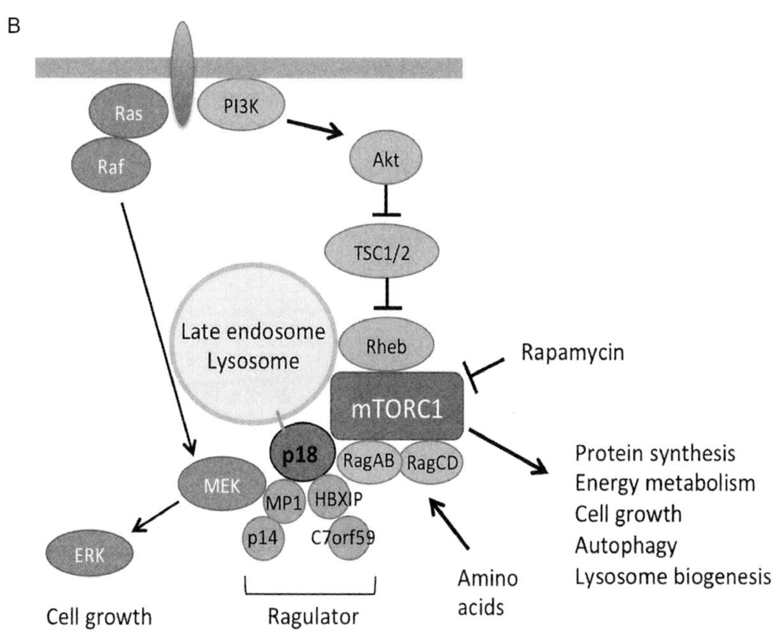

Figure 15.1 (A) Amino acid sequence of human p18. G2 is a myristoylation site. C3 and C4 are palmitoylation sites. L21–L23 is a diLeu motif. Y40 is a phosphorylation site. (B) Cell signaling pathways anchored to late endosome/lysosomes via p18/LAMTOR1. (For color version of this figure, the reader is referred to the online version of this chapter.)

pheochromocytoma cells and mouse embryonic fibroblasts) (Nada et al., 2009; Fig. 15.2).

2.1. Fractionation of DRMs

1. Plate cells in two 14 cm culture dishes at a concentration of 2.5×10^6 cells/plate, and grow them to ~70–80% confluence.
2. Wash both plates twice with 5 ml of ice-cold phosphate-buffered saline (PBS), and completely remove PBS.
3. To one plate, add 650 μl of 0.25% Triton X-100 in solution A (50 m*M* Tris–HCl (pH 7.4), 150 m*M* NaCl, 1 m*M* EDTA, 10 μg/ml aprotinin,

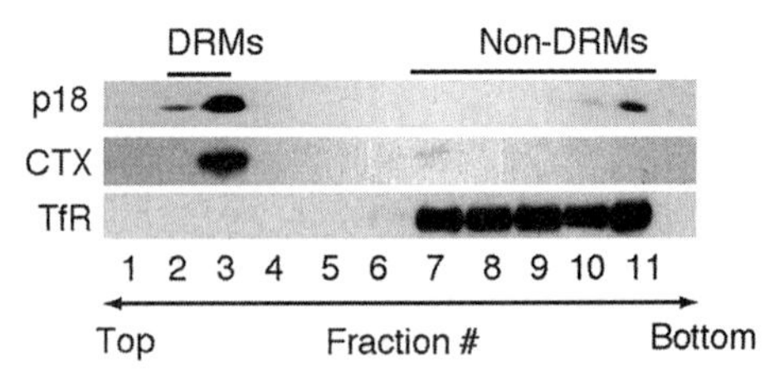

Figure 15.2 Identification of p18 in DRMs (Nada et al., 2009).

10 μg/ml leupeptin, 1 m*M* phenylmethylsulfonyl fluoride, 1 m*M* sodium orthovanadate, and 50 m*M* NaF), and collect the cells using a cell scraper. The concentration of Triton X-100 must be optimized for each cell type.

4. Add the cell lysate from the first plate to the second washed plate, collect the cells, and transfer the pooled lysate (1 ml) to a 2 ml tube.
5. After solubilization by rotating the tube for 1 h at 4 °C, mix the lysate with an equal volume (1 ml) of 85% sucrose in solution A.
6. Place the resulting 42.5% sucrose mixture (2 ml) at the bottom of a 13 × 51 mm ultracentrifuge tube (Beckman, #344057), and sequentially overlay 2.5 ml of 35% sucrose in solution A and 1 ml of 5% sucrose in solution A.
7. Centrifuge at 100,000 × *g* in a Beckman SW50.1 rotor for 16 h at 4 °C.
8. Collect 11 fractions (0.5 ml each) from the top of the gradient using a Pipetman. DRMs are recovered at the interface between the 35% and 5% sucrose layers (Fig. 15.2).

2.2. Detection of p18 by immunoblotting

1. Mix an aliquot of each fraction with SDS-PAGE sample buffer and resolve by SDS-PAGE. Detect p18 by immunoblotting with anti-LAMTOR1/p18 antibody (Cell Signaling Technology). As markers for microdomains and non-microdomain membranes, detect ganglioside GM1 and transferrin receptor, using HRP-conjugated cholera toxin B subunit (CTX) and antitransferrin receptor antibodies (Santa Cruz Biotechnology), respectively.
2. To detect p18-associated proteins, solubilize DRMs by adding 2% octyl-β-D-glucoside (ODG) and 1% NP-40, and precipitate the p18 complex with anti-LAMTOR1/p18 antibody or, when a tagged p18 is expressed, an antibody against the tag sequence (Fig. 15.2).

3. INTRACELLULAR LOCALIZATION OF p18

p18 is specifically localized to LE/Lys (Nada et al., 2009; Sancak et al., 2010; Takahashi et al., 2012). The localization of p18 in cultured cells has been determined by immunocytochemistry and expression of exogenous fluorescent fusion proteins, using LAMP-1 and Rab7 as markers for LE/Lys.

3.1. Immunocytochemistry

Localization of endogenous p18 can be determined by staining cells with antibodies against p18 and several LE/Lys marker proteins.

1. Seed the cells onto fibronectin-coated coverslips and grow them to ~10–30% confluence.
2. Wash the cells with PBS and fix them for 10 min with 4% paraformaldehyde in PBS.
3. After washing with PBS, permeabilize the cells with 50 μg/ml digitonin in PBS (Dig-PBS).
4. Incubate the cells with Blocking One® (Nacalai Tesque, Japan) for 1 h.
5. After washing the cells once with Dig-PBS, incubate them for 1 h at room temperature or overnight at 4 °C with primary antibody diluted 100-fold with 50% Can Get Signal® solution 1 (Toyobo Co., Ltd.) in Dig-PBS.

Primary antibodies

p18: Anti-LAMTOR1/C11orf59 (D11H6)® rabbit monoclonal antibody (Cell Signaling Technology).

Late endosome/lysosome: Anti-LAMP-1 rat monoclonal antibody (1D4B, Santa Cruz Biotechnology).

Late endosome: Anti-Rab7 (D95F2) XP® rabbit monoclonal antibody (Cell Signaling Technology).

Lysosome: Anti-Cathepsin D goat polyclonal antibody (G-19, Santa Cruz Biotechnology).

6. Wash the cells three times with Dig-PBS (10 min per wash).
7. Incubate the cells for 1 h at room temperature with Alexa Fluor® 488 (or 594)-conjugated secondary antibody diluted 1000-fold with 50% Can Get Signal® solution 2 (Toyobo Co., Ltd.) in Dig-PBS.
8. Wash the cells three times with Dig-PBS (10 min per wash) and once for 30 min.

9. Mount the specimen with ProLong® Gold antifade reagent (Invitrogen).
10. Observe the specimen using an Olympus IX81 confocal microscope controlled by Fluoview FV1000 software.
11. Localization of p18 can also be observed by immunostaining cells expressing HA-, FLAG-, or Myc-tagged p18 using a specific antibody against the tag sequence (Fig. 15.3A).

3.2. Expression of fluorescent fusion proteins

1. Vector construction: Clone p18 cDNA into pmKO1-N1 (pEGFP-N1 vector carrying a monomeric Kusabira-Orange 1 gene instead of EGFP), pEGFP-N1, or pmCherry-N1 (Clontech). In previous work, cDNAs for Rab7, Rab5, Rab11, Rab4, MP1, and p14 were cloned into the pEGFP-C1 or pmKO1-N1 vector (Nada et al., 2009).
2. Seed the cells onto fibronectin-coated coverslips and grow them in Dulbecco's modified Eagle medium (DMEM) supplemented with 10% FBS to ~10–30% confluence.
3. Transfect cells with the expression vector using Lipofectamine 2000 (Invitrogen) and grow them for 1–2 days.
4. Wash the cells with PBS and fix them for 10 min in 4% paraformaldehyde in PBS.
5. Mount the specimen with ProLong® Gold antifade reagent (Invitrogen).
6. Observe the specimen using an Olympus IX81 confocal microscope controlled by Fluoview FV1000 software (Fig. 15.3B).
7. For live cell imaging, seed cells onto a 35 mm diameter glass bottom dish coated with fibronectin in DMEM (phenol red-free), and monitor them using an Olympus IX71 microscope equipped with a CoolSNAP HQ camera (Roper Scientific, Trenton, NJ) controlled by MetaMorph software (Universal Imaging, West Chester, PA).

4. IDENTIFICATION OF p18-INTERACTING PROTEINS

p18-interacting proteins have been identified, using pull-down methods, from p18-deficient (p18$^{-/-}$) cells expressing FLAG- or Strep-tagged p18. To date, p14–MP1 (LAMTOR2-3) complex, which is known as a scaffold for MEK1 (Nada et al., 2009; Teis et al., 2006), RagAB/CD

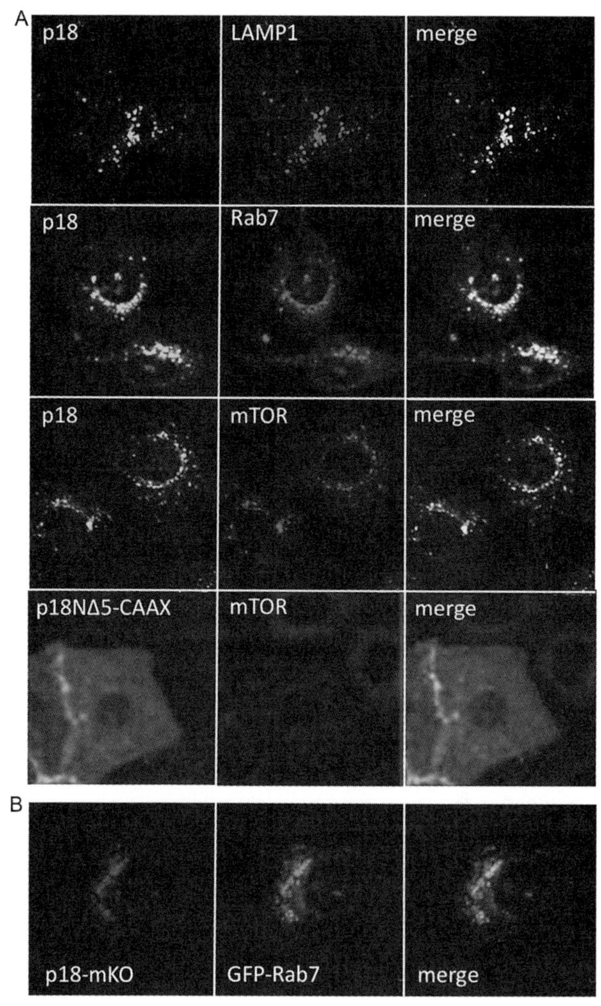

Figure 15.3 (A) Intracellular localization of p18. HA-tagged wild-type p18 or p18NΔ5-CAAX was expressed in p18$^{-/-}$ cells, and the cells were stained with anti-HA antibody. p18NΔ5-CAAX is localized to the plasma membrane as a result of deletion of the N-terminal acylation sites as well as addition of a CAAX motif at its C-terminus. LAMP-1 is a marker of both late endosomes and lysosomes. Rab7 is used as a specific marker of late endosomes. The results show that mTOR is localized to LE/Lys in a manner that depends on p18. (B) Analysis of p18 localization by expression of fluorescent fusion proteins (p18-mKO and GFP-Rab7). (For color version of this figure, the reader is referred to the online version of this chapter.)

GTPases (Sancak et al., 2010), V-ATPase (Zoncu et al., 2011), HBXIP, and C7orf59 (Bar-Peled, Schweitzer, Zoncu, & Sabatini, 2012) have been identified:

1. Vector construction for FLAG-tagged p18: clone p18 cDNA containing a C-terminal FLAG sequence into a retroviral pCX4 vector and stably transfected in $p18^{-/-}$ cells.
2. Plate cells transfected with FLAG-tagged p18 or FLAG vector in three culture dishes (14 cm) and grow them to ~70–80% confluence.
3. Separate DRMs by the method described in Section 2.1, and solubilize proteins in DRMs by adding an equal volume of 2× ODG buffer (40 m*M* Tris–HCl (pH 7.4), 300 m*M* NaCl, 2 m*M* EDTA, 2% NP-40, 4% ODG, 20 μg/ml aprotinin, 20 μg/ml leupeptin, 2 m*M* Na_3VO_4, 100 m*M* NaF, and 10% glycerol).
4. After rotating the tube for 1 h at 4 °C, preclear the solution by incubating with Protein G–Sepharose® (GE Healthcare) (40 μl/1 ml solution) for 15 min at 4 °C, followed by centrifugation at 8000 × *g* for 1 min.
5. To the precleared solution (1 ml), add 40 μl of Anti-FLAG® M2 Affinity Gel (Sigma) equilibrated with 1 × ODG buffer, and incubate with rotation for 3 h at 4 °C.
6. Spin down the gel by centrifugation at 8000 × *g* for 1 min, and wash it for three times each with 1 ml of 1 × ODG buffer.
7. To the pelleted gel, add 50 μl of 1 × SDS sample buffer and boil for 3 min.
8. Separate the proteins by SDS-PAGE (14% T or 8% T) and stain the gel using the MS Silver Staining Kit (Bio-Rad).
9. For mass spectrometry, excise protein bands that are detected only in FLAG-tagged p18-expressing cells and digest them *in situ* with Trypsin Gold (Promega). Analyze the digested samples by LC-ESI–MS/MS (LTQ Orbitrap Velos+ETD, Thermo Scientific). Proteins can then be identified by a database search using Mascot Daemon (Matrix Science) (Fig. 15.4A).

5. FUNCTIONAL ANALYSIS OF p18

p18 is a component of a "Ragulator" complex that binds RagAB/CD to recruit mTORC1 to LE/Lys (Sancak et al., 2010). Also, a recent report has demonstrated that Ragulator serves as a GEF for the Rag GTPases that signal amino acid levels to mTORC1 (Bar-Peled et al., 2012). These findings suggest that p18 is a crucial regulator of mTORC1 function. The

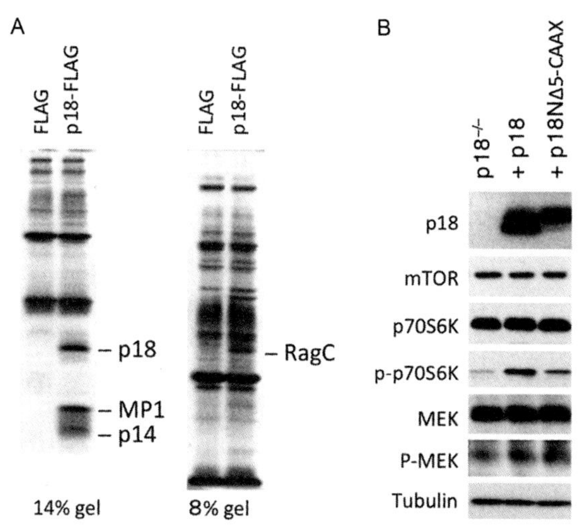

Figure 15.4 (A) Identification of p18-binding proteins by a pull-down method. FLAG-tagged p18 was stably expressed in p18$^{-/-}$ cells, and the p18 complex was purified from DRMs using an anti-FLAG antibody. Proteins specifically bound to p18 were identified by LC–MS/MS analysis. (B) Effects of p18 loss on the mTORC1 signaling pathway were analyzed by immunoblotting with the indicated antibodies. Function of S6K, an mTORC1 target, is attenuated upon loss of p18 from LE/Lys.

activity level of the p18-mTORC1 pathway can be assessed by monitoring the phosphorylation status of ribosomal protein S6 kinase (p70S6K) and 4E-BP1, a negative regulator of elF-4E. Phosphorylation of these mTORC1 target molecules leads to promotion of cell growth by activating protein synthesis.

Furthermore, the p18-mediated pathway is critical for lysosome biogenesis (Guillaumot et al., 2010; Takahashi et al., 2012; Teis et al., 2006). The loss of p18 greatly affects the distribution and maturation of lysosomes.

5.1. Monitoring the phosphorylation status of mTORC1 targets

Analyses of the phosphorylation status of S6K in p18$^{-/-}$ cells and p18$^{-/-}$ cells reexpressing p18 (p18$^{-/-}$+p18) or p18NΔ5-CAAX (p18$^{-/-}$+ p18NΔ5-CAAX) revealed that p18 on LE/Lys is essential for activation of the mTORC1 pathway (Fig. 15.4B). The p18-mTORC1 pathway can also be activated by the stimulation with growth-promoting factors, for example, insulin and IGFs.

1. Plate the cells onto 35 mm culture dishes and grow them to ~70–80% confluence.
2. After washing with PBS, lyse the cells in ODG buffer (250 μl) and clear the lysates by centrifugation at 15,000 rpm for 10 min.
3. Separate the proteins by SDS-PAGE and transfer them to nitrocellulose (or polyvinylidene difluoride) membrane.
4. Block the membrane with Blocking One-P® (Nacalai Tesque) for 30 min at room temperature, and wash it with Tris-buffered saline containing 0.1% Tween 20 (T-TBS).
5. Incubate the membrane for 1 h at room temperature or overnight at 4 °C with anti-phospho-p70S6K(Thr389) antibody (Cell Signaling Technology) and/or anti-p70S6K antibody (Cell Signaling Technology) diluted 1000-fold with 50% Can Get Signal® solution 1 (Toyobo Co., Ltd.) in T-TBS.

   Other antigens useful for assessing the activity of the p18-mTORC1 pathway include Rictor, Raptor, mTOR, mTORpS2448, 4E-BP1, 4E-BP1pT37/46, AKT, and AKTpS473.
6. After washing with T-TBS, incubate the membrane for 30 min at room temperature with HRP-conjugated secondary antibody diluted 10,000-fold with 50% Can Get Signal® solution 2 (Toyobo Co., Ltd.) in T-TBS.
7. Develop the washed membrane using an enhanced chemiluminescence detection system (Pierce). Quantify signal using a LAS 4000 mini (GE Healthcare).

5.2. Analysis of lysosome function

The loss of p18 greatly affects the localization, size, and number of LAMP-1-positive LE/Lys. In p18$^{-/-}$ cells, more small LE/Lys are diffusely distributed throughout the cytoplasm (Takahashi et al., 2012), suggesting that the p18-mTORC1 pathway is involved in the regulation of biogenesis and/or maturation of LE/Lys (Fig. 15.5A).

The role of the p18–mTORC1 pathway in lysosomal degradation was examined using Alexa Fluor® 488-conjugated bovine serum albumin (Alexa488-BSA) and DQ Red BSA as fluorogenic tracers (Takahashi et al., 2012). Alexa488-BSA is endocytosed, and its fluorescence decays following protein processing. By contrast, DQ Red BSA generates fluorescent products only when it is hydrolyzed in lysosomes (Vazquez and Colombo, 2009). The results demonstrated that lysosomal function, as

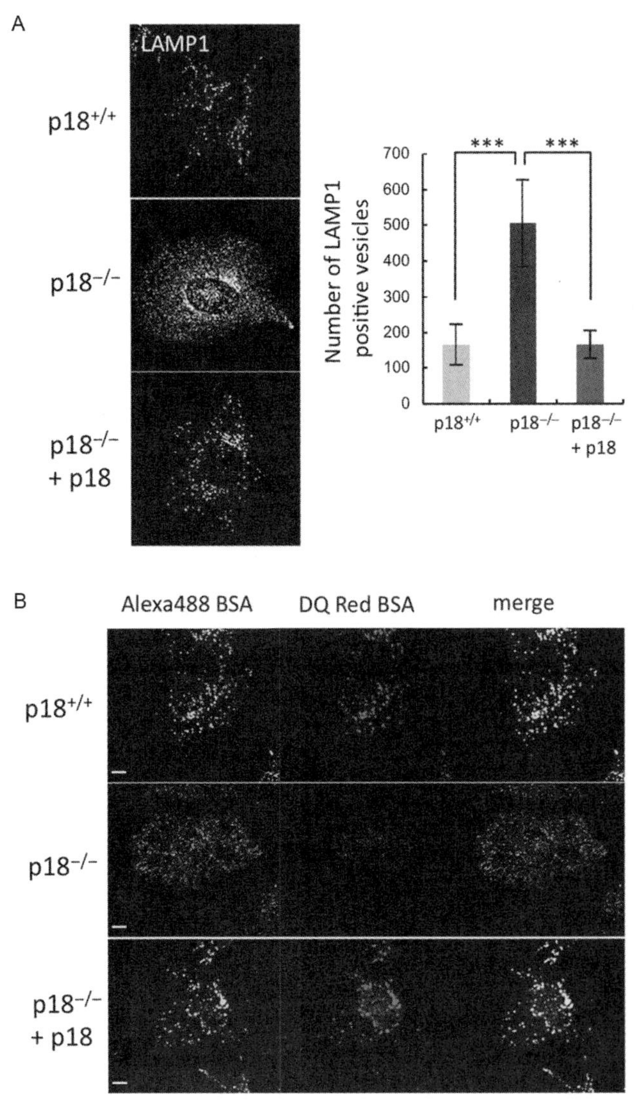

Figure 15.5 (A) Changes in distribution, size, and numbers of LE/Lys upon loss of p18, investigated using immunostaining for LAMP-1. Smaller-sized LE/Lys are diffusely distributed in the cytoplasm of p18$^{-/-}$ cells. (B) The indicated cells were cultured in the presence of Alexa488-BSA and DQ Red BSA, and localization of Alexa488-BSA and the fluorescent degradation products of DQ Red BSA were imaged. Degradation of DQ Red BSA was suppressed by loss of p18. (For color version of this figure, the reader is referred to the online version of this chapter.)

reflected by degradation of DQ Red BSA, is suppressed by the loss of p18 (Fig. 15.5B).

1. Culture cells on fibronectin-coated coverslips in media containing both Alexa488-BSA and DQ Red BSA (10 μg/ml each) for 30 min.
2. Wash the cells twice with media to remove free tracer.
3. Incubate the culture for 0 or 1 h, wash the cells with PBS, and fix them with 4% formaldehyde in PBS.
4. Mount the specimens with ProLong® Gold antifade reagent (Invitrogen).
5. Observe signals for Alexa488-BSA and the fluorescent degradation products of DQ Red BSA using an Olympus IX81 confocal microscope controlled by Fluoview FV1000 software (Fig. 15.5).

5.3. Analysis of p18$^{-/-}$ embryos

The *in vivo* function of p18 was verified by generating p18$^{-/-}$ mice. The mutant embryos died due to growth retardation at the egg cylinder stage. Defects were most evident in the visceral endoderm (VE), in which p18 is abundantly expressed. In control VE cells, larger LAMP-1-positive LE/Lys (Zheng et al., 2006), the so-called "giant" lysosomes, accumulated on the apical sides of cells. In mutant cells, however, relatively small LAMP-1-positive vesicles accumulated beneath the apical membrane. Electron microscopy confirmed that giant lysosomal structures were present in control cells; however, smaller lysosome-like structures with amorphous substances were abundant in the cytoplasm of mutant cells. These observations indicate that p18 is crucial for organization or biogenesis of LE/Lys-related organelles, even *in vivo* (Fig. 15.6). Therefore, it is likely that loss of p18 abrogates the critical functions of VE, that is, nutrient uptake, digestion, and delivery to epiblasts (Bielinska, Narita, & Wilson, 1999); these effects may be responsible for embryonic lethality.

5.3.1 Immunohistochemistry

1. Remove embryos (E6.5), immerse them in a 4% paraformaldehyde–4% sucrose solution in 0.1 *M* phosphate buffer (PB; pH 7.2), and fix them overnight at 4 °C.
2. Embed the samples in optimal cutting temperature compound (Miles, Elkhart, IN) after cryoprotection with 15% and 30% sucrose solutions, and then cut them into 10 μm sections using a cryostat (CM3050, Leica).
3. Place the sections on silane-coated glass slides and stain with hematoxylin and eosin (HE).

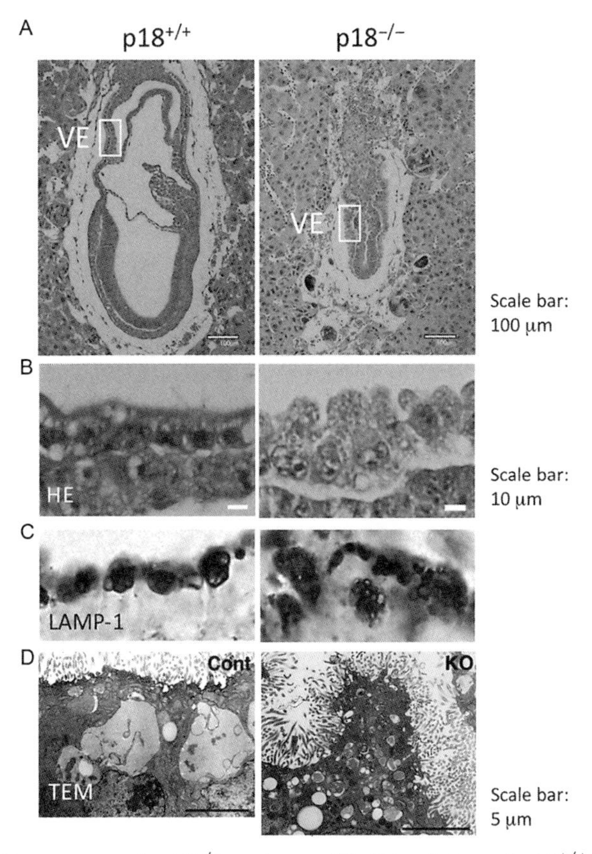

Figure 15.6 Phenotypes of $p18^{-/-}$ embryos. (A) HE staining of $p18^{+/+}$ and $p18^{-/-}$ embryos. Locations of visceral endoderm are indicated by white boxes. (B) Magnified views of VE. (C) Immunohistochemical analysis for LAMP-1. (D) Transmission electron microscopy (TEM) analysis. (For color version of this figure, the reader is referred to the online version of this chapter.)

4. For immunostaining, block the specimens with Blocking One®, and then incubate with anti-LAMP-1 antibody (1:100) in 50% Can Get Signal® solution 1 in Blocking One®.
5. Incubate the washed specimens with biotinylated goat anti-rat IgG antibody for 1 h, and then with peroxidase-conjugated streptavidin (Vector Laboratories) for 1 h at room temperature. Stain the specimens using the VECTASTAIN Elite ABC standard kit (Vector Labs) (Koike et al., 2003; Yagi et al., 2007).

5.3.2 Transmission electron microscopy

1. Immerse the embryos in 2% glutaraldehyde–2% paraformaldehyde, buffered with 0.1 *M* PB (pH 7.2), and then fix them overnight (or longer) at 4 °C.
2. Postfix the embryos with 1% OsO_4 in 0.1 *M* PB, block-stain with a 2% aqueous solution of uranyl acetate, dehydrate them with a graded series of ethanol, and embed in Epon 812.
3. Cut ultrathin sections using an ultramicrotome (Ultracut N, Reichert-Nissei), stain with uranyl acetate and lead citrate, and observe on a Hitachi H-7100 electron microscope (Koike et al., 2003).

6. SUMMARY AND PERSPECTIVE

The *in vitro* and *in vivo* data have revealed the crucial role of p18/LAMTOR1 on LE/Lys in the mTORC1 pathway. Notably, it became evident that, in addition to its vital roles in regulating cell growth and energy homeostasis, the p18–mTORC1 pathway also plays important roles in regulating lysosomal biogenesis and function. These findings suggest that the LE/Lys is a critical site for the operation of the p18–mTORC1 pathway and that there may be a functional link between control of cell growth and lysosomal functions. However, the molecular mechanisms that define the specific localization of p18 to LE/Lys are yet to be elucidated. Moreover, the mechanisms underlying the p18–mTORC1-mediated regulation of LE/Lys function remain thoroughly unknown. Because mTORC1 activity is required for these processes, there may be as-yet-unidentified targets of mTORC1 that are required for the regulation of lysosomal functions. Furthermore, the potential contribution of p18 to the MAPK pathway needs to be clarified. Searching for specific targets of the p18-mediated MAPK pathway would further define the roles of this pathway. Further studies of the p18-mediated pathway will provide clues to understanding the functional link between the endosome system and cell signaling pathways.

ACKNOWLEDGMENTS

We thank C. Oneyama for the advice on the gene-expression experiment, Y. Koreeda and A. Kawai for technical support in generating p18-knockout mice, and T. Akagi for the pCX4 vector. LC–MS/MS analysis was performed in the DNA-chip Development Center for Infectious Diseases (RIMD, Osaka University). This work was supported in part by the Uehara Foundation and a Grant-in-Aid for Scientific Research from the Ministry of Education, Culture, Sports, Science, and Technology of Japan.

REFERENCES

Bar-Peled, L., Schweitzer, L. D., Zoncu, R., & Sabatini, D. M. (2012). Ragulator is a GEF for the rag GTPases that signal amino acid levels to mTORC1. *Cell*, *150*, 1196–1208.

Bielinska, M., Narita, N., & Wilson, D. B. (1999). Distinct roles for visceral endoderm during embryonic mouse development. *International Journal of Developmental Biology*, *43*, 183–205.

Bohn, G., Allroth, A., Brandes, G., Thiel, J., Glocker, E., Schaffer, A. A., et al. (2007). A novel human primary immunodeficiency syndrome caused by deficiency of the endosomal adaptor protein p14. *Nature Medicine*, *13*, 38–45.

Guillaumot, P., Luquain, C., Malek, M., Huber, A. L., Brugiere, S., Garin, J., et al. (2010). Pdro, a protein associated with late endosomes and lysosomes and implicated in cellular cholesterol homeostasis. *PLoS One*, *5*, e10977.

Hoshino, D., Koshikawa, N., & Seiki, M. (2010). A p27(kip1)-binding protein, p27RF-Rho, promotes cancer metastasis via activation of RhoA and RhoC. *Journal of Biological Chemistry*, *286*, 3139–3148.

Koike, M., Shibata, M., Ohsawa, Y., Nakanishi, H., Koga, T., Kametaka, S., et al. (2003). Involvement of two different cell death pathways in retinal atrophy of cathepsin D-deficient mice. *Molecular and Cellular Neuroscience*, *22*, 146–161.

Malek, M., Guillaumot, P., Huber, A. L., Lebeau, J., Petrilli, V., Kfoury, A., et al. (2012). LAMTOR1 depletion induces p53-dependent apoptosis via aberrant lysosomal activation. *Cell Death & Disease*, *3*, e300.

Nada, S., Hondo, A., Kasai, A., Koike, M., Saito, K., Uchiyama, Y., et al. (2009). The novel lipid raft adaptor p18 controls endosome dynamics by anchoring the MEK-ERK pathway to late endosomes. *EMBO Journal*, *28*, 477–489.

Sancak, Y., Bar-Peled, L., Zoncu, R., Markhard, A. L., Nada, S., & Sabatini, D. M. (2010). Ragulator-Rag complex targets mTORC1 to the lysosomal surface and is necessary for its activation by amino acids. *Cell*, *141*, 290–303.

Takahashi, Y., Nada, S., Mori, S., Soma-Nagae, T., Oneyama, C., & Okada, M. (2012). The late endosome/lysosome-anchored p18-mTORC1 pathway controls terminal maturation of lysosomes. *Biochemical and Biophysical Research Communications*, *417*, 1151–1157.

Teis, D., Taub, N., Kurzbauer, R., Hilber, D., de Araujo, M. E., Erlacher, M., et al. (2006). p14-MP1-MEK1 signaling regulates endosomal traffic and cellular proliferation during tissue homeostasis. *Journal of Cell Biology*, *175*, 861–868.

Teis, D., Wunderlich, W., & Huber, L. A. (2002). Localization of the MP1-MAPK scaffold complex to endosomes is mediated by p14 and required for signal transduction. *Developmental Cell*, *3*, 803–814.

Vazquez, C. L., & Colombo, M. I. (2009). Assays to assess autophagy induction and fusion of autophagic vacuoles with a degradative compartment, using monodansylcadaverine (MDC) and DQ-BSA. *Methods in Enzymology*, *452*, 85–95.

Yagi, R., Waguri, S., Sumikawa, Y., Nada, S., Oneyama, C., Itami, S., et al. (2007). C-terminal Src kinase controls development and maintenance of mouse squamous epithelia. *EMBO Journal*, *26*, 1234–1244.

Zheng, B., Tang, T., Tang, N., Kudlicka, K., Ohtsubo, K., Ma, P., et al. (2006). Essential role of RGS-PX1/sorting nexin 13 in mouse development and regulation of endocytosis dynamics. In: *Proceedings of the National Academy of Sciences of the United States of America*, *103*, 16776–16781.

Zoncu, R., Bar-Peled, L., Efeyan, A., Wang, S., Sancak, Y., & Sabatini, D. M. (2011). mTORC1 senses lysosomal amino acids through an inside-out mechanism that requires the vacuolar H(+)-ATPase. *Science*, *334*, 678–683.

Zoncu, R., Efeyan, A., & Sabatini, D. M. (2010). mTOR: From growth signal integration to cancer, diabetes and ageing. *Nature Reviews Molecular Cell Biology*, *12*, 21–35.

CHAPTER SIXTEEN

Vascular Endothelial Growth Factor A-Stimulated Signaling from Endosomes in Primary Endothelial Cells

Gareth W. Fearnley[*], Gina A. Smith[*], Adam F. Odell[*], Antony M. Latham[*], Stephen B. Wheatcroft[†], Michael A. Harrison[‡], Darren C. Tomlinson[§], Sreenivasan Ponnambalam[*,1]

[*]Endothelial Cell Biology Unit, School of Molecular & Cellular Biology, University of Leeds, Leeds, United Kingdom
[†]Division of Diabetes and Cardiovascular Research, Faculty of Medicine & Health, University of Leeds, Leeds, United Kingdom
[‡]School of Biomedical Sciences, University of Leeds, Leeds, United Kingdom
[§]Biomedical Health Research Centre & Astbury Centre for Structural Molecular Biology, University of Leeds, Leeds, United Kingdom
[1]Corresponding author: e-mail address: s.ponnambalam@leeds.ac.uk

Contents

Methods in Enzymology, Volume 535
ISSN 0076-6879
http://dx.doi.org/10.1016/B978-0-12-397925-4.00016-X

Abstract

The vascular endothelial growth factor A (VEGF-A) is a multifunctional cytokine that stimulates blood vessel sprouting, vascular repair, and regeneration. VEGF-A binds to VEGF receptor tyrosine kinases (VEGFRs) and stimulates intracellular signaling leading to changes in vascular physiology. An important aspect of this phenomenon is the spatiotemporal coordination of VEGFR trafficking and intracellular signaling to ensure that VEGFR residence in different organelles is linked to downstream cellular outputs. Here, we describe a series of assays to evaluate the effects of VEGF-A-stimulated intracellular signaling from intracellular compartments such as the endosome–lysosome system. These assays include the initial isolation and characterization of primary human endothelial cells, performing reverse genetics for analyzing protein function; methods used to study receptor trafficking, signaling, and proteolysis; and assays used to measure changes in cell migration, proliferation, and tubulogenesis. Each of these assays has been exemplified with studies performed in our laboratories. In conclusion, we describe necessary techniques for studying the role of VEGF-A in endothelial cell function.

1. OVERVIEW

1.1. Receptor–ligand trafficking and signaling

The family of vascular endothelial growth factors (VEGFs) regulates different aspects of mammalian vascular physiology. The founding member of this family, VEGF-A, is synthesized and secreted as an *N*-glycosylated homodimer. Most of the downstream effects of VEGF-A bioactivity, such as vascular permeability, cell proliferation, cell migration, cell survival, and smooth muscle relaxation, result from binding to the VEGFR2 receptor tyrosine kinase and subsequent activation of downstream signaling pathways (Koch, Tugues, Li, Gualandi, & Claesson-Welsh, 2011).

VEGFR2 activation involves *trans*-autophosphorylation of six to seven specific cytoplasmic domain tyrosine residues creating phosphotyrosine epitopes, for example, pY1175 (Koch et al., 2011; Roskoski, 2007). Such post-translational modifications create binding sites for different enzymes and adaptors that contain phosphotyrosine-binding or Src homology 2 domains (Roskoski, 2007). For example, the recruitment of PLCγ1 to activated VEGFR2 stimulates hydrolysis of plasma membrane phosphatidylinositol 4,5-bisphosphate (PIP_2) to diacylglycerol (DAG) and inositol 1,4,5-triphosphate (IP_3) (Koch et al., 2011; Roskoski, 2007). The production of IP_3

activates IP_3 receptors on the sarcoplasmic reticulum, leading to a rise in cytosolic calcium ion levels. DAG activation of protein kinase C (PKC) activates the mitogen-activated protein kinase (MAPK) pathway. PKC also phosphorylates MEK, resulting in phosphorylation and activation of p42/44 MAPK (ERK1/2), transcriptional upregulation of angiogenic genes, and increased cell proliferation. The Shb adaptor molecule also binds to activated and tyrosine-phosphorylated VEGFR2, stimulating phosphatidylinositol 3-kinase to activate downstream master regulator and serine/threonine protein kinase, Akt. Activation of endothelial Akt triggers activation of endothelial nitric oxide synthase (eNOS) promoting nitric oxide production, vascular permeability, and vasodilation (Koch et al., 2011; Roskoski, 2007). Nitric oxide stimulates the expression and transcriptional activity of the HIF-1 transcription activator, which upregulates VEGF-A mRNA synthesis and provides a further link between eNOS and angiogenesis (Karar & Maity, 2011). Another regulatory protein, T-cell-specific adaptor (TSAd), can be recruited to VEGFR2 via the VEGFR2-pY1175 epitope and stimulates recruitment and activation of the c-Src proto-oncogene and tyrosine kinase, resulting in increased endothelial cell migration and vascular permeability. Sequential activation of CDC42, p38 MAPK, and heat-shock protein 27 following recruitment to the VEGFR2-pY1214 epitope stimulates cell migration and actin remodeling. Generation of the VEGFR2-pY1214 epitope is also linked to increased focal adhesion turnover (Koch et al., 2011; Roskoski, 2007).

Following VEGF-A activation, VEGFR2 is endocytosed and trafficked through the endosome–lysosome system followed by recycling back to the plasma membrane or degradation. A model for VEGFR2 trafficking was suggested proposing that VEGFR2 undergoes constitutive clathrin-dependent endocytosis with recycling from early endosomes (Bruns et al., 2010). Upon VEGFR2 binding to VEGF-A_{165a}, the rate of endocytosis is increased as VEGFR2 phosphorylation, ubiquitination, and proteolysis are linked to trafficking into late endosomes and lysosomes. Rab GTPases are members of the Ras superfamily of GTPases and regulate many aspects of membrane trafficking and organelle fusion. Internalization of VEGFR2 from the plasma membrane into early endosomes requires Rab5a (Jopling et al., 2009). Depletion of Rab5a increases VEGFR2 phosphorylation and MAPK signaling. Additionally, overexpressing the Rab5a GTPase-deficient and constitutively active Q79L mutant causes accumulation of VEGFR2 in enlarged endosomes (Bruns, Bao, Walker, & Ponnambalam, 2009; Jopling et al., 2009). Another member of the Rab family, Rab7a, is involved in the transport of VEGFR2 and other proteins from early to late endosomes (Bruns et al., 2009). Depletion of Rab7a decreases the level of detectable phosphorylated VEGR2 but increases MAPK signaling (Bruns et al., 2009; Jopling et al., 2009). Overexpression of either the Rab7a GDP-bound dominant-negative mutant (T22N) or a GTP-bound constitutively active mutant (Q67L) causes the accumulation of VEGFR2 within late endosomes (Bruns et al., 2009; Jopling et al., 2009). Additionally, depletion of Rab5a or Rab7a showed that Rab GTPase levels had either a stimulatory or an inhibitory effect on endothelial cell migration, respectively (Bruns et al., 2009; Jopling et al., 2009). Redirection of activated VEGFR2 from early to late endosomes suggests that specific compartments could regulate VEGFR2 signaling; this view is strengthened by the fact that a MAPK scaffold associates with endosomes and is required for full ERK1/2 activation (Santambrogio, Valdembri, & Serini, 2011). VEGFR2 undergoes constitutive endocytosis and recycling at a rate of 0.14 min^{-1} (Santambrogio et al., 2011) and is delivered into intracellular EEA1-positive, Rab5a-positive early endosomes before being transported back to the plasma membrane by a Rab4-dependent regulatory step in recycling endosomes (Gampel et al., 2006; Santambrogio et al., 2011). This novel endosomal recycling compartment is independent of another recycling pathway involving Rab11-positive endosomes but is dependent on c-Src tyrosine kinase activity (Gampel et al., 2006).

1.2. Vascular models

Primary endothelial cells are one of the few cell types that express relatively high levels of VEGFR2, and their capacity for both physiological and pathological angiogeneses makes them ideal model systems to study endothelial receptor–ligand complex regulation. VEGFR2–VEGF-A activation and signaling modulates endothelial cell migration, proliferation, tubulogenesis, and apoptosis. All of these outcomes can be replicated using different techniques and assays for the growth of defined primary endothelial cells used to study signaling pathways and effectors involved in a variety of cellular responses.

2. ENDOTHELIAL CELL CHARACTERIZATION

2.1. Introduction

Primary human umbilical vein endothelial cells (HUVECs) provide a physiologically relevant cell type for studying VEGF-A. Here, we describe methods for isolating and validating HUVECs. HUVECs can be retrieved from fresh (~3–18 h old) umbilical cords (~20 cm in length) by digestion of the lumen of the large umbilical vein with collagenase, purified and cultured at 37 °C in a humidified 5% (v/v) CO_2 atmosphere (Howell et al., 2004; Jaffe, Nachman, Becker, & Minick, 1973). HUVECs exhibit a characteristic "cobblestone" morphology (Fig. 16.1A) and can be validated by labeling for the endothelial-specific marker proteins platelet endothelial cell adhesion molecule (PECAM-1/CD31), vascular endothelial (VE)-cadherin, and von Willebrand factor (VWF) (Fig. 16.1B–D) and VEGFR2 (Fig. 16.1E) using immunofluorescence microscopy or flow cytometry. VEGFR2 expression is characteristic of endothelial cells, whereas VEGFR1 is more widely expressed (Fig. 16.1F).

2.2. Isolation and validation of endothelial cells (HUVECs)

2.2.1 Isolation of HUVECs from umbilical cords

1. The umbilical cord contains two arteries and one vein; the vein has the largest diameter. Cannulate the vein with a blunt-ended sterile needle attached to a 20 ml syringe. Flush with prewarmed (37 °C) PBS (phosphate-buffered saline) containing penicillin (100 units/ml), streptomycin (100 μg/ml), and amphotericin B (50 μg/ml) until all blood and clots are removed.
2. Using a hemostat, clamp the umbilical cord at one end and perfuse the vein with ~10 ml of serum-free MCDB131 medium (Invitrogen, Amsterdam, the Netherlands), containing 0.1% (w/v) type II-S collagenase from *Clostridium histolyticum* (Sigma-Aldrich, Poole, United Kingdom), until full before clamping the other end.
3. Incubate the clamped cord at 37 °C for 20 min.
4. Unclamp the vein carefully at one end, placing it within a sterile 50 ml screw cap plastic centrifuge tube. Unclamp the other end, and using a 20 ml syringe as before, flush extensively with PBS containing penicillin (100 units/ml), streptomycin (100 μg/ml), and amphotericin B (50 μg/ml). Collect all dislodged cellular material (up to a total volume of 50 ml) within the sterile tube.
5. Pellet cells by centrifugation at 200 × *g* for 5 min.
6. Resuspend cell pellet in 6–8 ml of prewarmed (37 °C) endothelial cell growth medium (ECGM) supplemented with 2% (w/v) fetal calf serum (FCS), recombinant epidermal growth factor (EGF, 5 ng/ml), hydrocortisone (0.2 μg/ml), recombinant basic fibroblast growth factor (bFGF, 10 ng/ml), recombinant long insulin-like growth factor 1

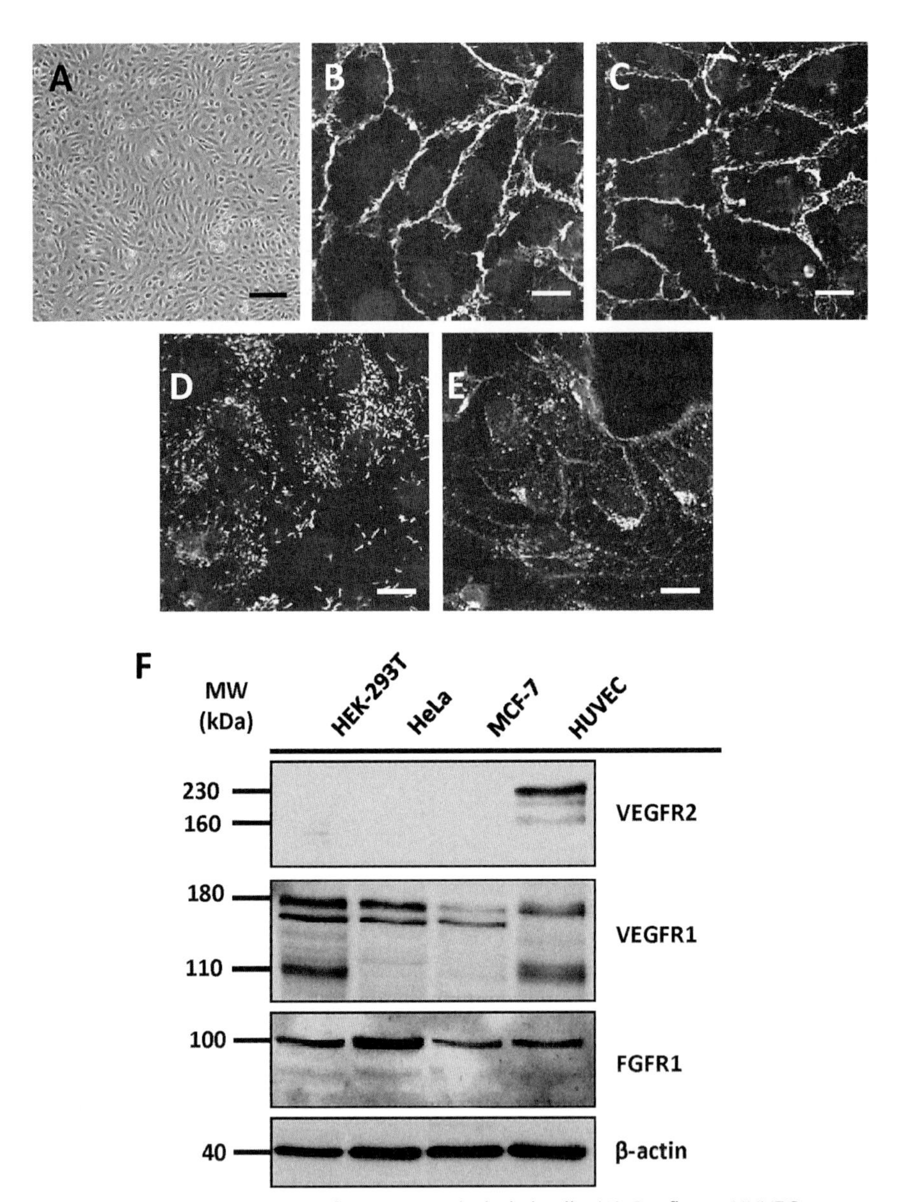

Figure 16.1 Characterization of primary endothelial cells. (A) Confluent HUVEC monolayer grown on gelatin-coated plastic processed for microscopy. Phase contrast picture shown at 10 × magnification. Bar, 60 μm. For immunofluorescence microscopy, confluent HUVECs were labeled with (B) anti-PECAM-1 (CD31), (C) anti-VE-cadherin, (D) anti-von Willebrand factor (VWF), and (E) anti-VEGFR2 antibodies as endothelial markers (green). The nucleus is labeled with DAPI (blue). Bar, 10 μm. (F) Total cell lysates (30 μg per lane) from HEK-293T, HeLa, MCF-7, and HUVECs were probed by immunoblotting using antibodies specific for human VEGFR1, VEGFR2, FGFR1 extracellular domains, or β-actin. Detection was carried out using HRP-conjugated secondary antibodies followed by enhanced chemiluminescence. (For interpretation of the references to color in this figure legend, the reader is referred to the online version of this chapter.)

(IGF-1, 20 ng/ml), ascorbic acid (1 μg/ml), and heparin (22.5 μg/ml) (PromoCell, Heidelberg, Germany).

7. Seed cells into a 75 cm^2 tissue culture flask precoated with 0.1% (w/v) porcine skin gelatin (PSG) (37 °C for 30 min) and incubate at 37 °C overnight.
8. Gently wash the cells four times with PBS to remove any nonadherent cells and replace with 10 ml fresh ECGM.
9. Replace ECGM growth medium every 2–3 days.

2.2.2 Endothelial cell passage

Passage HUVECs every 5–7 days upon cells reaching 70–90% confluence. Do not split HUVECs more than 1:3 per passage as splitting too sparsely causes cell cycle arrest. Only use HUVECs that have been cultured for 0–5 passages as after this point, the cells start to senesce, arrest, and downregulate expression of key endothelial-specific proteins:

1. Aspirate cell culture medium and wash cells twice with sterile PBS.
2. Add 1 ml TrypLE™ (Invitrogen) and incubate at 37 °C for 3 min (or until all cells have become dislodged).
3. Gently tap the side of the flask to remove any adherent cells.
4. Quench trypsinization with 5 ml MCDB131 medium containing 10% (v/v) FCS.
5. Transfer cells to a 50 ml centrifuge tube and pellet cells by centrifugation at 200 × *g* for 5 min.
6. Aspirate supernatant and resuspend cells in required amount of ECGM.
7. Seed cells into a gelatin-coated 75 cm^2 tissue culture flask, single or multiwell dishes.

2.2.3 Validation of endothelial cells using immunofluorescence microscopy

1. Seed HUVECs onto thickness #1.5, gelatin-coated, 13 mm diameter round glass coverslips (VWR Scientific, Lutterworth, United Kingdom). Culture these HUVECs for 1–5 days to the desired cell confluency, that is, seeding $1–2 \times 10^4$ cells will remain subconfluent for 1–2 days for cell proliferation studies, whereas seeding $8–10 \times 10^4$ cells will reach confluences after 1–2 days for signaling and cell–cell adhesion studies. *Note*: VE-cadherin levels are dependent on cell confluence. For optimal VE-cadherin staining, the cells must be close to 100% confluent (Odell, Hollstein, Ponnambalam, & Walker, 2012).

2. Aspirate media and add 300 μl of prewarmed chemical fixative: 10% (v/v) formalin (fixation and permeabilization) or 3% (w/v) paraformaldehyde (fixation only). Incubate coverslips at 37 °C for 5 min.
3. Wash coverslips three times with 500 μl PBS.
4. Permeabilize fixed cells in 1 ml 0.2% (v/v) Triton X-100 in PBS at room temperature for 4 min. *Note*: do not permeabilize cells if wanting to detect cell surface receptors only.
5. Wash coverslips three times with 500 μl PBS.
6. Incubate cells in 5% (w/v) bovine serum albumin (BSA) in PBS at room temperature for 1 h.
7. Wash coverslips three times with 500 μl PBS.
8. Incubate coverslips with primary antibody diluted in 1% (w/v) BSA in PBS at room temperature overnight (16–24 h).
9. Wash coverslips three times with 500 μl PBS.
10. Incubate with cross-purified, species-specific fluorescent-conjugated secondary antibodies (Invitrogen or Jackson ImmunoResearch, West Grove, United States) in PBS containing 1 μg/ml 4′,6-diamidino-2-phenylindole (DAPI) at room temperature for 2–3 h.
11. Wash coverslips three times with 500 μl PBS.
12. Mount onto microscope slides using Fluoromount-G (SouthernBiotech, Birmingham, United States) or equivalent mounting medium.

2.2.4 Validation of endothelial cells using flow cytometry

1. Detach HUVECs from tissue culture plastic using collagenase digestion. Rinse cells twice with prewarmed PBS followed by incubation with 1 ml of PBS containing 0.1% (w/v) collagenase type II-S and 5 m*M* EDTA at 37 °C for 20 min.
2. Transfer detached cells to a 1.5 ml microcentrifuge tube and pellet cells by centrifugation at 200 × *g* at 4 °C for 5 min.
3. Wash cells twice with ice-cold PBS repeating centrifugation at 200 × *g* at 4 °C for 5 min.
4. Block nonspecific binding sites by adding 1 ml 0.5% (v/v) fish skin gelatin (Sigma-Aldrich) in PBS on ice for 20 min.
5. Centrifuge at 200 × *g* at 4 °C for 5 min.
6. Remove supernatant and rinse in 500 μl ice-cold PBS, centrifuged as before.
7. Incubate cells with primary antibody (specific for human CD31, VE-cadherin, or VWF) diluted in PBS containing 0.1% (w/v) BSA and 1 m*M* sodium azide for 1 h on ice.

8. Wash three times with ice-cold PBS using repeated centrifugation at $200 \times g$ at 4 °C for 5 min.
9. Incubate cells with labeled fluorescent species-specific secondary antibodies diluted in 0.1% (w/v) BSA, 1 mM sodium azide, and PBS on ice for 1 h.
10. Wash cells three times with 500 μl ice-cold PBS by centrifugation at $200 \times g$ at 4 °C for 5 min.
11. Fix cells in PBS containing 1% (w/v) paraformaldehyde, 2% (w/v) glucose, and 0.02% (w/v) sodium azide.
12. Analyze samples using a Fortessa™ flow cytometer (Becton Dickinson, Oxford, United Kingdom). Set gates to distinguish positively stained cells from negative controls and analyze 1×10^4 events per experiment.

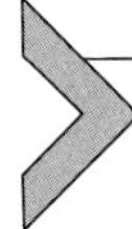

3. Reverse Genetics

3.1. Introduction

RNA interference (RNAi) is a commonly used method for studying the function of proteins. Here, we describe methods for performing RNAi in HUVECs. Rabs are Ras-related small GTPases that regulate VEGFR2 signaling and trafficking. Using RNAi, it is possible to significantly deplete Rab5a or Rab7a levels by 80% or more and trap membrane receptors such as VEGFR2 within early or late endosomes, respectively (Jopling et al., 2009).

3.2. Rab GTPase depletion using RNAi

Note: these working volumes are per 75 cm^2 flask; however, they can be scaled up or down.

1. For RNAi treatment of a single confluent 75 cm^2 flask, prepare a solution of 4 ml antibiotic- and serum-free Opti-MEM medium (Invitrogen) containing 240 pmol siRNA duplex and 32 μl Lipofectamine RNAiMAX transfection reagent (Invitrogen).
2. Invert briefly to mix components and incubate at room temperature for 20 min.
3. Detach HUVECs by trypsinization (see Section 2.2.2) and resuspend pellet at 2.5×10^5 cells/ml in antibiotic- and serum-free Opti-MEM. Add 8 ml of cell suspension ($\sim 2 \times 10^6$ cells) per gelatin-coated 75 cm^2 flask.

4. Add the siRNA/Lipofectamine mixture dropwise to the 75 cm^2 flask with gentle agitation.
5. Incubate at 37 °C for 6 h.
6. Remove Opti-MEM/siRNA media and replace with 10 ml prewarmed ECGM.
7. Culture cells at 37 °C for 48–72 h prior to assays for receptor function and endothelial cell responses.
8. Remove media and wash cells twice with prewarmed PBS. Add 1 ml of TrypLE™ and incubate at 37 °C for 3 min (or until all cells have become rounded and detached).
9. Gently tap the side of the flask to dislodge any remaining adherent cells.
10. Quench trypsinization by adding 5 ml of MCDB131 medium containing 20% (v/v) FCS.
11. Mix cells with trypan blue (Invitrogen) and determine cell number using a hemocytometer.
12. Transfer cells to a 50 ml centrifuge tube and pellet cells at 200 × *g* for 5 min.
13. Aspirate supernatant and resuspend cells at desired concentration in appropriate growth media.
14. Seed HUVECs into gelatin-coated tissue culture plates, multiwell plates, and coverslips for further experiments.
15. Process cells for analysis of VEGFR2 signaling and trafficking (Section 4) or endothelial cell responses (Sections 5 and 6).

3.3. Quantification of Rab GTPase depletion using RNAi

To quantify Rab GTPase depletion, protein levels are detected using immunoblotting. Take a sample containing 5×10^4–1×10^5 cells when seeding transfected HUVECs and compare them to nontransfected cells:

1. Pellet cells at 200 × *g* for 5 min. Remove and discard supernatant.
2. Add 250 μl 2% (w/v) SDS containing 1 m*M* protease inhibitor cocktail (Sigma-Aldrich) to lyse the cells. Gently invert the tube a few times to mix the lysate. This should become a highly viscous but clear solution.
3. Transfer cell lysate to 1.5 ml microcentrifuge tube.
4. Boil lysate at 95 °C for 5 min and sonicate for 3 s at 13–15 μ using a Soniprep 150 probe sonicator (Sanyo, Osaka, Japan).
5. Briefly centrifuge the cell lysate.

6. Determine protein concentration of samples using the bicinchoninic acid assay (BCA).
7. Aliquot 25–50 μg of cell lysate into a 1.5 ml microcentrifuge tube. Add required amount of 2× SDS-PAGE sample buffer (1 *M* Tris–HCl, pH 6.8, 4% (w/v) SDS, 20% (v/v) glycerol, 0.1% (w/v) bromophenol blue, and 4% (v/v) β-mercaptoethanol) to each sample.
8. Briefly centrifuge the cell lysate. Samples can now be stored at −20 °C before analysis.
9. Before performing immunoblot analysis of proteins, pierce the lid of the microcentrifuge tube (this stops build up of pressure and prevents sample loss) and boil the lysate at 92 °C for 5 min before brief centrifugation.
10. Subject cell lysate to denaturing SDS-PAGE on a 6–16% gradient gel at 120 V at room temperature for 1–2 h.
11. Transfer proteins onto reinforced 0.2 μm pore size nitrocellulose membrane at 300 mA at 4 °C for 3 h or at 30 mA at 4 °C overnight (16–24 h).
12. Incubate membranes briefly in Ponceau S (1 g/l Ponceau in distilled water containing 5% (v/v) glacial acetic acid). Rinse off excess dye using distilled water and check for the presence of polypeptides transferred onto membrane.
13. Rinse off Ponceau S stain using TBS-T (20 m*M* Tris pH7.6 and 137 m*M* NaCl containing 0.1% (v/v) Tween-20).
14. Block nonspecific antibody binding by incubating membrane in 5% nonfat milk in TBS-T at room temperature for 30 min.
15. Remove blocking solution and briefly rinse in TBS-T.
16. Incubate membrane with primary antibody (anti-Rab5a or anti-Rab7a) in 1% (w/v) BSA and 1% (w/v) sodium azide in TBS-T at 4 °C overnight.
17. Discard primary antibody and wash three times in TBS-T (10 min per wash).
18. Incubate membrane with species-specific conjugated-HRP secondary antibody in 1% (w/v) BSA and 1% (w/v) sodium azide in TBS-T at room temperature for 1–2 h.
19. Discard secondary antibody and wash three times in TBS-T (10 min per wash).
20. Invert membrane onto enhanced chemiluminescence solution for 1 min.

21. Visualize signal from immunoblot using a sensitive CCD-based imaging workstation, for example, Fujifilm LAS-3000 (Fuji, Japan) with analysis software, for example, AIDA (Fuji, Japan) (Fig. 16.2).
22. Quantify band intensity using AIDA analysis software and compare nontransfected to transfected cells to determine the % Rab depletion. *Note*: Blotting for a housekeeping protein such as actin or tubulin acts as an internal control for variations in sample loading. Dividing the value for Rab GTPase intensity by the tubulin or actin intensity standardizes loading.

4. VEGFR TRAFFICKING, SIGNALING, AND PROTEOLYSIS

4.1. Introduction

In nonstimulated endothelial cells, VEGFR2 localizes in early endosomes, the Golgi, and plasma membrane (Bruns et al., 2009; Jopling, Howell, Gamper, & Ponnambalam, 2011; Manickam et al., 2011). Intensity and duration of VEGF-A stimulation determines the relative distribution of intracellular VEGFR2 and the proportion trafficked from early to late

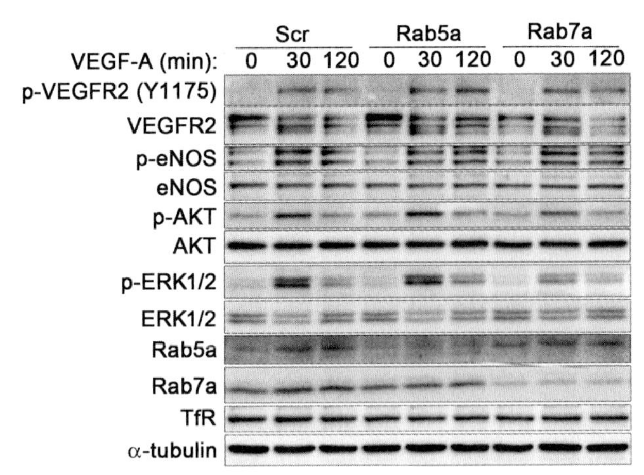

Figure 16.2 Rab GTPase depletion using RNAi on endothelial cells. HUVECs were reverse-transfected with scrambled (Scr), Rab5a- or Rab7a-specific siRNA duplexes as described. Cells were grown for 48 h prior to serum starvation and stimulation with 20 ng/ml VEGF-A for the indicated times. 30 μg of total protein was fractionated by SDS-PAGE prior to immunoblotting with the indicated antibodies to non-phosphorylated and phosphorylated (p-) proteins. Note the enhanced signaling output and increased VEGFR2 levels evident following Rab5a depletion.

endosomes. VEGFR2 trafficking to different cellular compartments influences the signaling outcome; endocytosis is required for Akt and ERK (extracellular signal-regulated kinases) activation, whereas p38 MAPK is activated by surface VEGFR2 only (Chen et al., 2010; Lampugnani, Orsenigo, Gagliani, Tacchetti, & Dejana, 2006; Sawamiphak et al., 2010).

4.2. VEGFR2 trafficking and localization

4.2.1 Cell surface biotinylation

1. Stimulate or treat HUVECs cultured on gelatin-coated 6-well plates as appropriate, place on ice, and wash twice with ice-cold PBS containing 2 mM $MgCl_2$ and 2 mM $CaCl_2$.
2. Incubate cells with 0.3 mg/ml EZ-Link Sulfo-NHS-LC-Biotin (Thermo Fisher, Waltham, United States) in PBS containing 2 mM $MgCl_2$ and 2 mM $CaCl_2$ on ice for 45 min with gentle agitation.
3. Quench biotinylation in TBS (20 mM Tris pH 7.6 and 137 mM NaCl) and lyse cells on ice for 1 h in KSHM lysis buffer (20 mM HEPES pH 7.4, 140 mM KCl, 10 mM potassium acetate, 80 mM sucrose, 2 mM $MgCl_2$, 20 mM Na_2MoO_4, 1 mM Na_3VO_4, 1 mM NaF, and 0.5% (v/v) Triton X-100) followed by 2 washes with PBS.
4. Clear lysates by centrifugation at 16,000 × g at 4 °C for 30 min. Measure and standardize protein concentrations in cell lysates as previously described.
5. Incubate 150 μg of cell lysate with 40 μl of packed NeutrAvidin–agarose beads on a rotating wheel at 4 °C for 3 h.
6. Briefly centrifuge samples to pellet NeutrAvidin–agarose beads, aspirate supernatant, and gently wash in KHSM lysis buffer. Repeat twice more.
7. Add 2 × reducing sample buffer and elute proteins by heat denaturation at 95 °C for 5 min prior to SDS-PAGE and immunoblotting.

4.2.2 Internalization and recycling assays

1. Culture HUVECs until confluent on gelatin-coated 10 cm dishes.
2. Serum starve cells in MCDB131 containing 0.2% (w/v) BSA for 3 h before placing on ice.
3. Wash cells three times with ice-cold PBS containing 2 mM $MgCl_2$ and 2 mM $CaCl_2$.
4. Incubate cells at 4 °C for 30 min with 0.2 mg/ml cleavable NHS-SS-Biotin (Thermo Fisher, Waltham, United States) in PBS containing 2 mM calcium chloride and 2 mM magnesium chloride. Wash cells once with TBS followed by two washes in PBS. All washes performed at 4 °C.

5. Lyse control samples (total biotinylation) using lysis buffer containing 1% (v/v) NP-40, 50 m*M* Tris pH 7.5, 150 m*M* NaCl, and protease inhibitor cocktail (Sigma-Aldrich) or chemically fix the samples for microscopy (see Step 10).
6. Centrifuge cell lysates at 16,000 × *g* at 4 °C for 30 min. Discard pellet and store lysates on ice.
7. Incubate remaining samples at 37 °C in serum-free MCDB131 medium for varying lengths of time to allow internalization of biotinylated cell surface receptor–ligand complexes.
8. Remove exposed cell surface biotin label using three sequential 10 min incubations in buffer containing 100 m*M* sodium 2-mercaptoethansulfonate (MESNA), 50 m*M* Tris pH 8.6, 100 m*M* NaCl, 1 m*M* EDTA, and 0.2% (w/v) BSA.
9. Quench the reaction using 120 m*M* iodoacetamide in PBS.
10. Fix and process cells for microscopy (see Section 2.2.3) or lyse in an isotonic NP-40 lysis buffer.
11. For receptor recycling analysis, subject cells to a second incubation at 37 °C for 20 min in serum-free MCDB131 medium prior to a second round of MESNA washes.
12. Lyse cells, centrifuge lysates, and determine protein concentrations using the BCA assay.
13. Adjust cell lysates such that the protein concentrations are identical for each experimental condition.
14. For each 150 μg of cell lysate, add 40 μl of packed NeutrAvidin–agarose beads (Thermo Fisher, Waltham, United States), place on a rotating wheel, and incubate at 4 °C overnight. This promotes binding of biotinylated proteins to the NeutrAvidin–agarose beads.
15. Centrifuge each lysate briefly to pellet the beads. Wash beads three times in lysis buffer, and resuspend in reducing 2 × sample buffer. Incubate lysates at 95 °C for 5 min to denature proteins prior to analysis by SDS-PAGE and immunoblotting (Section 3.3).

4.2.3 Direct receptor recycling assay

1. Culture HUVECs on gelatin-coated coverslips until confluent and serum starve in MCDB131 medium for 3 h prior to treatment.
2. Incubate cells at 37 °C for 1 h with primary antibody (in PBS) or at 4 °C as a negative control.
3. After incubation, chill on ice and remove bound cell surface antibody by acid washing cells twice in ice-cold MCDB131 medium adjusted to pH 2.0.

4. Wash cells twice in ice-cold MCDB131 medium at standard pH.
5. Incubate cells with a species-specific fluorescent-conjugated secondary antibody (in PBS) at 37 °C for 1 h. Remove surface-bound antibody by acid wash as previously described.
6. Fix cells and visualize VEGFR2 localization by immunofluorescence microscopy. Only proteins that have recycled at least 1.5 times are visible by microscopy.

4.3. Analysis of VEGF-A-stimulated signaling events

Changes in signaling pathways can be monitored by detection of posttranslational modifications (e.g., phosphorylation) on key signaling proteins (VEGFR2, Akt, eNOS, and MAPK) whose total levels are also simultaneously quantified using SDS-PAGE and immunoblot analysis:

1. Culture HUVECs until confluent in gelatin-coated 6-well plates.
2. Aspirate media and wash once in PBS.
3. Serum starve cells in MCDB131 medium containing 0.2% (w/v) BSA at 37 °C for 3 h.
4. Stimulate cells with 25 ng/ml VEGF-A for desired time course.
5. Lyse cells in 100 μl lysis buffer (2% (w/v) SDS, phosphatase inhibitor cocktail, and protease inhibitor cocktail in PBS).
6. Transfer lysates into 1.5 ml microcentrifuge tubes.
7. Boil lysates at 92 °C for 5 min. Sonicate for 3 s using a tip probe sonicator set at 10–15 μ.
8. Briefly centrifuge cell lysates.
9. Determine protein concentration using the BCA assay.
10. Aliquot 25–50 μg of cell lysate into a 1.5 ml microcentrifuge tube. Add required amount of 2 × SDS-PAGE sample buffer to each sample.
11. Briefly centrifuge the cell lysate. Samples can now be stored at −20 °C before analysis.
12. Load 25–50 μg of protein lysate on a 10% SDS-PAGE gel and carry out immunoblot analysis as previously described (see Section 3.3) using primary antibodies against desired signaling node.
13. Visualize signal from immunoblot using a sensitive CCD-based imaging workstation, for example, Fujifilm LAS-3000 (Fuji, Japan) with analysis software, for example, AIDA (Fuji, Japan) (Fig. 16.2).
14. Quantify band intensity using analysis software and compare VEGF-A-stimulated protein levels to nonstimulated control levels. *Note*: Blotting for a housekeeping protein such as actin or tubulin acts as an internal control for variations in sample loading. Dividing the value for protein of interest intensity by the tubulin or actin intensity standardizes loading.

5. ENDOTHELIAL CELL RESPONSES

5.1. Introduction

VEGF-A modulates vasculogenesis and angiogenesis by regulating specific endothelial cell responses such as cellular proliferation, migration, and viability. Cellular proliferation is regulated by the ERK, p38, and Akt signaling nodes (Horowitz & Seerapu, 2012; Liu et al., 2006). These pathways are activated downstream of VEGFR2 upon stimulation with VEGF-A (Olsson, Dimberg, Kreuger, & Claesson-Welsh, 2006). Disruption of VEGFR2 endosomal signaling and proteolysis was shown to regulate the levels of ERK1/2 and Akt activation, respectively (Bruns et al., 2009, 2010; Jopling et al., 2009), which impacts on cellular outputs such as cell proliferation. A simple and effective assay for evaluating cell proliferation is by monitoring the incorporation of the pyrimidine 5-bromo-2′-deoxyuridine (BrdU) analog into newly synthesized DNA (instead of thymidine) using a nonradioactive, ELISA-like colorimetric assay.

The early stages of angiogenesis and vasculogenesis depend heavily on endothelial cell migration (Carmeliet, 2005; Schmidt, Brixius, & Bloch, 2007). Our lab has shown that disruption of VEGFR2 trafficking and proteolysis in the endosome–lysosome system perturbs VEGF-A-stimulated cell migration (Bruns et al., 2009; Jopling et al., 2009). Additionally, VEGF-A also regulates endothelial cell viability and apoptosis (Gerber et al., 1998). This section provides a subset of cellular assays to determine the effect that blocking endosomal signaling has on cell outcome.

5.2. Cell proliferation

Note: We recommend using a cell proliferation ELISA kit (Roche Diagnostics, Burgess Hill, United Kingdom). This protocol is optimized for the use of this cell proliferation ELISA kit and reagents provided within. The length of incubation times may need optimization for nonkit reagents:

1. Seed 2×10^3 transfected or nontransfected cells per well of a gelatin-coated 96-well plate and culture overnight.
2. Stimulate endothelial cells with VEGF-A for desired times.
3. Incubate cells with 10 μM BrdU for 2–24 h; above 8 h is best for HUVECs as they exhibit relatively slow proliferation. Generally, use multiple wells (3–5) per experiment.

4. Remove media by inverting the plate and gently tapping on some tissue paper.
5. Fix cells in 200 μl per well FixDenat (ethanol-based fixative and DNA denaturing solution) at room temperature for 30 min.
6. Remove FixDenat solution thoroughly by flicking and tapping the plate as previously described.
7. Add 100 μl per well of anti-Brdu-POD (anti-Brdu primary antibody plus species-specific conjugated-HRP secondary antibody) working solution and incubate at room temperature for 90 min. *Note*: if using an anti-Brdu primary antibody plus species-specific conjugated-HRP secondary antibody, perform Steps 8–9 after primary and secondary antibody incubation.
8. Remove antibody conjugate by flicking and tapping the plate as previously described.
9. Wash by adding 200 μl per well of wash solution (PBS).
10. Remove wash solution (PBS) by flicking and tapping the plate as previously described.
11. Repeat Steps 9–10 twice more to give a total of three washes.
12. Add 100 μl per well of substrate solution (3,3′,5,5′-tetramethylbenzidine) and incubate at room temperature for 5–30 min until color change is sufficient for photometric detection. Measure absorbance routinely after 10–30 min at 370 or 650 nm using a multiwell plate absorbance reader.
13. Stop reaction by adding 25 μl per well of stop solution (1 *M* H_2SO_4) and incubate at room temperature for 1 min while mixing thoroughly.
14. Measure the absorbance immediately at 450 nm using a multiwell plate absorbance reader.

5.3. Cell migration

The endothelial cell migration assay was carried out using 8 μm pore size polyester membrane (Transwell) inserts from BD Biosciences (Oxford, United Kingdom):

1. Place Transwell inserts into a 24-well tissue culture plate. Aliquot 400 μl of 0.1% (w/v) PSG per well and place Transwell filter into well. Add 100 μl 0.1% (w/v) PSG to the inside chamber of the Transwell filter. Leave at 37 °C for 1 h.
2. Gently remove PSG and rinse both well and Transwell inserts once with PBS. Take care not to damage the Transwell membrane.

3. Aspirate all traces of liquid and place inside a warm incubator until completely dry.
4. Set up desired chemotactic gradient by placing 400 μl of control medium (containing chemokine or growth factor) per well of a new 24-well plate.
5. Carefully place precoated Transwell insert into well ensuring no air is trapped between the membrane and the liquid.
6. Seed 6×10^4 transfected or nontransfected endothelial cells in100 μl total volume into the center of each Transwell insert. Ensure the cells have been resuspended in control media lacking the chemokine or growth factor under study.
7. Allow endothelial cells to migrate across the Transwell membrane (toward the chemokine or growth factor) for 18–24 h.
8. Gently remove media from inside the Transwell insert by aspiration and transfer inserts into a fresh well containing 400 μl 10% (v/v) formalin. Allow chemical fixation to occur at room temperature for 5 min.
9. Rinse Transwell filters three times by gentle submersion into a beaker of PBS.
10. Stain endothelial cells present on the membrane by placing the Transwell inserts into fresh wells containing 400 μl of crystal violet solution (filter-sterilized solution of 0.2% (w/v) crystal violet and 20% (v/v) methanol in distilled water) at room temperature for 30 min.
11. Remove Transwell filter and rinse by gentle submersion into a beaker of PBS.
12. Remove cells from the upper (internal) side of the filter using a cotton bud.
13. Leave filters at room temperature overnight and allow to dry completely.
14. View the membrane using a digital microscope system with low-power (4 × and 10 ×) and high-power (60 ×) objective lenses. Collect 3–5 random field images (containing no more than 100 cells per image to reduce counting errors) per membrane (Fig. 16.3A).
15. Average the number of migrated cells per field and express them as a fold or percentage increase or decrease compared to the number of migrated cells under control conditions in the absence of the cytokine (Fig. 16.3B).

5.4. Apoptosis and cell cycle analysis

Flow cytometry is a convenient technique to evaluate endothelial cell apoptosis or DNA content. In contrast to previous techniques that detect cell

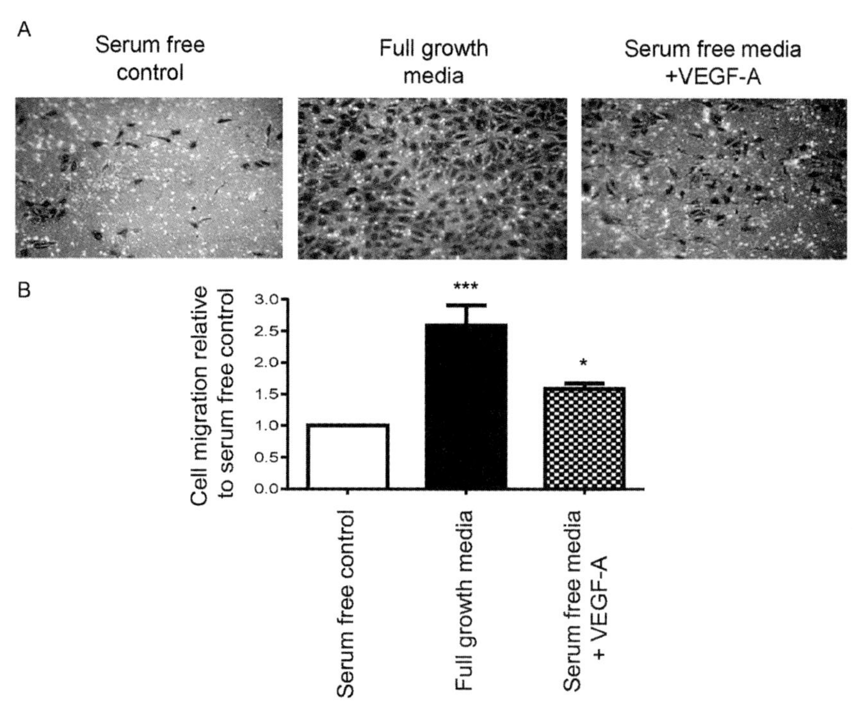

Figure 16.3 VEGF-A-stimulated endothelial cell migration. HUVECs were seeded into Transwell filters and migration occurred over 24 h in full-growth media or serum-free media without or with VEGF-A_{165a} (25 ng/ml). (A) Cells were fixed with formalin and stained with crystal violet. The number of migrated cells was counted from three random field images per experiment and an average taken. (B) Quantification of migration assay showed that HUVEC stimulation with exogenous VEGF-A_{165a} (25 ng/ml) promotes ~1.5-fold increase in endothelial migration when compared to nonstimulated controls. Error bars denote ±SEM ($n=3$). $^{*}p<0.05$; $^{***}p<0.001$; all values compared against nonstimulated control. (For color version of this figure, the reader is referred to the online version of this chapter.)

surface membrane receptors using antibody-based labeling, probes are used to label nonprotein molecules. In one case, a phospholipid, phosphatidylserine (PS), is exposed at the extracellular/exposed leaflet of the plasma membrane lipid bilayer upon mammalian cell commitment to apoptosis or programmed cell death (Fadok et al., 1992). This can be monitored using annexin V, which binds to PS in the presence of calcium ions (Koopman et al., 1994). The cellular DNA content is also substantially altered during states such as interphase, mitosis, or apoptosis. This change in DNA content can also be monitored using flow cytometry by monitoring the binding of a fluorescent DNA-binding dye to endothelial cell DNA.

5.4.1 Apoptosis assay

1. Seed nontransfected or transfected endothelial cells into gelatin-coated 6-well plates and culture until ~70% confluent.
2. Stimulate endothelial cells with ligand in 1 ml total volume of media. Incubate for desired times, for example, 0–72 h.
3. At the end of the treatment period, remove cell culture media and store on ice.
4. Add 250 μl well TrypLE™ (Invitrogen) and incubate at 37 °C for 3 min or until all cells have detached.
5. Resuspend cells in original media and transfer cells to 1.5 ml centrifuge cells.
6. Pellet cells at 200 × *g* at 4 °C for 5 min and discard supernatant carefully.
7. Wash cells in 500 μl ice-cold binding buffer (10 m*M* HEPES pH 7.5, 140 m*M* NaCl, and 2.5 m*M* $CaCl_2$).
8. Pellet cells at 200 × *g* at 4 °C for 5 min and discard supernatant carefully.
9. Resuspend cells in 500 μl ice-cold binding buffer.
10. Add FITC-conjugated chicken liver annexin V (or equivalent), isolated and labeled as previously described (Boustead, Brown, & Walker, 1993), to a final concentration of 10 μg/ml of labeled FITC-annexin V. Resuspend gently by pipetting and incubate at room temperature in the dark for 20 min.
11. Pellet cells at 200 × *g* at 4 °C for 5 min and discard supernatant carefully.
12. Resuspend labeled cells in ice-cold binding buffer.
13. Add DNA labeling dye DAPI to a final concentration of 2 μg/ml immediately prior to analysis on the flow cytometer.
14. Analyze labeled cells using a flow cytometer set up to detect DAPI (360 nm excitation and 460 nm emission) on the *y*-axis and FITC on *x*-axis (490 excitation and 520 emission). *Note*: To reduce the appearance of nonspecific binding, FITC channel may have to be left-shifted to bring live cells into bottom-left quadrant (~220 V). HUVECs characteristically display low forward scatter (~80 V).

5.4.2 Cell cycle and genomic DNA analysis

1. Seed nontransfected or transfected HUVECs into gelatin-coated 6-well plates and culture until ~70% confluent.

2. Stimulate endothelial cells with ligand in 1 ml total volume of media. Incubate for desired times, for example, 0–72 h.
3. Aspirate media and add 250 μl TrypLE™ to each well. Incubate at 37 °C for 3 min or until all cells have detached from the plastic surface.
4. Quench trypsinization by adding 1 ml of MCDB131 containing 20% (v/v) FCS.
5. Determine cell number using hemocytometer. Transfer cell suspension to 1.5 ml microcentrifuge tube.
6. Pellet cells at 200 × *g* at 4 °C for 5 min.
7. Add ice-cold 70% ethanol dropwise under vortexing to resuspend cells to a final concentration of $\sim 1 \times 10^6$ cells per ml.
8. Immediately place cell suspension at −20 °C overnight. Such samples can be stored in this way for a few weeks before flow cytometry analysis.
9. Before carrying out flow cytometry, pellet the cells by centrifugation at 200 × *g* at 4 °C for 5 min.
10. Aspirate ethanol fixative and resuspend the cell pellet in 500 μl PBS by gentle pipetting. Recentrifuge as before and resuspend in 500 μl PBS. Repeat centrifugation step as before.
11. Aspirate supernatant and resuspend pellet in 0.5 ml of 100 μg/ml ribonuclease and 50 μg/ml propidium iodide in PBS. Incubate at 37 °C for 3 h.
12. Pellet cells by centrifugation at 200 × *g* at room temperature for 5 min.
13. Carefully remove and discard supernatant. Gently resuspend cell pellet in 1 ml of PBS.
14. Run samples on a Fortessa flow cytometer set to run at a low flow rate (12–60 μl/min) and analyze data using ModFit software (Becton Dickinson).

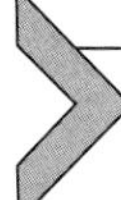

6. TUBULOGENESIS

6.1. Introduction

A functional assay that evaluates endothelial cell capacity to form vascular tubes *in vitro* is the endothelial–fibroblast coculture assay (Bishop et al., 1999). In this assay, primary endothelial cells (HUVECs) are seeded on a confluent layer of normal human dermal foreskin fibroblasts and cultured for 7–10 days, depending on growth conditions, media, and treatments.

One important advantage of this assay is that other assays employing biological matrices (e.g., collagen, Matrigel) also promote growth of multicellular/tubular structures of nonvascular cells, for example, fibroblasts and epithelial cells. However, this coculture assay appears restricted in its specificity in promoting endothelial tube formation (tubulogenesis) within a heterogeneous cellular population (Beilmann, Birk, & Lenter, 2004; Donovan, Brown, Bishop, & Lewis, 2001). Tubulogenesis is an essential feature in the phenomenon of angiogenesis. This organotypic angiogenesis assay has been used to evaluate the action of small-molecule tyrosine kinase inhibitors on VEGF-A-stimulated responses (Kankanala et al., 2012; Latham et al., 2012).

6.2. Organotypic angiogenesis assay

1. Culture primary human foreskin fibroblasts (PromoCell) in Q333 medium (PAA Laboratories, Pasching, Austria) until ~70% confluent in either a 75 cm^2 flask or a 10 cm dish.
2. Aspirate the medium and rinse cells briefly twice with PBS.
3. Add 1 ml of TrypLE™ and incubate at 37 °C for 3 min (or until all cells have become rounded and dislodged).
4. Quench trypsinization with 5 ml of MCDB131 medium containing 20% (v/v) FCS.
5. Transfer cell suspension into a sterile 50 ml plastic centrifuge tube. Pellet cells at $200 \times g$ at room temperature for 5 min.
6. Aspirate solution and resuspend cell pellet in 24 ml of Q333 growth medium.
7. Seed 500 μl of fibroblast suspension into each well of a gelatin-coated 48-well plate.
8. Culture fibroblasts until they become confluent (24–48 h).
9. Detach nontransfected or transfected HUVECs and resuspend at $\sim 1 \times 10^4$ cells/ml in media containing Q333/ECGM (1:1).
10. Aspirate the growth medium from the fibroblasts.
11. Add 500 μl of the HUVEC cell suspension (~5000 cells) to each well (*day* 1). Culture cells overnight (16–24 h).
12. Aspirate the growth media and add 500 μl ECGM (*day* 2). Culture overnight (16–20 h).

13. Aspirate growth medium, and add fresh 500 μl ECGM containing growth factor (e.g., VEGF-A) and/or compounds, drugs, etc., depending on experimental conditions (*day* 3). Replace this medium with the exact experimental conditions every 2–3 days until the end of the 7–10-day period.
14. Aspirate growth medium and add 200 μl 10% (v/v) formalin at room temperature for 10–20 min to chemically fix the cells.
15. Aspirate fixative and briefly rinse twice with PBS.
16. Block nonspecific antibody binding by adding 500 μl 1% (w/v) BSA in PBS at room temperature for 30 min.
17. Aspirate solution and add 250 μl of mouse anti-PECAM1 antibody (0.4 μg/ml; Santa Cruz Biotechnology, United States) in 1% (w/v) BSA in PBS overnight.
18. Wash sample with three rinses of 500 μl PBS.
19. Incubate fixed cells with 250 μl secondary HRP-conjugated antibody per well. A species-specific HRP-conjugated secondary antibody (1–10 μg/ml) is diluted into 1% (w/v) BSA in PBS.
20. Wash sample three times with 500 μl PBS.
21. Stain endothelial tubules by adding 150 μl of a 3,3′-diaminobenzidine (DAB)/urea/hydrogen peroxide development solution (Sigma-Aldrich).
22. Allow HRP activity to proceed and color to develop by incubating at room temperature for 15–20 min. *Note*: orange color begins to develop immediately and is maximal after 15–20 min.
23. Stop reaction by aspirating the substrate and adding 500 μl PBS.
24. Analyze samples using an inverted microscope connected to a digital camera with phase contrast optics. Collect images of 3–5 random fields per experiment (Fig. 16.4A).
25. Using NIH ImageJ or similar quantification software, analyze endothelial tube profiles. Automated batch analyses of tubule dimensions can be performed using an open-source software package called AngioQuant (www.cs.tut.fi/sgn/csb/angioquant). For each field, count number of branch points and measure total endothelial tubule length. Calculate the average endothelial tubule branch points and average total tubule length per experiment. Compare these values to negative (e.g., lacking VEGF-A) and positive (plus VEGF-A) controls (Fig. 16.4B).

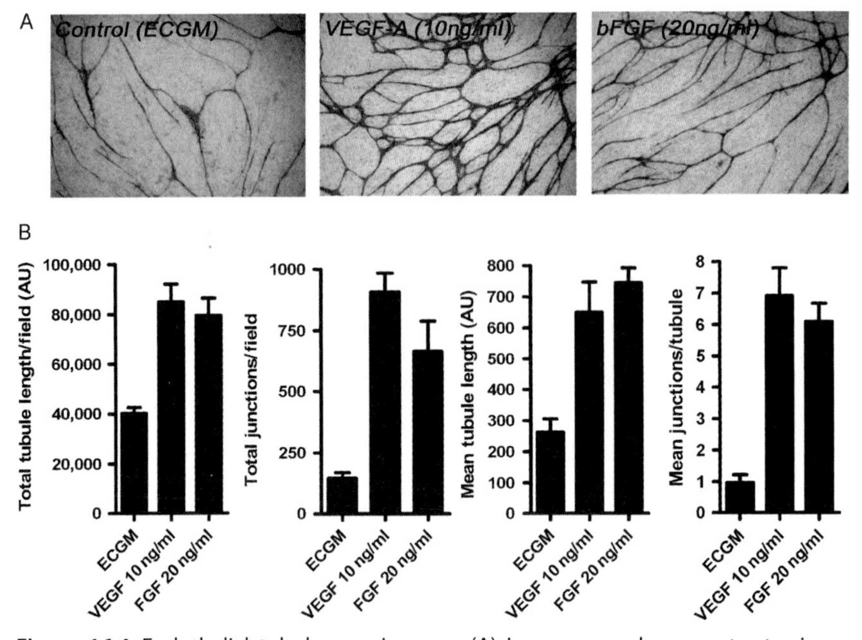

Figure 16.4 Endothelial tubulogenesis assay. (A) Low-power phase contrast microscopy of antibody-based HRP staining of endothelial-specific proteins to monitor endothelial tube formation in fibroblast coculture. This was carried out in either control growth medium (EGCM), ECGM plus VEGF-A (10 ng/ml, 0.22 n*M*), or ECGM plus bFGF (20 ng/ml, 1.25 n*M*). (B) Quantification of mean tubule length or number of tubule junctions under the different growth conditions. Error bars indicate ±SEM.

7. SUMMARY

The VEGF-A cytokine was originally identified by its ability to modulate vascular function in mammals. In the past 20 years, the complexity of this gene family has been vastly increased by the discovery of multiple genes in mammals (VEGF-A, VEGF-B, VEGF-C, VEGF-D, and PlGF), viral orthologs (VEGF-E), and snake venoms (VEGF-F) (Koch et al., 2011; Ponnambalam & Alberghina, 2011; Ruiz de Almodovar, Lambrechts, Mazzone, & Carmeliet, 2009). Many of these genes encode multiple splice variants whose functions remain obscure. Genetic ablation of VEGFs is invariably lethal during early embryogenesis, making it difficult to evaluate gene function in different tissues and organs. Of note, VEGFs have been increasingly implicated in epithelial and neuronal function, suggesting that these cytokines have a much wider regulatory function than anticipated. This highlights the need for studies within this area using different cell-based systems to decode the biological activity of the VEGF superfamily.

There is increasing evidence that VEGF-A-dependent signaling and proangiogenic outcomes are regulated by receptor–ligand trafficking and processing through the endosome–lysosome system. Regulators of receptor-mediated endocytosis (Bruns et al., 2010; Ewan et al., 2006), endosome-linked Rab GTPases (Gampel et al., 2006; Jopling et al., 2009; Reynolds et al., 2009), and endosome-associated ubiquitination machinery (Ewan et al., 2006; Hasseine et al., 2007) regulate VEGF-A-stimulated downstream signaling pathways. Ablation of specific regulators linked to different membrane compartments in which a VEGFR complex is located could delineate specific signaling and cell response outcomes associated with temporal and spatial distribution of the receptor–ligand complex. In this way, one can determine how different biological signals are generated in time and space upon receptor binding to a specific ligand, for example, VEGF-A.

A major challenge is to evaluate the role of a complex series of enzymes including GTPases and ubiquitin ligases in modulating response to VEGF-A through VEGFR2. Different studies suggest a role for Rab5a in regulating VEGFR2 trafficking through early endosomes (Jopling et al., 2009), whereas evidence suggests a role for Rab4a and/or Rab11a in recycling from endosomes (Gampel et al., 2006; Reynolds et al., 2009). The Rab7a GTPase associated with late endosomes regulates VEGFR2 trafficking toward final degradation in lysosomes (Jopling et al., 2009). Intriguingly, VEGFR2 is also trafficked slowly through the secretory pathway with evidence for a SNARE-regulated mechanism in controlling its transit through the Golgi apparatus (Manickam et al., 2011). The role of ubiquitination in VEGFR2 function is contradictory: although early studies suggested a role for the E3 ubiquitin ligase c-Cbl (Duval, Bedard-Goulet, Delisle, & Gratton, 2003), more recent studies suggest roles for different E3 ubiquitin ligases (Bruns et al., 2010; Meyer et al., 2011; Murdaca et al., 2004). Notably, another receptor tyrosine kinase such as epidermal growth factor receptor (EGFR/ErbB1) exhibits both mono- and polyubiquitination upon ligand binding and activation (Haglund & Dikic, 2012; Haglund et al., 2003; Huang & Sorkin, 2005; Sorkin & Goh, 2009). Different endosome-associated ubiquitin ligases and deubiquitinases could thus fine-tune cellular responses to a receptor–ligand complex depending on its residence time within a specific compartment and final proteolysis of the ubiquitinated membrane receptor.

The VEGFR–VEGF axis is of much interest in understanding vascular function but is also implicated in tumor neovascularization, arterial repair, and regeneration after heart attacks and strokes. Our work provides a suite of techniques and assays to stimulate further work in this area and related fields such as neurobiology. Such assays can be easily adapted toward studies in primary and transformed epithelial, neuronal, and immune cells, systems that are likely to respond to other VEGF splice variants or family members. Increasingly, VEGF dysfunction is implicated in conditions such as amyotrophic lateral sclerosis (motor neuron disease), multiple sclerosis, and Alzheimer's disease (Ponnambalam & Alberghina, 2011). Deciphering the role of a specific VEGFR–VEGF complex in a cellular and tissue-specific context will shed light not only on basic mechanisms of receptor signaling and function but also on possible therapeutic strategies in a wide variety of ailments.

ACKNOWLEDGMENTS

This work was supported by a Heart Research UK PhD studentship (G. W. F.), a British Heart Foundation PhD studentship (G. A. S), and British Heart Foundation project grant (S. B. W.).

REFERENCES

Beilmann, M., Birk, G., & Lenter, M. C. (2004). Human primary co-culture angiogenesis assay reveals additive stimulation and different angiogenic properties of VEGF and HGF. *Cytokine*, *26*, 178–185.

Bishop, E. T., Bell, G. T., Bloor, S., Broom, I. J., Hendry, N. F., & Wheatley, D. N. (1999). An in vitro model of angiogenesis: Basic features. *Angiogenesis*, *3*, 335–344.

Boustead, C. M., Brown, R., & Walker, J. H. (1993). Isolation, characterization and localization of annexin-v from chicken liver. *The Biochemical Journal*, *291*, 601–608.

Bruns, A. F., Bao, L., Walker, J. H., & Ponnambalam, S. (2009). VEGF-A-stimulated signalling in endothelial cells via a dual receptor tyrosine kinase system is dependent on co-ordinated trafficking and proteolysis. *Biochemical Society Transactions*, *37*, 1193–1197.

Bruns, A. F., Herbert, S. P., Odell, A. F., Jopling, H. M., Hooper, N. M., Zachary, I. C., et al. (2010). Ligand-stimulated VEGFR2 signaling is regulated by co-ordinated trafficking and proteolysis. *Traffic*, *11*, 161–174.

Carmeliet, P. (2005). Angiogenesis in life, disease and medicine. *Nature*, *438*, 932–936.

Chen, T. T., Luque, A., Lee, S., Anderson, S. M., Segura, T., & Iruela-Arispe, M. L. (2010). Anchorage of VEGF to the extracellular matrix conveys differential signaling responses to endothelial cells. *The Journal of Cell Biology*, *188*, 595–609.

Donovan, D., Brown, N. J., Bishop, E. T., & Lewis, C. E. (2001). Comparison of three in vitro human 'angiogenesis' assays with capillaries formed in vivo. *Angiogenesis*, *4*, 113–121.

Duval, M., Bedard-Goulet, S., Delisle, C., & Gratton, J. P. (2003). Vascular endothelial growth factor-dependent down-regulation of Flk-1/KDR involves Cbl-mediated ubiquitination—Consequences on nitric oxide production from endothelial cells. *The Journal of Biological Chemistry*, *278*, 20091–20097.

Ewan, L. C., Jopling, H. M., Jia, H., Mittar, S., Bagherzadeh, A., Howell, G. J., et al. (2006). Intrinsic tyrosine kinase activity is required for vascular endothelial growth factor receptor 2 ubiquitination, sorting and degradation in endothelial cells. *Traffic, 7*, 1270–1282.

Fadok, V. A., Voelker, D. R., Campbell, P. A., Cohen, J. J., Bratton, D. L., & Henson, P. M. (1992). Exposure of phosphatidylserine on the surface of apoptotic lymphocytes triggers specific recognition and removal by macrophages. *Journal of Immunology, 148*, 2207–2216.

Gampel, A., Moss, L., Jones, M. C., Brunton, V., Norman, J. C., & Mellor, H. (2006). VEGF regulates the mobilization of VEGFR2/KDR from an intracellular endothelial storage compartment. *Blood, 108*, 2624–2631.

Gerber, H. P., McMurtrey, A., Kowalski, J., Yan, M. H., Keyt, B. A., Dixit, V., et al. (1998). Vascular endothelial growth factor regulates endothelial cell survival through the phosphatidylinositol 3′-kinase Akt signal transduction pathway—Requirement for Flk-1/KDR activation. *The Journal of Biological Chemistry, 273*, 30336–30343.

Haglund, K., & Dikic, I. (2012). The role of ubiquitylation in receptor endocytosis and endosomal sorting. *Journal of Cell Science, 125*, 265–275.

Haglund, K., Sigismund, S., Polo, S., Szymkiewicz, I., Di Fiore, P. P., & Dikic, I. (2003). Multiple monoubiquitination of RTKs is sufficient for their endocytosis and degradation. *Nature Cell Biology, 5*, 461–466.

Hasseine, L. K., Murdaca, J., Suavet, F., Longnus, S., Giorgetti-Peraldi, S., & Van Obberghen, E. (2007). Hrs is a positive regulator of VEGF and insulin signaling. *Experimental Cell Research, 313*, 1927–1942.

Horowitz, A., & Seerapu, H. R. (2012). Regulation of VEGF signaling by membrane traffic. *Cellular Signalling, 24*, 1810–1820.

Howell, G. J., Herbert, S. P., Smith, J. M., Mittar, S., Ewan, L. C., Mohammed, M., et al. (2004). Endothelial cell confluence regulates Weibel–Palade body formation. *Molecular Membrane Biology, 21*, 413–421.

Huang, F. T., & Sorkin, A. (2005). Growth factor receptor binding protein 2-mediated recruitment of the RING domain of Cbl to the epidermal growth factor receptor is essential and sufficient to support receptor endocytosis. *Molecular Biology of the Cell, 16*, 1268–1281.

Jaffe, E. A., Nachman, R. L., Becker, C. G., & Minick, C. R. (1973). Culture of human endothelial cells derived from umbilical veins—Identification by morphologic and immunological criteria. *The Journal of Clinical Investigation, 52*, 2745–2756.

Jopling, H. M., Howell, G. J., Gamper, N., & Ponnambalam, S. (2011). The VEGFR2 receptor tyrosine kinase undergoes constitutive endosome-to-plasma membrane recycling. *Biochemical and Biophysical Research Communications, 410*, 170–176.

Jopling, H. M., Odell, A. F., Hooper, N. M., Zachary, I. C., Walker, J. H., & Ponnambalam, S. (2009). Rab GTPase regulation of VEGFR2 trafficking and signaling in endothelial cells. *Arteriosclerosis, Thrombosis, and Vascular Biology, 29*, 1119–1124.

Kankanala, J., Latham, A. M., Johnson, A. P., Homer-Vanniasinkam, S., Fishwick, C. W. G., & Ponnambalam, S. (2012). A combinatorial in silico and cellular approach to identify a new class of compounds that target VEGFR2 receptor tyrosine kinase activity and angiogenesis. *British Journal of Pharmacology, 166*, 737–748.

Karar, J., & Maity, A. (2011). PI3K/AKT/mTOR pathway in angiogenesis. *Frontiers in Molecular Neuroscience, 4*, 51.

Koch, S., Tugues, S., Li, X., Gualandi, L., & Claesson-Welsh, L. (2011). Signal transduction by vascular endothelial growth factor receptors. *The Biochemical Journal, 437*, 169–183.

Koopman, G., Reutelingsperger, C. P. M., Kuijten, G. A. M., Keehnen, R. M. J., Pals, S. T., & Vanoers, M. H. J. (1994). Annexin-v for flow cytometric detection of phosphatidylserine expression on b-cells undergoing apoptosis. *Blood, 84*, 1415–1420.

Lampugnani, M. G., Orsenigo, F., Gagliani, M. C., Tacchetti, C., & Dejana, E. (2006). Vascular endothelial cadherin controls VEGFR-2 internalization and signaling from intracellular compartments. *The Journal of Cell Biology*, *174*, 593–604.

Latham, A. M., Odell, A. F., Mughal, N. A., Issitt, T., Ulyatt, C., Walker, J. H., et al. (2012). A biphasic endothelial stress-survival mechanism regulates the cellular response to vascular endothelial growth factor A. *Experimental Cell Research*, *318*, 2297–2311.

Liu, Z. J., Xiao, M., Balint, K., Soma, A., Pinnix, C. C., Capobianco, A. J., et al. (2006). Inhibition of endothelial cell proliferation by Notch1 signaling is mediated by repressing MAPK and PI3K/Akt pathways and requires MAML1. *The FASEB Journal*, *20*, 1009–1011.

Manickam, V., Tiwari, A., Jung, J.-J., Bhattacharya, R., Goel, A., Mukhopadhyay, D., et al. (2011). Regulation of vascular endothelial growth factor receptor 2 trafficking and angiogenesis by Golgi localized t-SNARE syntaxin 6. *Blood*, *117*, 1425–1435.

Meyer, R. D., Srinivasan, S., Singh, A. J., Mahoney, J. E., Gharahassanlou, K. R., & Rahimi, N. (2011). Pest motif serine and tyrosine phosphorylation controls vascular endothelial growth factor receptor 2 stability and downregulation. *Molecular and Cellular Biology*, *31*, 2010–2025.

Murdaca, J., Treins, C., Monthouel-Kartmann, M. N., Pontier-Bres, R., Kumar, S., Van Obberghen, E., et al. (2004). Grb10 prevents Nedd4-mediated vascular endothelial growth factor receptor-2 degradation. *The Journal of Biological Chemistry*, *279*, 26754–26761.

Odell, A. F., Hollstein, M., Ponnambalam, S., & Walker, J. H. (2012). A VE-cadherin-PAR3-α-catenin complex regulates the Golgi localization and activity of cytosolic phospholipase A(2)α in endothelial cells. *Molecular Biology of the Cell*, *23*, 1783–1796.

Olsson, A. K., Dimberg, A., Kreuger, J., & Claesson-Welsh, L. (2006). VEGF receptor signalling—In control of vascular function. *Nature Reviews Molecular Cell Biology*, *7*, 359–371.

Ponnambalam, S., & Alberghina, M. (2011). Evolution of the VEGF-regulated vascular network from a neural guidance system. *Molecular Neurobiology*, *43*, 192–206.

Reynolds, A. R., Hart, I. R., Watson, A. R., Welti, J. C., Silva, R. G., Robinson, S. D., et al. (2009). Stimulation of tumor growth and angiogenesis by low concentrations of RGD-mimetic integrin inhibitors. *Nature Medicine*, *15*, 392–400.

Roskoski, R., Jr. (2007). Vascular endothelial growth factor (VEGF) signaling in tumor progression. *Critical Reviews in Oncology/Hematology*, *62*, 179–213.

Ruiz de Almodovar, C., Lambrechts, D., Mazzone, M., & Carmeliet, P. (2009). Role and therapeutic potential of VEGF in the nervous system. *Physiological Reviews*, *89*, 607–648.

Santambrogio, M., Valdembri, D., & Serini, G. (2011). Increasing traffic on vascular routes. *Molecular Aspects of Medicine*, *32*, 112–122.

Sawamiphak, S., Seidel, S., Essmann, C. L., Wilkinson, G. A., Pitulescu, M. E., Acker, T., et al. (2010). Ephrin-B2 regulates VEGFR2 function in developmental and tumour angiogenesis. *Nature*, *465*, 487–491.

Schmidt, A., Brixius, K., & Bloch, W. (2007). Endothelial precursor cell migration during vasculogenesis. *Circulation Research*, *101*, 125–136.

Sorkin, A., & Goh, L. K. (2009). Endocytosis and intracellular trafficking of ErbBs. *Experimental Cell Research*, *315*, 683–696.

CHAPTER SEVENTEEN

Assessment of Internalization and Endosomal Signaling: Studies with Insulin and EGF

Barry I. Posner[*,†,1], **John J.M. Bergeron**[*,†]
[*]Department of Medicine, McGill University, Montreal, Quebec, Canada
[†]Department of Cell Biology, McGill University, Montreal, Quebec, Canada
[1]Corresponding author: e-mail address: barry.posner@mcgill.ca

Contents

Abstract

Endosomes are isolated from rat liver using high-speed centrifugation through sucrose density gradients. They are distinguishable from Golgi elements, with which they coisolate, by their capacity to concentrate internalized protein ligands (viz., insulin and epidermal growth factor (EGF)) in receptor-bound intact form. Endosomal signaling to relevant substrates can be readily shown for insulin and EGF receptor tyrosine kinases (RTKs), respectively. Both RTKs undergo dephosphorylation in endosomes. This can be inhibited by the powerful phosphotyrosine phosphatase inhibitors—the peroxovanadium compounds. *In vivo* administration of these compounds has been shown to activate selectively the endosomal insulin receptor kinase and promote signaling. Taken together, these observations constitute the basis for the hypothesis for which there is now ample evidence. Furthermore, a substantial body of work has documented the importance of endosomal signaling for growth, development, and disease.

Methods in Enzymology, Volume 535
ISSN 0076-6879
http://dx.doi.org/10.1016/B978-0-12-397925-4.00017-1

1. INTRODUCTION

Early studies on the uptake of protein ligands into cells concluded that they were internalized into lysosomes where they underwent degradation and clearance from the cell (Terris & Steiner, 1976). In the late 1970s, we and others demonstrated the uptake and concentration of internalized ligands into nonlysosomal structures (Bergeron, Sikstrom, Hand, & Posner, 1979; Josefsberg, Posner, Patel, & Bergeron, 1979). In our initial studies, we described the internalization of insulin and epidermal growth factor (EGF) into subcellular fractions/structures previously designated as Golgi elements (Posner, Patel, Verma, & Bergeron, 1980). It soon became clear that these fractions were heterogeneous and contained "unique" vesicles, distinguishable from Golgi elements (Khan, Posner, Khan, & Bergeron, 1982; Khan, Posner, Verma, Khan, & Bergeron, 1981), which concentrated the internalized ligands (Kay, Khan, Posner, & Bergeron, 1984). The term endosome (EN) was soon thereafter introduced to describe those subcellular constituents accumulating internalized ligands such as insulin (Marsh, Bolzau, & Helenius, 1983).

In this chapter, we present methods demonstrating that these endosomal structures are a critical site for cellular signaling. Our studies have focused on insulin and EGF, but there has been a host of other work demonstrating the generality of this mechanism for peptide hormones and peptides involved in promoting growth and differentiation (Posner & Laporte, 2010).

2. ENDOSOMAL SIGNALING BY INSULIN

The following summarize the steps involved in preparing a combined endosomal fraction (C-EN) from rat liver. The fraction is derived from both a light mitochondrial and microsomal pellet and is enriched in endosomal elements but contains other components as well. However, it was shown that this preparation is free of plasma membrane (PM) contamination (Bergeron, Posner, Josefsberg, & Sikstrom, 1978) and it will simply be referred to as ENs. PM is prepared by a standard procedure (Hubbard, Wall, & Ma, 1983) that is not described here.

2.1. Preparation of a rat liver endosomal fraction

1. Sprague Dawley rats weighing 150–200 g each are used in all these studies. They are fed *ad libitum* and fasted overnight before the experiment.

The animals are anesthetized by ether, and insulin (0–150 μg/100 g bwt) is injected into the portal or jugular vein in 0.4 ml phosphate-buffered saline (PBS).

2. Rats are killed by decapitation at the noted times after insulin injection (Fig. 17.1) and exsanguinated, and the livers are quickly placed into ice-cold 0.25 *M* sucrose in buffer B (4 m*M* imidazole, pH 7.5, containing protease inhibitors 1 m*M* benzamidine, 5 m*M* iodoacetamide, 1 m*M* phenylmethylsulfonyl fluoride (PMSF), 1000 KIU/ml of aprotinin

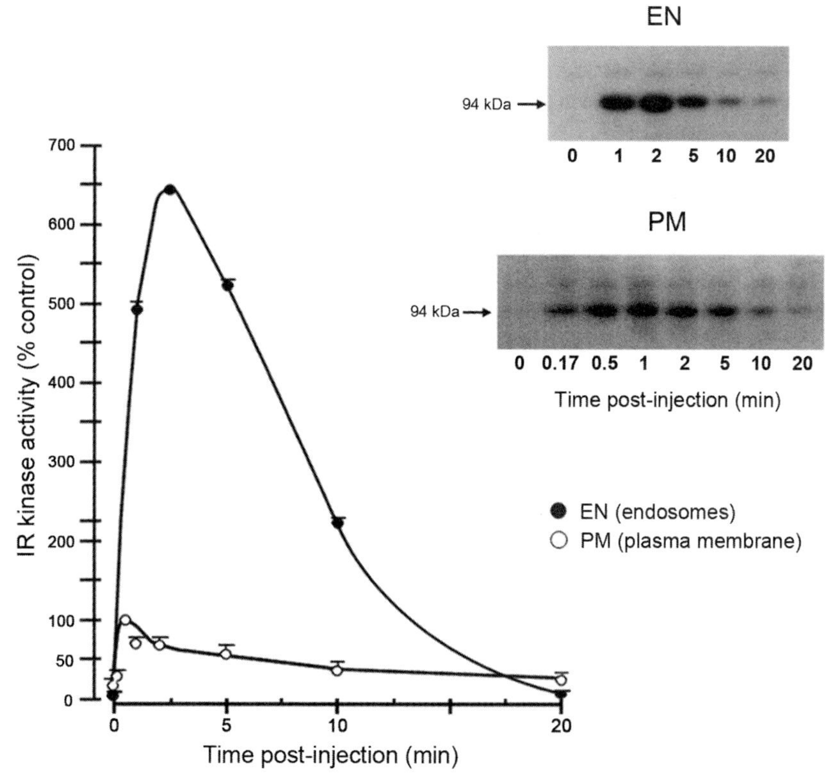

Figure 17.1 Time course of IRK autophosphorylation in PM and ENs. Rats were injected with insulin (1.5 μg/100 g bwt) or buffer alone. ENs and PM were prepared (Section 2.1) from two rats at each time point. They were subjected to autophosphorylation and solubilization after which the immunoprecipitated IRK was resolved on SDS-PAGE followed by radioautography and quantitative densitometry of the 94 kDa (β-subunit) band as described (Section 2.2.2). ^{32}P labeling is expressed as a percent of that in the IRK of PM prepared at 2 min postinjection of insulin (set to 100%). *This research was originally published in Khan et al. (1989), © The American Society for Biochemistry and Molecular Biology.* (For color version of this figure, the reader is referred to the online version of this chapter.)

and phosphatase inhibitors 2 mM NaF, 20 mM sodium molybdate, and 10 μM sodium orthovanadate).

3. The minced liver is homogenized (5 ml/g liver) with five up-and-down strokes of a Potter–Elvehjem homogenizer (clearance 0.15–0.23 mm); and the homogenate is centrifuged at $3200 \times g_{av}$ for 10 min (Beckman 60 Ti rotor).
4. The nuclear mitochondrial pellet is discarded and the supernatant centrifuged at $200{,}000 \times g_{av}$ for 30 min (Beckman 50.2 rotor) to obtain the combined light mitochondrial–microsomal pellet (L+P) that is resuspended (one stroke in a Dounce homogenizer, type B) in 1.15 M sucrose (5 ml/g liver) in buffer B.
5. The 1.15 M suspension is placed beneath a discontinuous sucrose gradient comprised of 1.0 and 0.6 M sucrose layers each dissolved in buffer B. The C-EN is obtained at the 1.0/0.6 M interface following centrifugation at $80{,}000 \times g_{av}$ for 195 min (Beckman SW 27 rotor).

In some studies (Section 3), a Golgi-EN (GE) fraction is prepared by flotation in a linear sucrose gradient as described previously (Bergeron, Rachubinski, Sikstrom, Posner, & Paiement, 1982). Since as with C-ENs, we are concerned only with the EN content of the GE fraction, we will refer to them as well simply as ENs. As with C-ENs, this fraction contains Golgi elements but is free of PM.

2.2. Determination of insulin receptor kinase activity and content in ENs

In this section, we describe the techniques used to measure insulin receptor kinase (IRK) activation in PM and ENs. The description starts with earlier technologies and the methodology shifts to the more efficient use of specific antibodies for Western blotting as these reagents became available.

2.2.1 Kinase activity after lectin purification of IRK

1. ENs are suspended (2 mg protein/ml) in 50 mM HEPES, pH 7.6, containing 10 mM $MgCl_2$, 0.1%BSA, 1 mM PMSF, bacitracin (100 U/ml), aprotinin (1000 KIU/ml), and 1% Triton X-100 and shaken vigorously for 1 h at 4 °C.
2. Insoluble material is removed by centrifugation at $100{,}000 \times g_{av}$ for 60 min (Beckman SW 60 Ti rotor). The supernatant containing solubilized material (10 mg/2 ml) is applied to a 2 ml wheat germ agglutinin–Sepharose column at 4 °C, which had been previously washed

extensively with 50 mM HEPES, pH 7.6, 10 mM $MgCl_2$, and 0.1% Triton X-100.

3. The eluate is recycled over the column four times, and the column is washed with buffer containing proteases and phosphatase inhibitors prior to eluting glycoproteins with 0.3 M N-acetyl-D-glucosamine as described previously (Khan, Savoie, Bergeron, & Posner, 1986).
4. IRK, corresponding to 10–20 fmol of insulin binding (50 μl of eluate) as measured with ^{125}I-insulin (Khan et al., 1986), is incubated in 90 μl of 20 mM HEPES, 8 mM $MnCl_2$, 10 mM $MgCl_2$, 100 μM $ZnSO_4$, 50 mM NaF, 10 μM Na vanadate, 270 μM dithiothreitol (DTT), 20 mM β-glycerophosphate, and BSA (10 μg/ml). Phosphorylation is initiated by adding 10 μl [γ-^{32}P]ATP to a final concentration of 50 μM (14 μCi/nmol) and the incubation is continued for 15 min at room temperature.
5. To measure IRK autophosphorylation, the reaction is stopped by adding 50 μl 6% SDS "stopping buffer" prior to subjecting to SDS-polyacrylamide gel electrophoresis (PAGE: 4% stacking and 10% resolving gel) under reducing conditions. This is followed by radioautography at −70 °C using enhancing screens and quantitative densitometry of the bands at 94 kDa as described (Khan et al., 1989; see Fig. 17.1).
6. To measure IRK extrinsic kinase activity, the incubation of Step 4 includes 200 μg of poly(Glu:Tyr) (4:1) and excludes DTT and β-glycerophosphate in a final volume of 160 μl. It is terminated by spotting 70 μl aliquots on chromatography paper (1 × 1 in.) followed by washing in 10% trichloroacetic acid–10 mM Na pyrophosphate, rinsing in 100% ethanol before scintillation counting as described (Khan et al., 1986).

2.2.2 IRK autophosphorylation in intact endosomal fractions

1. Autophosphorylation in an intact cell fraction is carried out by incubating ENs (30–150 μg protein) in 0.25 M sucrose buffer otherwise identical to that described in Step 4 of Section 2.2.1.
2. The reaction is terminated by adding 50 μl of 150 mM HEPES (pH 7.4), containing 4.5% Triton X-100, 3% Na deoxycholate, 0.3% SDS, 30 mM EDTA, 3 mM PMSF, 3 mM Na vanadate, 0.3 mM $ZnSO_4$, 120 mM β-glycerophosphate, 3% BSA, and 150 mM ATP followed by vigorous shaking at 4 °C for 1 h.
3. The solubilized membranes are diluted fourfold and then incubated with anti-insulin receptor antibody against the IRK β-subunit (according to

provider's specifications) for 4 h at 4 °C to generate an IRK–antibody complex.

4. This is followed by incubation with protein A–Sepharose (4 mg) for 1 h.
5. The immunoprecipitate is washed twice with 50 m*M* HEPES (pH 7.4), 0.1% Triton X-100, and 0.1% SDS followed by SDS-PAGE under reducing conditions (Khan et al., 1989).
6. Prior to radioautography and quantitative densitometry, the gels are treated with 1 *M* KOH for 2 h at 55 °C to remove alkali-labile serine and threonine phosphorylation.

2.2.3 Analysis of IRK activation by measuring IRK and phosphotyrosine (PY) content

We initially employed anti-IRK and anti-PY antibodies prepared in our own laboratory (Burgess et al., 1992). Suitable antibodies, along with instructions on how to employ them, are now commercially available:

1. Equal amounts of EN protein from insulin-treated and control rats are suspended in 3× Laemmli sample buffer (6.9% SDS, 30% glycerol, 300 m*M* DTT, and 1.1 *M* Tris–HCL, pH 6.8) (Bevan et al., 1995) and boiled for 5 min.
2. The samples are subjected to SDS-PAGE under reducing conditions, and phosphoproteins are transferred from SDS gels to nitrocellulose membranes by a standard method (Burgess et al., 1992); and membranes were then incubated with blocking solution for 1 h at room temperature.
3. For IR Western blots, the blocking buffer is PBS containing 4% powdered milk; and for PY Western blots, it is PBS containing 20% fetal calf serum.
4. The blocking solution is replaced with 50 ml of solution containing the primary antibody (anti-IRK or anti-PY) and gently rocked at room temperature for 2 h.
5. This was followed by 3–10 min washes with 50 ml of 1% Tween 20 before the blots were transferred to 50 ml of ^{125}I-labeled secondary antibody (5×10^5 dpm/transferred lane) and incubated for 1 h at room temperature followed by three additional washes.
6. The nitrocellulose membranes were mounted on Whatman 3MM paper and air-dried prior to radioautography at −70 °C and subsequent densitometry of the 94 kDa IRK β-subunit (Bevan et al., 1995).

Using the earlier methods, we were able to demonstrate that the IRK internalized into ENs and was highly activated therein (Fig. 17.1).

2.3. Dephosphorylation of PY–IRK in ENs

These studies followed the discovery that pervanadate activated the IRK in the absence of insulin (Kadota et al., 1987) and the synthesis of stable peroxovanadium compounds (pVs; Posner et al., 1994). They demonstrate that the pVs are powerful inhibitors of phosphotyrosine phosphatases (PTPs), which play a role in modulating IRK activation in ENs.

2.3.1 Phosphorylation in intact endosomal fractions

1. Rat hepatic ENs are prepared 2 min after the injection of insulin (1.5 μg/100 g bwt) or buffer alone and are suspended (25 μg protein) in 90 μl of solution at a final concentration of 50 m*M* HEPES (pH 7.4), 150 m*M* KCl, 5 m*M* NaCl, 5 m*M* $MgCl_2$, and 1 m*M* DTT in the presence or absence of pV as noted in Fig. 17.2A.

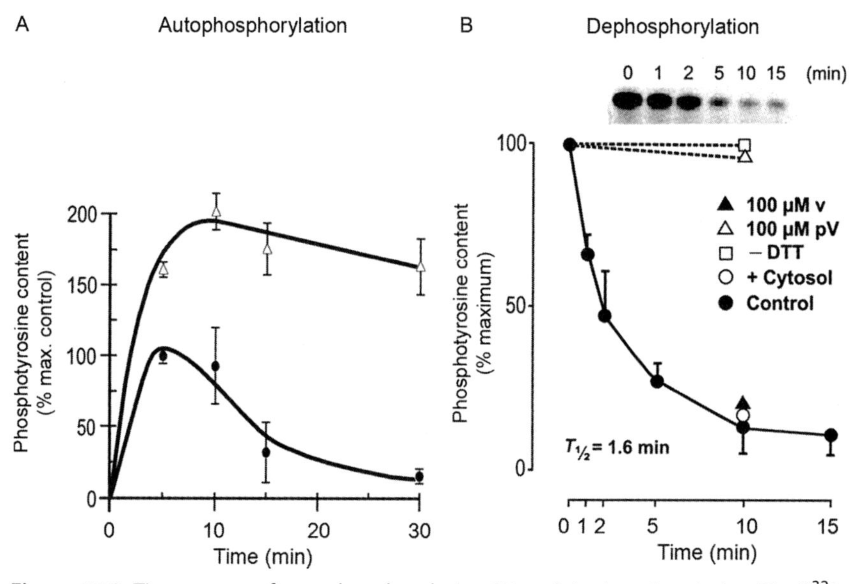

Figure 17.2 Time course of autophosphorylation (A) and dephosphorylation (B) of ^{32}P-labeled endosomal IRK receptors. Rats were injected with insulin (1.5 μg/100 g bwt), and 2 min later, ENs were prepared (Section 2.1) and suspended in assay buffer containing [γ-^{32}P] ATP with (Δ) or without(•) 100 μ*M* pV. (A) At the noted times, ENs were solubilized and the IRK was immunoprecipitated and subjected to SDS-PAGE, radioautography, and quantitative densitometry of the 94 kDa (β-subunit) band as described (Section 2.3.1). (B) After 5 min of incubation, autophosphorylation was terminated (zero time) by adding EDTA/ATP (Section 2.3.2). At the noted times thereafter, ENs were solubilized and the IRK processed for quantitative densitometry as described in (A). *This research was originally published in Faure, Baquiran, Bergeron, and Posner (1992), © The American Society for Biochemistry and Molecular Biology.*

2. To determine ^{32}P labeling of the IRK β-subunit in intact ENs, phosphorylation is initiated by adding 10 μl [γ-^{32}P] ATP.
3. (30 μCi/nmol) at a final concentration of 25 μ*M*, and incubating the mixture at 37 °C.
4. At various times thereafter (Fig. 17.2A), the reaction is terminated by adding 50 μl ice-cold solubilizing buffer (4.5% Triton X-100, 150 m*M* HEPES (pH 7.4), 150 m*M* EDTA, 6 m*M* orthovanadate, 3 m*M* PMSF, and 3 m*M* benzamidine) followed by shaking for 1 h at 4 °C.
5. The solubilized membranes are diluted by adding 290 μl of ice-cold 50 m*M* HEPES and 10 μl of anti-IRK and incubated at 4 °C for 18 h prior to mixing with 50 μl of ice-cold protein A–Sepharose (2 mg).
6. Immunoprecipitates are generated and subjected to SDS-PAGE and radioautography as noted in Section 2.2.2 (Steps 4–6).

2.3.2 Dephosphorylation of ^{32}P-labeled IRK

1. ENs are prepared and suspended and autophosphorylation is initiated as in Section 2.3.1 (Steps 1 and 2).
2. After 5 min incubation at 37 °C, phosphorylation is stopped by adding 10 μl of prewarmed (37 °C) stopping buffer containing 500 m*M* HEPES (pH 7.4), 100 m*M* EDTA, and 5 m*M* unlabeled ATP.
3. The incubation is continued at 37 °C, and at different times thereafter (Fig. 17.2B), dephosphorylation is terminated by processing each mixture as described in Section 2.3.1 (Steps 3–5).

The studies depicted in Fig. 17.2A clearly demonstrate that IRK autophosphorylation in intact endosomal structures is a dynamic process. Thus, phosphorylation in control membranes reached a peak and subsequently declined in parallel with a decline in the ATP concentration in the reaction mixture (Faure et al., 1992). But in the presence of pV, a powerful PTP inhibitor, this decline was not observed as the dephosphorylation process had been abrogated. This is illustrated in Fig. 17.2B where dephosphorylation of the IRK is permitted to occur after peak labeling with ^{32}P. The addition of pV but not orthovanadate completely inhibits IRK dephosphorylation. The same results are elicited using anti-PY antibodies instead of [γ-^{32}P] ATP (Faure et al., 1992).

2.4. Activation of endosomal IRK is sufficient for insulin action

In this section, we demonstrate that the insulin response can be entrained by activating the IRK in the complete absence of insulin. Further, the IRK in

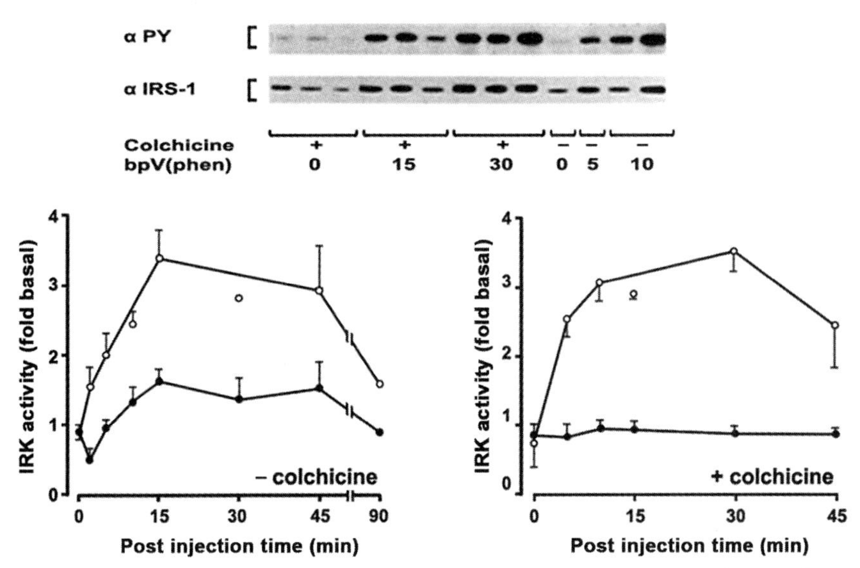

Figure 17.3 Time course of IRK activation in PM (•) and ENs (○) by bpV(phen) with (+) or without (−) colchicine pretreatment. Rats were injected with colchicines or buffer and 1 h later with bpV(phen) (0.6 μmol/100 g bwt). At the noted times, hepatic PM and ENs were prepared and IRK activity assayed as described (Section 2.2.1). Cytosols were prepared and IRS-1 was immunoprecipitated and subjected to SDS-PAGE, immunoblotting, radioautography, and quantitative densitometry (Section 2.4). Colchicine pretreatment blocked bpV(phen)-induced activation of IRK in PM without affecting that in ENs (right panel). Nevertheless, IRS-1 underwent tyrosine phosphorylation indicating a direct endosomal origin for insulin signaling. *This research was originally published in Bevan et al. (1995), © The American Society for Biochemistry and Molecular Biology.*

ENs can be activated in the absence of prior activation at the cell surface (i.e., in PM) (Fig. 17.3).

The preparation of endosomal IRK is carried out in rats anesthetized and injected via jugular vein with colchicine (25 μmol/100 g bwt) in 0.9% saline or saline alone 1 h prior to the injection of bpV(phen) (0.6 μmol/100 g bwt) prepared as previously described (Bevan et al., 1995). As seen in Fig. 17.3, ENs and PM are prepared as described in Section 2.1; and IRK activity is determined as described in Sections 2.2.1 and 2.2.2. IRS-1 tyrosine phosphorylation is measured to determine activation of the insulin signaling pathway (White, 1994):

1. An aliquot of liver homogenate, prepared as in Section 2.1, is subjected to centrifugation at 200,000 $\times g_{av}$ for 45 min (SW40 Beckman rotor) to yield a supernatant constituting the cytosolic fraction.
2. Cytosolic protein (15 mg in 1.1 ml) is incubated with 1% Triton X-100 (final concentration) for 1 h at 4 °C and centrifuged at 12,000 $\times g_{av}$ for 5 min (Eppendorf microcentrifuge) to remove insoluble material.
3. IRS-1 antibody (initially provided by M. F.White (Bevan et al., 1995) but now commercially available) is added in 10 µl volume, and the incubation is continued for another 4 h.
4. Protein A–Sepharose is added as a 50% slurry (150 µl) after preequilibration in 50 m*M* HEPES (pH 7.4), 150 m*M* NaCl, and 2 m*M* Na orthovanadate; and the mixture is shaken for 1 h.
5. After centrifugation as in Step 2, the pellet is rinsed three times with buffer (Step 4) containing 1% Triton X-100 and 0.1% SDS followed by boiling in 210 µl of Laemmli buffer (Bevan et al., 1995).
6. The samples (70 µl) are subjected to SDS-PAGE (7.5% gel), transferred to immobilon-P membranes, and immunoblotted with anti-PY and anti-IRS-1 antibodies for radioautography and quantitative densitometry as described in Section 2.2.3.

By using colchicine to block receptor recycling, we could show that the administration of bpV(phen) activated the IRK in ENs exclusively (Fig. 17.3, bottom right). In this circumstance, IRS-1 tyrosine phosphorylation was clearly observed (top panel) indicating effective signaling by the endosomal IRK. In addition, it was shown that bpV(phen), though it could not access skeletal muscle when given *in vivo*, was nevertheless capable of lowering blood glucose levels indicating that a hepatic effect to decrease glucose production was probably responsible (Bevan et al., 1995).

3. ENDOSOMAL SIGNALING BY EPIDERMAL GROWTH FACTOR

In parallel with the studies on the IRK, we demonstrated the endosomal localization of activated EGF receptor (EGFR) (Kay et al., 1986) and its SHC substrate (Wada, Lai, Posner, & Bergeron, 1992) as judged by their respective levels of PY content. Subsequently, we assessed the compartmentalization of the activated EGFR in liver parenchyma by quantitative subcellular fractionation. The isolation of the fractions and the characterization of the reagents have been described in detail (Di Guglielmo, Baass, Ou, Posner, & Bergeron, 1994). The following describes how we proceeded to quantitate EGFR compartmentalization after the administration of EGF (10 μg/100 g bwt) into the hepatic portal vein:

1. Calculation of protein yields. This is done using the Bradford method for protein assessment using a standard curve of bovine serum albumin (Bradford, 1976). From the wet weights of liver and the protein content of the fractions, then, the experimental yields are calculated.
2. All fractions are prepared as described (Di Guglielmo et al., 1994) and subjected to SDS-PAGE followed by immunoblotting and quantitative densitometry as described earlier.
3. The EGFR content in the total particulate fraction is taken as 100%. The EGFR concentration is deduced, from densitometric analysis, as 2.54 ± 0.26 arbitrary units/mg cell fraction protein.
4. The EGFR is largely localized at the PM in the control steady state where its concentration is determined experimentally as 12.2 ± 3.4 units/mg cell fraction protein.
5. From the recovery of the PM fraction (30.5%) and the yield (3.5 ± 0.8 mg protein/g liver), the receptor content of PMs per gram of liver is calculated (Fig. 17.4).
6. For ENs, the same is done where the yield of EGFR is 0.4 ± 0.09 mg protein/g liver. We estimated that the recovery of ENs from liver homogenates is 16%, leading us to deduce the amount of EGFR and pEGFR (immunoblotting with respective antibodies) at different times of receptor activation *in situ* (Fig. 17.4).
7. The same kind of analysis is carried out for SHC, pSHC, and GRB2.

Although it is well established that ligand binding to its cognate receptor (viz., the EGFR or IRK) takes place initially at the PM, quantitative subcellular fractionation, along with immunoblotting and quantitative densitometry, reveals that the highest abundance of activated EGFR is seen in hepatic ENs. This is equally true for downstream signaling partners (SHC and GRB2) of the EGFR. Notably, these observations were confirmed by visualizing, in living cells at 15 min after EGF stimulation, endosomal colocalization of both the EGFR and SHC (Colwill, Renewable Protein Binder Working Group, & Graslund, 2011).

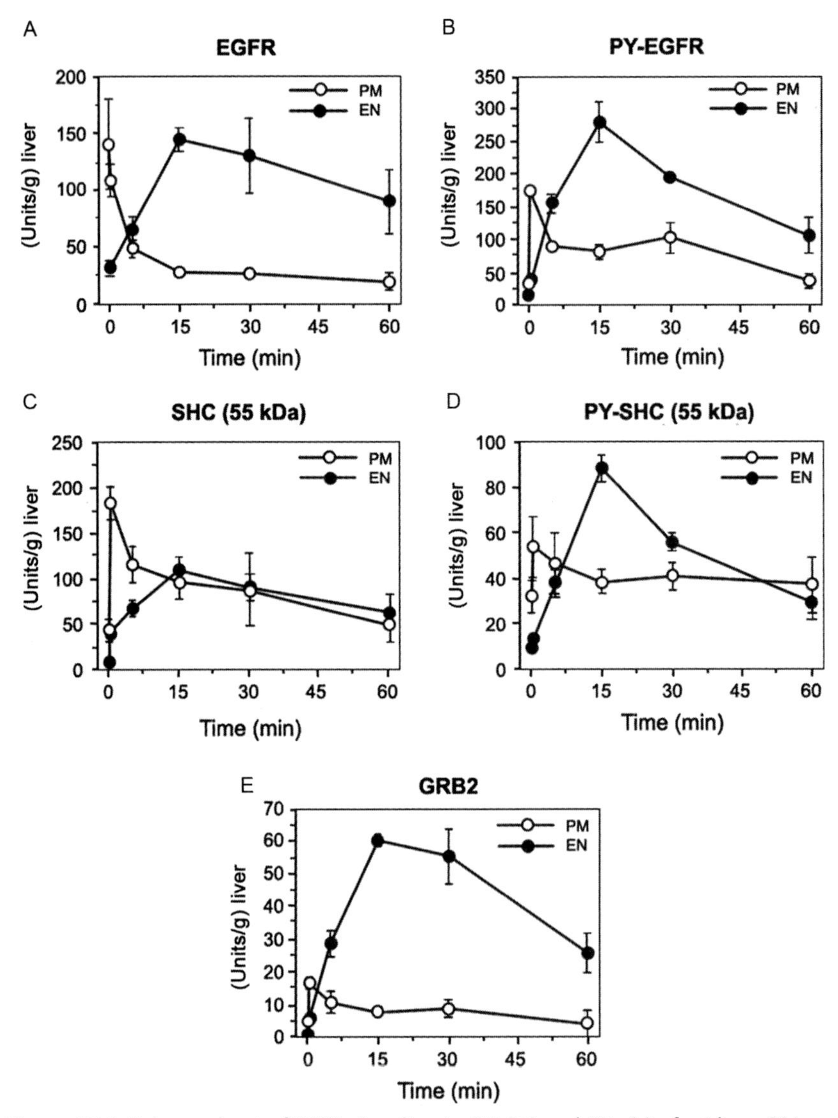

Figure 17.4 Balance sheet of EGFR signaling in PM (○) and ENs (•) of rat liver. PM and ENs were isolated from different rat livers (three separate experiments for each time point) at different times after the injection of EGF into the hepatic portal vein. By immunoblotting and quantitative densitometry, the EGFR (A) was lost from the PM with a $T_{1/2}$ of 1 min with a small burst of tyrosine-phosphorylated receptor (B) seen at 30 s. However, the highest abundance phosphotyrosine receptor was found in ENs at 15 min. The major substrate of the EGFR (SHC) (C) was recruited to the PM by 30 s, but tyrosine-phosphorylated SHC (D) was largely endosomal as was the adaptor protein GRB2 (E). Since the data are calculated for yields of the respective isolated organelles, then, taken together, the site of signal transduction quantitatively is largely endosomal. *This research was originally published in Di Guglielmo et al. (1994), © The Nature Publishing Group.*

4. SUMMARY

Since these early observations, it has become clear that endosomal signaling is central to the functioning and physiological consequences of multiple receptor systems (Hupalowska & Miaczynska, 2012; Posner & Laporte, 2010). Thus, endosomal signaling is important in development and embryogenesis (Kopan & Ilagan, 2009; Lund & Delotto, 2011). It also appears to play a critical role in neuronal development and function (Alsina, Ledda, & Paratcha, 2012) as well as the modulation of behavior (Beaulieu et al., 2008). Furthermore, evidence is accumulating for a role for endosomal signaling in the pathogenesis and modulation of disease states such as Alzheimer's disease (Rajendran et al., 2008) and cancer progression (Espada, Calvo, Diaz-Prado, & Medina, 2009), as well as in immunomodulation (Mullershausen et al., 2009).

REFERENCES

Alsina, F. C., Ledda, F., & Paratcha, G. (2012). New insights into the control of neurotrophic growth factor receptor signaling: Implications for nervous system development and repair. *Journal of Neurochemistry*, *123*, 652–661.

Beaulieu, J. M., Marion, S., Rodriguiz, R. M., Medvedev, I. O., Sotnikova, T. D., Ghisi, V., et al. (2008). A beta-arrestin 2 signaling complex mediates lithium action on behavior. *Cell*, *132*, 125–136.

Bergeron, J. J., Posner, B. I., Josefsberg, Z., & Sikstrom, R. (1978). Intracellular polypeptide hormone receptors. The demonstration of specific binding sites for insulin and human growth hormone in Golgi fractions isolated from the liver of female rats. *Journal of Biological Chemistry*, *253*, 4058–4066.

Bergeron, J. J., Rachubinski, R. A., Sikstrom, R. A., Posner, B. I., & Paiement, J. (1982). Galactose transfer to endogenous acceptors within Golgi fractions of rat liver. *The Journal of Cell Biology*, *92*, 139–146.

Bergeron, J. J., Sikstrom, R., Hand, A. R., & Posner, B. I. (1979). Binding and uptake of ^{125}I-insulin into rat liver hepatocytes and endothelium. An in vivo radioautographic study. *The Journal of Cell Biology*, *80*, 427–443.

Bevan, A. P., Burgess, J. W., Drake, P. G., Shaver, A., Bergeron, J. J., & Posner, B. I. (1995). Selective activation of the rat hepatic endosomal insulin receptor kinase. Role for the endosome in insulin signaling. *Journal of Biological Chemistry*, *270*, 10784–10791.

Bradford, M. M. (1976). A rapid and sensitive method for the quantitation of microgram quantities of protein utilizing the principle of protein–dye binding. *Analytical Biochemistry*, *72*, 248–254.

Burgess, J. W., Wada, I., Ling, N., Khan, M. N., Bergeron, J. J., & Posner, B. I. (1992). Decrease in beta-subunit phosphotyrosine correlates with internalization and activation of the endosomal insulin receptor kinase. *Journal of Biological Chemistry*, *267*, 10077–10086.

Colwill, K., Renewable Protein Binder Working Group, & Graslund, S. (2011). A roadmap to generate renewable protein binders to the human proteome. *Nature Methods*, *8*, 551–558.

Di Guglielmo, G. M., Baass, P. C., Ou, W. J., Posner, B. I., & Bergeron, J. J. (1994). Compartmentalization of SHC, GRB2 and mSOS, and hyperphosphorylation of Raf-1 by EGF but not insulin in liver parenchyma. *EMBO Journal, 13*, 4269–4277.

Espada, J., Calvo, M. B., Diaz-Prado, S., & Medina, V. (2009). Wnt signalling and cancer stem cells. *Clinical & Translational Oncology: Official Publication of the Federation of Spanish Oncology Societies and of the National Cancer Institute of Mexico, 11*, 411–427.

Faure, R., Baquiran, G., Bergeron, J. J., & Posner, B. I. (1992). The dephosphorylation of insulin and epidermal growth factor receptors. Role of endosome-associated phosphotyrosine phosphatase(s). *Journal of Biological Chemistry, 267*, 11215–11221.

Hubbard, A. L., Wall, D. A., & Ma, A. (1983). Isolation of rat hepatocyte plasma membranes. I. Presence of the three major domains. *The Journal of Cell Biology, 96*, 217–229.

Hupalowska, A., & Miaczynska, M. (2012). The new faces of endocytosis in signaling. *Traffic, 13*, 9–18.

Josefsberg, Z., Posner, B. I., Patel, B., & Bergeron, J. J. (1979). The uptake of prolactin into female rat liver. Concentration of intact hormone in the Golgi apparatus. *Journal of Biological Chemistry, 254*, 209–214.

Kadota, S., Fantus, I. G., Deragon, G., Guyda, H. J., Hersh, B., & Posner, B. I. (1987). Peroxide(s) of vanadium: A novel and potent insulin-mimetic agent which activates the insulin receptor kinase. *Biochemical and Biophysical Research Communications, 147*, 259–266.

Kay, D. G., Khan, M. N., Posner, B. I., & Bergeron, J. J. (1984). In vivo uptake of insulin into hepatic Golgi fractions: Application of the diaminobenzidine-shift protocol. *Biochemical and Biophysical Research Communications, 123*, 1144–1148.

Kay, D. G., Lai, W. H., Uchihashi, M., Khan, M. N., Posner, B. I., & Bergeron, J. J. (1986). Epidermal growth factor receptor kinase translocation and activation in vivo. *Journal of Biological Chemistry, 261*, 8473–8480.

Khan, M. N., Baquiran, G., Brule, C., Burgess, J., Foster, B., Bergeron, J. J., et al. (1989). Internalization and activation of the rat liver insulin receptor kinase in vivo. *Journal of Biological Chemistry, 264*, 12931–12940.

Khan, M. N., Posner, B. I., Khan, R. J., & Bergeron, J. J. (1982). Internalization of insulin into rat liver Golgi elements. Evidence for vesicle heterogeneity and the path of intracellular processing. *Journal of Biological Chemistry, 257*, 5969–5976.

Khan, M. N., Posner, B. I., Verma, A. K., Khan, R. J., & Bergeron, J. J. (1981). Intracellular hormone receptors: Evidence for insulin and lactogen receptors in a unique vesicle sedimenting in lysosome fractions of rat liver. *In: Proceedings of the National Academy of Sciences of the United States of America, 78*, 4980–4984.

Khan, M. N., Savoie, S., Bergeron, J. J., & Posner, B. I. (1986). Characterization of rat liver endosomal fractions. In vivo activation of insulin-stimulable receptor kinase in these structures. *Journal of Biological Chemistry, 261*, 8462–8472.

Kopan, R., & Ilagan, M. X. (2009). The canonical Notch signaling pathway: Unfolding the activation mechanism. *Cell, 137*, 216–233.

Lund, V. K., & Delotto, R. (2011). Regulation of Toll and Toll-like receptor signaling by the endocytic pathway. *Small GTPases, 2*, 95–98.

Marsh, M., Bolzau, E., & Helenius, A. (1983). Penetration of Semliki Forest virus from acidic prelysosomal vacuoles. *Cell, 32*, 931–940.

Mullershausen, F., Zecri, F., Cetin, C., Billich, A., Guerini, D., & Seuwen, K. (2009). Persistent signaling induced by FTY720-phosphate is mediated by internalized S1P1 receptors. *Nature Chemical Biology, 5*, 428–434.

Posner, B. I., Faure, R., Burgess, J. W., Bevan, A. P., Lachance, D., Zhang-Sun, G., et al. (1994). Peroxovanadium compounds. A new class of potent phosphotyrosine phosphatase inhibitors which are insulin mimetics. *Journal of Biological Chemistry, 269*, 4596–4604.

Posner, B. I., & Laporte, S. A. (2010). Cellular signalling: Peptide hormones and growth factors. *Progress in Brain Research*, *181*, 1–16.
Posner, B. I., Patel, B., Verma, A. K., & Bergeron, J. J. (1980). Uptake of insulin by plasmalemma and Golgi subcellular fractions of rat liver. *Journal of Biological Chemistry*, *255*, 735–741.
Rajendran, L., Schneider, A., Schlechtingen, G., Weidlich, S., Ries, J., Braxmeier, T., et al. (2008). Efficient inhibition of the Alzheimer's disease beta-secretase by membrane targeting. *Science*, *320*, 520–523.
Terris, S., & Steiner, D. F. (1976). Retention and degradation of ^{125}I-insulin by perfused livers from diabetic rats. *The Journal of Clinical Investigation*, *57*, 885–896.
Wada, I., Lai, W. H., Posner, B. I., & Bergeron, J. J. (1992). Association of the tyrosine phosphorylated epidermal growth factor receptor with a 55-kD tyrosine phosphorylated protein at the cell surface and in endosomes. *Journal of Cell Biology*, *116*, 321–330.
White, M. F. (1994). The IRS-1 signaling system. *Current Opinion in Genetics & Development*, *4*, 47–54.

CHAPTER EIGHTEEN

Quantitative Proteomic Analysis of Compartmentalized Signaling Networks

Maria Hernandez-Valladares[2], Veronica Aran[2], Ian A. Prior[1]
Division of Cellular and Molecular Physiology, Institute of Translational Medicine, University of Liverpool, Liverpool, United Kingdom
[1]Corresponding author: e-mail address: iprior@liverpool.ac.uk

Contents

Abstract

Ras proteins operate predominantly from the plasma membrane; however, they have also been localized to most intracellular compartments. Various functions and signaling outputs have been ascribed to endomembranous Ras although systematic comparison and measurement of potential outputs have not yet been carried out. We describe the methodology for isolating and measuring compartment-specific signaling networks using quantitative proteomics. This approach reveals the potential of a subcellular platform for supporting specific outputs and will inform subsequent studies of endogenous isoform-specific Ras signaling.

[2] These authors contributed equally to this work.

Methods in Enzymology, Volume 535
ISSN 0076-6879
http://dx.doi.org/10.1016/B978-0-12-397925-4.00018-3

1. INTRODUCTION

Ras proteins are small-molecular-weight GTPases that regulate key signaling pathways downstream of cell surface growth factor receptors. Oncogenic mutations are found in 16% of human cancers and result in aberrant cell proliferation, survival, and transformation (Prior, Lewis, & Mattos, 2012). Three almost identical, ubiquitously expressed isoforms (HRAS, KRAS, and NRAS) are not biologically redundant. This is believed to be at least in part due to differential trafficking and localization of each isoform resulting in overlapping but distinctive distributions within the plasma membrane and endomembrane compartments (Henis, Hancock, & Prior, 2009). The advent of improved imaging technologies has enabled careful characterization of the dynamic association of Ras isoforms with intracellular organelles such as the ER/Golgi, endosome, and mitochondria; however, there is considerable debate over the functional relevance of these Ras pools (Fehrenbacher, Bar-Sagi, & Philips, 2009; Omerovic, Laude, & Prior, 2007). One strategy for investigating this phenomenon involves ectopic expression of Ras mutants targeted to specific subcellular locations (Chiu et al., 2002; Matallanas et al., 2006). This approach reveals the potential for each cellular compartment to sustain specific signaling and phenotypic consequences that can then be followed up in studies of endogenous isoform-specific Ras signaling. In this chapter, we will describe the compartmentalized Ras constructs and cell lines that we have developed together with our strategy for measuring the global phosphorylation-based signaling outputs from each location.

2. COMPARTMENTALIZED RAS CONSTRUCTS

To target Ras to specific locations, we utilized a GFP-tagged Ras-acceptor construct where the NRas C-terminal hypervariable region (HVR; amino acids 166–189) was deleted and replaced with a Xho1 restriction site to facilitate addition of novel membrane-targeting sequences by DNA subcloning (Laude & Prior, 2008). This generates a generic Ras construct that will have no confounding influence of isoform-specific localization information. A selection of organelle-targeting motifs for ER/Golgi, Golgi, endomembrane, and mitochondria are cloned in frame to the C-terminus that are separated from the Ras protein by a 3-amino acid (ARA) linker sequence (Fig. 18.1A for details; Barr, Nakamura, & Warren, 1998; Horie, Suzuki, Sakaguchi, & Mihara, 2002; Laude & Prior, 2008;

A

	GFP	Ras	Membrane targetting motif
N-Ras	GFP	1 N-Ras G domain 164	YRMKKLNSSDDGTQGCMGLPCVVM 189
ER/Golgi-Ras	GFP	1 N-Ras G domain 164	ARA-AAAAAAAAAAAAAAGCMGLPCVVM 180 189
Golgi-Ras	GFP	1 N-Ras G domain 164	ARA- 784 GM130 986
Mito-Ras	GFP	1 N-Ras G domain 164	ARA- 170 OMP25 206
Endo-Ras	GFP	1 N-Ras G domain 164	ARA- 274 FENS1-FYVE 358 274 FENS1-FYVE 358

B

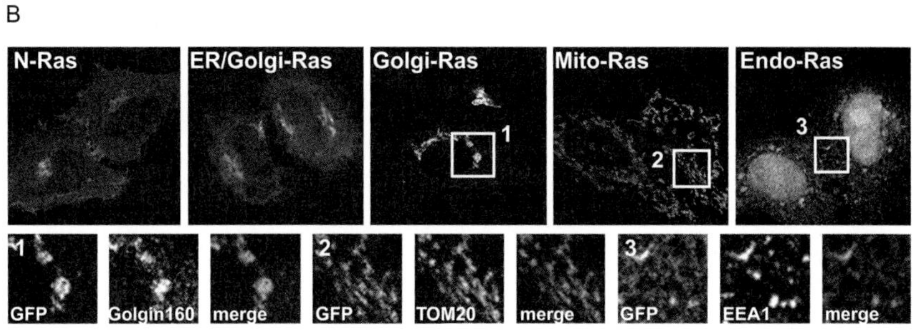

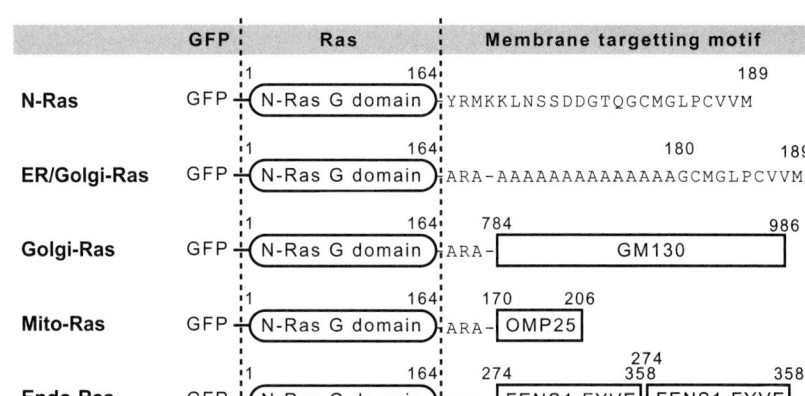

Figure 18.1 Organelle-targeted Ras chimeras. (A) GFP-Ras was subcloned to replace the C-terminal hypervariable region for an organelle-targeting motif. (B) HeLa cells transiently transfected with targeted Ras chimeras show specific localization to their target organelle. Endomembrane-Ras has a more complex distribution but clear colocalization with the endosomal marker EEA1 can be observed. (For color version of this figure, the reader is referred to the online version of this chapter.)

Ridley et al., 2001). Both wild-type and constitutively active (G12V) variants were produced. Figure 18.1B illustrates the highly specific targeting associated with each construct. Only endomembranous Ras displays a heterogeneous localization that includes not only endosomes but also the nucleus and cytosol—compartments not typically associated with Ras signaling and therefore presumed to be silent in our assays.

3. STABLE CELL LINE GENERATION

To generate stable cell lines for compartmentalized Ras study, we opted to use the Invitrogen Flp-In system. This avoids some of the problems associated with transfection and antibiotic-based stable cell line selection where the gene insertion occurs randomly in the cell genome. Consequently, this frequently results in variable gene expression between cell lines and limited ability to compare between clones. For the Flp-In system, host cells haboring an FRT (Flp recombinase target site) are required. The gene of interest, in a vector also possessing an FRT site, is then transfected into these cells and Flp recombinase-mediated exchange results in the insertion of the plasmid containing the gene of interest into the cell genome (O'Gorman, Fox, & Wahl, 1991). The advantage of this approach is that there is a single insertion site within the genome reducing the possible variability of off-target effects and resulting in highly similar expression levels between clones. We generated both HeLa S3 and NIH3T3 cells stably expressing wild-type and G12V mutant compartmentalized Ras variants. We mainly use the stable cell lines in phenotypic and cell signaling assays. For phosphoproteomic analysis of compartmentalized Ras function, we opted to use both transient transfection and Flp-In cell strategies. This was in part because long-term expression of active Ras mutants typically results in negative feedback pathways downregulating many Ras signaling pathways (Omerovic, Clague, & Prior, 2010).

3.1. Cloning and cell transfection

1. The compartmentalized Ras genes were cloned into pEF5/FRT/V5 *via* topoisomerase reaction according to manufacturer's instructions. An example is given later with NIH3T3 cell generation.
2. Gene constructs were transfected together with the vector pOG44 carrying the Flp recombinase:
   2.1 NIH3T3 Flp-In cells (Invitrogen) were split to the following density 1×10^5 cells/well in a 6-well plate resulting in ~50% confluency the following day for transfection.
   2.2 0.9 μg pOG44 + 0.1 μg pEF5/FRT/V5-compartmentalized Ras were transfected using the TransIT-3T3 transfection kit (Mirus Bio) according to manufacturer's instructions. Cells were incubated in DMEM (Invitrogen), 10% FBS for 48 h before splitting into a 10 cm dish.

Note: Some cells are more amenable to Flp-In technology than others; Flp-In HeLa S3 cells (made in-house) produced at least 10 × more colonies than NIH3T3. Routinely, we plate 10% and 90% of the transfected cells into 10 cm dishes to ensure that at the point that colonies formed, we could choose the plate where colonies had not merged with each other.

3.2. Clonal cell selection, isolation, and verification

1. Cell selection is initiated once cells have attached. For NIHI3T3, we used 200 μg/ml hygromycin B (concentration will vary for each cell line and has to be determined empirically). Cells are cultured for approximately 20 days to allow clonal colonies to form; the media is changed every 3 days with hygromycin included.
2. Colony picking is performed once visible white spots can be seen with the naked eye. The media is aspirated and cells washed twice with 5 ml of PBS. One milliliter of ice-cold 0.05% trypsin is added to cover the cells and then immediately aspirated off. Using a 10 μl pipette, 10–20 μl of ice-cold trypsin is added directly on top of a target colony and detachment of the colony is facilitated by scraping with the pipette tip. The detached cells are pipetted into a 24-well plate well containing 1 ml of media, and a 1 ml pipette is used to repeatedly pipette to generate a homogenous suspension.
3. Cell clones were maintained in 200 μg/ml hygromycin media while being expanded up for archival freezing and testing using fluorescence imaging for GFP and Western blotting for the presence of the appropriate Ras construct.

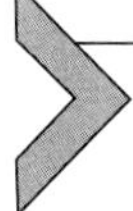

4. PROTEOMIC ANALYSIS OF COMPARTMENT-SPECIFIC SIGNALING

We are interested in quantifying relative changes of protein abundance and protein phosphorylation status as readouts for the influence of compartment-specific Ras on cell state and cell signaling. To achieve this, we use stable isotopic labeling in cell culture (SILAC)-based quantitative proteomics approaches that allow direct ratiometric comparison of these parameters between cell lines and experimental conditions (Harsha, Molina, & Pandey, 2008). Detailed methods are provided later; but briefly, we typically performed triplexed experiments with each cell line labeled with isotopic variants of arginine and lysine; cell lysates are generated, mixed together 1:1:1, trypsinized, and fractionated; and then, enrichment strategies for pSer-, pThr-, and pTyr-modified peptides are applied (Dephoure & Gygi, 2011; Rogers & Foster, 2009). Methods to detect phosphopeptides in the mass spectrometer and analyze the spectra have also been described elsewhere (Boersema, Mohammed, & Heck, 2009; Cox & Mann, 2008; Cox et al., 2009). The isotopic labeling allows subsequent deconvolution of the experimental conditions due to mass shifts in the mass spectrometry spectra; the relative intensity of each peak allows ratiometric analysis of the response of each experimental condition. An overview of the workflow can be seen in Fig. 18.2.

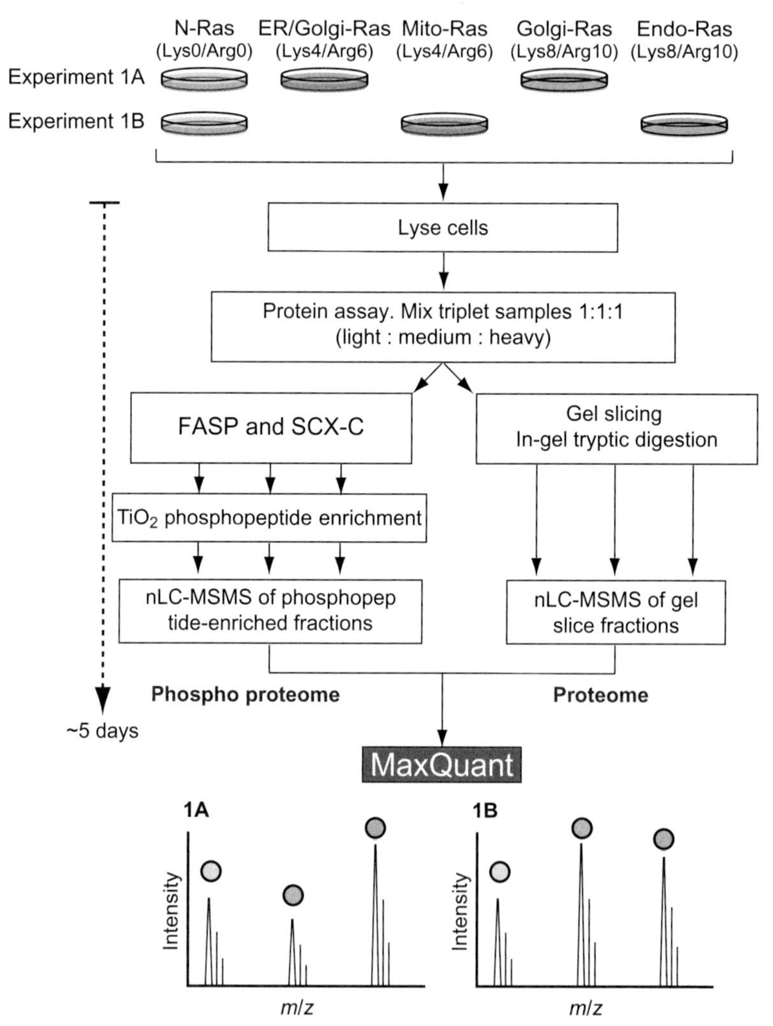

Figure 18.2 Workflow for a large-scale phosphopeptide preparation. FASP stands for filter-aided sample preparation (Wisniewski, Zougman, Nagaraj, & Mann, 2009), SCXC for strong cation exchange chromatography, and TiO_2-C for TiO_2 chromatography. (For color version of this figure, the reader is referred to the online version of this chapter.)

4.1. Cell culture and lysis

1. Cells are labeled with light, medium, or heavy isotopic variants of arginine and lysine until complete metabolic incorporation has been achieved (typically $\geq$7 cell doublings). For HeLa S3 cells, we used DMEM, 10% dialyzed FBS (Dundee Cell Products) containing 200 mg/l L-proline, L-lysine (Lys0) and L-arginine (Arg0), L-lysine-$^{2}H_4$ (Lys4) and L-arginine-$U^{13}C_6$ (Arg6), or L-lysine-U-$^{13}C_6$-$^{15}N_2$ (Lys8) and L-arginine-U-$^{13}C_6$-$^{15}N_4$ (Arg10), at final concentration of 84 mg/l for arginine and 146 mg/l for lysine amino acids. For each compartment-specific Ras HeLa S3 cell line, we prepared 3 × 15 cm dishes, which were 80–90% confluent on the day of the experiment. The large-scale culture needs to be able to generate 6–7 mg of lysate per cell line.
2. Cells were lysed in the following buffer: 50 m*M* Tris–HCl, pH 7.5, 100 m*M* NaCl, 50 m*M* NaF, 1% (w/v) NP-40, 0.1% (w/v) sodium deoxycholate, 1 m*M* EDTA, 1 m*M* EGTA, mammalian protease inhibitors (SIGMA, added according manufacturers), 5 m*M* glycerol phosphate, 2 m*M* sodium orthovanadate, and PhosSTOP phosphatase inhibitor cocktail tablets (Roche, one tablet added per 10 ml of lysis buffer). To lyse the cells, remove the medium from the dishes and wash 2 × with ice-cold PBS. Carefully aspirate the remaining PBS, add 3 ml lysis buffer per 15 cm dish, and rock the dishes on ice for 10 min. Check lysis has worked using a phase contrast microscope (nuclei should remain intact with this lysis buffer).
3. Lysates are microfuged at 13,000 rpm, 4 °C for 10 min; the supernatant is removed and assayed for protein content (e.g., BCA assay). Based on the protein assay, the supernatants are then mixed 1:1:1.
4. Supernatant proteins are precipitated by adding four volumes of cold (−20 °C) acetone and incubating overnight at −20 °C.

4.2. Protein digestion by the filter-aided sample preparation (FASP) method

The FASP method developed by Wisniewski et al. (2009) uses the membrane in a filter device as a reaction surface to allow washing, chemical modification, and proteolysis of proteins. Cell lysates in urea buffer are reduced using DTT and then concentrated and washed in standard filtration devices, reduced thiols are then alkylated, and finally, the proteins are trypsinized. The peptides generated can then be eluted through the membrane for further enrichment and analysis. Although the FASP method is compatible with the use of NP-40 (Wisniewski, Zielinska, & Mann, 2011), we prefer to eliminate *a priori* this detergent from the protein sample by precipitation with acetone to avoid the difficult removal of NP-40 micelles:

1. Centrifuge the protein precipitates at $300 \times g$ for 1 min. Carefully decant the acetone, remove remainder with a pipette, and allow sample to dry at room temperature for 5–10 min. *Note*: Do not overdry or this will result in pellets that are difficult to resuspend.
2. Resuspend at 10 mg/ml with 8 *M* urea and 0.1 *M* Tris–HCl, pH 8.5. Vortex and/or pipette up and down gently to dissolve the protein pellets. Avoid making foam. If necessary, sonicate at room temperature but avoid heating the sample.
3. Add DTT to 0.1 *M* and incubate the samples at room temperature for 30 min.
4. Pour the dissolved samples into 30k Amicon Ultra-15 ml filtration devices (Millipore) previously washed with 5 ml of the urea–Tris solution. We recommend using one filtration device per 10 mg of lysate.
5. Spin the devices down at the maximum recommended speed (see Millipore instructions) at room temperature until the concentrate reaches $\leq$2 ml. Discard the flow-through.
6. Add 10 ml of the urea–Tris solution, mix well, and spin again to concentrate to a $\leq$2 ml final volume per device. Discard the flow-through. Repeat this step one more time.
7. To alkylate the sample, mix the concentrate with 2–3 ml of 50 m*M* chloroacetamide prepared in urea–Tris solution, and incubate in the dark at room temperature for 30 min. Spin the devices to concentrate to 1–2 ml.
8. Add 10 ml of the urea–Tris solution, mix, and spin down to concentrate to a $\leq$2 ml. Discard the flow-through. Repeat this step one more time.
9. Add 10 ml of 50 m*M* ammonium bicarbonate (ambic), mix well, and spin down to concentrate to a $\leq$2 ml. Discard flow-through. Repeat this step two more times.
10. The sample is now ready for protein digestion using trypsin. Take 10–15 μl of the concentrate for SDS-PAGE analysis. Add 100 μg of Trypsin Gold (Promega) dissolved in ambic solution per filtration device, mix well, and incubate in a wet chamber at 37 °C overnight.
11. The following day, place the filters in new 50 ml Falcon tubes. Spin down for 15 min to collect the peptides. Retain 10–15 μl of the digested concentrate for SDS-PAGE analysis.

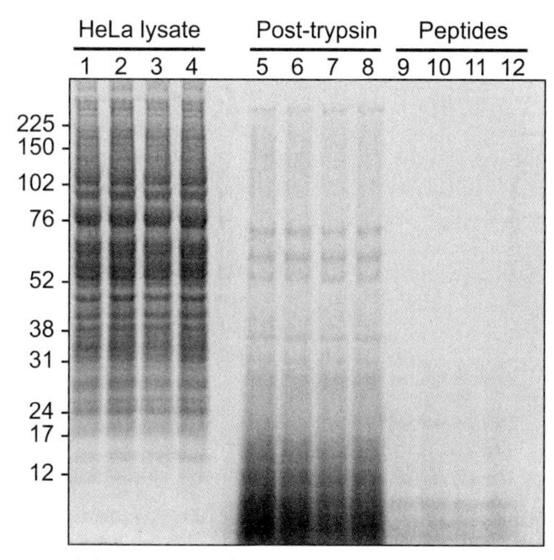

Figure 18.3 Efficiency of the FASP method. Assessed by SDS-PAGE analysis on a 4–12% gradient gel. Lanes 1–4, four concentrates of HeLa S3 cytoplasmic lysates before trypsin digestion; lanes 5–8, same concentrates after trypsin digestion and before peptide elution; lanes 9–12, filtrated peptides after spinning and two additional washing/spinning steps.

12. Wash the filters with 2 ml of ambic solution and spin down for 10 min. Repeat this step one more time.
13. Take 25–30 μl of the filtrate for SDS-PAGE analysis. Analyze the three samples on a 4–12% gradient gel to check digestion efficiency (see Fig. 18.3).
14. Measure the peptide concentration in the filtrate by reading absorbance at 280 nm and assuming that a 1 mg/ml peptide solution has 1.1 AU. We usually obtain 7–8 mg of peptides from a 20 mg protein sample.
15. Acidify the peptides to 1% (v/v) TFA. Check if pH is ~3 with pH strips. The peptides are now ready for desalting.

4.3. Desalting of peptides

Before peptide fractionation, the acidified filtrates need to be salt-free.

1. Use a 20 cc Oasis HLB SPE cartridge (Waters) to desalt the peptides from 10 mg of protein. Follow the manufacturer's recommendations for solvents and volumes.
2. Split the eluted volume in protein low-binding tubes (Eppendorf). In a vacuum drier, dry the tubes at full vacuum. Dried tubes can be stored in the freezer if desired until next step.

Optional: to check the binding capacity of the cartridges, keep 100–200 μl of the flow-through and perform phosphopeptide enrichment as describe later and LC–MS/MS analysis.

4.4. Peptide fractionation by strong cation exchange chromatography

To separate complex peptide mixtures before the phosphopeptide enrichment step, strong cation exchange chromatography (SCXC) has proven to be a robust tool that elutes peptides according to their solution state charge (Beausoleil et al., 2004):

1. Before performing SCXC, prepare the following buffers:
   - Binding buffer: 5 m*M* ammonium formate in 20% acetonitrile, pH 2.7 (by adding TFA).
   - Elution buffer: 200 m*M* ammonium formate in 20% acetonitrile, pH 2.7 (by adding TFA).
   - Nucleic acid cleaning buffer: 50 m*M* K_2HPO_4, 1 *M* NaCl, pH 7.5.
   - Storage buffer: 0.2 *M* sodium acetate in 20% ethanol.

   All the buffers should be filtered before use. The binding and elution buffers should be adjusted to pH 2.7 before adding 20% acetonitrile.
2. Dried peptides are dissolved with 2–2.5 ml of 30% acetonitrile pH 2.7 (remember to pH the solution before adding acetonitrile). Spin down any insoluble material.
3. Place a 1 ml S-Resource column (GE Healthcare Life Sciences) onto your liquid chromatography system. Wash the pumps and the injection loop. Equilibrate the column with five column volumes (CV) of binding buffer. We recommend working at a flow rate of 1 ml/min during all the chromatographic process.
4. Load the peptides onto the column.
5. Wash the column with 5–10 CV of binding buffer and collect the unbound fraction in 1.5 ml protein low-binding tubes or in a 96-well protein low-binding plate (Eppendorf). This flow-through is usually enriched with singly positive charged mono-phosphopeptides.
6. To elute multiply charged phosphopeptides, apply a three-step salt gradient consisting of 20 CV from 0% to 50% elution buffer (Olsen et al., 2006; Olsen & Macek, 2009), 5 CV from 50% to 100% elution buffer, and 5 CV from 100% elution buffer.
7. Wash the column with 10 CV of nucleic acid cleaning buffer. Then wash the column with 5 CV of filtered water and 5 CV of storage buffer. Store the column at 4 °C.

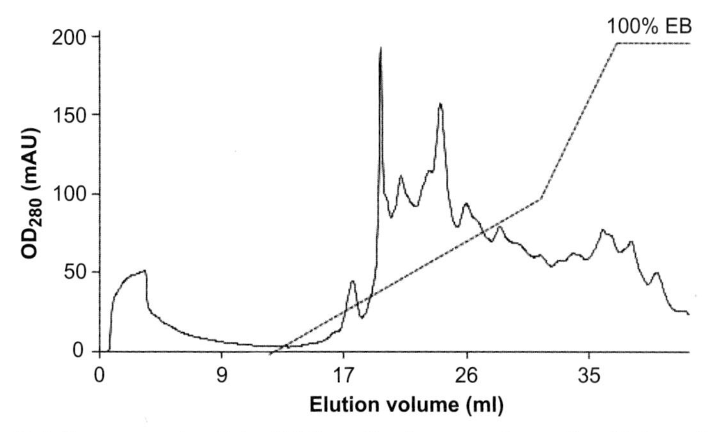

Figure 18.4 Representative HeLa SCXC profile. An example profile of 20 mg HeLa S3 cytoplasmic lysate after trypsin digestion in FASP devices using an S-Resource column as cation exchanger. The dashed line indicates the three-step salt gradient. EB stands for elution buffer.

8. In a vacuum drier, dry the flow-through and elution fractions at full vacuum. Dried tubes can be stored in the freezer if desired until the next step.

A typical SCXC separation of peptides from a HeLa S3 protein extract is shown in Fig. 18.4.

4.5. Phosphopeptide enrichment by TiO_2 chromatography (TiO_2-C)

Immobilized metal ion affinity chromatography (IMAC) was first introduced by Andersson and Porath (1986) to identify phosphorylated proteins. The technique was later widely used for enrichment of phosphopeptides prior to mass spectrometric analysis (Figeys et al., 1998; Li & Dass, 1999; Neville et al., 1997). Despite the impressive enrichment rate, enriched phosphopeptide samples were "contaminated" with unphosphorylated peptides containing multiple acidic amino acid residues that copurified in IMAC (Thingholm, Jensen, Robinson, & Larsen, 2008). As an efficient alternative to IMAC, TiO_2-C and the use of 2,5-dihydroxybenzoic acid, phthalic acid, or glycolic acid in the binding buffer have proven to be more selective for phosphopeptide capturing minimizing the binding of unphosphorylated peptides (Larsen, Thingholm, Jensen, Roepstorff, & Jorgensen, 2005; Pinkse, Uitto, Hilhorst, Ooms, & Heck, 2004; Thingholm, Jorgensen, Jensen, & Larsen, 2006). In the succeeding text is a protocol for TiO_2-C using glycolic acid in the binding buffer by Körner with minor variations (Dulla, Daub, Hornberger, Nigg, & Korner, 2010; Sugiyama et al., 2007):

1. Add 100 μl of 80 g/l glycolic acid (SIGMA) in 80% (v/v) acetonitrile and 2% (v/v) TFA (GA solution) to each tube containing dried peptides from SCXC flow-through and elution fractions. Vortex for 15–20 s and then spin at 13,000 rpm for 5 min to remove any insoluble material.
2. To prepare TiO_2 microcolumns, place a small disk of C8-silica membrane (Varian GmbH) into a 200 μl pipette tip to make a plug at the bottom of the tip. Prepare as many as dried tubes containing peptides. Stir the TiO_2 slurry prepared by dissolving 100 mg of TiO_2 beads (GL Sciences) in 1 ml of 80% (v/v) acetonitrile and 0.2% (v/v) TFA. Keep stirring while dispensing 20–30 μl of the slurry into the tip. Spin onto the disk at 4000 rpm for 30 s. The microcolumns can be easily spun using tube adapters (GL Sciences) in a protein low-binding tube.
3. Wash the microcolumns with 60 μl of 0.6% (w/v) NH_4OH solution by spinning at 4000 rpm for 2 min. Empty the tubes if necessary so that the tip bottom does not touch the different solutions.
4. Wash the microcolumns with 60 μl of GA solution. Spin at 4000 rpm for 2 min. Repeat the wash with 90 μl of GA solution to assure equilibration of the microcolumns. Spin at 4000 rpm for 3 min. Discard the flow-through.
5. Transfer the dissolved peptides into the microcolumns and spin at 2000 rpm for 5 min. To enhance the binding, apply the flow-through again to the same microcolumn, flick a bit the microcolumn to shake the beads, and spin again. To check the binding efficiency, collect the flow-through and desalt with StageTips before LC–MS/MS analysis (Rappsilber, Mann, & Ishihama, 2007).
6. Wash the microcolumns with 60 μl of the GA solution by spinning at 2000 rpm for 5 min.
7. Wash the microcolumns with 60 μl of 80% (v/v) acetonitrile and 0.2% (v/v) TFA by spinning at 2000 rpm for 5 min.
8. Wash the microcolumns with 60 μl of 20% (v/v) acetonitrile by spinning at 4000 rpm for 2 min.
9. Place the microcolumns in new protein low-binding tubes. Elute phosphopeptides with 60 μl 0.6% (w/v) NH_4OH by spinning at 1000 rpm for 15 min. If elution is not completed, spin at 1200–1500 rpm until it is done. Add 60 μl of 60% (v/v) acetonitrile to the microcolumns to elute phosphopeptide trapped in the C8-silica disk. Spin at 2000 rpm for 5 min.

10. Place again the microcolumns in new protein low-binding tubes. Elute remaining phosphopeptides with 60 μl of 1% (w/v) pyrrolidine by spinning at 1000 rpm for 15 min. Add again 60 μl of 60% (v/v) acetonitrile to the microcolumns and spin at 2000 rpm for 5 min.
11. In a vacuum drier, dry the tubes at full vacuum. Dried tubes can be stored in the freezer until LC–MS/MS analysis if desired.
12. Dissolve the dried phosphopeptide pellets with 15–20 μl of 0.5% (v/v) formic acid. Vortex for 15–20 s. Spin the tubes at 13,000 rpm for 5 min to remove any insoluble material. The samples are now ready for mass spectrometric analysis.

Optionally, dissolved phosphopeptides can be desalted using StageTips before loading into the LC–MS/MS system. However, we found this step unnecessary because of the desalting procedure setup in the LC–MS/MS protocol.

4.6. In-gel proteome digestion

To correlate phosphorylation events with protein expression levels, the proteome is also analyzed by LC–MS/MS. The proteome is prepared by fractionation by SDS-PAGE and trypsin digestion of the sliced gel bands. We typically identify 2000–2400 proteins by LC–MS/MS with a 90 min elution gradient from 150 μg of a HeLa S3 proteome sample loaded on a 4–12% gradient gel (Invitrogen) and sliced into 24 gel bands.

4.7. Mass spectrometric analysis of phosphopeptides

LC–MS/MS equipment and modes of operation will vary from facility to facility. Later is a brief overview of how we run our samples based on the specification and capacity of available equipment.

About 5 μl of each sample is fractionated by nanoscale C18 high-performance liquid chromatography on a Waters nanoACQUITY UPLC system coupled to an LTQ Orbitrap XL (Thermo Fisher) fitted with a Proxeon nanoelectrospray source. TiO_2-C enriched phosphopeptides from the SCXC elution fractions are loaded onto a 5 cm × 180 μm trap column (BEH-C18 Symmetry; Waters) and resolved on a 25 cm × 75 μm column using a 21 min linear gradient of 0–37.5% (v/v) acetonitrile in 0.1% (v/v) formic acid at a flow rate of 400 nl/min (65 °C). For the small number of TiO_2-C enriched phosphopeptides from the SCXC flow-through fractions that contained many phosphopeptides, we used 121 min linear gradients. The mass spectrometer acquired full MS survey scans in the Orbitrap ($R = 30{,}000$; m/z range 300–2000) and performed MSMS on the top 5 multiply charged ions in the linear ion quadruple ion trap after fragmentation using collision-induced dissociation (30 ms at 35% energy). Dynamic exclusion of 180 s ($n=1$) was applied to full MS ions previously selected for MSMS. Raw MW peak list files were processed with the MaxQuant software suite (Cox et al., 2009); examples of further types of data analysis can be seen in our previous work (Omerovic, Hammond, Prior, & Clague, 2012).

5. SUMMARY

A phosphoproteomic workflow for large-scale preparations that include cells lysis, protein digestion by the FASP procedure, desalting, SCXC, TiO_2-C, and proteome analysis by SDS-PAGE/in gel digestion is described here in detail. In the last few years, the phosphoproteomics field has made a rapid progress in developing a variety of highly specific and sensitive methods to characterize the impact of phosphorylation events in signaling pathways. The phosphoproteomic workflow consists of several laborious techniques that aim to identify most of the phosphopeptides from complex samples by MS.

We have suggested a phosphoproteomic workflow that best works in our laboratory. However, variations of this workflow (e.g., filter-based affinity capture and elution, hydrophobic interaction chromatography, and in-solution digestion) determined by availability of equipment, expertise, and resources have proved to give excellent results in other laboratories (Boersema, Mohammed, & Heck, 2008; McNulty & Annan, 2008; Olsen & Macek, 2009; Rush et al., 2005; Thingholm, Jensen, & Larsen, 2009; Wisniewski, Nagaraj, Zougman, Gnad, & Mann, 2010). Moreover, the combination of two or more workflows seems to identify the largest number of phosphorylation sites described so far (Olsen et al., 2010).

In summary, the approach that we have described gives a comprehensive snapshot of the proteome and signaling status of a cell line. Our strategy to overexpress compartment-specific Ras proteins reveals the potential of a platform to support a particular type of signaling or phenotypic response. The next challenge is to see to what extent the insights generated correlate with endogenous isoform-specific Ras responses.

ACKNOWLEDGMENTS

The authors thank J. R. Wiśniewski, G. Palmisano, and T. Geiger for sharing further details of their phosphoproteomic protocols. Our research is funded by North West Cancer Research, the BBSRC (BB/G018162), and the Wellcome Trust (WT085201).

REFERENCES

Andersson, L., & Porath, J. (1986). Isolation of phosphoproteins by immobilized metal (Fe^{3+}) affinity chromatography. *Analytical Biochemistry, 154*, 250–254.

Barr, F. A., Nakamura, N., & Warren, G. (1998). Mapping the interaction between GRASP65 and GM130, components of a protein complex involved in the stacking of Golgi cisternae. *The EMBO Journal, 17*, 3258–3268.

Beausoleil, S. A., Jedrychowski, M., Schwartz, D., Elias, J. E., Villen, J., Li, J., et al. (2004). Large-scale characterization of HeLa cell nuclear phosphoproteins. *Proceedings of the National Academy of Sciences of the United States of America, 101*, 12130–12135.

Boersema, P. J., Mohammed, S., & Heck, A. J. (2008). Hydrophilic interaction liquid chromatography (HILIC) in proteomics. *Analytical and Bioanalytical Chemistry, 391*, 151–159.

Boersema, P. J., Mohammed, S., & Heck, A. J. (2009). Phosphopeptide fragmentation and analysis by mass spectrometry. *Journal of Mass Spectrometry, 44*, 861–878.

Chiu, V. K., Bivona, T., Hach, A., Sajous, J. B., Silletti, J., Wiener, H., et al. (2002). Ras signalling on the endoplasmic reticulum and the Golgi. *Nature Cell Biology, 4*, 343–350.

Cox, J., & Mann, M. (2008). MaxQuant enables high peptide identification rates, individualized p.p.b.-range mass accuracies and proteome-wide protein quantification. *Nature Biotechnology, 26*, 1367–1372.

Cox, J., Matic, I., Hilger, M., Nagaraj, N., Selbach, M., Olsen, J. V., et al. (2009). A practical guide to the MaxQuant computational platform for SILAC-based quantitative proteomics. *Nature Protocols, 4*, 698–705.

Dephoure, N., & Gygi, S. P. (2011). A solid phase extraction-based platform for rapid phosphoproteomic analysis. *Methods, 54*, 379–386.

Dulla, K., Daub, H., Hornberger, R., Nigg, E. A., & Korner, R. (2010). Quantitative site-specific phosphorylation dynamics of human protein kinases during mitotic progression. *Molecular and Cellular Proteomics, 9*, 1167–1181.

Fehrenbacher, N., Bar-Sagi, D., & Philips, M. (2009). Ras/MAPK signaling from endomembranes. *Molecular Oncology, 3*, 297–307.

Figeys, D., Gygi, S. P., Zhang, Y., Watts, J., Gu, M., & Aebersold, R. (1998). Electrophoresis combined with novel mass spectrometry techniques: Powerful tools for the analysis of proteins and proteomes. *Electrophoresis, 19*, 1811–1818.

Harsha, H. C., Molina, H., & Pandey, A. (2008). Quantitative proteomics using stable isotope labeling with amino acids in cell culture. *Nature Protocols, 3*, 505–516.

Henis, Y. I., Hancock, J. F., & Prior, I. A. (2009). Ras acylation, compartmentalization and signaling nanoclusters (Review). *Molecular Membrane Biology, 26*, 80–92.

Horie, C., Suzuki, H., Sakaguchi, M., & Mihara, K. (2002). Characterization of signal that directs C-tail-anchored proteins to mammalian mitochondrial outer membrane. *Molecular Biology of the Cell, 13*, 1615–1625.

Larsen, M. R., Thingholm, T. E., Jensen, O. N., Roepstorff, P., & Jorgensen, T. J. (2005). Highly selective enrichment of phosphorylated peptides from peptide mixtures using titanium dioxide microcolumns. *Molecular and Cellular Proteomics, 4*, 873–886.

Laude, A. J., & Prior, I. A. (2008). Palmitoylation and localisation of RAS isoforms are modulated by the hypervariable linker domain. *Journal of Cell Science, 121*, 421–427.

Li, S., & Dass, C. (1999). Iron(III)-immobilized metal ion affinity chromatography and mass spectrometry for the purification and characterization of synthetic phosphopeptides. *Analytical Biochemistry, 270*, 9–14.

Matallanas, D., Sanz-Moreno, V., Arozarena, I., Calvo, F., Agudo-Ibanez, L., Santos, E., et al. (2006). Distinct utilization of effectors and biological outcomes resulting from site-specific Ras activation: Ras functions in lipid rafts and Golgi complex are dispensable for proliferation and transformation. *Molecular and Cellular Biology, 26*, 100–116.

McNulty, D. E., & Annan, R. S. (2008). Hydrophilic interaction chromatography reduces the complexity of the phosphoproteome and improves global phosphopeptide isolation and detection. *Molecular and Cellular Proteomics*, 7, 971–980.

Neville, D. C., Rozanas, C. R., Price, E. M., Gruis, D. B., Verkman, A. S., & Townsend, R. R. (1997). Evidence for phosphorylation of serine 753 in CFTR using a novel metal-ion affinity resin and matrix-assisted laser desorption mass spectrometry. *Protein Science*, *6*, 2436–2445.

O'Gorman, S., Fox, D. T., & Wahl, G. M. (1991). Recombinase-mediated gene activation and site-specific integration in mammalian cells. *Science*, *251*, 1351–1355.

Olsen, J. V., Blagoev, B., Gnad, F., Macek, B., Kumar, C., Mortensen, P., et al. (2006). Global, in vivo, and site-specific phosphorylation dynamics in signaling networks. *Cell*, *127*, 635–648.

Olsen, J. V., & Macek, B. (2009). High accuracy mass spectrometry in large-scale analysis of protein phosphorylation. *Methods in Molecular Biology*, *492*, 131–142.

Olsen, J. V., Vermeulen, M., Santamaria, A., Kumar, C., Miller, M. L., Jensen, L. J., et al. (2010). Quantitative phosphoproteomics reveals widespread full phosphorylation site occupancy during mitosis. *Science Signaling*, *3*, ra3.

Omerovic, J., Clague, M. J., & Prior, I. A. (2010). Phosphatome profiling reveals PTPN2, PTPRJ and PTEN as potent negative regulators of PKB/Akt activation in Ras-mutated cancer cells. *The Biochemical Journal*, *426*, 65–72.

Omerovic, J., Hammond, D. E., Prior, I. A., & Clague, M. J. (2012). A global snap-shot of the influence of endocytosis upon EGF receptor signaling output. *Journal of Proteome Research*, *11*, 5157–5166.

Omerovic, J., Laude, A. J., & Prior, I. A. (2007). Ras proteins: Paradigms for compartmentalised and isoform-specific signalling. *Cellular and Molecular Life Sciences*, *64*, 2575–2589.

Pinkse, M. W., Uitto, P. M., Hilhorst, M. J., Ooms, B., & Heck, A. J. (2004). Selective isolation at the femtomole level of phosphopeptides from proteolytic digests using 2D-NanoLC-ESI-MS/MS and titanium oxide precolumns. *Analytical Chemistry*, *76*, 3935–3943.

Prior, I. A., Lewis, P. D., & Mattos, C. (2012). A comprehensive survey of Ras mutations in cancer. *Cancer Research*, *72*, 2457–2467.

Rappsilber, J., Mann, M., & Ishihama, Y. (2007). Protocol for micro-purification, enrichment, pre-fractionation and storage of peptides for proteomics using StageTips. *Nature Protocols*, *2*, 1896–1906.

Ridley, S. H., Ktistakis, N., Davidson, K., Anderson, K. E., Manifava, M., Ellson, C. D., et al. (2001). FENS-1 and DFCP1 are FYVE domain-containing proteins with distinct functions in the endosomal and Golgi compartments. *Journal of Cell Science*, *114*, 3991–4000.

Rogers, L. D., & Foster, L. J. (2009). Phosphoproteomics—Finally fulfilling the promise? *Molecular BioSystems*, *5*, 1122–1129.

Rush, J., Moritz, A., Lee, K. A., Guo, A., Goss, V. L., Spek, E. J., et al. (2005). Immunoaffinity profiling of tyrosine phosphorylation in cancer cells. *Nature Biotechnology*, *23*, 94–101.

Sugiyama, N., Masuda, T., Shinoda, K., Nakamura, A., Tomita, M., & Ishihama, Y. (2007). Phosphopeptide enrichment by aliphatic hydroxy acid-modified metal oxide chromatography for nano-LC–MS/MS in proteomics applications. *Molecular and Cellular Proteomics*, *6*, 1103–1109.

Thingholm, T. E., Jensen, O. N., & Larsen, M. R. (2009). Analytical strategies for phosphoproteomics. *Proteomics*, *9*, 1451–1468.

Thingholm, T. E., Jensen, O. N., Robinson, P. J., & Larsen, M. R. (2008). SIMAC (sequential elution from IMAC), a phosphoproteomics strategy for the rapid separation of monophosphorylated from multiply phosphorylated peptides. *Molecular and Cellular Proteomics*, *7*, 661–671.

Thingholm, T. E., Jorgensen, T. J., Jensen, O. N., & Larsen, M. R. (2006). Highly selective enrichment of phosphorylated peptides using titanium dioxide. *Nature Protocols*, *1*, 1929–1935.

Wisniewski, J. R., Nagaraj, N., Zougman, A., Gnad, F., & Mann, M. (2010). Brain phosphoproteome obtained by a FASP-based method reveals plasma membrane protein topology. *Journal of Proteome Research*, *9*, 3280–3289.

Wisniewski, J. R., Zielinska, D. F., & Mann, M. (2011). Comparison of ultrafiltration units for proteomic and N-glycoproteomic analysis by the filter-aided sample preparation method. *Analytical Biochemistry*, *410*, 307–309.

Wisniewski, J. R., Zougman, A., Nagaraj, N., & Mann, M. (2009). Universal sample preparation method for proteome analysis. *Nature Methods*, *6*, 359–362.

CHAPTER NINETEEN

Separation of Magnetically Isolated TNF Receptosomes from Mitochondria

Vladimir Tchikov, Jürgen Fritsch, Stefan Schütze[1]
Institute of Immunology, Christian-Albrechts-University of Kiel, Kiel, Germany
[1]Corresponding author: e-mail address: schuetze@immunologie.uni-kiel.de

Contents

Abstract

We previously demonstrated that tumor necrosis factor receptor-1 (TNF-R1) initiates distinct TNF signaling pathways depending on the localization of the receptor. While TNF-R1 at the plasma membrane transmits proinflammatory and antiapoptotic signals, internalized TNF-R1 forms signaling endosomes (TNF receptosomes) that transmit

Methods in Enzymology, Volume 535
ISSN 0076-6879
http://dx.doi.org/10.1016/B978-0-12-397925-4.00019-5

proapoptotic signals. These findings were obtained by a novel technique for the isolation of morphologically intact endocytic vesicles containing magnetically labeled TNF-R1 complexes using a high-gradient, free-flow magnetic chamber. Since intact mitochondria appeared to be a major contaminating organelle in these preparations, we subsequently included a second purification step by iodixanol density centrifugation to obtain a mitochondria-free receptosome preparation.

1. INTRODUCTION

The classical model of signal transduction involves cell-surface receptors that are activated after binding to their ligands and transmit intracellular signals to generate secondary messengers. However, it has become clear that certain signaling pathways require receptor internalization for proper signal transduction (i.e., EGF-R, Trk, NGF-R, insulin-R, FGF-R1, Tf-R, IL5-R, AMPA-R, Frizzled, FcεRI, CXCR4, and Met; Cadigan, 2010; Lei, Mazumdar, & Martinez-Moczygemba, 2011; Shenoy, 2007; Sigismund et al., 2012). Thus, a more sophisticated picture emerged suggesting that endocytosis orchestrates cell signaling by coupling and integrating different cascades on the surface of endocytic vesicles (reviewed by McPherson, Kay, & Hussain, 2001; Miaczynska, Pelkmans, & Zerial, 2004; Murphy, Padilla, Hasdemir, Cottrell, & Bunnett, 2009; Scita & Di Fiore, 2010; Sigismund et al., 2012; Sorkin & Von Zastrow, 2002; Teis & Huber, 2003).

The essential role of tumor necrosis factor receptor-1 (TNF-R1) internalization for initiation of proapoptotic signaling was initially demonstrated by us using pharmacological inhibitors (Schütze et al., 1999), by deletion of a region termed "TNF-R1 internalization domain" (Schneider-Brachert et al., 2004), or by transducing cells with the adenoviral protein 14.7K (Schneider-Brachert et al., 2006; Schütze, Tchikov, & Schneider-Brachert, 2008).

In order to characterize the functional role of receptosomes, we developed a system based on ligand-specific immunomagnetic labeling of cell-surface receptors and selective purification of morphologically and functionally intact receptosomes in a patented free-flow magnetic chamber (Schütze, Tchikov, & Kabelitz, 2003, Tchikov & Schütze, 2008; Tchikov, Schütze, & Krönke, 1999; Tchikov, Winoto-Morbach, Krönke, Kabelitz, & Schütze, 2001).

This immunomagnetic separation approach was instrumental to identify TNF receptosomes as "death-signaling vesicles" (Schneider-Brachert et al.,

2004) and allowed the discovery of a novel adenoviral immune escape mechanism (Schneider-Brachert et al., 2006). It also led to the identification of CARP-2, an endosome-associated E3 ubiquitin ligase for RIP, which is involved in the regulation of antiapoptotic NF-κB signaling (Liao et al., 2008), and to the identification of the complex formation of the ESCRT proteins Alix and ALG-2 with procaspase-8 and TNF-R1, regulating TNF-induced apoptosis (Mahul-Mellier et al., 2008). Furthermore, the immunomagnetic separation technique allowed the identification of the enzyme riboflavin kinase (RFK) as a novel TNF-R1 death-domain adaptor protein, coupling the receptor to NADPH oxidase (Yazdanpanah et al., 2009), and the coupling of TNF-R1 via the polycomb group protein EED to neutral sphingomyelinase (Philipp et al., 2010).

Our studies showed that during intracellular trafficking, TNF-R1-containing endosomes fuse with endolysosomal compartments leading to the formation of multivesicular bodies. These fusion events are mandatory for the proteolytic activation of endolysosomal enzymes like A-SMase by a caspase-8 and caspase-7 cascade and finally cathepsin D and the proapoptotic protein BID. These events trigger the mitochondrial amplification loop of apoptosis (Edelmann et al., 2011; Schütze & Schneider-Brachert, 2009; Schütze et al., 2008). The results of our investigations on the compartmentalization of TNF-R1 signaling pathways are summarized in Fig. 19.1.

The immunomagnetic separation method was additionally applied to characterize the internalization and maturation of CD95 receptosomes. In this case, a biotinylated anti-CD95 antibody as ligand and streptavidin-coated magnetic beads were used for separation in the high-gradient magnetic chamber (Feig, Tchikov, Schütze, & Peter, 2007; Lee et al., 2006). Also, the death-inducing signaling complex (DISC) recruitment to internalized TRAIL receptors was demonstrated using this method (Lemke et al., 2010).

After successful application of the magnetic separation device for the isolation and functional characterization of TNF-, CD95-, and TRAIL receptosomes, we subsequently adopted this approach for the immunomagnetic isolation of soluble proteins and intact organelles from total cell lysates (Tchikov, Fritsch, Kabelitz, & Schütze, 2010). Here, we isolated subcellular organelles at various stages of vesicular maturation by coupling antibodies specific for signature proteins of endosome trafficking and fusion, like Rab5 and Vti1b, to magnetic nanobeads. These vesicles contain internalized TNF receptors (TNF-R1) in early endosomes and activated TNF-R1-associated signaling complexes like caspase-8 and cathepsin D in late endosomes/multivesicular organelles.

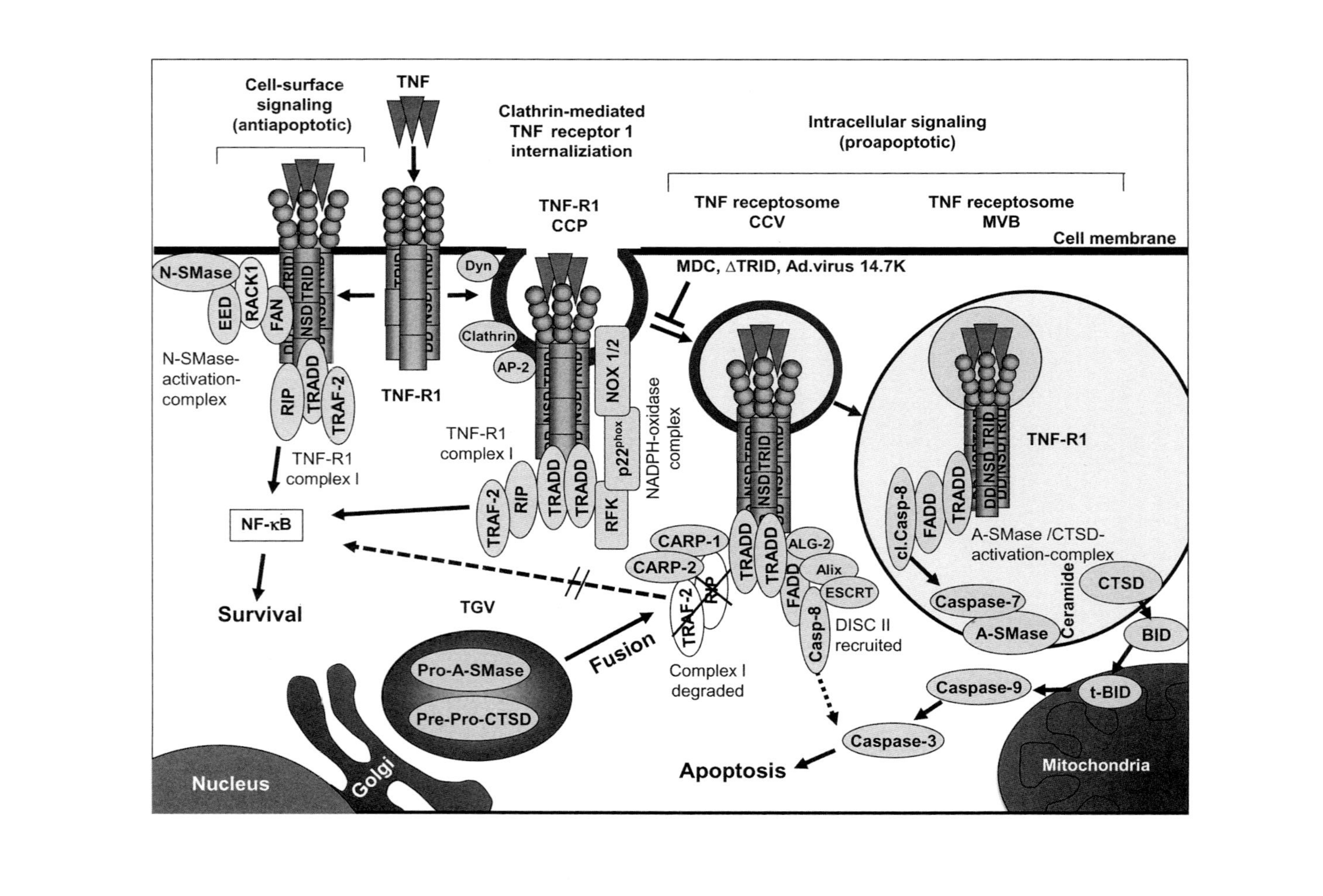

Cell-surface signaling (antiapoptotic)
TNF
Clathrin-mediated TNF receptor 1 internaliziation
Intracellular signaling (proapoptotic)
TNF-R1 CCP
TNF receptosome CCV
TNF receptosome MVB
Cell membrane
N-SMase
EED
RACK1
FAN
TRID
NSD
DD
Dyn
MDC, ΔTRID, Ad.virus 14.7K
Clathrin
AP-2
NOX 1/2
N-SMase-activation-complex
RIP
TRADD
TRAF-2
TNF-R1
p22phox
NADPH-oxidase complex
TNF-R1 complex I
TNF-R1 complex I
RFK
TNF-R1
NF-κB
CARP-1
CARP-2
ALG-2
Alix
ESCRT
cl.Casp-8
FADD
A-SMase /CTSD-activation-complex
Ceramide
CTSD
Caspase-7
A-SMase
BID
Survival
TGV
Fusion
Casp-8
DISC II recruited
Complex I degraded
Pro-A-SMase
Pre-Pro-CTSD
Caspase-9
t-BID
Caspase-3
Apoptosis
Nucleus
Golgi
Mitochondria

Figure 19.1 TNF receptosomes as "death-signaling vesicles." After TNF binding, an initial complex of adaptor proteins is formed at the "death domain" (DD) of cell-surface TNF-R1. It consists of TRADD, RIP-1, and TRAF-2 (termed *complex I*), which is capable of transducing the signal to NF-κB activation and cell survival. Also, at the cell surface, a distinct complex consisting of the proteins FAN, RACK1, and EED associates to the NSD of TNF-R1, coupling and signaling to the membrane-bound enzyme neutral sphingomyelinase (N-SMase) (the *N-SMase activation complex*). Within minutes after ligand binding, TNF-R1 is internalized in a clathrin/AP-2/dynamin-dependent manner, recruiting the *NADPH oxidase complex* (RFK, $p22^{phox}$, and NOX1/2) to the DD-adaptor protein TRADD. During intracellular trafficking, the E3 ubiquitin ligases CARP-1 and CARP-2 are recruited upon endosomal fusion, and these molecules lead to the K48 ubiquitination-mediated proteasomal degradation of RIP-1 and complex I and, by this, to the termination of NF-κB signaling. Concomitantly, the adaptor protein FADD binds to TRADD at TNF-R1, and the ESCORT proteins Alix and ALG-2 associate with TNF-R1 receptosomes and mediate the recruitment of caspase-8 to the TRADD/FADD complex, forming the DISC, also termed *complex II*. Activated caspase-8 can then induce apoptosis by direct activation of the executioner caspase-3. Alternatively, TNF receptosomes fuse with *trans*-Golgi vesicles containing pro-A-SMase and pre-pro-cathepsin D (CTSD) to form multivesicular endosomes. Here, activated caspase-8 and caspase-7 sequentially activate A-SMase, generating the lipid second messenger ceramide that binds to and activates the protease CTSD. CTSD translocates through the multivesicular membrane and cleaves the bcl-family member BID in the cytosol to generate tBID that eventually induces the mitochondrial apoptosis cascade involving cytochrome c release and caspase-9 and caspase-3 activation. (For color version of this figure, the reader is referred to the online version of this chapter.)

The free-flow, high-gradient magnetic system can also be successfully applied to isolate intracellular phagosomes after infection of cells with pathogens like mycobacteria, listeria, and chlamydia (Steinhäuser et al., 2013). This novel approach facilitates the fast purification of intact bacteria-containing phagosomes and allows the comprehensive biochemical characterization of gram-positive bacteria as well as mycobacteria-containing phagosomes from as few as 2×10^7 primary cells of different origin. In this study, labeling of bacteria was performed by using lipobiotin as target for streptavidin-coated MACS MicroBeads. The labeling procedure did not influence the uptake of the labeled bacteria nor did it induce unspecific signaling in phagosomes (Steinhäuser et al., 2013).

In addition to whole membrane compartments, the immunomagnetic approach can be applied for the isolation of rare soluble proteins like NOD-2 from primary γδ-T cells, too (Marischen et al., 2011).

One major problem in the preparation of a distinct subcellular compartment like signaling endosomes (receptosomes) is the fact that they are not located isolated within the cell. They rather form contact sites with many other cytoplasmic organelles and can interact with each other through stable junctions where the two membranes are kept in close contact (Elbaz & Schuldiner, 2011; Rassow, 2011; Reinehr & Haeussinger, 2008; Rowland & Voeltz, 2012; Toulmay & Prinz, 2011; West et al., 2011).

In addition to physiological organelle interactions *in vivo*, mechanical homogenization of cells may create artificial aggregates of organelles or even create hybrid membrane vesicles (Salomon, Janssen, & Neefjes, 2010). These heterogeneous membrane preparations mostly contain contaminations from mitochondria (Waugh, Chu, Clayton, Minogue, & Hsuan, 2011). Thus, to obtain information solely related to the compartment of interest, for example, TNF receptosomes, new approaches are required to enhance the purity of the preparations.

In this chapter, we describe a protocol in which receptosomes are purified by a two-step procedure: a ligand-specific immunomagnetic approach as the first step subsequently followed by ultracentrifugation of the magnetic isolates on an iodixanol density barrier to separate contaminating mitochondria from receptosomes in the second step.

2. PHYSICAL BACKGROUND

2.1. Magnetically labeled receptosomes are "light vesicles"

Based on an average 1:1 ratio of lipids to protein content of biological membranes (Dupuy & Engelman, 2008; Takamori et al., 2006) and mean densities of 1.02 g/ml for lipid (Koenig & Gawrisch, 2005) and 1.38 g/ml for protein (Fischer, Polikarpov, & Craievich, 2004; Quillin & Matthews, 2000), the receptosome membrane should have a density of 1.20 g/ml. Given a sphere-like receptosome of 200 nm in diameter surrounded by the 8 nm double membrane, the membrane and internal compartment composes 24% and 76% of the total receptosome volume, respectively (Fig. 19.2A(a)). With densities of 1.20 g/ml for the membrane and 1.006 g/ml for the internal compartment (equal to the internalization medium), the mean density of the whole receptosome is calculated by $0.24 \times 1.20 + 0.76 \times 1.006 = 1.053$ g/ml. This density corresponds to a 5% iodixanol solution. Thus, receptosomes can be considered as "light vesicles."

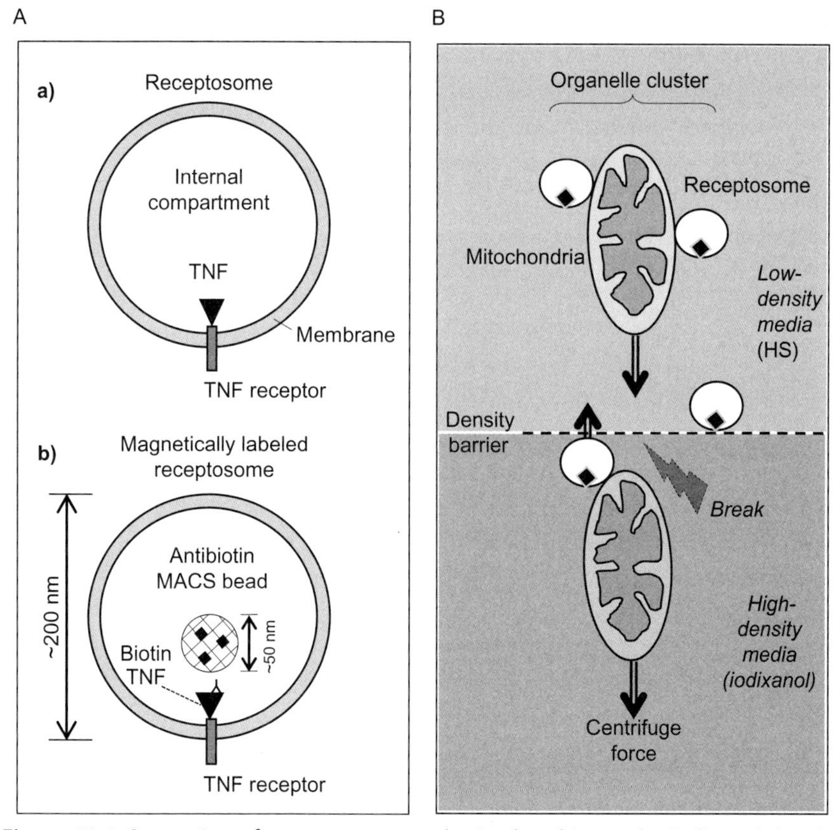

Figure 19.2 Separation of receptosomes and mitochondria on the iodixanol density barrier. (A) Properties of unlabeled receptosomes (a) and receptosomes labeled with antibiotin antibody-coupled iron oxide–dextran nanobeads (MACS MicroBeads) bound to biotinTNF (b). Black diamonds indicate iron oxide crystals embedded in the dextran matrix of the nanobead. (B) Separation of magnetic receptosomes and mitochondria at the iodixanol density barrier during centrifugation.

After magnetic labeling of the TNF receptors, internalized receptosomes include MACS MicroBeads inside (Fig. 19.2A(b)). The beads contain iron oxide (e.g., magnetite) nanocrystals embedded into a dextran matrix (Kantor, Gibbons, Miltenyi, & Schmitz, 1998). At densities of 5.2 g/cm^3 for magnetite crystals and 1.38 g/cm^3 for dextran and with the volumes of 26% for iron oxide and 74% for dextran, we calculated $0.26 \times 5.2 + 0.74 \times 1.38 = 2.37$ g/cm^3 for the density of the magnetic bead.

The difference in densities of the bead and the internal endosomal compartment is great ($2.37 - 1.006 = 1.364$ g/cm^3), but since the 50 nm bead occupies only a small part in the 200 nm receptosome, which is $(50\ \text{nm}/200\ \text{nm})^3 = 0.016$ or 1.6% (Fig. 19.2A(b)), the total effect in the shift of density of the whole receptosome caused by the bead is only small ($1.364 \times 0.016 = 0.022$ g/ml). The density of the bead-loaded receptosome was calculated as $0.022 + 1.053 = 1.075$ g/ml. This value is equal to a 9% iodixanol solution.

Compared to the small and light receptosomes, mitochondria are relatively heavy organelles (Dieteren et al., 2011; Srere, 1980). In an iodixanol gradient, mitochondria will float at a density of 1.14–1.16 g/ml, corresponding to a 24% iodixanol–sucrose solution.

2.2. The density barrier

The density barrier is defined as the interface between low- and high-density media, in our case the homogenization solution (HS) and the iodixanol solution (Fig. 19.2B). By choosing the iodixanol barrier density between 9% (receptosomes) and 24% (mitochondria), we expected an optimal separation of receptosomes and mitochondria by ultracentrifugation on the 16% iodixanol barrier in which receptosomes will be floating on the barrier, while mitochondria will pass through the barrier. The high centrifugal force will disrupt the bonds between receptosomes and mitochondria at the barrier.

3. REAGENTS AND CELLS

3.1. Reagents

Biotinylated TNF (biotinTNF) was obtained from Fluorokine Kit (R&D Systems, Wiesbaden, Germany), and antibiotin MACS MicroBeads were from Miltenyi Biotec GmbH. The 60% iodixanol solution was from OptiPrep™ (Axis-Shield PoC AS, Oslo, Norway), and the Application Sheets are available at http://www.axis-shield-density-gradient-media.

com. The HS was from the manufacturer of OptiPrep and contains 0.25 *M* sucrose in 10 m*M* triethanolamine–acetic acid buffer, pH 7.8 (TEA buffer, for details, see the Application Sheet S06 from the OptiPrep manual). The Protease Inhibitors Set was from Roche Diagnostics (Mannheim, Germany). Primary antibodies were obtained against TNF-R1 (H5; Santa Cruz Biotech, Inc.), TRADD (monoclonal; BD PharMingen), FADD (monoclonal; BD Transduction Lab.), caspase-8 (C-15; Scaffidi et al., 1997), cathepsin D (Calbiochem GmbH, Germany), and clathrin and Vti1b (BD Transduction Lab.). Secondary antibodies were horseradish peroxidase (HRP)-conjugated rabbit anti-mouse and goat anti-rabbit antibodies (Dianova GmbH, Germany). Streptavidin conjugated with HRP was obtained from GE Healthcare, UK. Protein measurements were performed with BCA Protein Assay Reagents (Pierce, USA).

3.2. Cultivation and preparation of cells

1. As an example, we used U937 cells, obtained from ATCC and maintained in CLICK's RPMI culture medium (Biochrom, Berlin, Germany) supplemented with 5% fetal calf serum.
2. Wash 10^8 cells for two times by centrifugation with 30 ml of cold phosphate-buffered saline (PBS) at 100 × *g*, 10 min. Resuspend the cell pellet in a total volume of 250 μl cold PBS; transfer the cell suspension into a fresh tube and store at 4 °C.

3.3. Protocols

Figure 19.3 shows the flow scheme of the two-step separation protocol, first involving immunomagnetic separation of TNF receptosomes from total lysates in the magnetic chamber (Steps 1–6) and second the purification of isolated receptosomes from contaminating mitochondria by iodixanol density centrifugation (Step 7).

4. PROTOCOL 1: MAGNETIC SEPARATION

Antibiotin MACS MicroBeads are used for magnetic labeling of biological material that is prelabeled with biotinylated primary antibodies or ligands. In this case, biotinTNF bound to the TNF receptor is recognized by a monoclonal antibiotin antibody coupled to 50 nm magnetic nanoparticles (antibiotin MACS MicroBeads).

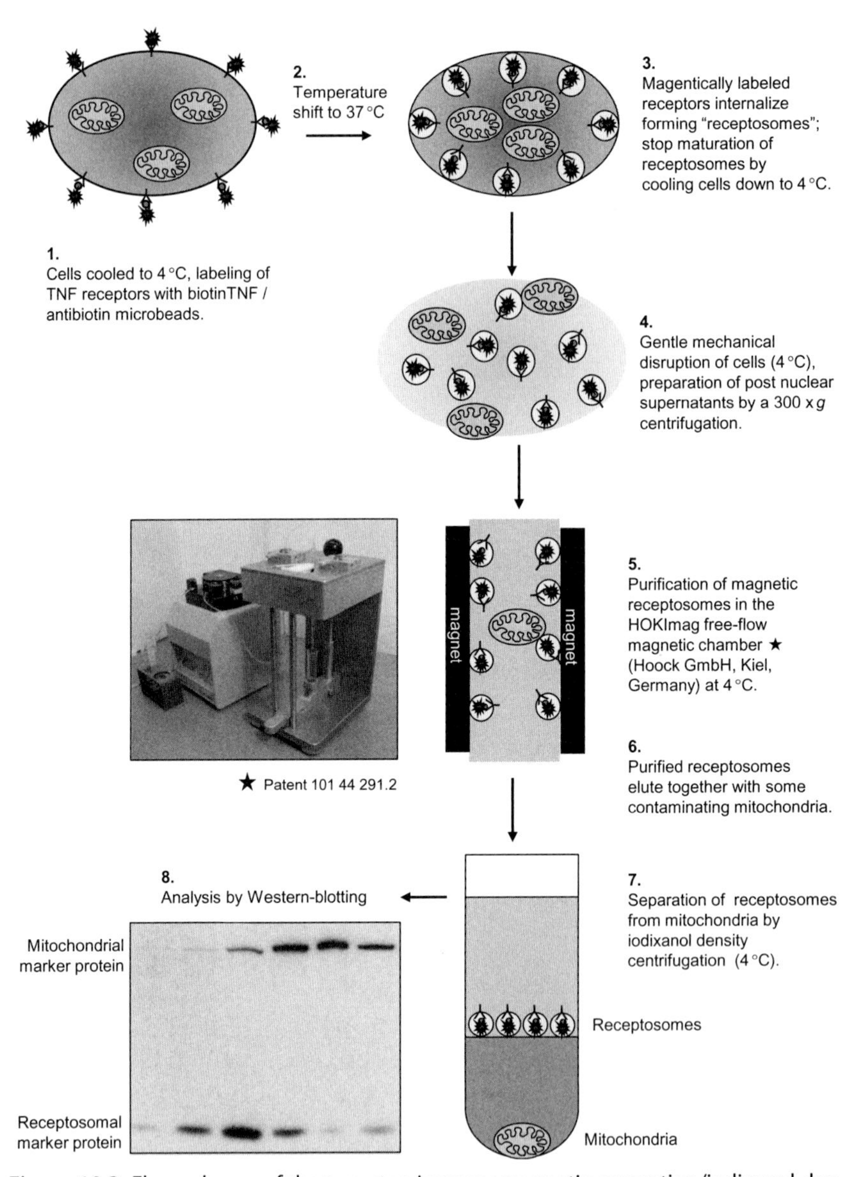
2.
Temperature
shift to 37 °C
3.
Magentically labeled
receptors internalize
forming "receptosomes";
stop maturation of
receptosomes by
cooling cells down to 4 °C.
1.
Cells cooled to 4 °C, labeling of
TNF receptors with biotinTNF /
antibiotin microbeads.
4.
Gentle mechanical
disruption of cells (4 °C),
preparation of post nuclear
supernatants by a 300 x g
centrifugation.
magnet
magnet
5.
Purification of magnetic
receptosomes in the
HOKImag free-flow
magnetic chamber ★
(Hoock GmbH, Kiel,
Germany) at 4 °C.
★ Patent 101 44 291.2
6.
Purified receptosomes
elute together with some
contaminating mitochondria.
8.
Analysis by Western-blotting
7.
Separation of receptosomes
from mitochondria by
iodixanol density
centrifugation (4 °C).
Mitochondrial
marker protein
Receptosomes
Receptosomal
marker protein
Mitochondria

Figure 19.3 Flow scheme of the two-step immunomagnetic separation/iodixanol density barrier centrifugation protocol. See Chapters 4 and 5 for details. (For color version of this figure, the reader is referred to the online version of this chapter.)

4.1. Labeling of TNF receptors with biotinTNF and antibiotin MACS MicroBeads (Step 1 in Fig. 19.3)

1. Add 100 μl of biotinTNF to the cells prepared as described before in Step 2 of Section 3.2, mix carefully, and incubate for 1 h on ice to label TNF receptors on the cell plasma membrane.
2. Wash unbound biotinTNF from the cells for at least two times by centrifugation with 30 ml of cold PBS for 10 min at 100 × *g*.
3. Resuspend the cell pellet again in a total volume of 250 μl and add 100 μl of antibiotin MACS MicroBeads solution to the cell suspension. Incubate as earlier for another hour to couple the magnetic beads to the biotinylated TNF/TNF receptor complex at 4 °C.
4. Wash unbound MACS MicroBeads from the cells for two times by centrifugation as earlier with 30 ml of cold PBS at 100 × *g* for 10 min, and resuspend the cell pellet in a total volume of 500 μl cold FCS-free RPMI medium, and then, store it on ice.

4.2. Internalization and formation of receptosomes (Step 2 in Fig. 19.3)

Synchronized internalization of magnetically labeled biotinTNF/TNF receptor complexes from the cell surface is induced by shifting the incubation temperature from 4 to 37 °C:

1. Transfer and resuspend the labeled cells in 20 ml of 37 °C prewarmed FCS-free medium followed by incubation for different time points.
2. Stop the internalization and maturation of receptosomes by adding 20 ml of ice-cold medium into the incubation tube. Sediment cells by centrifugation at 4 °C at 100 × *g* for 10 min.

4.3. Homogenization of cells (Step 4 in Fig. 19.3)

1. Resuspend cells in 30 ml of precooled HS and centrifuge at 150 × *g* for 10 min. Discard the supernatant and resuspend the cells in 350 μl of HS supplemented with protease inhibitors from the Protease Inhibitors Set (Roche Diagnostics, Mannheim, Germany) supplemented with 12.5 U/ml Benzonase (Merck, Darmstadt, Germany).
2. For homogenization, transfer the cells into the homogenization device described previously (Tchikov & Schütze, 2008), and vortex the sample with glass beads for 5 min on ice. Collect the homogenate in a separate tube and centrifuge at 300 × *g* for 5 min. Transfer the supernatant again into a separate tube. Resuspend the pellet in 350 μl of HS, and repeat the

procedure for three times. Collect supernatants into one tube and finally clear by centrifugation as earlier. The collected supernatants (i.e., lysate) are subjected to the following magnetic separation.

Alternatively, and depending on the cell type, homogenization may also be performed by using steel beads instead of glass beads as earlier or by ultrasonication: In this case, cell sediments are resuspended in a total volume of 800 μl HS and 12.5 U Benzonase (Merck). Cells are lysed at 4 °C using a Branson W-450 sonication device equipped with a cup resonator (G. Heinemann, Schwäbisch Gmünd, Germany).

4.4. Magnetic separation of the lysate (Steps 5 and 6 in Fig. 19.3)

The magnetic device developed in our lab (HOKImag, now manufactured by Hoock GmbH, Kiel, Germany) is used for the purification of magnetically labeled receptosomes from the lysate (Fig. 19.4A). The device contains the magnetic system providing an especially configurated inhomogeneous 2-T magnetic field in the gap between the magnetic poles.

Separation of magnetically labeled material is performed in free-flow plastic tubes of 5 mm in diameter and with 0.1 mm thin walls. Unlike MACS columns that are filled with iron balls crucial for generating the magnetic force inside the columns (Gruetzkau & Radbruch, 2010), our columns are empty inside and the magnetic separation is driven by the magnetic force generated by especially designed external magnetic poles. With these columns, clogging of aggregates is completely excluded. To minimize contaminations at the wall of the tube with nonmagnetic material, the internal surface of the tubes is silanized with PlusOne Repel-Silane ES (GE Healthcare GmbH, Germany). Silanization of the tubes also facilitates the release of the magnetic fraction from the tube walls after magnetic separation when the column is removed from the magnetic field.

For the separation procedure, the column is placed in the gap of the magnetic chamber and connected to the peristaltic pump to allow for the loading of the lysate (pumping from the right to the left) or generating the washing flow after the separation (pumping from the left to the right). The device is stored at 4 °C. The pump (i.e., Minipuls 3 Peristaltic Pump, Gilson Inc., USA) is supplemented with a Tygon R3607 tube (ID 0.89 mm, orange–orange, VWR Intern. GmbH, Germany, Cat. No. 224-2238).

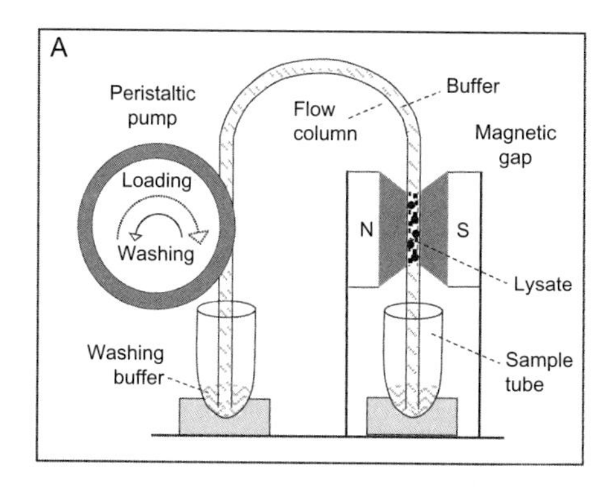

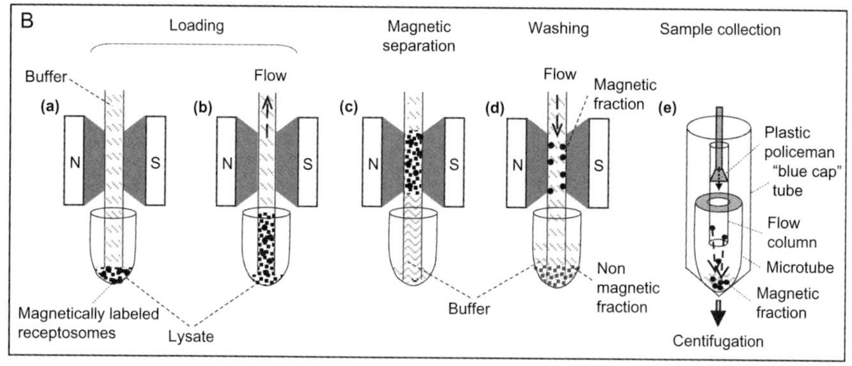

Figure 19.4 Immunomagnetic separation of TNF receptosomes in the high-gradient, free-flow magnetic chamber. (A) The magnetic separation device contains the magnetic system providing an inhomogeneous 2-T magnetic field in the gap between magnetic poles (N–S). The thin-wall flow column is placed in the gap and connected to the peristaltic pump for developing the loading flow of the lysate (pumping from the right to the left) or the washing flow (pumping from the left to the right). (B) Loading of the chamber (a, b), magnetic separation of the material (c), washing (d), and collection of samples by centrifugation (e). See protocols 1–8 of Section 4.4.

Assembling the column

1. Place the flow column into the magnetic gap of the magnetic device.
2. Place the right tip of the flow column in the right tube containing wash buffer (HS). Completely fill the flow column with HS by pumping from right to left. No air breaks should be visible in the flow system (Fig. 19.4A).
3. Prepare 750 μl sample lysate in a fresh 14-ml tube (Falcon tube, 14 ml polystyrene round-bottom 17 x 100 mm style; BD Labware, USA) and immerse the right tip of the flow column within (Fig. 19.4B(a)).

4. Pump the sample upward into the flow column (2 rpm × 5 min), leaving the rest of the buffer at the bottom of the tube to avoid pumping air bubbles (Fig. 19.4B(b)).

Loading the sample

5. Replace the sample tube with a fresh tube preloaded with 600 μl of supporting buffer (HS supplemented with 5% iodixanol). Pump 500 μl of the buffer upward (2 rpm × 5 min) until the lysate is localized within the magnetic gap (Fig. 19.4B(c)).
6. Incubate the lysate in the magnetic field for 1 h at steady state (Fig. 19.4B(c)).

Washing

7. Pump the washing buffer through the flow column (from the left to the right) at low speed (0.4 rpm × 25 min) for removing the unbound lysate from the magnetic gap. Continue washing at higher speed (2 rpm × 30 min) (Fig. 19.4B(d)).

Sample collection

8. Remove the tube containing the nonmagnetic fraction from the separation device and place a fresh 1.5-ml microtube below the separation column. Label the flow column at the top of the magnetic gap. Cut off the lower part of the separation column at the bottom of the magnetic gap. Remove the separation column from the magnetic gap and cut off the upper part of the column at a distance of about 5 mm upward to the upper label. Since this will disrupt the contact of the column to the pump, the magnetic fraction will elute from the column into the sample microtube. Most of the magnetic fraction will be immediately released from the column. Additionally, scraping the walls of the column with a plastic policeman (or use the plunger of a 1-ml syringe) followed by a short-time centrifugation (e.g., 3500 × g for 1 min) of the collecting device (supplied by the manufacturer) placed in a 50-ml centrifuge tube (Cellstar® 50-ml tube; "blue cap," Greiner Bio-One GmbH, Germany) will quantitatively collect the magnetic fraction into the microtube.

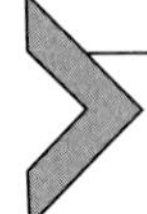

5. PROTOCOL 2: SEPARATION BY IODIXANOL DENSITY BARRIER CENTRIFUGATION

OptiPrep is a 60% weight per volume (w/v) solution of iodixanol in water produced by Axis-Shield PoC (Oslo, Norway) for the preparation of density gradients. The manufacturer recommends to prepare a 50% working solution (WS) from the original OptiPrep product and then to mix the WS with the HS for obtaining the desired density (see product description).

5.1. Preparation of the density barrier solution

1. To prepare a 50% (w/v) iodixanol WS, mix 5 volumes of OptiPrep with 1 volume of the HS.
2. Prepare the 16% (w/v) iodixanol–sucrose solution (density 1.107 g/ml) by mixing 3.2 ml of the WS with 6.8 ml of HS according to the recommendations and the density table from the manufacturer.
3. Load 2 ml of the 16% iodixanol solution on the bottom of the 4-ml centrifuge tube (Ultra-Clear™ Centrifuge Tube, 11 × 60 mm; Beckman Instruments, Inc., CA, USA).

5.2. Loading sample on the density barrier and ultracentrifugation

1. Adjust the volume of the magnetic fraction prepared earlier with HS to 2 ml.
2. Carefully overlay the magnetic fraction on top of the density barrier.
3. Centrifuge the sample at 150,000 × *g* for 1 h in a swing-out rotor (SW60Ti; Beckman Coulter, Krefeld, Germany) without brake.
4. Carefully collect fractions from the top to the bottom accordingly to the scheme in Fig. 19.5 (A = 650 μl, B = 800 μl, fractions from C to G are 250 μl each, H = 800 μl, and I and J = 250 μl each), and store them frozen for BCA measurement and the following Western blot analysis. Alternatively perform the chemical fixation for electron microscopy.
5. The fractions are characterized by Western blotting, shown in Figs. 19.5 and 19.7 with 3 μg of protein loaded on each lane.

6. RESULTS

6.1. Separation of receptosomes from mitochondria on the density barrier

During centrifugation, the magnetically isolated material loaded on the density barrier is fractionated into two major distinct fractions. Most of the analyzed proteins are recovered in the floating fraction at the barrier (fractions D and E) and a minor amount at the bottom of the tube (fraction J) (Fig. 19.5B). The Western blot analysis revealed that fractions D–F mainly

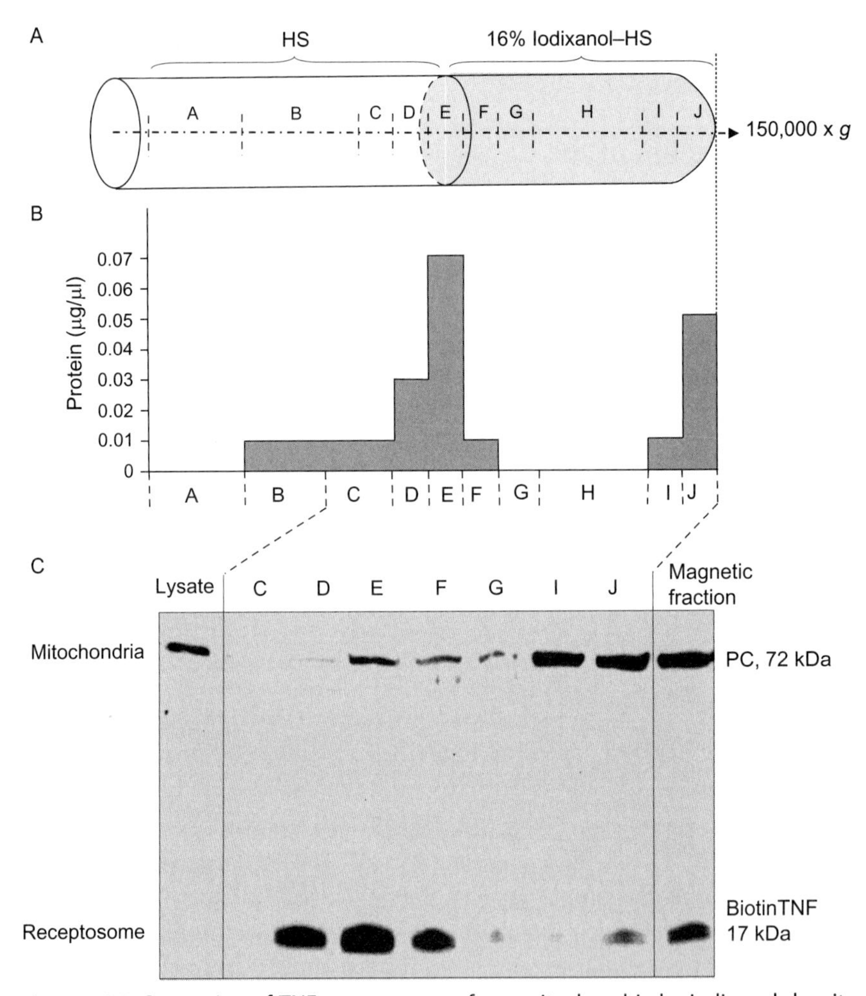

Figure 19.5 Separation of TNF receptosomes from mitochondria by iodixanol density centrifugation. (A) Scheme of the centrifuge tube containing sample in HS and a 16% iodixanol-HS solution. Fractionation of the material is indicated by letters A–J. (B) Distribution of protein in the fractions after centrifugation. (C) Western blotting of the fractions, showing the distribution of the mitochondrial marker protein pyruvate carboxylase (PC) and biotinTNF as marker for TNF receptosomes.

contained TNF receptosomes as indicated by the detection of biotinTNF. Fractions I and J mainly contained mitochondria, indicated by the presence of pyruvate carboxylase (PC) (Fig. 19.5C).

Samples from floating fraction E and pellet fraction J were also analyzed by electron microscopy. This revealed intact TNF receptosomes

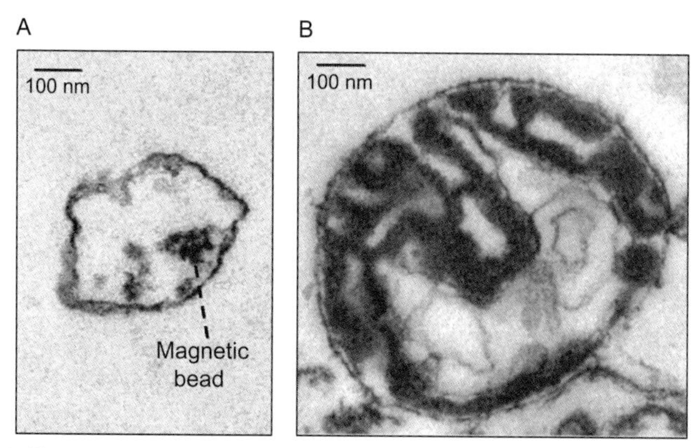

Figure 19.6 Electron microscopy of an intact TNF receptosome (A) floating on the density barrier and a mitochondrion (B) from the pellet after centrifugation.

characterized by containing the MACS MicroBeads inside the vesicle in the floating fraction (Fig. 19.6A) and mitochondria characterized by the electron-dense internal matrix in the pellet (Fig. 19.6B).

6.2. Characterization of the fractions

The fractions from the iodixanol centrifugation were analyzed for the distribution of proteins characteristic of TNF receptosomes, for example, biotinTNF; TNF-R1; the DISC proteins TRADD, FADD, and caspase-8; signature proteins of endocytosis and endosomal trafficking like clathrin (internalization), the SNARE-protein Vti1b (fusion with *trans*-Golgi compartments), cathepsin D (fusion with lysosomal compartments); as well as mitochondrial marker proteins like cox-IV and PC (Fig. 19.7).

In line with the result shown in Fig. 19.5, the mitochondrial proteins are almost exclusively found in the pellet fraction J. In the floating fractions D, E, and F, a strong enrichment of the TNF-R1 and the ligand biotinTNF is apparent compared to the total lysate fraction, indicating successful isolation of TNF-R1 receptosomes. Within these fractions, also the trafficking marker proteins clathrin, Vti1b, and cathepsin D are highly enriched. Vti1b and mature and cleaved forms of cathepsin D indicate fusion between TNF receptosomes and cathepsin D containing transport vesicles in line with our previous observations (Schneider-Brachert et al., 2004) (see also Fig. 19.1). The presence of the DISC proteins TRADD, FADD, and caspase-8 in the same fractions further characterizes the successful preparation of biologically active and intact TNF receptosomes.

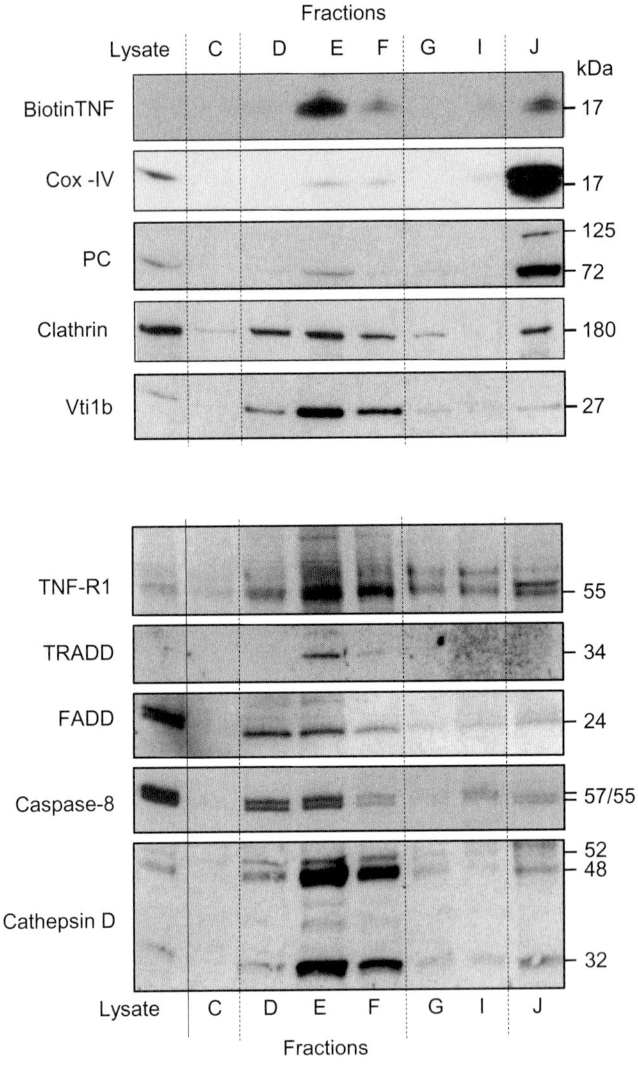

Figure 19.7 Western blot analysis of isolates from the iodixanol density centrifugation. Fractions from the iodixanol density centrifugation were separated by SDS-PAGE and blotted for the proteins indicated. The upper panel shows the separation from the biotinTNF signal mainly in the floating fraction E from mitochondria, indicated by the marker proteins cox-IV and pyruvate carboxylase (PC) in the pellet fraction J (upper panel). In line with the distribution of biotinTNF, the TNF-R1 and the TNF-R1-associated proteins TRADD, FADD, and caspase-8 are highly enriched in the floating fractions D–F (lower panel), indicating the presence of intact TNF receptosomes. The signature proteins of vesicular trafficking and maturation, clathrin, Vti1b, and cathepsin D, are also present in the same fractions. (For color version of this figure, the reader is referred to the online version of this chapter.)

Some residual TNF receptosome "signaling proteins" at the bottom fraction can be explained by "a leakage" of receptosomes through the barrier, which is also indicated by faint signals from biotinTNF and TNF-R1 at the bottom fraction J. This "leakage" may occur when the morphology of the receptosomes is destroyed, resulting in enhanced density of the organelle by iodixanol flowing into the vesicle.

However, compared to the signals from the receptosome fraction, the leakage signals are poor, indicating some 10% loss of receptosomes from the barrier fraction to the bottom fraction.

The cox-IV pattern indicates only negligible amounts of mitochondrial contaminations in the receptosome fraction. This proves the superior purity of TNF receptosomes isolated on the density barrier.

7. CONCLUSIONS

Traditionally, isolation of organelles from cell lysates is widely performed by separating different compartments based on their differences in density by density gradient centrifugation. The technical limitations of these techniques are still evident (Mathias, 1966), and density gradient separation should rather be termed an enrichment than an isolation of a specific organelle (Schmidt et al., 2009). Mitochondria seem to be a general contamination in organelle separation by density gradients (Waugh et al., 2011). Therefore, new separation techniques are needed that can reduce the heterogeneity of isolated materials.

One way to add specificity to the preparations is the immunospecific labeling of organelles, using antibodies against organelle-specific antigens and isolation of the vesicles by magnetic force (Tchikov et al., 2010; Tchikov & Schütze, 2008). Immunomagnetic sorting has been successfully used to purify Golgi vesicles (Mura, Becker, Orellana, & Wolff, 2002), endosomes (Perrin-Cocon, Marche, & Villiers, 1999), lysosomes (Diettrich, Mills, Johnson, Hasilik, & Winchester, 1998), nuclei (Kausch, Owen, Narayanswami, & Bruce, 1999), mitochondria (Tang, Zhao, Liang, Zhang, & Wang, 2012), and plasma membranes (Lawson et al., 2006). We extended this technology to the purification of TNF-R1-, CD95-, and TRAIL receptosomes by using either biotinylated ligands or receptor-specific antibodies (Edelmann et al., 2011; Feig et al., 2007; Lee et al., 2006; Lemke et al., 2010; Liao et al., 2008; Philipp et al., 2010; Schneider-Brachert et al., 2006, 2004; Yazdanpanah et al., 2009). Commercially available magnetic separation systems suffered either from using large magnetic beads, not suitable for isolation of endosomes, or from separation columns that contain an iron matrix inside. The inner matrix creates a huge surface that provokes contaminations and occlusion caused by the viscous lysates. Therefore, we have developed a novel free-flow, high-gradient magnetic chamber system that generates a highly inhomogeneous magnetic field inside a column without matrix and that allows to use small magnetic beads of 50 nm in diameter. The device (HOKImag) is now produced by the company Hoock GmbH (Kiel, Germany).

During our investigations, we noticed that our immunomagnetic receptosome preparations were still contaminated with mitochondria. We thus included a second purification step by using an independent physical principle—the differences in density between both receptosomes as "light vesicles" and mitochondria as "heavy vesicles"—and made use of the physical shear forces at the interface between low-density media and high-density media. The centrifugation of the magnetic fraction obtained from the first-step magnetic separation on an iodixanol density barrier resulted in an almost complete removal of mitochondria and, by this, in a highly purified and morphological intact receptosome fraction. This can be used for further biochemical investigations including analysis of the proteome of receptosomes.

ACKNOWLEDGMENTS

We thank Casimir Malanda for excellent technical assistance. This work was supported by grants from the Deutsche Forschungsgemeinschaft (DFG) Collaborative Research Center 877, Project B1, by the DFG-Project SCHU 733/9-2, and by the Schleswig-Holstein Cluster of Excellence given to S. S.

REFERENCES

Cadigan, K. M. (2010). Receptor endocytosis: Frizzled joins the ubiquitin club. *EMBO Journal*, *29*, 2099–2100.

Dieteren, C., Gielen, S., Nijtmans, L., Smeitink, L., Swarts, H., Brock, R., et al. (2011). Solute diffusion is hindered in the mitochondrial matrix. In: *Proceedings of the National Academy of Sciences of the United States of America*, *108*, 8657–8662.

Diettrich, O., Mills, K., Johnson, A. W., Hasilik, A., & Winchester, B. G. (1998). Application of magnetic chromatography to the isolation of lysosomes from fibroblasts of patients with lysosomal storage disorders. *FEBS Letters*, *441*, 369–372.

Dupuy, A., & Engelman, D. (2008). Protein area occupancy at the center of the red blood cell membrane. In: *Proceedings of the National Academy of Sciences of the United States of America*, *105*, 2848–2852.

Edelmann, B., Bertsch, U., Tchikov, V., Winoto-Morbach, S., Perrotta, C., Jakob, M., et al. (2011). Caspase-8 and caspase-7 sequentially mediate proteolytic activation of acid sphingomyelinase in TNF-R1 receptosomes. *EMBO Journal*, *30*, 379–394.

Elbaz, Y., & Schuldiner, M. (2011). Staying in touch: The molecular era of organelle contact sites. *Trends in Biochemical Sciences, 36*, 616–623.
Feig, C., Tchikov, V., Schütze, S., & Peter, M. E. (2007). Palmitoylation of CD95 facilitates formation of SDS-stable receptor aggregates that initiate apoptosis signaling. *EMBO Journal, 26*, 221–231.
Fischer, H., Polikarpov, I., & Craievich, A. (2004). Average protein density is a molecular-weight-dependent function. *Protein Science, 13*, 2825–2828.
Gruetzkau, A., & Radbruch, A. (2010). Small but mighty: How the MACS-technology based on nanosized superparamagnetic particles has helped to analyze the immune system within the last 20 years. *Cytometry Part A, 77A*, 643–647.
Kantor, A. B., Gibbons, I., Miltenyi, S., & Schmitz, J. (1998). In D. Recktenwald & A. Radbruch (Eds.), *Cell separation methods and applications. Part IV: Magnetic Methods* (p. 153). New York, Hong Kong: Marcel Dekker, Inc.
Kausch, A. P., Owen, T. P., Jr., Narayanswami, S., & Bruce, B. D. (1999). Organelle isolation by magnetic immunoabsorption. *BioTechniques, 26*, 336–343.
Koenig, B., & Gawrisch, K. (2005). Specific volumes of unsaturated phosphatidylcholines in the liquid crystalline lamellar phase. *Biochimica et Biophysica Acta, 1715*, 65–70.
Lawson, E. L., Clifton, J. G., Huang, F., Li, X., Hixson, D. C., & Josic, D. (2006). Use of magnetic beads with immobilized monoclonal antibodies for isolation of highly pure plasma membranes. *Electrophoresis, 27*, 2747–2758.
Lee, K. H., Feig, C., Tchikov, V., Schickel, R., Hallas, C., Schütze, S., et al. (2006). The role of receptor internalization in CD95 signaling. *EMBO Journal, 25*, 1009–1023.
Lei, J. T., Mazumdar, T., & Martinez-Moczygemba, M. (2011). Three lysine residues in the common beta chain of the IL-5 receptor are required for JAK-dependent receptor ubiquitination, endocytosis and signaling. *Journal of Biological Chemistry, 286*, 40091–40103.
Lemke, J., Noack, A., Adam, D., Tchikov, V., Bertsch, U., Röder, C., et al. (2010). TRAIL signaling is mediated by DR4 in pancreatic tumor cells despite the expression of functional DR5. *Journal of Molecular Medicine, 88*, 729–740.
Liao, W., Xiao, Q., Tchikov, V., Fujita, K., Yang, W., Wincovitch, S., et al. (2008). CARP-2 is an endosome-associated ubiquitin ligase for RIP and regulates TNF-induced NF-kappaB activation. *Current Biology, 18*, 641–649.
Mahul-Mellier, A. L., Strappazon, F., Petiot, A., Chatellard-Causse, C., Torch, S., Blot, B., et al. (2008). Alix and ALG-2 are involved in tumor necrosis factor receptor-1 induced cell death. *Journal of Biological Chemistry, 283*, 34954–34965.
Marischen, L., Wesch, D., Oberg, H. H., Rosenstiel, P., Trad, A., Shomali, M., et al. (2011). Functional expression of NOD2 in human peripheral blood γδ T-cells. *Scandinavian Journal of Immunology, 74*, 126–134.
Mathias, A. (1966). Separation of subcellular particles. *British Medical Bulletin, 22*, 146–152.
McPherson, P. S., Kay, B. K., & Hussain, N. K. (2001). Signaling on the endocytic pathway. *Traffic, 2*, 375–384.
Miaczynska, M., Pelkmans, L., & Zerial, M. (2004). Not just a sink: Endosomes in control of signal transduction. *Current Opinion in Cell Biology, 16*, 400–406.
Mura, C. V., Becker, M. I., Orellana, A., & Wolff, D. (2002). Immunopurification of Golgi vesicles by magnetic sorting. *Journal of Immunological Methods, 260*, 263–271.
Murphy, J. E., Padilla, B. E., Hasdemir, B., Cottrell, G. S., & Bunnett, N. W. (2009). Endosomes: A legitimate platform for the signaling train. In: *Proceedings of the National Academy of Sciences of the United States of America, 106*, 17615–17622.
Perrin-Cocon, L. A., Marche, P. N., & Villiers, C. L. (1999). Purification of intracellular compartments involved in antigen processing: A new method based on magnetic sorting. *Biochemical Journal, 338*, 123–130.
Philipp, S., Puchert, M., Adam-Klages, S., Tchikov, V., Winoto-Morbach, S., Mathieu, S., et al. (2010). The polycomb group protein EED couples TNF-receptor 1 to neutral

sphingomyelinase. In: *Proceedings of the National Academy of Sciences of the United States of America*, *107*, 1112–1117.
Quillin, M., & Matthews, B. (2000). Accurate calculation of the density of proteins. *Acta Crystallographica*, *D56*, 791–794.
Rassow, J. (2011). Helicobacter pylori vacuolating toxin A and apoptosis. *Cell Communication and Signalling*, *9*, 26–35.
Reinehr, R., & Haeussinger, D. (2008). CD95 ligation and intracellular membrane flow. *Biochemistry Journal*, *413*, e11–e12.
Rowland, A., & Voeltz, G. (2012). Endoplasmic reticulum–mitochondria contacts: Function of the junction. *Nature Reviews Molecular Cell Biology*, *13*, 607–615.
Salomon, I., Janssen, H., & Neefjes, J. (2010). Mechanical forces used for cell fractionation can create hybrid membrane vesicles. *International Journal of Biological Sciences*, *6*, 649–654.
Scaffidi, C., Medema, J. P., Krammer, P. H., & Peter, M. E. (1997). FLICE is predominantly expressed as two functionally active isoforms, caspase-8/a and caspase-8/b. *Journal of Biological Chemistry*, *272*, 26953–26958.
Schmidt, H., Gelhaus, C., Lucius, R., Nebendahl, M., Leippe, M., & Janssen, O. (2009). Enrichment and analysis of secretory lysosomes from lymphocyte populations. *BMC Immunology*, *10*, 41–52.
Schneider-Brachert, W., Tchikov, V., Merkel, O., Jakob, M., Hallas, C., Kruse, M. L., et al. (2006). Inhibition of TNF receptor 1 internalization by adenovirus 14.7K as a novel immune escape mechanism. *Journal of Clinical Investigation*, *116*, 2901–2913.
Schneider-Brachert, W., Tchikov, V., Neumeyer, J., Jakob, M., Winoto-Morbach, S., Held-Feindt, J., et al. (2004). Compartmentalization of TNF receptor 1 signaling: Internalized TNF receptosomes as death signaling vesicles. *Immunity*, *21*, 415–428.
Schütze, S., Machleidt, T., Adam, D., Schwandner, R., Wiegmann, K., Kruse, L. M., et al. (1999). Inhibition of receptor internalization by monodansylcadaverine selectively blocks p55 TNF receptor death domain signaling. *Journal of Biological Chemistry*, *274*, 10203–10212.
Schütze, S., & Schneider-Brachert, W. (2009). Impact of TNF-R1 and CD95 internalization on apoptotic and antiapoptotic signaling. *Results and Problems in Cell Differentiation*, *49*, 63–85.
Schütze, S., Tchikov, V., Kabelitz, D., & Krönke, M. (2003). German patent DE 101 44 291.
Schütze, S., Tchikov, V., & Schneider-Brachert, W. (2008). Regulation of TNFR1 and CD95 signaling by receptor compartmentalization. *Nature Reviews Molecular Cell Biology*, *9*, 655–662.
Scita, G., & Di Fiore, P. P. (2010). The endocytic matrix. *Nature*, *463*, 464–473.
Shenoy, S. K. (2007). Seven-transmembrane receptors and ubiquitination. *Circulation Research*, *100*, 1142–1154.
Sigismund, S., Confalonieri, S., Ciliberto, A., Polo, S., Scita, G., & Di Fiore, P. P. (2012). Endocytosis and signaling: Cell logistics shape the eukaryotic cell plan. *Physiological Reviews*, *92*, 273–366.
Sorkin, A., & Von Zastrow, M. (2002). Signal transduction and endocytosis: Close encounters of many kinds. *Nature Reviews Molecular Cell Biology*, *3*, 600–614.
Srere, P. A. (1980). The infrastructure of the mitochondrial matrix. *Trends in Biochemical Sciences*, *5*, 120–121.
Steinhäuser, C., Heigl, U., Tchikov, V., Schwarz, J., Gutsmann, T., Seeger, K., et al. (2013). Lipid-labeling facilitates a novel magnetic isolation procedure to characterize of pathogen-containing phagosomes. *Traffic*, *14*, 321–336.
Takamori, S., Holt, M., Stenius, K., Lemke, E., Gronborg, M., Riedel, D., et al. (2006). Molecular anatomy of a trafficking organelle. *Cell*, *127*, 831–846.
Tang, B., Zhao, L., Liang, R., Zhang, Y., & Wang, L. (2012). Magnetic nanoparticles: An improved method for mitochondrial isolation. *Molecular Medicine Reports*, *5*, 1271–1276.

Tchikov, V., Fritsch, J., Kabelitz, D., & Schütze, S. (2010). Immunomagnetic isolation of subcellular compartments. *Methods in Microbiology*, *37*, 21–34.

Tchikov, V., & Schütze, S. (2008). Immunomagnetic isolation of tumor necrosis factor receptosomes. *Methods in Enzymology*, *442*, 101–123.

Tchikov, V., Schütze, S., & Krönke, M. (1999). Comparison between immunofluorescence and immunomagnetic techniques of cytometry. *Journal of Magnetism and Magnetic Materials*, *194*, 242.

Tchikov, V., Winoto-Morbach, S., Krönke, M., Kabelitz, D., & Schütze, S. (2001). Adhesion of immunomagnetic particles targeted to antigens and cytokine receptors on tumor cells determined by magnetophoresis. *Journal of Magnetism and Magnetic Materials*, *225*, 285–293.

Teis, D., & Huber, L. A. (2003). The odd couple: Signal transduction and endocytosis. *Cellular and Molecular Life Sciences*, *60*, 2020–2033.

Toulmay, A., & Prinz, W. (2011). Lipid transfer and signaling at organelle contact sites: The tip of the iceberg. *Current Opinion in Cell Biology*, *23*, 458–463.

Waugh, M., Chu, E., Clayton, E., Minogue, S., & Hsuan, J. (2011). Detergent-free isolation and characterization of cholesterol-rich membrane domains from *trans*-Golgi network vesicles. *Journal of Lipid Research*, *52*, 582–589.

West, A., Brodsky, I., Rahner, C., Woo, D., Erdjument-Bromage, H., Tempst, P., et al. (2011). TLR signalling augments macrophage bactericidal activity through mitochondrial ROS. *Nature*, *472*, 476–479.

Yazdanpanah, B., Wiegmann, K., Tchikov, V., Krut, O., Pongratz, C., Schramm, M., et al. (2009). Riboflavin kinase couples TNF Receptor 1 to NADPH oxidase. *Nature*, *460*, 1159–1163.

CHAPTER TWENTY

Deubiquitinases and Their Emerging Roles in β-Arrestin-Mediated Signaling

Sudha K. Shenoy[1]
Department of Medicine, Duke University Medical Center, Durham, North Carolina, USA
[1]Corresponding author: e-mail address: sudha@receptor-biol.duke.edu

Contents

Abstract

The two homologous mammalian proteins called β-arrestin1 (also known as arrestin2) and β-arrestin2 (also called arrestin3) are now widely accepted as endocytic and signaling adaptors for G protein-coupled receptors (GPCRs), growth factor receptors, and ion channels. The sustained interactions of β-arrestins with activated GPCRs have been shown to correlate with the agonist-induced ubiquitination on distinct domains in the β-arrestin molecule. Additionally, ubiquitination of β-arrestin promotes its interaction with proteins that mediate endocytosis (e.g., clathrin) and signaling (e.g., c-RAF). Recent studies have demonstrated that deubiquitination of β-arrestin by specific deubiquitinating enzymes (DUBs) acts as an important regulatory mechanism, which determines the stability of β-arrestin–GPCR binding and fine-tunes β-arrestin-dependent

Methods in Enzymology, Volume 535
ISSN 0076-6879
http://dx.doi.org/10.1016/B978-0-12-397925-4.00020-1

signaling to downstream kinases. Accordingly, ubiquitination/deubiquitination of β-arrestin can serve as an on/off switch for its signaling and endocytic functions. Moreover, by regulating the stability and localization of signalosomes, deubiquitination of β-arrestins by DUBs imparts spatial and temporal resolution in GPCR signaling.

1. INTRODUCTION

The two highly homologous nonvisual arrestins (β-arrestin1 or arrestin2 and β-arrestin2 or arrestin3), which regulate most G protein-coupled receptors (GPCRs), were originally discovered for their role in interdicting the coupling of G proteins to GPCRs (Lefkowitz et al., 1992). However, within the past decade, this view has changed and novel functions and protein interactions of β-arrestins have been discovered (DeWire, Ahn, Lefkowitz, & Shenoy, 2007). Detailed analyses of the interaction of β-arrestin with proteins of the endocytic machinery (e.g., clathrin, AP2, NSF, and ARF6), signal transduction machinery (e.g., c-RAF, ASK-1, ERK1/2, JNK3, and c-Src, AKT), ubiquitination/deubiquitination pathway (E3 ubiquitin ligases such as Mdm2 and Nedd4 and deubiquitinases such as USP33), and other regulatory proteins (PDE4D and PP2A) have indicated that β-arrestins can function as specialized adaptors, scaffolds, and signal transducers upon GPCR activation (Beaulieu & Caron, 2008; DeWire et al., 2007; Gurevich & Gurevich, 2010; Houslay, 2009; Kendall & Luttrell, 2009; Min & Defea, 2011; Reiter, Ahn, Shukla, & Lefkowitz, 2012; Shenoy & Lefkowitz, 2011).

β-Arrestins are homogenously distributed in the cytoplasm of quiescent cells; however, upon agonist stimulation of GPCRs and subsequent phosphorylation of the receptor cytoplasmic domains by the serine–threonine kinases called GRKs (G protein-coupled receptor kinases), β-arrestins rapidly translocate to the plasma membrane and bind the agonist-activated receptors (Barak, Ferguson, Zhang, & Caron, 1997). Receptor interaction triggers a series of conformational changes of the β-arrestin molecule itself, allowing it to engage various kinases leading to specific signaling outcomes and physiological responses (Gurevich & Gurevich, 2004; Shenoy & Lefkowitz, 2011). The stability of activated GPCR–β-arrestin complexes and their cotrafficking to endosomes are defined by both the nature of the phosphorylation barcode on cytoplasmic domains of GPCRs and the ubiquitination status of β-arrestin (Nobles et al., 2011; Oakley, Laporte,

Holt, Barak, & Caron, 2001; Perroy, Pontier, Charest, Aubry, & Bouvier, 2004; Shenoy et al., 2007; Shenoy & Lefkowitz, 2005).

Ubiquitination is a universal posttranslational modification that was discovered in the context of protein degradation by the cellular proteases called 26S proteasomes (Hershko & Ciechanover, 1998). However, it is now evident that ubiquitination has an equally important role to mark membrane proteins for endocytic trafficking and lysosomal sorting (Mukhopadhyay & Riezman, 2007). Additionally, ubiquitination tags substrate proteins to alter subcellular localization, modulate interactions with other proteins, and regulate enzymatic activity.

Deubiquitinating enzymes (aka deubiquitinases or DUBs) are proteases that reverse ubiquitination (Pickart & Rose, 1985; Wilkinson, 2000). DUBs hydrolyze the isopeptide bond between the amino group of a substrate protein's lysine residue and the carboxyl group of the last residue (glycine) in ubiquitin. DUBs can also dissolve the covalent bond formed between an internal lysine of a ubiquitin unit and glycine of an adjoining ubiquitin unit within a polyubiquitin chain. The human genome contains about 80 functional DUBs divided into five subfamilies based on their catalytic domains, of which the ubiquitin-specific protease (USP) subclass consisting of >50 cysteine proteases is the largest. The USP enzymes rarely catalyze generic deubiquitination but display substrate-specific activity and act as checkpoints in both endocytic and signaling pathways (Reyes-Turcu, Ventii, & Wilkinson, 2009).

The crucial roles of ubiquitination and deubiquitination are well illustrated by the studies on the trafficking and signaling mechanisms of the β_2 adrenergic receptor (β_2AR) and its association with β-arrestin2 (Shenoy & Lefkowitz, 2003; Shenoy, McDonald, Kohout, & Lefkowitz, 2001; Shenoy et al., 2009). Agonist stimulation leads to ubiquitination of the β_2AR and of β-arrestin2, which is recruited to the activated receptor. Ubiquitination of the β_2AR is mediated by the E3 ubiquitin ligase NEDD4, which in turn is recruited by β-arrestin2 (Han, Kommaddi, & Shenoy, 2013; Shenoy et al., 2008). This modification on the β_2AR is required for degradation of the receptor in lysosomes, but not for its internalization. During postendocytic sorting, ubiquitinated β_2ARs are deubiquitinated by two homologous USPs: USP20 and USP33. This deubiquitination allows the receptor to be recycled and resensitized at the plasma membrane for a new round of activation and trafficking cycle (Berthouze, Venkataramanan, Li, & Shenoy, 2009).

Agonist-induced ubiquitination of β-arrestin2 has distinct functional outcomes: ubiquitinated β-arrestin2 promotes internalization of activated GPCRs, forms a tight complex and cotraffics with internalized GPCRs into endosomes, and facilitates prolonged activation of the mitogen-activated protein kinases, ERK 1 and 2, on signaling endosomes or signalosomes (Shenoy et al., 2007, 2001). If β-arrestin is not ubiquitinated (or is deubiquitinated), it rapidly dissociates from the GPCR and does not localize to endosomes and signalosomes are destabilized (Shenoy et al., 2009). Accordingly, when β-arrestin2 binds the activated β_2AR, it is rapidly polyubiquitinated by the E3 ubiquitin ligase MDM2. This ubiquitination is, however, promptly reversed by the DUB USP33 and the deubiquitinated β-arrestin2 dissociates from the internalizing β_2AR complex. This chapter focuses on the interaction of β-arrestin2 and USP33 and describes the assays used to determine the specific effects of USP33 on the trafficking of β-arrestin2 and on ERK signaling. These assays could be easily adapted to characterize other DUBs and associated signaling adaptor(s) or receptors.

2. INTERACTION OF β-ARRESTIN2 WITH THE DEUBIQUITINASE USP33

USP33 was identified as a β-arrestin-binding protein in a yeast two-hybrid screen, and the interaction was confirmed in mammalian cells by coimmunoprecipitation assays (Shenoy et al., 2009). While examining the kinetics of agonist-dependent association of endogenously expressed USP33 with β-arrestin2, characteristic binding patterns were revealed (Fig. 20.1). Upon stimulation of the β_2AR with isoproterenol, a robust increase in the association of β-arrestin2 and USP33 was detected. However, vasopressin (V_2R agonist) promoted dissociation of β-arrestin2 and USP33. This differential binding pattern is most likely the underlying molecular mechanism for the differential kinetics of β-arrestin2 ubiquitination obtained for these two GPCRs: transient ubiquitination (rapid deubiquitination) upon β_2AR stimulation versus sustained ubiquitination (no deubiquitination) upon V_2R stimulation. The detailed steps to assay the kinetics of β-arrestin–USP33 interaction are described later.

2.1. β_2AR-stimulated association of β-arrestin2 and USP33

1. HEK-293 cells (purchased from American Type Culture Collection, ATCC) were maintained at 37 °C in 95% air and 5% CO_2 and were

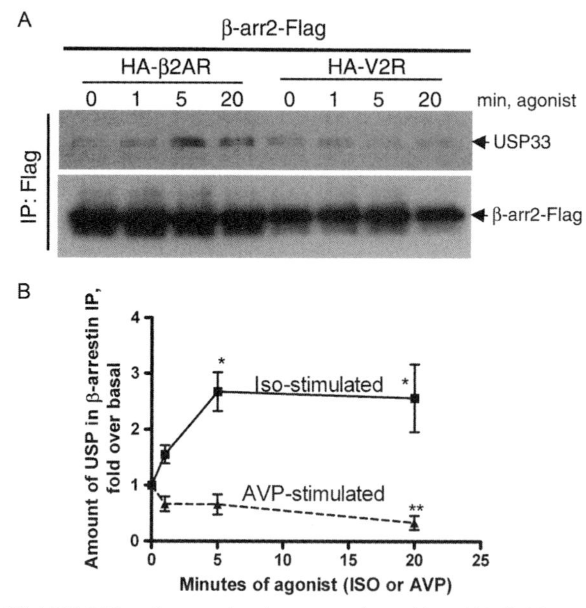

Figure 20.1 (A) HEK-293 cells transiently expressing either HA-β_2AR or HA-V_2R along with β-arrestin2-FLAG were stimulated with respective agonists for the indicated times, β-arrestins immunoprecipitated, and the IP probed with USP33 antisera (top panel) and anti-FLAG M2 antibody (bottom panel). (B) The graph shows quantification of USP33 in β-arrestin precipitates obtained from three independent experiments. The four time points within each binding curve were analyzed by one-way ANOVA. In each case, stimulated samples were significantly different from unstimulated samples; $^{*}p<0.05$, $^{**}p<0.01$. *Adapted from Shenoy et al. (2009).*

cultured in Eagle's minimum essential medium (MEM) supplemented with 10% fetal bovine serum (FBS) and 1% penicillin–streptomycin (pen–strep) solution.

2. HEK-293 cells grown to 40–50% confluence on four 100 mm dish were transfected with two pcDNA plasmid constructs (2 μg HA-β_2AR and 1.5 μg βarr2-FLAG per dish). Twenty-four hours later, the cells were trypsinized from the four dishes, mixed together, and replated on four dishes. This helps obtain an even distribution of transfected cells on the four individual plates. Forty-eight hours after the original transfection step, the cells were replenished with unsupplemented MEM containing 0.1% BSA and 10 m*M* HEPES, pH 7.5, for 1 h and stimulated with 1 μ*M* isoproterenol dissolved in sterile water or phosphate-buffered saline (PBS) for 1, 5, and 20 min at 37 °C. Isoproterenol solution was freshly prepared and used immediately to prevent oxidation. If

prolonged stimulation is desired, isoproterenol should be dissolved in 0.1 m*M* ascorbic acid. One dish was treated with vehicle alone (designated as nonstimulated or NS).

3. At the end of stimulation, all dishes were placed on ice. The media was aspirated and cells scraped into 0.8 mL of ice-cold lysis buffer (50 m*M* HEPES, pH 7.5, 0.5% IGEPAL® CA-630, 250 m*M* NaCl, 2 m*M* EDTA, 10% (w/v) glycerol, 10 m*M* *N*-ethylmaleimide (NEM) and antiproteases (1 m*M* sodium orthovanadate, 1 m*M* sodium fluoride, 1 m*M* phenylmethylsulfonyl fluoride, leupeptin (5 μg/mL), aprotinin (5 μg/mL), pepstatin A (1 μg/mL), benzamidine (100 μ*M*), and soybean trypsin inhibitor (10 μg/mL)). The NEM solution was freshly prepared by dissolving the chemical in 100% ethanol.
4. The cell lysates from the earlier step were clarified by centrifugation (15,000 rpm) in a microcentrifuge placed at 4 °C. In general, the protein concentration of the extracts was ~1 μg/μL. An aliquot of lysates was set aside for lysate immunoblotting. The remainder of each sample (~800 μg protein) was immunoprecipitated with anti-FLAG M2 affinity agarose. The immunoprecipitation samples were rotated at 4 °C overnight. Next, the proteins that nonspecifically stick to the agarose beads were removed by repeated centrifugation (high speed for 1 min) and washed with lysis buffer three times. After the final wash, 25 μL of 2× SDS sample buffer (Laemmli buffer) was added to the IP beads. The bound protein was eluted by incubation at room temperature for 10 min. The eluted samples were separated on gradient SDS polyacrylamide gels (4–20%, Invitrogen); a prestained molecular weight ladder was included in one well. USP33 displays a mobility of ~110 kDa and β-arrestin2 migrates at ~52 kDa. Next, protein bands from the gel were transferred to nitrocellulose membrane (0.2 μm pore size, Bio-Rad) for serial immunoblotting: first with an anti-USP33 antibody, after which the blots were stripped with Restore™ Western Blot Stripping Buffer (Thermo Scientific Pierce Protein Biology Products) and then reblotted with anti-FLAG M2 antibody. Each primary antibody was followed by incubation with appropriate secondary antibody conjugated to horseradish peroxidase. The signals were detected using SuperSignal® West Pico reagent (Pierce) and exposing the membrane to X-ray films or capturing digitally with a chemiluminescence detection system.

   Note: standard Western blot methodology described elsewhere is applicable for the earlier step (Kang, Tian, & Benovic, 2013).

2.2. V_2R-stimulated dissociation of β-arrestin2 and USP33

The experimental procedures were identical to the steps in Section 2.1 with the following changes:

- Cells were transfected with a HA-V_2R plasmid instead of HA-$β_2$AR.
- Cells were stimulated with 1 μ*M* V_2R agonist (AVP or arginine vasopressin peptide was dissolved in sterile water and stored as aliquots at −80 °C).

3. ASSAY OF USP33 ACTIVITY *In Vitro*

Detection of agonist-stimulated ubiquitination of β-arrestin requires the addition of 10 m*M* NEM for biochemical assays. NEM alkylates sulfhydryl groups and blocks the enzymatic activity of DUBs, which are cysteine proteases, thus preserving the ubiquitination on β-arrestin2. In order to assess deubiquitination of β-arrestin2 by cognate DUBs, an *in vitro* assay with purified proteins is a useful approach to complement cellular assays with DUB overexpression. This section will describe the protocols to purify active USP33 from COS-7 cells and to test its enzymatic activity on generic substrates (i.e., polyubiquitin chains) and on ubiquitinated β-arrestin2 isolated by immunoprecipitation (Fig. 20.2).

3.1. Purification of USP33

1. COS-7 cells (purchased from American Type Culture Collection, ATCC) were maintained at 37 °C in 95% air and 5% CO_2 and were cultured in Dulbecco's modified Eagle's medium (DMEM) supplemented with 10% FBS and 1% pen–strep solution.
   Note: HEK-293 cells can also be used for DUB purification.
2. On day 1, COS-7 cells were plated on six 150 mm dishes. On day 2, cells were transfected with 5 μg plasmid DNA (HA-USP33) per dish. Twenty-four to forty-eight hours posttransfection, cells were collected in lysis buffer containing protease inhibitors (as listed in Section 2.1). However, EDTA and NEM should not be added to the buffer as they inhibit DUBs.
3. The cell extracts were centrifuged and the clarified supernatant mixed with anti-HA affinity agarose beads. Lysates prepared from the six dishes transfected with HA-USP33 were transferred into one 15 mL conical polypropylene tube and set up as a single immunoprecipitation.

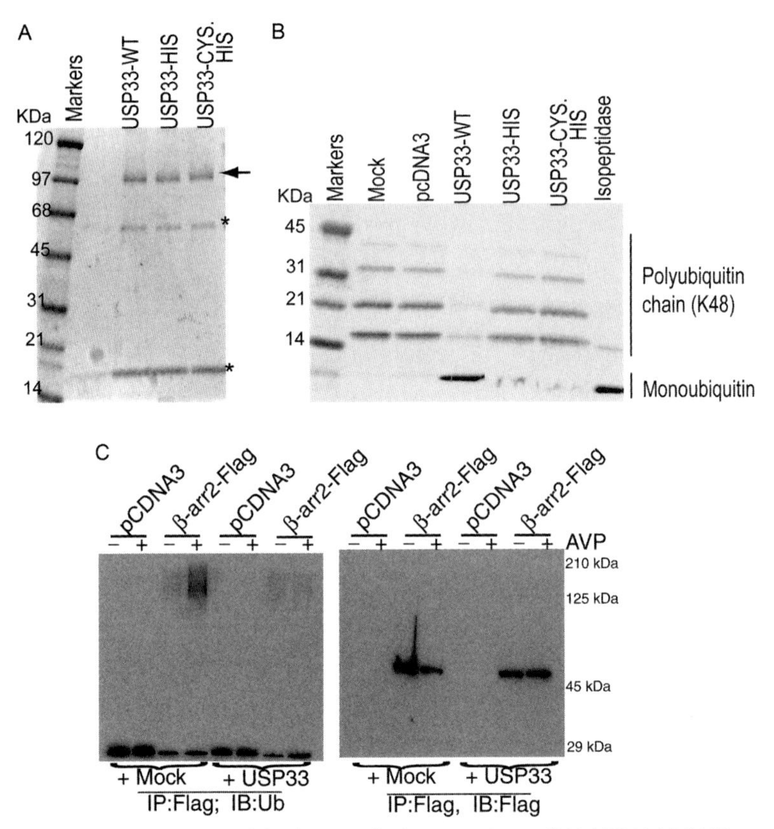

Figure 20.2 (A) Coomassie gel displays purified preparations of HA-USP33: WT, HIS mutant, and CYS.HIS mutant. The arrow indicates mobility of HA-USP33. Asterisks indicate the heavy- and light-chain IgG bands that coelute from the HA affinity beads. (B) Enzymatic activity measured by *in vitro* DUB assay. After purification, USP33 WT and mutants (as indicated) were incubated at 37 °C with the polyubiquitin chain (K48). The isopeptidase T enzyme is known to cleave polyubiquitin chains to yield monoubiquitin and represents a positive control. (C) FLAG immunoprecipitates isolated before or after 1 μM AVP stimulation from HEK-293 cells that transiently express the V2R with either pcDNA3 or β-arrestin2-FLAG are treated with a mock purification (first four lanes in each panel) or purified HA-USP33 (last four lanes in each panel). After incubation, samples are subjected to Western blot analyses with an antiubiquitin antibody (left panel) and reprobed with an anti-FLAG antibody (right panel). *Adapted from Berthouze et al. (2009) and Shenoy et al. (2009).* (For color version of this figure, the reader is referred to the online version of this chapter.)

Thirty microliters of HA agarose suspension were added for every 1000 μg of lysate proteins and the IP rotated overnight at 4 °C. After the IP incubation, repeated centrifugation and wash steps with the lysis buffer were performed to eliminate nonspecific protein interactions. After the final wash, elution buffer (50 mM Tris–HCl, pH 8.0) supplemented with protease inhibitors and HA peptide (60–100 μg peptide per mL) were added and the samples incubated on 4 °C rotator for 1 h. For the sample setup described here, 2 mL of elution buffer can be used. After this elution step, samples were centrifuged at 3000 × *g* for 5 min, and the supernatant was collected into a new sterile tube. The elution step was repeated once and the eluates were combined.

Note: HA peptide is dissolved in sterile water (5 mg/mL) and stored as aliquots at −80 °C.

4. The eluted protein was concentrated with an ultrafiltration spin column (50,000 molecular weight cutoff). About 600 μg of HA-USP33 (1 μg/μL) is obtained from a single purification setup as described earlier. One, two, and five micrograms of the purified sample were analyzed by SDS–PAGE and Coomassie blue staining. USP33 should be easily visualized as a major band at ~110 kDa (Fig. 20.2A).
5. A polyubiquitin chain depolymerization assay was performed to check the activity of isolated USP33. For this, multiubiquitin chains without any monoubiquitin were used as substrates: polyubiquitin chains with 2–7 ubiquitin units, which are linked through lysine 48 or 63 linkages, are suitable substrates (catalogs # UC-230 and #UC-330, Boston Biochem, Inc.). Ten to twenty nanomoles of purified USP33 were mixed with 100 nmol of ubiquitin chains in 50 m*M* Tris–HCl, pH 8.0, 1 m*M* DTT, and 5 m*M* $MgCl_2$ and incubated at 37 °C for 2 h. The reactions were terminated by adding SDS sample buffer and analyzed by SDS–PAGE and Coomassie blue staining. Disappearance of polyubiquitin and appearance of monoubiquitin will indicate that the purified USP33 is an active DUB (Fig. 20.2B).
6. USP33 is a cysteine protease and its enzymatic activity relies on the thiol group of a cysteine in the active site (Komander, Clague, & Urbe, 2009). Deprotonation of this cysteine (C214 in HA-tagged murine USP33) is assisted by histidine (H683 in HA-USP33), which is polarized by a conserved aspartate residue. These three residues make up the catalytic triad. In order to confirm that a loss of enzymatic activity of USP33 would prevent depolymerization of polyubiquitinated chains, a USP33 mutant in which the conserved cysteine and histidine residues of the catalytic domain are altered ($USP33^{CYS:HIS}$) should also be tested in parallel. The same purification procedure described for wild-type USP33 is applicable to obtain purified USP33 mutant(s) (see Fig. 20.2).

Note: The depolymerization assay as described earlier is a simple and quick protocol to test the quality and DUB activity of purified USPs, but it is not suitable if the kinetics or rate of enzymatic activity is desired. This can be measured by following real-time deubiquitination as described elsewhere (Nicholson et al., 2008).

3.2. Deubiquitination of β-arrestin *in vitro*

1. HEK-293 cells (COS-7 cells can also be used) were transfected with HA-V_2R and β-arrestin2-FLAG plasmid constructs. Twenty-four to forty-eight hours posttransfection, the culture media was replaced with unsupplemented media containing 0.1% BSA and 10 m*M* HEPES for 1 h after which cells were stimulated with vehicle or 1 μ*M* AVP for 15 min. β-Arrestin2 immunoprecipitates were isolated using methods described in Section 2.1. After the third wash with lysis buffer, extra wash steps were performed with ice-cold Tris–HCl, pH 8.0, supplemented with protease inhibitors. Any residual EDTA and NEM in the isolated β-arrestin IP samples were removed by the extra washes.
2. The IP beads were carefully suspended in 50 μL Tris–HCl and each IP was divided into two equal portions: One is a control sample that is not exposed to DUB activity, while the other is subjected to *in vitro* deubiquitination. To the latter, 2 μg of purified USP33 that has been confirmed for DUB activity (as in Section 3.1) was added. The samples were incubated at 30 °C for 15 min and subsequently washed with lysis buffer supplemented with protease inhibitors and NEM. The wash buffer was removed carefully and 25 μL SDS sample buffer was added. The reaction samples were then analyzed by western blotting (serial immunoblots for ubiquitin and β-arrestin2) followed by chemiluminescence detection.

   Note: Ubiquitin antibodies vary in their sensitivity and have to be tested on standard substrates. In our laboratory, we test the ubiquitin antibody by immunoblotting serial dilutions of monomeric ubiquitin as well as polyubiquitin chains that are commercially available. We also include a control IP (cell lysates prepared from untransfected cells and immunoprecipitated with either anti-FLAG or anti-HA affinity agarose beads), to serve as a negative control. A smear in the negative control indicates nonspecific background; however, this is not necessarily from a batch of antibody of poor quality and could be attributed to cross contamination. Such background problem can be rectified by freshly preparing SDS sample buffers with sterile water and by using sterile or autoclaved tips.

4. EFFECTS OF USP33 OVEREXPRESSION

Both G proteins and β-arrestins function as GPCR signal transducers, and each links the activated receptor to intracellular effectors. Furthermore, the G protein- and β-arrestin-dependent pathways can operate independently of each other (DeWire et al., 2007). ERK signaling, which promotes cell growth and survival through phosphorylation of cytosolic and nuclear targets, has been shown to be activated in a β-arrestin-dependent manner. G protein-mediated and β-arrestin-mediated ERK pathways can be separated into spatially and temporally distinct components. In general, G proteins mediate a rapid phase of ERK signaling, whereas β-arrestins regulate a sustained phase (DeWire et al., 2007). Additionally, sustained phospho-ERK (pERK) activity on endosomes is correlated with stable ubiquitination of β-arrestin (Shenoy et al., 2007).

When USP33 activity is increased by overexpression, β-arrestin2 is prone to rapid deubiquitination. This reduces both the magnitude and extent of ERK activity and prevents β-arrestin2 from cotrafficking with internalized GPCRs that are known to engage in a stable complex with β-arrestins (e.g., the V_2R; see Fig. 20.3). The inhibitory effect of USP33 on trafficking requires its enzymatic activity because overexpression of $USP33^{CYS:HIS}$ mutant does not block the normal endosomal recruitment of β-arrestin2 upon V_2R stimulation. In contrast, overexpressed wild-type USP33 attenuates ERK activation and blocks recruitment of β-arrestin2 to endosomes (Fig. 20.3).

The experimental procedures that were used to define the effects of USP33 on ERK signaling and trafficking of β-arrestin2 are described below.

4.1. Effect on β-arrestin-dependent ERK activation

1. HEK-293 cells plated on 100 mm dishes were transfected with HA-V_2R plasmid along with either pcDNA3 vector or HA-USP33/pcDNA3. Two micrograms of each plasmid were used and two dishes for each transfection were set up. Twenty-four to forty-eight hours later, cells from each transfection were subcultured into multiple 6-well or 12-well dishes; equal numbers of cells per well were carefully

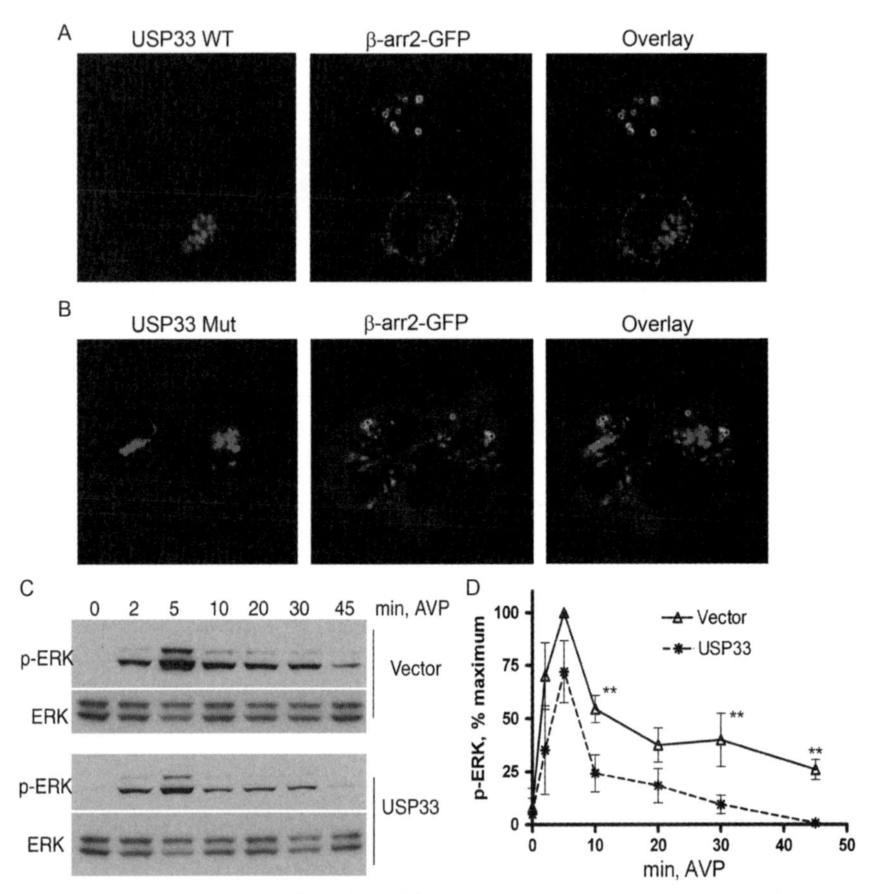

Figure 20.3 (A) HEK-293 cells with stable HA-V_2R expression were transiently transfected with β-arrestin2-GFP and RFP-USP33 and stimulated with 1 μ*M* AVP. The panel shows two cells, one with no detectable RFP-USP33 that has the expected endosomal recruitment of β-arrestin2 and the other cell that expresses USP33 in which β-arrestin recruitment to the endosomes is inhibited. (B) Confocal micrographs show normal endosomal distribution of β-arrestin2-GFP upon AVP stimulation in cells overexpressing RFP-USP33$^{CYS:HIS}$. (C) HEK-293 cells transiently expressing HA-V_2R with either vector or USP33 were stimulated with 1 μ*M* AVP for the indicated times. Equivalent amounts of whole cell lysates were analyzed for pERK and ERK content. (D) The graph represents quantification of pERK bands from four independent experiments. $^{**}p<0.01$ two-way ANOVA, $n=4$. *Adapted from Shenoy et al. (2009).* (For color version of this figure, the reader is referred to the online version of this chapter.)

dispersed, and cells were plated such that each dish has one well of each transfection to be used for a single time point of agonist stimulation. This method of plating prevents disturbance of cells that correspond to different time points, which would happen if cells for multiple time points are plated on the same multiwell dish. Cells were plated to assess

ERK activation at 2, 5, 10, 20, 30, and 45 min of agonist stimulation. Cells were also plated to include wells with no stimulation and one well with cells that were lysed and used for determining protein concentration.

2. Prior to agonist stimulation, cells were replenished with unsupplemented media for 4 h (overnight, serum deprivation is not required, but can be used). At the end of starvation, first, the cells from wells set aside for protein determination were analyzed. After aspiration of media, 150 μL lysis buffer supplemented with protease inhibitors was added. Cell extracts were collected and clarified by centrifugation and protein concentration was measured. Next, cells in the assay plates were stimulated by adding 1 μ*M* AVP and incubating at 37 °C for the desired times.

   Note: dilute the stock such that the agonist can be dispensed as 10 μL drops with a repeat pipettor (Eppendorf).
3. At the end of stimulation, the media was quickly aspirated and 150 μL of 2 × SDS sample buffer was added. The cell extracts were collected in labeled microfuge tubes. Using a 26 gauge syringe needle, a hole was carefully pierced at the center of the lid. The lysates were boiled for 5 min on a heat block set at 100 °C.
4. The protein concentration of each transfected sample was estimated using the protein measurements from Step 2. Because transfected cells are dispensed equally in all wells, the measurements obtained for cells from the "protein well" of each transfection setup would correlate with that of the lysates collected in 2 × SDS buffer. Twenty to twenty-five micrograms of lysates were used for SDS–PAGE and western blotting.
5. The lysates were analyzed by SDS–PAGE and immunoblotted with an anti-pERK antibody (Cell Signaling Technology, 1:2000 dilution; this antibody specifically recognizes dual phosphorylation at threonine-183 and tyrosine-185 of ERK). After this, the membrane was stripped and immunoblotted with anti-ERK IgG (rabbit polyclonal, Millipore, 1:2000 dilution).
6. The band signals were quantified by densitometry; the pERK signal from each lane normalized to the corresponding ERK signal. The data were imported into either Microsoft Excel or GraphPad PRISM for generating graphs and further analysis.

   Note: care should be taken not to use band signals that are saturated for quantification, but to use light exposures of films or appropriate captures of digital scan.

4.2. Effect on trafficking of β-arrestin

1. HEK-293 cells plated on 100 mm dishes were transfected with HA-V_2R (2 μg), β-arrestin2-GFP (1 μg), and pcDNA vector or USP33 (2 μg). Twenty hours after transfection, the transfected cells were subcultured on glass-bottomed confocal dishes that have been coated with collagen. Rat tail collagen solution was prepared by dissolving 10 mg lyophilized collagen into 10 mL sterile PBS with calcium and magnesium containing 50 μL 10 N acetic acid at room temperature overnight. Each 35 mm glass-bottomed dish was coated with rat tail collagen solution to ensure tight adherence of the cells and to prevent cells from lifting off of the plate during wash steps. One milliliter of the collagen solution was added to the plate and spread around to coat evenly. The collagen was retrieved into a sterile tube and reused for coating remaining plates. After removing the collagen, the dishes were air-dried in the tissue culture hood for 15 min and then washed twice with PBS before plating cells.
2. The monolayers of transfected cells on glass-bottomed dishes were serum-starved by replacing culture media with unsupplemented media (as earlier) for 1–2 h at 37 °C. Next, the cells were stimulated with vehicle or 1 μ*M* AVP for 10 min. At the end of stimulation, the media was aspirated and fixing solution was added (5% formaldehyde diluted in sterile PBS). The dishes were incubated at room temperature for 15–30 min. The fixed cells were washed three times with sterile PBS.
3. The fixed cells were next permeabilized with 0.1% Triton (mixed in a 2% BSA solution prepared in sterile PBS) for 20 min at room temperature. Longer permeabilization affects the morphology of endosomes and is not recommended. Permeabilized cells were washed once with sterile PBS.
4. Serial immunostaining was performed to help visualization of HA-V_2R and USP33. Each incubation step with antibody listed later was followed by three washes of the cellular monolayers with sterile PBS. First, the permeabilized cells were incubated with anti-USP33 (rabbit polyclonal antibody, Bethyl Laboratories, 1:300 dilution) at 4 °C overnight. Following PBS washes, anti-rabbit IgG conjugated to Alexa 594 (Invitrogen, 1:500 dilution) was added. The dishes were covered with aluminum foil and incubated at room temperature for 1 h. Next, the secondary antibody solution was aspirated, and cells were washed three times with sterile PBS and incubated with the second primary antibody, a monoclonal anti-HA IgG (12CA5, 1:300 dilution), at 4 °C overnight.

Following the PBS washes, anti-mouse IgG conjugated to Alexa 633 was added and the dishes were incubated at room temperature for 1 h. After three washes with PBS, the cells were covered with 1 mL PBS and the dishes were stored protected from direct light. Although the fixed immunostained samples can be stored at 4 °C for a few days, the fluorescence deteriorates with time; therefore, image acquisition with a confocal microscope was carried out immediately.

5. Confocal images were acquired by adapting settings that allow multitrack sequential excitation (488, 568, and 633 nm) and emission (515–540 nm for GFP, 585–615 nm for Alexa 594, and 650 nm for Alexa 633) filter sets. Each scan was set for 1 μm optical slice and either a 40 × or 100 × oil objective was used. The detector gain and amplifier offset were adjusted for individual wavelength channel to minimize saturation. 1024 × 1024 frame size and scan speed of 4 were used to collect the final image. All dishes from one experiment were examined in parallel and the images for individual treatments were collected using the same acquisition settings. The confocal image was exported as a tagged image file (TIF file without any compression) for further processing with Adobe Photoshop software.

   Note: Image processing should not involve any changes to the original image. If any increase or decrease of brightness and contrast is required to visualize details, this change should be applied to the entire image and steps taken should be included in the methods section of the resulting manuscript.

In the earlier set of experiments, USP33 plasmid can be substituted with a red fluorescent protein-tagged USP33 (RFP-USP33, as in Fig. 20.3), in which case only HA-V_2R visualization would require immunostaining described earlier (primary antibody, 12CA5, and secondary anti-mouse IgG conjugated to Alexa 633). Additionally, trafficking of GPCR and/or β-arrestin can be analyzed by coexpressing the inactive forms of USP33 ($USP33^{CYS}$, $USP33^{HIS}$, or $USP33^{CYS:HIS}$).

5. EFFECTS OF USP33 KNOCKDOWN

While USP33 overexpression blocked the cointernalization of β-arrestin with activated V_2R on endosomes and curtailed ERK activation mediated by β-arrestin, knockdown of endogenously expressed USP33 using small interfering RNA (siRNA) produced reciprocal effects.

Depletion of USP33 completely inhibited the deubiquitination of β-arrestin2 observed after stimulating the β_2AR for 10 min (Fig. 20.4A). Additionally, in cells transfected with siRNA targeting USP33, β-arrestin2 formed a tight complex with the activated β_2AR and cointernalized with the receptor into endosomes (Fig. 20.4B). The stabilization of β-arrestin2 ubiquitination also correlated with the protracted time course of ERK activation induced by isoproterenol stimulation (Fig. 20.4C and D). These findings strongly suggest that USP33 acts as an endogenous inhibitor and fine-tunes β-arrestin2-dependent signaling in response to β_2AR stimulation.

The following siRNA oligonucleotides were chemically synthesized as double-stranded 19-nucleotide duplexes with two nucleotide 3′ dTdT overhangs (Thermo Scientific or Dharmacon) in a deprotected and desalted form. Lyophilized siRNA was dissolved in the recommended buffer and stored as 20 μg aliquots frozen at −20 °C. The siRNA sequences (sense, 5′–3′) that target human USP33 are USP33-1 CAAUGUUAAUUCAGGAUGA, USP33-2 GGCUUGGAUCUUCAGCCAU, and USP33-3 GAUCAUG UGGCGAAGCAUA. In addition, parallel transfection with a control nontargeting sequence (UUCUCCGAACGUGUCACGU) was also characterized. All the three USP33 siRNA listed earlier resulted in complete knockdown of the USP33 protein as shown in Fig. 20.4E.

5.1. Effect on ubiquitination of β-arrestin

1. HEK-293 cells plated on 100 mm dishes were transfected with HA-β_2AR (1 μg) and β-arrestin2-FLAG (2 μg) plasmids along with siRNA oligos (either control or USP33-targeting) using Lipofectamine™ 2000 or GeneSilencer® (both yield similar efficiency of transfection). Fifteen to twenty micrograms siRNA were used for each 100 mm dish. siRNA was mixed with 700 μL unsupplemented MEM (containing no additives); 20–24 μL Lipofectamine™ reagent was mixed in 300 μL MEM in a separate microfuge vial. The two solutions were then mixed by transferring the Lipofectamine™ mixture into the tube containing siRNA. The mixture was incubated at room temperature for 10 min and then added to the cells. Prior to this addition, the cells were replenished with 4 mL of warm unsupplemented MEM. After 3 h, 5 mL of MEM supplemented with 20% FBS and 2% pen–strep was added to the transfected cells. Forty-eight hours after transfection, cells were further divided or used as such to test effects of protein knockdown. GeneSilencer® transfection involved a similar protocol: 50 μL of the GeneSilencer® transfection reagent was

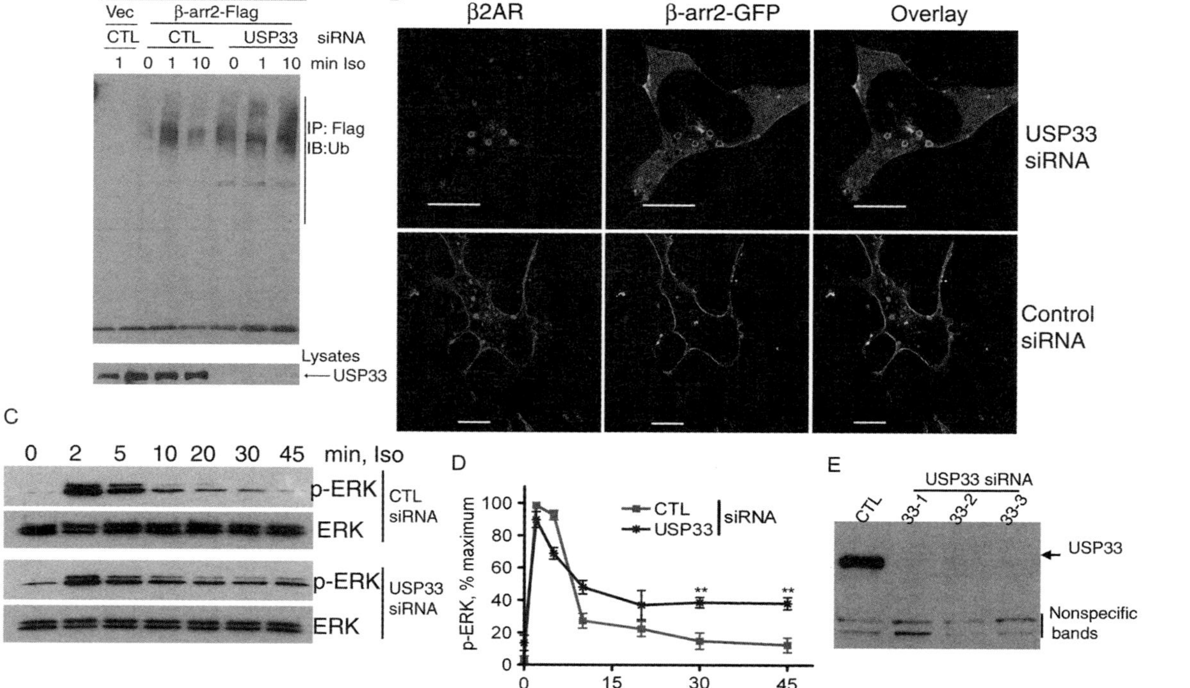

Figure 20.4 (A) HEK-293 cells were transfected with HA-β_2AR, β-arrestin2-FLAG, and either control or USP33 siRNA. FLAG immunoprecipitates were isolated after Iso stimulation for the indicated times and probed with an antiubiquitin antibody. The lower panel shows USP33 levels in control and USP knockdown samples. (B) Confocal micrographs show distribution of FLAG-β_2AR (red) and β-arrestin2-GFP (green) in USP33-depleted cells (top row) and control cells (bottom row) in agonist-stimulated HEK-293 cells. Scale bars, 10 μm. (C) Western blots of pERK and ERK in response to a time course of Iso (100 n*M*) activation, detected in HEK-293 cells with stable β_2AR expression (2 pmol/mg protein) and with indicated siRNA transfections. (D) This graph represents quantification of pERK signals from three independent experiments. $^{**}p < 0.01$, control versus USP33, two-way ANOVA. (E) A representative blot for USP33 levels is displayed showing western blot analysis of 25 μg of lysate protein from each siRNA transfection. *Adapted from Shenoy et al. (2009)*. (For interpretation of the references to color in this figure legend, the reader is referred to the online version of this chapter.)

added to 300 μL of MEM, while RNA was mixed with 240 μL of siRNA diluent and 180 μL of MEM. Both solutions were allowed to stand for 5 min at room temperature and then mixed together. After further incubation for 10 min, the entire transfection mixture was added to cells. The remaining steps were similar to the transfection setup with Lipofectamine™.

2. Stimulation and harvesting of cells and immunoprecipitation of β-arrestin2 were performed as described in Section 2.1. Immunoblotting was performed to detect polyubiquitination with an antiubiquitin IgG, followed by serial blotting for USP33 and β-arrestin2.

5.2. Effect on trafficking of β-arrestin

1. HEK-293 cells on 100 mm dishes were transfected with HA-β_2AR (2 μg) and β-arrestin2-GFP (1 μg) plasmids along with siRNA oligos (either control or USP33-targeting) using Lipofectamine™ 2000 as described in Section 5.1.
2. 24 h after transfection, cells were split on to glass-bottomed confocal dishes as described in Section 4.2.
3. Monolayers of cells on confocal dishes were serum-starved by replacing the culture media with unsupplemented media (as in Section 4.2) for 1–2 h at 37 °C. Next, the cells were stimulated with vehicle or 1 μ*M* isoproterenol for 20 min. At the end of stimulation, the cells were processed by fixing, permeabilizing, immunostaining, and imaging as described in Section 4.2. For the detection of β_2AR, anti-β_2AR polyclonal IgG (H-20, Santa Cruz Biotechnology) was used as a primary antibody, and anti-rabbit IgG conjugated with Alexa 594 was used as the secondary antibody. Colocalization of β-arrestin2-GFP and internalized β_2AR was detected only in cells with USP33 knockdown (Fig. 20.4B).

5.3. Effect on β-arrestin-dependent ERK activation

1. HEK-293 cells on 100 mm dishes were transfected with HA-β_2AR and siRNA oligos (either control or USP33-targeting) using Lipofectamine™ 2000 as described in Section 5.1. Forty-eight hours later, cells from each transfection were subcultured into multiple 6-well or 12-well dishes, and ERK activation induced by 100 n*M* isoproterenol at 2, 5, 10, 20, 30, and 45 min was measured as described in Section 4.1. ERK activation at the later time points corresponds to the β-arrestin-dependent phase of signaling in these cells, and this is significantly increased upon USP33 knockdown (Fig. 20.4C and D).

6. SUMMARY

β-Arrestin-dependent signaling is a recently established paradigm in GPCR-initiated pathways; how β-arrestin signaling is initiated and regulated remains largely undefined. Deubiquitination by USP33 regulates the timing as well as localization of β-arrestin-dependent MAPK activity and thus USP33 functions as an inhibitor of β-arrestin-mediated signaling. DUBs are also beginning to be appreciated for their role in the regulation of intracellular trafficking of a growing list of GPCRs. The assays described in this chapter could help elucidate the regulatory roles of DUBs and identify novel substrates involved in GPCR signaling.

ACKNOWLEDGMENTS

I thank Dr S. Rajagopal for critical reading of the chapter. This work was supported by an NIH Grant HL 080525 (S. K. S.) and a Grant-in-Aid from the American Heart Association (S. K. S.).

REFERENCES

Barak, L. S., Ferguson, S. S., Zhang, J., & Caron, M. G. (1997). A beta-arrestin/green fluorescent protein biosensor for detecting G protein-coupled receptor activation. *The Journal of Biological Chemistry*, *272*, 27497–27500.

Beaulieu, J. M., & Caron, M. G. (2008). Looking at lithium: Molecular moods and complex behaviour. *Molecular Interventions*, *8*, 230–241.

Berthouze, M., Venkataramanan, V., Li, Y., & Shenoy, S. K. (2009). The deubiquitinases USP33 and USP20 coordinate beta2 adrenergic receptor recycling and resensitization. *EMBO Journal*, *28*, 1684–1696.

DeWire, S. M., Ahn, S., Lefkowitz, R. J., & Shenoy, S. K. (2007). Beta-arrestins and cell signaling. *Annual Review of Physiology*, *69*, 483–510.

Gurevich, V. V., & Gurevich, E. V. (2004). The molecular acrobatics of arrestin activation. *Trends in Pharmacological Sciences*, *25*, 105–111.

Gurevich, V. V., & Gurevich, E. V. (2010). Custom-designed proteins as novel therapeutic tools? The case of arrestins. *Expert Reviews in Molecular Medicine*, *12*, e13.

Han, S. O., Kommaddi, R. P., & Shenoy, S. K. (2013). Distinct roles for beta-arrestin2 and arrestin-domain-containing proteins in beta(2) adrenergic receptor trafficking. *EMBO Reports*, *14*, 164–171.

Hershko, A., & Ciechanover, A. (1998). The ubiquitin system. *Annual Review of Biochemistry*, *67*, 425–479.

Houslay, M. D. (2009). Arrestin times for developing antipsychotics and beta-blockers. *Science Signaling*, *2*, pe22.

Kang, D. S., Tian, X., & Benovic, J. L. (2013). Beta-arrestins and G protein-coupled receptor trafficking. *Methods in Enzymology*, *521*, 91–108.

Kendall, R. T., & Luttrell, L. M. (2009). Diversity in arrestin function. *Cellular and Molecular Life Sciences*, *66*, 2953–2973.

Komander, D., Clague, M. J., & Urbe, S. (2009). Breaking the chains: Structure and function of the deubiquitinases. *Nature Reviews Molecular Cell Biology*, *10*, 550–563.

Lefkowitz, R. J., Inglese, J., Koch, W. J., Pitcher, J., Attramadal, H., & Caron, M. G. (1992). G-protein-coupled receptors: Regulatory role of receptor kinases and arrestin proteins. *Cold Spring Harbor Symposia on Quantitative Biology*, *57*, 127–133.

Min, J., & Defea, K. (2011). Beta-arrestin-dependent actin reorganization: Bringing the right players together at the leading edge. *Molecular Pharmacology*, *80*, 760–768.

Mukhopadhyay, D., & Riezman, H. (2007). Proteasome-independent functions of ubiquitin in endocytosis and signaling. *Science*, *315*, 201–205.

Nicholson, B., Leach, C. A., Goldenberg, S. J., Francis, D. M., Kodrasov, M. P., Tian, X., et al. (2008). Characterization of ubiquitin and ubiquitin-like-protein isopeptidase activities. *Protein Sciences*, *17*, 1035–1043.

Nobles, K. N., Xiao, K., Ahn, S., Shukla, A. K., Lam, C. M., Rajagopal, S., et al. (2011). Distinct phosphorylation sites on the beta(2)-adrenergic receptor establish a barcode that encodes differential functions of beta-arrestin. *Science Signaling*, *4*, ra51.

Oakley, R. H., Laporte, S. A., Holt, J. A., Barak, L. S., & Caron, M. G. (2001). Molecular determinants underlying the formation of stable intracellular g protein-coupled receptor-beta-arrestin complexes after receptor endocytosis*. *The Journal of Biological Chemistry*, *276*, 19452–19460.

Perroy, J., Pontier, S., Charest, P. G., Aubry, M., & Bouvier, M. (2004). Real-time monitoring of ubiquitination in living cells by BRET. *Nature Methods*, *1*, 203–208.

Pickart, C. M., & Rose, I. A. (1985). Ubiquitin carboxyl-terminal hydrolase acts on ubiquitin carboxyl-terminal amides. *Journal of Biological Chemistry*, *260*, 7903–7910.

Reiter, E., Ahn, S., Shukla, A. K., & Lefkowitz, R. J. (2012). Molecular mechanism of beta-arrestin-biased agonism at seven-transmembrane receptors. *Annual Review of Pharmacology and Toxicology*, *52*, 179–197.

Reyes-Turcu, F. E., Ventii, K. H., & Wilkinson, K. D. (2009). Regulation and cellular roles of ubiquitin-specific deubiquitinating enzymes. *Annual Review of Biochemistry*, *78*, 363–397.

Shenoy, S. K., Barak, L. S., Xiao, K., Ahn, S., Berthouze, M., Shukla, A. K., et al. (2007). Ubiquitination of beta-arrestin links seven-transmembrane receptor endocytosis and ERK activation. *The Journal of Biological Chemistry*, *282*, 29549–29562.

Shenoy, S. K., & Lefkowitz, R. J. (2003). Trafficking patterns of beta-arrestin and G protein-coupled receptors determined by the kinetics of beta-arrestin deubiquitination. *The Journal of Biological Chemistry*, *278*, 14498–14506.

Shenoy, S. K., & Lefkowitz, R. J. (2005). Receptor-specific ubiquitination of beta-arrestin directs assembly and targeting of seven-transmembrane receptor signalosomes. *The Journal of Biological Chemistry*, *280*, 15315–15324.

Shenoy, S. K., & Lefkowitz, R. J. (2011). Beta-arrestin-mediated receptor trafficking and signal transduction. *Trends in Pharmacological Sciences*, *32*, 521–533.

Shenoy, S. K., McDonald, P. H., Kohout, T. A., & Lefkowitz, R. J. (2001). Regulation of receptor fate by ubiquitination of activated beta 2-adrenergic receptor and beta-arrestin. *Science*, *294*, 1307–1313.

Shenoy, S. K., Modi, A. S., Shukla, A. K., Xiao, K., Berthouze, M., Ahn, S., et al. (2009). Beta-arrestin-dependent signaling and trafficking of 7-transmembrane receptors is reciprocally regulated by the deubiquitinase USP33 and the E3 ligase Mdm2. In: *Proceedings of the National Academy of Sciences of the United States of America*, *106*, 6650–6655.

Shenoy, S. K., Xiao, K., Venkataramanan, V., Snyder, P. M., Freedman, N. J., & Weissman, A. M. (2008). Nedd4 mediates agonist-dependent ubiquitination, lysosomal targeting, and degradation of the {beta}2-adrenergic receptor. *The Journal of Biological Chemistry*, *283*, 22166–22176.

Wilkinson, K. D. (2000). Ubiquitination and deubiquitination: Targeting of proteins for degradation by the proteasome. *Seminars in Cell and Developmental Biology*, *11*, 141–148.

CHAPTER TWENTY-ONE

TSLP Expression Induced via Toll-Like Receptor Pathways in Human Keratinocytes

Toshiro Takai[*,1], Xue Chen[*,†], Yang Xie[*,‡], Anh Tuan Vu[*,§,¶], Tuan Anh Le[*,§,||], Hirokazu Kinoshita[*,§], Junko Kawasaki[*,§], Seiji Kamijo[*], Mutsuko Hara[*], Hiroko Ushio[*], Tadashi Baba[#], Keiichi Hiramatsu[#], Shigaku Ikeda[*,§], Hideoki Ogawa[*,§], Ko Okumura[*]

[*]Atopy (Allergy) Research Center, Juntendo University Graduate School of Medicine, Tokyo, Japan
[†]Department of Dermatology, Peking University People's Hospital, Beijing, China
[‡]Department of Dermatology, The 3rd Affiliated Hospital of Sun Yat-sen University, Guangzhou, China
[§]Department of Dermatology and Allergology, Juntendo University Graduate School of Medicine, Tokyo, Japan
[¶]Quyhoa National Leprosy-Dermatology Hospital, Quynhon, Vietnam
[||]Department of Dermatology and Allergology, Institute of Clinical Medical and Pharmaceutical Sciences 108, Hanoi, Vietnam
[#]Department of Microbiology and Infection Control Science, Juntendo University Graduate School of Medicine, Tokyo, Japan
[1]Corresponding author: e-mail address: t-takai@juntendo.ac.jp

Contents

Methods in Enzymology, Volume 535
ISSN 0076-6879
http://dx.doi.org/10.1016/B978-0-12-397925-4.00021-3

Abstract

The skin epidermis and mucosal epithelia (airway, ocular tissues, gut, and so on) are located at the interface between the body and environment and have critical roles in the response to various stimuli. Thymic stromal lymphopoietin (TSLP), a cytokine expressed mainly by epidermal keratinocytes (KCs) and mucosal epithelial cells, is a critical factor linking the innate response at barrier surfaces to Th2-skewed acquired immune response. TSLP is highly expressed in skin lesions of atopic dermatitis patients. Here, we describe on Toll-like receptor (TLR)-mediated induction of TSLP expression in primary cultured human KCs, placing emphasis on experimental methods used in our studies. Double-stranded RNA (TLR3 ligand), flagellin (TLR5 ligand), and diacylated lipopeptide (TLR2–TLR6 ligand) stimulated human KCs to express TSLP and *Staphylococcus aureus* membranes did so via the TLR2–TLR6 pathway. Atopic cytokine milieu upregulated the TLR-mediated induction of TSLP. Culturing in the absence of glucocorticoid before stimulation enhanced the TSLP expression. Extracellular double-stranded RNA induced TSLP via endosomal acidification- and NF-κB-dependent pathway. Specific measurement of the long *TSLP* transcript, which contributes to the production of the TSLP protein, rather than total or the short transcript is useful for accurate detection of functional human *TSLP* gene expression. The results suggest that environment-, infection-, and/or self-derived TLR ligands contribute to the initiation and/or amplification of Th2-type skin inflammation including atopic dermatitis and atopic march through the induction of TSLP expression in KCs and include information helpful for understanding the role of the gene–environment interaction relevant in allergic diseases.

1. INTRODUCTION

Thymic stromal lymphopoietin (TSLP) is an IL-7-like cytokine initially identified in the culture supernatant of a thymic stromal cell line. Highly expressed in the epidermis in skin lesions of atopic dermatitis patients, TSLP was subsequently found to be a critical factor linking responses at interfaces between the body and environment (skin, airway, gut, ocular tissues, and so on) to Th2 responses (Liu, 2006; Takai, 2012). TSLP-activated dendritic cells secrete Th2-recruiting chemokines but not IL-12 and induce naïve T cells to differentiate into inflammatory Th2 cells producing IL-4, IL-5, IL-13, and TNF-α through OX40 ligand.

TSLP is expressed mainly by keratinocytes (KCs) and epithelial cells at barrier surfaces and constitutively expressed in the thymus and intestinal epithelial cells. TSLP expression in the epidermis, epithelium, and submucosa in skin, airway, and ocular tissues plays a critical role in the pathogenesis of allergic diseases. At barrier interfaces, environmental stimuli such as allergen sources, viruses, microbes, helminths, diesel exhaust, cigarette smoke, and chemicals trigger TSLP production, resulting in initiation of the sensitization process and the exacerbation of allergic diseases (Takai, 2012; Takai & Ikeda, 2011). Proinflammatory cytokines, Th2-related cytokines, and IgE contribute to TSLP production, indicating an amplification cycle for the Th2 response (Takai, 2012). Disturbance of epidermal homeostasis triggers also TSLP production, and increase in TSLP concentrations in the epidermis induces the onset of Th2 cytokine-associated inflammation, that is, "atopic march" (Spergel, 2005; Takai, 2012).

Studies of environmental, endogenous, transcriptional, and posttranscriptional regulatory mechanisms of TSLP expression will contribute to the elucidation of the pathogenesis of allergic diseases and other TSLP-related disorders and to the development of new approaches in prevention and therapy (Takai, 2012; Takai et al., 2009). Toll-like receptors (TLRs) are one of the innate receptor families that recognize pathogens or pathogen-derived products (Kawai & Akira, 2010). TLRs are expressed in human KCs (Kollisch et al., 2005; Lebre et al., 2007). We reported that double-stranded RNA (dsRNA) (TLR3 ligand) (Kinoshita et al., 2009; Vu et al., 2011), flagellin (TLR5 ligand) (Le et al., 2011), and diacylated lipopeptide (TLR2–TLR6 ligand) stimulated primary cultured human KCs to express TSLP and *Staphylococcus aureus* membranes did so in a manner dependent on TLR2 and TLR6 (Vu et al., 2010). Here, we describe on the TLR-mediated induction of TSLP expression in primary cultured human KCs, placing emphasis on experimental methods used in our studies.

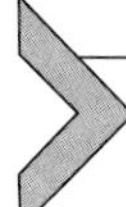

2. REAGENTS FOR STIMULATION OR TREATMENT OF HUMAN KCs

2.1. TLR ligands

Activity of TLR ligands sometimes differs among manufacturers and/or batches. The following TLR ligands and concentrations were used for stimulation of KCs in our studies: 1 μg/mL or 0.01–100 μg/mL polyI:C (TLR3 ligand) (GE Healthcare/Amersham, Buckinghamshire, United Kingdom),

1 or 0.01–1 μg/mL (S,R)-(2,3-bispalmitoyloxypropyl)-Cys-Gly-Asp-Pro-Lys-His-Pro-Lys-Ser-Phe (TLR2–TLR6 ligand) (FSL-1; InvivoGen, San Diego, CA), 5 μg/mL (*S*)-[2,3-bis(palmitoyloxy)-(2-RS)-propyl]-*N*-palmitoyl-(R)-Cys-(*S*)-Ser-(*S*)-Lys4-OH, 3HCl (TLR2–TLR1 ligand) (Pam_3CSK_4; Calbiochem, San Diego, CA), 20 μg/mL peptidoglycan from *S. aureus* (TLR2 ligand), 20 μg/mL LPS from *Escherichia coli* serotype 0111: B4 (TLR4 ligand) (Sigma, St Louis, MO), 20 or 1–100 ng/mL flagellin purified from *Salmonella typhimurium* strain 14028 (TLR5 ligand) (Alexis Biochemicals, San Diego, CA), 20 μg/mL loxoribine (TLR7 ligand), 10 μg/mL CL097 (TLR7/8 ligand) (InvivoGen, San Diego, CA), and 5 μmol/L human CpG (TLR9 ligand) (Hycult Biotechnology, Uden, The Netherlands).

2.2. Cytokines

The following recombinant human cytokines and concentrations were used for stimulation of KCs in our studies: 20 ng/mL TNF-α, 10 ng/mL IL-1α, 100 ng/mL IL-4, 100 ng/mL IL-13, 100 ng/mL IFN-γ, 100 ng/mL IL-10, 10 ng/mL TGF-β, 100 ng/mL IL-17, 100 ng/mL IL-22 (R&D Systems, Minneapolis, MN), 10 ng/mL (2000 U/mL) IFN-α, 100 ng/mL (2000 U/mL) IFN-β (PeproTech, London, United Kingdom), and 50 ng/mL TGF-α (Calbiochem).

2.3. Other reagents

The following reagents were used for stimulation or treatment of KCs in our studies, and concentrations effective and/or used in our studies will be described later. Endosomal acidification inhibitor bafilomycin A1 (Sigma); hydrocortisone (Kurabo, Osaka, Japan) and other glucocorticoids (dexamethasone, prednisolone, betamethasone, fluticasone propionate, and clobetasol propionate); calcineurin inhibitors (FK506 monohydrate/ tacrolimus and cyclosporin A); and vitamin D receptor agonists, calcitriol (1,25-dihydroxyvitamin D3, the active form of vitamin D3) and its analog MC903 (calcipotriol; Dovonex) (Sigma), were dissolved in dimethyl sulfoxide (Sigma). They are stored at −20 °C and were diluted with medium just prior to use.

2.4. *S. aureus*

2.4.1 Subcellular fractionation of S. aureus

S. aureus strain MW2, which was isolated from a patient with fatal septicemia and septic arthritis, was subjected to subcellular fractionation (subcellular-secreted, cell-wall, membranous, and cytoplasmic fractions). MW2 was grown overnight in tryptic soy broth (Becton, Dickinson, and Company, Sparks, MD) at 37 °C and subjected to subcellular fractionation, basically as described (Baba & Schneewind, 1996). Briefly, the cell wall was enzymatically solubilized by lysostaphin in a buffer osmotically equivalent to the cytoplasm after removal of the culture medium as an extracellular fraction. Isolation of the cell-wall fraction by centrifugation was then followed by the resuspension of protoplasts in the membrane buffer. Freeze–thaw cycles were repeated five times to disturb the integrity of the cytoplasmic membrane that was recovered by subsequent ultracentrifugation. The supernatant after ultracentrifugation was then isolated as the cytoplasm. The membrane was resuspended in the membrane buffer by serial passage through a 27-gauge syringe needle, and the suspension was used for experiments.

In our study, concentrations of the subcellular fractions were expressed in units per milliliter (1 U corresponds to the sample prepared from 1 mL of overnight bacterial culture of approximately 3×10^9 bacterial cells) (Vu et al., 2010).

2.4.2 *Preparation of heat-killed* S. aureus

D-YSD1 and D-ISK1 were *S. aureus* strains isolated from patients with atopic dermatitis.

S. aureus strains MW2, D-YSD1, and D-ISK1 were grown in tryptic soy broth at 37 °C to the stationary phase. Cell numbers were determined by measurement of optical density at 562 nm with Novaspec II (Amersham, Amersham, United Kingdom). The *S. aureus* cells were harvested by centrifugation, followed by heat inactivation (70 °C for 60 min).

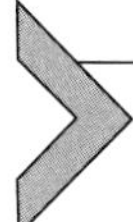

3. CELL CULTURE AND STIMULATION OF HUMAN KCs

3.1. Human KCs

In our studies, primary human KCs from infant foreskins were purchased from Cascade Biologics (Portland, OR). We recommend using primary KCs at the second to fifth passages (see Section 3.2). The KC responsiveness sometimes differs among donors. We recommend examining several donors first. After finding a responsive donor, reproducibility of the results in the donor and then that in some of the other donors should be examined.

3.2. Cell culture in serum-free medium

In our studies on TSLP release, KCs were cultured in a serum-free medium, EpiLife-KG2 (Invitrogen, Carlsbad, CA), added with supplements (0.1 ng/mL epidermal growth factor, 10 μg/mL insulin, 0.5 μg/mL hydrocortisone, 50 μg/mL gentamicin, 50 ng/mL amphotericin B, and 0.4% vol/vol bovine brain pituitary extract) (Kurabo). KCs were detached from the bottom of the dish/flask by incubating the monolayer of the cells with prewarmed trypsin–EDTA solution for 5–10 min at room temperature. Trypsin-neutralizing solution (Cambrex, Walkersville, MD) was added to the cell suspension, and cells were formed into pellets by centrifugation at 1000 rpm for 5 min. After this, they were resupended in culture medium, and the cell number was counted by trypan blue exclusion to determine the viability. KCs should be passaged before they reach 100% confluence (80% or less). The cells expanded by culturing in 9 cm dishes can be stored in liquid N_2.

3.3. Stimulation in serum-free medium

Each of the procedures of KC treatment before stimulation in our previous studies has minor differences. Typical one we recommend is as follows: KCs were seeded at 8000–10,000 cells/well in flat-bottomed 96-well microculture plates (for ELISA), 25,000–30,000 in 24-well plates, or 60,000–70,000 cells/well in 12-well plates (for PCR and multiple-cytokine ELISA) and cultured until they reached 90–100% confluence, and then, the medium was changed to fresh medium without the addition of hydrocortisone. After further cultivation for 24 h or longer, cells were stimulated with TLR ligands, cytokines, and so on in the absence of hydrocortisone. As hydrocortisone, a glucocorticoid, is supplied as one of the supplement bottles (Kurabo) (see Section 3.2), medium with the supplements except for hydrocortisone can be prepared. The removal of hydrocortisone is an important step to enhance the level of TSLP expression (Le et al., 2009, 2010).

3.4. Stimulation of KCs cultured in medium containing fetal bovine serum (FBS)

In some experiments in our studies, KCs were cultured with FBS as follows. Cells were cultured until they reached 70–90% confluence, and the medium was changed to one with the supplements and 5% FBS and renewed every 3 days. At day 9, the medium was replaced with that not containing hydrocortisone and FBS. At day 10, the KCs were stimulated in fresh medium without hydrocortisone and FBS. Culture supernatants for ELISA were recovered at day 11 or day 12 (24 or 48 h after stimulation). In KCs cultured with 5% FBS, higher levels of long TSLP transcript can be induced, and detectable TSLP was released even by stimulation with cytokines (see Section 6).

4. ANALYTICAL METHODS

4.1. Cytokine ELISA

Concentrations of cytokines and chemokines were measured with ELISA kits (DuoSet, R&D Systems, Minneapolis, MN). Culture supernatants were collected at 24–48 h after the stimulation and their diluted samples (e.g., 1:2 for TSLP and less than 1:20 for IL-8) were subjected to ELISA. In our studies, minimum detection limits for TSLP and IL-8 in the supernatants were 1.9–7.8 and 156 pg/mL, respectively.

4.2. Quantitative real-time PCR (qPCR)

At 3–9 h after the stimulation, total RNA was extracted from cells with RNeasy Plus Micro Kit (QIAGEN, Hilden, Germany). cDNA was synthesized with SuperScript II reverse transcriptase (Invitrogen) and random primers. qPCR was performed with the TaqMan method with an ABI7500 (Applied Biosystems, Piscataway, NJ). The mRNA level was normalized to the gene expression of β-actin and was shown relative to the control group.

The total level of the *TSLP* transcript (total *TSLP*) was analyzed by using a probe set (Hs00263639_m1; Applied Biosystems), which does not distinguish between the long and short forms. The long form of *TSLP* was detected using a long-form-specific probe set (Hs01572933_m1; Applied Biosystems). Specific measurement of the long *TSLP* transcript, which contributes to the production of the TSLP protein, rather than total or the short transcript is useful for accurate detection of functional human *TSLP* gene expression (see Section 6).

4.3. Transfection of KCs with small interfering RNAs (siRNA)

Transfection of siRNAs for target genes can induce specific knockdown of expression of target genes. siRNAs designed for possible specific inhibition are commercially available from several manufacturers. We used Stealth siRNAs (Invitrogen, Carlsbad, CA). Briefly, KCs were transfected with siRNAs for target genes. Transfection reagents (such as Lipofectamine 2000 and Lipofectamine RNAiMAX) (Invitrogen) were mixed with siRNA solution and OPTI-MEM (Gibco BRL, Gaithersburg, MD). After incubation, basal medium without the supplements was added and the solution was added to each of the wells. After cultivation with siRNAs for an optimum period, KCs were stimulated. Specific knockdown of gene expression of target genes and effect to the TSLP gene expression were analyzed by qPCR.

Nonspecific inhibition of expression of genes other than the target gene occurred in inappropriate experimental conditions or in the use of some of the designed siRNAs. Optimization of the experimental conditions and trial use of at least three different siRNAs for each of the target genes are necessary. Reproducibility of the results should be carefully checked by using at least two different siRNAs. The amount and ratio of reagents and the timing in each step should be optimized for each of the cases. The three procedures used in our previous studies are described below in detail.

4.3.1 TLR5

KCs at 60–70% confluence in 12-well tissue culture plates were transfected with the following Stealth siRNAs using Lipofectamine 2000: TLR5-siRNA1, 5′-AAUUCAACUUCCCAAAUGAAGGAUG-3′; TLR5-siRNA2, 5′-UCAGAUGGCUAAAUACUCCUGGUGG-3′; control siRNA1 (scrambled sequence of TLR5-siRNA1), 5′-AAUGGUCAACC CUUAAACAAGUAUG-3′; and control siRNA2 (scrambled sequence of TLR5-siRNA2), 5′-UCAGGAGGGUAUCUAAUCAUCCUGG-3′ (Le et al., 2011). Lipofectamine 2000 (4 μL) was mixed with 2 μL of a 20 μ*M* siRNA solution and 100 μL of OPTI-MEM. After incubation for 30 min at room temperature, a total of 500 μL of basal medium without the supplements was added and the solution (600 μL) was added to each of the wells. After cultivation with siRNAs for 24 h, the medium was changed to hydrocortisone-free medium. After further cultivation for 24 h, KCs were stimulated with flagellin. The expression of mRNA at 8 h after stimulation was analyzed by qPCR (see Section 5.2).

4.3.2 TLR2, TLR6, and TLR1

KCs were transfected with siRNAs for TLR1, TLR2, and TLR6 and were stimulated with FSL-1 or the *S. aureus* membranous fraction as follows (Vu et al., 2010). Cells were cultured until 70–80% confluent; then, the medium was changed to basal medium without the supplements, and the cells were incubated for 8 h. KCs were transfected with the following Stealth siRNAs using Lipofectamine 2000: TLR1-siRNA1, 5′-AGACC UUGCUGAUAUUCAAAUGAGC-3′; TLR2-siRNA3, 5′-UUCAGAG UGAGCAAAGUCUCUCCGG-3′; TLR6-siRNA2, 5′-UGGGAAUGC UGUUCUGUGGAAUGGG-3′; and TLR6-siRNA3, 5′-AAUAAGUCC GCUGCGUCAUGAGAGC-3′. We also used control siRNAs with a similar guanine–cytosine (GC) content to the TLR siRNAs (Stealth RNAi Universal Negative Controls LO#2 for 35–45% GC and Med#2 for 45–55% GC, Invitrogen). Lipofectamine 2000 (2 μL) was mixed with 2 μL of a 20 μ*M* siRNA solution and 100 μL of OPTI-MEM. After incubation for 30 min at room temperature, basal medium without the supplements was added to a total volume of 600 μL. Then, 600 or 400 μL of solution was added per well to the 12-well or 24-well plates, respectively. After cultivation with siRNAs overnight, the medium was changed to hydrocortisone-free medium. After further cultivation for 24 h, KCs were stimulated with FSL-1 or the *S. aureus* membranous fraction in hydrocortisone-free medium. The expression of mRNA at 4 h after stimulation was analyzed by qPCR (see Section 5.3).

4.3.3 RelA and IRF3

KCs were transfected with siRNAs for transcription factors RelA and IRF3 and were stimulated with polyI:C as follows (Vu et al., 2011). KCs were cultured until 70–80% confluent, and then, the medium was changed to basal medium without the supplements and the cells were incubated for 8 h. KCs were transfected with the following Stealth siRNAs using Lipofectamine RNAiMAX (Invitrogen): Re1AsiRNA2 (Primer number HSS9161), 5′-UUUACGUUUCUCCUCAAUCCGGUGA-3′ (44% GC); IRF3-siRNA3 (HSS105507), 5′-AACUCAUCCGAAU GUCUUCCUGGG-3′ (48% GC); and control siRNAs with a similar GC content to the siRNAs (Stealth RNAi Universal Negative Controls LO#2 for 35–45% GC and Med#2 for 45–55% GC). 3 μL of Lipofectamine RNAiMAX and 3 μL of a 20 μ*M* siRNA solution were separately diluted in 50 μL of OPTI-MEM, respectively, and the dilutions were mixed, total volume of 100 μL. After incubation for 20 min at room temperature, 800 μL of basal medium without the supplements was added, total volume of 900 μL. Then, 600 or 400 μL of solution was added per well to the 12-well or 24-well plates, respectively. After cultivation with siRNAs for 16 h, the medium was changed to hydrocortisone-free medium. After further cultivation for 24 h, KCs were stimulated with polyI:C. The expression of mRNA at 6 and 9 h after stimulation was analyzed by qPCR (see Section 5.1).

4.4. Flow cytometry

TLR3 protein expression was analyzed as follows (Vu et al., 2011). KCs were cultured in flat-bottomed six-well cell plates until 90% confluent. The medium was then changed to that without hydrocortisone. After further culturing overnight, KCs were stimulated with polyI:C or cultured in the absence of polyI:C for 18 h in fresh medium without hydrocortisone. Then, cells were collected after being detached from the plates by treatment with trypsin. Subsequently trypsin was inactivated with trypsin neutralization solution (Kurabo) and the cells were washed. Fc receptors were blocked with Clear Back (MBL, Nagoya, Japan) at 4 °C for 20 min; then, cells were stained for TLR3 expression with phycoerythrin-conjugated mouse anti-human TLR3 monoclonal antibody (clone TLR3.7, mouse IgG1/κ, eBioscience, San Diego, CA) or phycoerythrin-conjugated mouse IgG1/κ isotype control monoclonal antibody (clone MOPC-21; BD Biosciences Pharmingen, San Diego, CA) (1 μg–antibody/5 × 10^5 cells/30 μL/test) at 4 °C for 30 min. In the staining for intracellular TLR3, BD Cytofix/Cytoperm reagents (BD Biosciences, San Jose, CA) were used according to the manufacturer's instructions (see Section 5.1).

In our unpublished results, we clearly detected protein expression of TLR2, but not TLR1 and TLR6, on human KCs by flow cytometry, possibly because the expression levels of TLR1 and TLR6 were lower and/or the antibodies we used against TLR1 and TLR6 might not have high affinity (see Section 5.3).

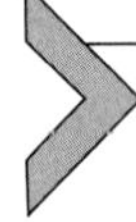

5. TLR-MEDIATED PATHWAYS OF TSLP INDUCTION IN HUMAN KCs

5.1. TLR3

ELISA of the culture supernatants showed that KCs treated with polyI:C, mimicking viral dsRNA, released detectable amounts of TSLP (Kinoshita et al., 2009). In the presence of hydrocortisone, the other ligands tested did not induce TSLP's release, whereas peptidoglycan and flagellin induced IL-8's release, although less effectively than polyI:C. Stimulation of KCs with polyI:C-induced gene expression of TSLP and other cytokines (TSLP, IFN-β, TNF-α, IL-6, and GM-CSF) and chemokines

(CXCL8/IL-8, CXCL9/MIG, CXCL10/IP-10, CXCL11/I-TAC, CCL2/MCP-1, CCL5/RANTES, CCL20/MIP-3α, and CCL22/MDC).

Glucocorticoids (frequently used at 1 μ*M*) suppress TSLP gene expression in human KCs (Le et al., 2009, 2010). Removal of hydrocortisone (300 n*M*), a glucocorticoid, from the medium prior to stimulation (see Section 3.4) enhanced the polyI:C-induced release and gene expression of TSLP. Calcineurin inhibitors showed no or much less inhibitory effect compared with glucocorticoids.

Endosomal TLR3 and cytosolic RIG-like receptors (RLRs; such as RIG-I and MDA5) and PKR have been reported to recognize dsRNA. The dsRNA sensors survey viral invasion or infection by sensing viral dsRNA to induce cellular responses (Matsumoto & Seya, 2008; McAllister & Samuel, 2009; Takeuchi & Akira, 2010), and recently, TLR3 has been suggested to sense self dsRNA released from damaged cells to induce inflammation (Cavassani et al., 2008; Lai et al., 2009). Stimulation of human KCs with polyI:C induced an upregulation of the gene expression of the four dsRNA sensors (Le et al., 2010; Vu et al., 2011). The relative level of mRNA and the increase caused by polyI:C were higher for the RLRs (RIG-I and MDA5) than for TLR3 and PKR. Although TLR3 showed the lowest mRNA expression among the four dsRNA sensors, analysis by flow cytometry showed the intracellular protein expression of TLR3 without stimulation to be enhanced after the polyI:C stimulation and weak cell-surface expression after the polyI:C stimulation (Vu et al., 2011).

Bafilomycin A1 (20–100 n*M*), which inhibits endosomal acidification to block the TLR3 pathway, blocked the dsRNA-induced expression of TSLP, IL-8, IFN-β, and other molecules including the dsRNA sensors, whereas it did not inhibit FSL-1 (TLR2–TLR6 ligand)-induced expression of TSLP and IL-8 (Vu et al., 2011). Our recent unpublished results also supported that TLR3 is the sensor for extracellular dsRNA in primary human KCs (our unpublished observations), whereas viral infection may stimulate cytosolic dsRNA sensors to induce TSLP (Kawasaki et al., 2011).

NF-κB activation has been considered essential for TSLP production (Lee & Ziegler, 2007; Takai, 2012; Zaph et al., 2007). Cellular signaling through TLR3 or RLRs similarly results in the activation of two transcription factors: NF-κB and IRF3 (Matsumoto & Seya, 2008; Takeuchi & Akira, 2010). We asked which of NF-κB and IRF3 contributes to the dsRNA-induced gene expression of TSLP in human KCs (Vu et al., 2011). In the analysis of siRNA transfection, the dsRNA-induced gene expression of TSLP depended on RelA, a component of NF-κB, but not

IRF3, similar to IL-8 but different from IFN-β, which depended on both IRF3 and RelA. The results indicate that endosomal acidification and the subsequent activation of NF-κB are necessary to sense extracellular dsRNA, suggesting the importance of the TLR3–NF-κB axis to trigger production of TSLP against extracellular dsRNA.

5.2. TLR5

Flagellin is the major structural protein of the flagella of gram-negative bacteria. Two receptors, the cell-surface TLR5 and the intracellular receptor Ipaf, have been reported to recognize flagellin (Hayashi et al., 2001; Miao et al., 2006). In the absence of hydrocortisone, stimulation of human KCs with flagellin induced the release of TSLP and gene expression of not only TSLP but also other proinflammatory molecules: cytokines (TSLP, TNF-α, IL-6, and GM-CSF), chemokines (CCL2, CCL5, CCL20, CCL27, CXCL8, and CXCL10), and an adhesion molecule (CD54/ICAM-1), suggesting importance of environmental flagellin as Th2 adjuvant in epicutaneous sensitization to indoor allergens (Le et al., 2011). Knockdown of TLR5 gene expression by siRNA reduced the flagellin-induced upregulation of TSLP gene expression in human KCs. Importantly, the presence of flagellin in house dust, administration of which via airway can cause TLR5-dependent allergic airway responses in mice, has been demonstrated (Wilson et al., 2012).

5.3. TLR2–TLR6

TLR2 forms a heterodimer with either TLR6 or TLR1 for the specific recognition of diacylated or triacylated lipoproteins/lipopeptides, respectively (Schenk, Belisle, & Modlin, 2009). The ratio of expression levels of mRNA for TLR1, TLR2, and TLR6 in unstimulated KCs was approximately 10:40:1 (Vu et al., 2010).

KCs treated with a synthetic diacylated lipopeptide, FSL-1, released detectable amounts of TSLP and IL-8 (Vu et al., 2010). Peptidoglycan and a synthetic triacylated lipopeptide, Pam_3CSK_4, induced the release of IL-8 but not TSLP at the concentrations tested. FSL-1 induced the gene or protein expression of not only TSLP but also other proinflammatory molecules similar to those induced by flagellin (see Section 5.2), whereas Pam_3CSK_4 induced lower levels or no significant upregulation of their expression.

S. aureus heavily colonizes the lesions of patient with atopic dermatitis and is known to trigger a worsening of the disease. Subcellular fractions of *S. aureus* and heat-killed whole bacterial cells of *S. aureus* also stimulated KCs to release TSLP (Vu et al., 2010). Knockdown of TLR2 or TLR6 gene expression by siRNA reduced the *S. aureus* membrane- and FSL-1-induced upregulation of TSLP gene expression, suggesting that ligands for the TLR2–TLR6 heterodimer in *S. aureus* membranes including diacylated lipoproteins/lipopeptides could promote Th2-type inflammation through TSLP production in KCs.

5.4. Cytokine milieu

Proinflammatory cytokines such as TNF-α and Th2 cytokines (IL-4 and IL-13) synergistically induce the release of TSLP in human skin explants (Bogiatzi et al., 2007).

In primary human KCs, Th2 cytokines, TNF-α, or type I IFNs (IFN-α and IFN-β) enhanced the polyI:C-induced release of TSLP, whereas Th1 (IFN-γ), Treg (TGF-β), or Th17 (IL-17A) cytokines suppressed it (Kinoshita et al., 2009). Th2/TNF cytokines or TGF-α (one of the ligands for epidermal growth factor receptor) enhanced the flagellin-induced release of TSLP (Le et al., 2011). Th2 cytokines and TNF-α upregulated FSL-1-induced release of TSLP. IFN-γ and TGF-β showed partial inhibition of the release of TSLP induced by FSL-1, the *S. aureus* membranous fraction, Th2/TNF cytokines, and the *S. aureus* membranous fraction plus Th2/TNF cytokines in the KCs cultured with FBS (Vu et al., 2010). IL-17A showed smaller inhibition of the release of TSLP induced by FSL-1 and Th2/TNF cytokines. The results suggest an amplification cycle between the TLR-mediated induction of TSLP and atopic cytokine milieu and a possible blockade of this cycle by other cytokine milieus.

6. LONG *TSLP* TRANSCRIPT RESPONSIBLE TO EXPRESSION OF THE TSLP PROTEIN

In primary human bronchial epithelial cells, two forms of *TSLP* transcripts (long and short) have been reported, and polyI:C upregulates the expression of long *TSLP* and induces the release of the TSLP protein (Harada et al., 2009). Single nucleotide polymorphisms in the promoter region of long *TSLP* were associated with asthma.

In primary human KCs, TLR ligands (polyI:C, flagellin, and FSL-1) or cytokines, a proinflammatory cytokine (TNF-α) and Th2 cytokines (IL-4 and IL-13), predominantly and/or markedly upregulated the gene expression of the long *TSLP* form but not the short form, indicating that the long *TSLP* contributes to the production of the TSLP protein (Xie, Takai, Chen, Okumura, & Ogawa, 2012). Without stimulation with TLR ligands or the atopic cytokine milieu, the expression level of the long form was strictly suppressed, while the short form was constitutively expressed or was further upregulated in culture conditions with overgrowth or with vitamin D receptor agonists, calcitriol or its analog MC903. Specific measurement of the long-form *TSLP*, but not conventional measurement of total *TSLP*, should be useful as the most accurate method of functional human *TSLP* gene expression.

7. SUMMARY

The skin epidermis and mucosal epithelia (airway, ocular tissues, gut, and so on) are located at the interface between the body and environment and have critical roles in the response to various stimuli (Takai, 2012; Takai & Ikeda, 2011). TSLP expressed mainly by epidermal KCs and mucosal epithelial cells links the innate response at barrier surfaces to Th2-skewed acquired immune response in allergic diseases. TSLP in the epidermis is important in the induction of the onset of Th2cytokine-associated inflammation, that is, "atopic march."

We studied TLR-mediated TSLP induction in primary cultured human KCs. DsRNA (TLR3 ligand) (Kinoshita et al., 2009; Vu et al., 2011), flagellin (TLR5 ligand) (Le et al., 2011), and diacylated lipopeptide (TLR2–TLR6 ligand) stimulated human KCs to express TSLP and *S. aureus* membranes did so in a manner dependent on TLR2 and TLR6 (Vu et al., 2010). Culturing in the absence of glucocorticoid before stimulation enhanced the TSLP expression (Le et al., 2009, 2010). Extracellular dsRNA induced TSLP via endosomal acidification- and NF-κB-dependent pathway, suggesting the importance of the TLR3–NF-κB axis to trigger production of TSLP against extracellular dsRNA. TLR5-mediated TSLP induction in KCs suggests the importance of environmental flagellin in house dust as Th2 adjuvant in epicutaneous sensitization to indoor allergens. *S. aureus* heavily colonizes the lesions of patient with atopic dermatitis and is known to trigger a worsening of the disease. *S. aureus*-induced TSLP expression via TLR2–TLR6 suggests that ligands for the TLR2–TLR6 heterodimer in *S. aureus* membranes including diacylated lipoproteins/lipopeptides could promote

Th2-type inflammation through TSLP production in KCs. Atopic cytokine milieu (IL-4, IL-13, TNF-α, and/or TGF-α) upregulated the TLR-mediated induction of TSLP, and others (IFN-γ, TGF-β, and IL-17A) inhibited it, suggesting an amplification cycle in Th2 inflammation and a possible blockade of this cycle by other cytokine milieus. Specific measurement of the long *TSLP* transcript, which contributes to the production of the TSLP protein, rather than total or the short transcript is useful for accurate detection of functional human *TSLP* gene expression (Xie et al., 2012). The results suggest that environment-, infection- and/or self-derived TLR ligands contribute to the initiation and/or amplification of Th2-type skin inflammation including atopic dermatitis and atopic march through the induction of TSLP expression in KCs and include information helpful for understanding the role of the gene–environment interaction relevant in allergic diseases.

ACKNOWLEDGMENTS

We gratefully acknowledge the contributions of Xiao-Ling Wang, Hendra Gunawan, Takasuke Ogawa, Takeshi Kato, and others to this work as well as Michiyo Matsumoto for secretarial assistance. This work was supported by the Ministry of Education, Culture, Sports, Science and Technology (MEXT)-supported Program for Strategic Research Foundation at Private Universities, Grants-in-Aid for Scientific Research from MEXT, the Takeda Science Foundation, and the Abbot Japan Allergy Research Award.

REFERENCES

Baba, T., & Schneewind, O. (1996). Target cell specificity of a bacteriocin molecule: A C-terminal signal directs lysostaphin to the cell wall of *Staphylococcus aureus*. *EMBO Journal*, *15*, 4789–4797.

Bogiatzi, S. I., Fernandez, I., Bichet, J. C., Marloie-Provost, M. A., Volpe, E., Sastre, X., et al. (2007). Cutting edge: Proinflammatory and Th2 cytokines synergize to induce thymic stromal lymphopoietin production by human skin keratinocytes. *Journal of Immunology*, *178*, 3373–3377.

Cavassani, K. A., Ishii, M., Wen, H., Schaller, M. A., Lincoln, P. M., Lukacs, N. W., et al. (2008). TLR3 is an endogenous sensor of tissue necrosis during acute inflammatory events. *Journal of Experimental Medicine*, *205*, 2609–2621.

Harada, M., Hirota, T., Jodo, A. I., Doi, S., Kameda, M., Fujita, K., et al. (2009). Functional analysis of the thymic stromal lymphopoietin variants in human bronchial epithelial cells. *American Journal of Respiratory Cell and Molecular Biology*, *40*, 368–374.

Hayashi, F., Smith, K. D., Ozinsky, A., Hawn, T. R., Yi, E. C., Goodlett, D. R., et al. (2001). The innate immune response to bacterial flagellin is mediated by Toll-like receptor 5. *Nature*, *410*, 1099–1103.

Kawai, T., & Akira, S. (2010). The role of pattern-recognition receptors in innate immunity: Update on Toll-like receptors. *Nature Immunology*, *11*, 373–384.

Kawasaki, J., Ushio, H., Kinoshita, H., Fukai, T., Niyonsaba, F., Takai, T., et al. (2011). Viral infection induces thymic stromal lymphopoietin (TSLP) in human keratinocytes. *Journal of Dermatological Science*, *62*, 131–134.

Kinoshita, H., Takai, T., Le, T. A., Kamijo, S., Wang, X. L., Ushio, H., et al. (2009). Cytokine milieu modulates release of thymic stromal lymphopoietin from human keratinocytes stimulated with double-stranded RNA. *Journal of Allergy and Clinical Immunology, 123*, 179–186.

Kollisch, G., Kalali, B. N., Voelcker, V., Wallich, R., Behrendt, H., Ring, J., et al. (2005). Various members of the Toll-like receptor family contribute to the innate immune response of human epidermal keratinocytes. *Immunology, 114*, 531–541.

Lai, Y., Di Nardo, A., Nakatsuji, T., Leichtle, A., Yang, Y., Cogen, A. L., et al. (2009). Commensal bacteria regulate Toll-like receptor 3-dependent inflammation after skin injury. *Nature Medicine, 15*, 1377–1382.

Le, T. A., Takai, T., Kinoshita, H., Suto, H., Ikeda, S., Okumura, K., et al. (2009). Inhibition of double-stranded RNA-induced TSLP in human keratinocytes by glucocorticoids. *Allergy, 64*, 1231–1232.

Le, T. A., Takai, T., Vu, A. T., Kinoshita, H., Chen, X., Ikeda, S., et al. (2011). Flagellin induces the expression of thymic stromal lymphopoietin in human keratinocytes via toll-like receptor 5. *International Archives of Allergy and Immunology, 155*, 31–37.

Le, T. A., Takai, T., Vu, A. T., Kinoshita, H., Ikeda, S., Ogawa, H., et al. (2010). Glucocorticoids inhibit double-stranded RNA-induced thymic stromal lymphopoietin release from keratinocytes in atopic cytokine milieu more effectively than tacrolimus. *International Archives of Allergy and Immunology, 153*, 27–34.

Lebre, M. C., van der Aar, A. M., van Baarsen, L., van Capel, T. M., Schuitemaker, J. H., Kapsenberg, M. L., et al. (2007). Human keratinocytes express functional Toll-like receptor 3, 4, 5, and 9. *Journal of Investigative Dermatology, 127*, 331–341.

Lee, H. C., & Ziegler, S. F. (2007). Inducible expression of the proallergic cytokine thymic stromal lymphopoietin in airway epithelial cells is controlled by NFkappaB. *Proceedings of the National Academy of Sciences of the United States of America, 104*, 914–919.

Liu, Y. J. (2006). Thymic stromal lymphopoietin: Master switch for allergic inflammation. *Journal of Experimental Medicine, 203*, 269–273.

Matsumoto, M., & Seya, T. (2008). TLR3: Interferon induction by double-stranded RNA including poly(I:C). *Advanced Drug Delivery Reviews, 60*, 805–812.

McAllister, C. S., & Samuel, C. E. (2009). The RNA-activated protein kinase enhances the induction of interferon-beta and apoptosis mediated by cytoplasmic RNA sensors. *Journal of Biological Chemistry, 284*, 1644–1651.

Miao, E. A., Alpuche-Aranda, C. M., Dors, M., Clark, A. E., Bader, M. W., Miller, S. I., et al. (2006). Cytoplasmic flagellin activates caspase-1 and secretion of interleukin 1beta via Ipaf. *Nature Immunology, 7*, 569–575.

Schenk, M., Belisle, J. T., & Modlin, R. L. (2009). TLR2 looks at lipoproteins. *Immunity, 31*, 847–849.

Spergel, J. M. (2005). Atopic march: Link to upper airways. *Current Opinion in Allergy and Clinical Immunology, 5*, 17–21.

Takai, T. (2012). TSLP expression: Cellular sources, triggers, and regulatory mechanisms. *Allergology International, 61*, 3–17.

Takai, T., & Ikeda, S. (2011). Barrier dysfunction caused by environmental proteases in the pathogenesis of allergic diseases. *Allergology International, 60*, 25–35.

Takai, T., Vu, A. T., Le, T. A., Kinoshita, H., Ushio, H., Ikeda, S., et al. (2009). Reply. *Journal of Allergy and Clinical Immunology, 124*, 864–865.

Takeuchi, O., & Akira, S. (2010). Pattern recognition receptors and inflammation. *Cell, 140*, 805–820.

Vu, A. T., Baba, T., Chen, X., Le, T. A., Kinoshita, H., Xie, Y., et al. (2010). *Staphylococcus aureus* membrane and diacylated lipopeptide induce thymic stromal lymphopoietin in keratinocytes via the Toll-like receptor 2-Toll-like receptor 6 pathway. *Journal of Allergy and Clinical Immunology, 126*, 985–993, 93.e1-3.

Vu, A. T., Chen, X., Xie, Y., Kamijo, S., Ushio, H., Kawasaki, J., et al. (2011). Extracellular double-stranded RNA induces TSLP via an endosomal acidification- and NF-κB-dependent pathway in human keratinocytes. *Journal of Investigative Dermatology, 131*, 2205–2212.

Wilson, R. H., Maruoka, S., Whitehead, G. S., Foley, J. F., Flake, G. P., Sever, M. L., et al. (2012). The Toll-like receptor 5 ligand flagellin promotes asthma by priming allergic responses to indoor allergens. *Nature Medicine, 18*, 1705–1710.

Xie, Y., Takai, T., Chen, X., Okumura, K., & Ogawa, H. (2012). Long *TSLP* transcript expression and release of TSLP induced by TLR ligands and cytokines in human keratinocytes. *Journal of Dermatological Science, 66*, 233–237.

Zaph, C., Troy, A. E., Taylor, B. C., Berman-Booty, L. D., Guild, K. J., Du, Y., et al. (2007). Epithelial-cell-intrinsic IKK-beta expression regulates intestinal immune homeostasis. *Nature, 446*, 552–556.

CHAPTER TWENTY-TWO

Endosomal Signaling by Protease-Activated Receptors

Neil Grimsey, Huilan Lin, JoAnn Trejo[1]
Department of Pharmacology, School of Medicine, University of California, La Jolla, California, USA
[1]Corresponding author: e-mail address: joanntrejo@ucsd.edu

Contents

Abstract

Protease-activated receptors (PARs) are a family of G protein-coupled receptors (GPCRs) that are uniquely activated by proteolysis. There are four members of the PAR family including: PAR1, PAR2, PAR3, and PAR4. PARs are expressed primarily in the cells of the vasculature and elicit cellular responses to coagulant and anticoagulant proteases. PAR1 exemplifies the unusual proteolytic mechanism of receptor activation. Thrombin binds to and cleaves the N-terminal exodomain of PAR1, generating a new N-terminus that functions as a tethered ligand. The N-terminal tethered ligand domain of PAR1 binds intramolecularly to the receptor to trigger transmembrane signaling and cannot diffuse away. Similar to other GPCRs, activation of PARs promotes coupling to heterotrimeric G proteins at the plasma membrane. After activation, PARs are rapidly internalized to endosomes and then sorted to lysosomes and degraded. Internalization functions to uncouple PARs from heterotrimeric G proteins at the cell surface. However, recent studies indicate that activated internalized PARs signal from endosomes through the recruitment of β-arrestins and potentially other pathways. Here, we provide an overview of methods and strategies used to examine endosomal signaling by PARs.

Methods in Enzymology, Volume 535
ISSN 0076-6879
http://dx.doi.org/10.1016/B978-0-12-397925-4.00022-5

1. INTRODUCTION

The idea that G protein-coupled receptors (GPCRs) can signal from endosomes was substantiated by studies showing that β-arrestins function as scaffolds to facilitate activation of MAPK signaling cascades on endosomes (Lefkowitz & Shenoy, 2005). DeFea et al. were the first to show that activation of PAR2 results in β-arrestin-mediated recruitment of a Raf-1 and ERK1/2 signaling complex on endosomes (DeFea et al., 2000; Dery, Thoma, Wong, Grady, & Bunnett, 1999). In subsequent work, we demonstrated that phosphorylation of the PAR2 C-tail is critical for stabilizing β-arrestin association and kinetics of ERK1/2 activation, but is not essential for receptor desensitization nor internalization (Ricks & Trejo, 2009; Stalheim et al., 2005). In examining PAR1 signaling, it has become clear that β-arrestins transiently associate with the receptor (Chen, Paing, & Trejo, 2004) and are unlikely to mediate signaling from endosomes, raising the possibility that other mechanisms exist. We have shown that activated PAR1 is internalized and sorted to early endosomes at a time that coincides with p38 activation (Fig. 22.1; Dores et al., 2012; Paing, Johnston, Siderovski, & Trejo, 2006), suggesting that p38 signaling may be initiated or sustained on endosomes. The majority of published studies have focused

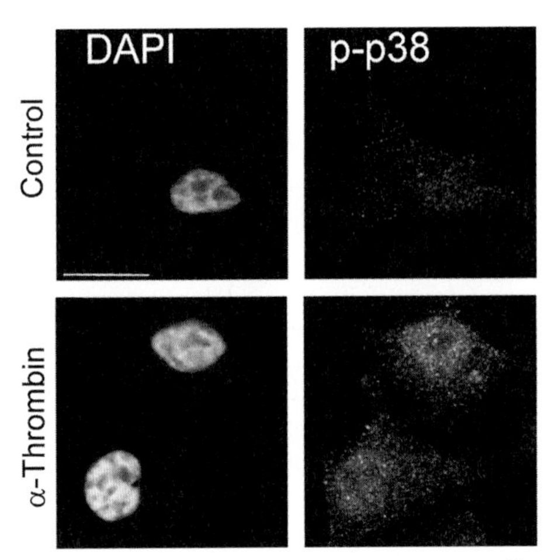

Figure 22.1 Thrombin-induced p38 phosphorylation. HeLa cells were serum-starved and either left untreated (control) or treated with 10 n*M* α-thrombin for 7 min at 37 °C. Cells were fixed, processed, and immunostained with antiphospho p38 antibody and imaged by confocal microscopy. Cells were counterstained with DAPI to image nuclei. Scale bar, 10 μm.

on PAR1 and PAR2 signaling and there is limited knowledge with regard to endosomal signaling by PAR3 or PAR4.

While previous publications have provided detailed methodologies for examining growth factor receptor endosomal signaling, here, we will provide an overview of methods used to examine endosomal signaling by protease-activated receptors (PARs) including imaging of p38, ERK1/2, and β-arrestins on endosomes and biochemical approaches to examine signaling complexes associated with PARs.

2. IMAGING OF P38, ERK1/2, AND PAR1 ON ENDOSOMES

To investigate the potential activation of p38 and ERK1/2 on endosomes following stimulation of PAR1, we have used immunofluorescence microscopy. In these assays, endogenous p38 and ERK1/2 are detected with antibodies. The colocalization with early endosomal antigen-1 (EEA1), a marker of early endosomes, and/or PAR1 is used to assess recruitment to endosomes. While we outline approaches for PAR1, similar strategies can be used to examine p38 and ERK1/2 activation following stimulation of other PARs. We describe procedures for HeLa cells, which are commonly used to examine endocytic trafficking and endosomal signaling, and human umbilical vein-derived EA.hy926 endothelial cells, which express endogenous PAR1 and PAR2.

2.1. Detection of p38 MAPK by fluorescence microscopy

1. Glass coverslips (12 mm circular No. 1, Chemglass Life Sciences #1760-012) are submerged in 100% ethanol, air dried, autoclaved, and then placed in each well of a 24-well plate. Coverslips are coated with 0.4 ml of 0.33 μg/ml fibronectin (Sigma Cat. #F1141) diluted in phosphate buffered saline (PBS), pH 7.4, for 30 min at room temperature (RT) before cells are plated.
2. HeLa cells expressing PAR1 are seeded at 3×10^4 cells/well of a 24-well plate in DMEM supplemented with 10% fetal bovine serum (FBS) on fibronectin-coated coverslips. Human EA.hy926 endothelial cells are seeded at 1.5×10^5 cells/well in 24-well plates with DMEM containing 10% FBS as described (Edgell, McDonald, & Graham, 1983). Cells are grown for 48 h to reach ~80% confluence.
3. Cells are then serum-starved by replacing growth medium with DMEM containing 10 m*M* HEPES, 1 mg/ml BSA, and 1 m*M* $CaCl_2$

(added just prior to use) (HeLa cells) or DMEM supplemented with 0.2% FBS (EA.hy926 cells).

4. After 24 h of serum starvation, cells are washed and incubated for an additional 3 h at 37 °C with starvation media and then stimulated with 10 n*M* α-thrombin (Enzyme Research Laboratories Cat. #HT 1002a) for various times at 37 °C.
5. After stimulation, cells are placed on ice, gently washed twice with cold PBS, and fixed in 4% paraformaldehyde (PFA) for 5 min on ice, followed by 12 min at RT, and then washed three times with PBS.
6. Coverslips are removed and placed with cells facing upward onto a clean surface and then incubated with 100 μl of blocking buffer (5% goat serum + 0.3% Triton™ X-100 diluted in PBS) for 60 min at RT and then washed twice with PBS at RT.
7. Coverslips are then incubated with a 100 μl of either antiphospho-p38 rabbit antibody (Cell Signaling Technology Cat. #4511) at 1:800 dilution, anti-p38 rabbit antibody (Cell Signaling Technology Cat. #9212) at 1:100 dilution, or anti-EEA1 monoclonal antibody (BD Biosciences Cat. #610457) at 1:1000 dilution in antibody dilution buffer (1% (w/v) BSA, 0.3% Triton™ X-100 reconstituted in PBS) and incubated overnight at 4 °C in a sealed humidified container.
8. Cells are washed three times with PBS at RT and then incubated with 100 μl of Alexa Fluor® 594 conjugated goat anti-rabbit antibody (Life Technologies Cat. #A-11012) or Alexa Fluor® 488 goat anti-mouse antibody (Life Technologies Cat. #A-11001) diluted at 1:500 in antibody dilution buffer for 1 h at RT in a humidified chamber kept in the dark.
9. Cells are then washed three times with 100 μl of PBS at RT with each wash left on the cells for 5 min before removing (second wash includes 0.5 mg/ml 4,6-diamidino-2-phenylindole (DAPI) reconstituted in PBS).
10. After the last wash, 14 μl of FluorSave™ Reagent (EMD Millipore Cat. #345789) is gently layered on the coverslip. The coverslip is then mounted on a glass microscope slide (Fisher Cat. #12-550-123) and allowed to dry overnight. Mounted coverslips can be stored for months at 4 °C.
11. Mounted coverslips are warmed to RT before imaging on an Olympus DSU-IX81 Spinning Disc Confocal imaging system fitted with a Plan Apo 60× oil objective (1.4 NA; Olympus) and a digital Hamamatsu Photonics ORCA-ER camera.

2.2. Detection of ERK1/2 by fluorescence microscopy

1. To detect the presence of ERK1/2 by fluorescence microscopy, follow Steps 1–4 in Section 2.1. After fixation with 4% PFA for 5 min on ice (Step 5 in Section 2.1), wash cells three times with ice-cold PBS.
2. Then add ice-cold 100% methanol (use enough to cover the cells completely) and incubate at −20 °C for 10 min. Cells are then washed twice with cold PBS.
3. After the second wash, the cells are incubated with 500 μl of blocking buffer containing 0.3% Triton™ X-100 and 5% goat serum diluted in PBS for 60 min at RT.
4. Cells are then washed twice with PBS at RT and incubated with antiphospho-ERK1/2 mouse antibody (Cell Signaling Technology Cat. #4372) or anti-ERK1/2 rabbit antibody (Cell Signaling Technology Cat. #9102) diluted 1:200 in antibody dilution buffer described in Step 7 of Section 2.1.
5. After primary ERK1/2 antibody incubation, follow Steps 8–11 in Section 2.1.

2.3. Detection of PAR1 and PAR2 on endosomes by fluorescence microscopy

To assess PAR1 or PAR2 colocalization with p38 and ERK1/2, cells are prelabeled with antibodies directed against epitope-tagged PARs expressed in HeLa cells or endogenous PARs expressed in endothelial cells. This approach can be used to determine if activated receptors colocalize with p38 and/or ERK1/2 signaling complexes on endosomes.

1. Follow Steps 1–3 of Section 2.1.
2. After 24 h of serum starvation, HeLa cells or EA.hy926 cells are washed and incubated for 3 h at 37 °C with DMEM containing 10 m*M* HEPES, 1 mg/ml BSA, and 1 m*M* $CaCl_2$ and then chilled on ice.
3. HeLa cells expressing either PAR1 or PAR2 containing an N-terminal FLAG or HA epitope tag are then labeled with M1 anti-FLAG mouse antibody (Sigma Cat #F3040) or anti-HA mouse antibody (Covance Cat. # MMS-101R), respectively, diluted to 1:500 in DMEM starvation buffer for 1 h on ice. Endothelial cells expressing endogenous PAR1 can be labeled with anti-PAR1 WEDE mouse antibody (Beckman Coulter Cat. #IM2085) diluted at 1:100 in DMEM starvation buffer. Endogenous PAR2 can be labeled with anti-PAR2 rabbit polyclonal antibody generously provided by Dr. Wolfram Ruf (The Scripps Research Institute, La Jolla, CA) diluted at 1:500 in DMEM starvation buffer.

4. Cells are then washed three times and stimulated with either 100 μ*M* TFLLRNPNDK (PAR1-specific agonist peptide), 100 μ*M* SLIGKV (PAR2-specific agonist), 10 n*M* α-thrombin, or 10 n*M* trypsin (Sigma Cat. #T-1426) for various times at 37 °C. Note that proteases cleave off the N-terminal FLAG and HA epitope tags so should be avoided if using anti-FLAG or anti-HA antibodies.
5. Cells are then placed on ice, washed with cold PBS, and processed for either p38 or ERK1/2 immunostaining as described in Sections 2.1 and 2.2. Note that the p38 immunostaining procedure will result in loss of PAR1 from the cell surface due to the Triton™ X-100 detergent. However, internalized PAR1 on endosomal structures is preserved.

3. β-ARRESTIN RECRUITMENT TO ENDOSOMES

β-Arrestins associate with activated GPCRs on endosomes and function as scaffolds to facilitate assembly of MAPK signaling complexes (Lefkowitz & Shenoy, 2005). The intracellular localization of β-arrestin-1-GFP or β-arrestin-2-GFP using microscopy is a simple method to assess GPCR-stimulated endosomal signaling. Using HeLa cells, we described methods to examine colocalization of β-arrestins with activated PAR2 (Stalheim et al., 2005). A similar strategy can be used to assess β-arrestin-GFP recruitment to thrombin-activated PAR1–PAR2 heterodimer in HeLa cells (Lin & Trejo, 2013). COS7 cells express low levels of β-arrestins and can be used as an alternative if a more robust cell model system is needed. Note that β-arrestin association with activated PAR1 is transient (Chen et al., 2004) and β-arrestins are not required for PAR1 internalization (Paing, Stutts, Kohout, Lefkowitz, & Trejo, 2002). Thus, this method cannot be used to detect β-arrestin association with activated PAR1.

3.1. Detection of β-arrestin-GFP by fluorescence microscopy

1. Glass coverslips (18 mm circular No. 1, Fisher #12-545-100) are treated as described in Step 1 of Section 2.1 and placed in 12-well dishes.
2. HeLa cells stably expressing a human PAR2 containing an N-terminal FLAG epitope (Stalheim et al., 2005) are seeded at 8×10^4 cells/well in 12-well dishes to achieve ~40% confluency (higher confluency will reduce transfection efficiency) and grown overnight in antibiotic-free DMEM containing 10% FBS.

3. Media is removed and replaced with 1000 μl of antibiotic-free DMEM containing 10% FBS.
4. For each well of a 12-well dish, the following is prepared. An aliquot 1.8 μl of 1 mg/ml PEI (Polysciences, Inc., Cat. #23966) is added to 100 μl of reduced serum Opti-MEM (Life Technologies Cat. #11058-021), gently mixed, and incubated for 10 min at RT. Then, 300 ng of cDNA plasmid encoding β-arrestin-1-GFP or β-arrestin-2-GFP is added to the PEI–Opti-MEM, gently mixed, and incubated at RT for 20 min. The ratio of PEI to plasmid used is optimized for each batch of 1 mg/ml PEI. Typically, ratios of 6:1 to 3:1 of PEI/plasmid result in ~60–80% transfection efficiency of HeLa cells after 48 h.
5. The final PEI, plasmid, and Opti-MEM mixture is then added dropwise to each well. After 24 h, wells are washed twice and left overnight in DMEM starvation buffer containing 10 m*M* HEPES, 1 mg/ml BSA, and 1 m*M* $CaCl_2$.
6. The next day, cells at ~80% confluency are washed twice with DMEM starvation buffer and then incubated for an additional 3 h at 37 °C in starvation media. Each well is then washed once with prechilled DMEM starvation buffer, placed on ice, and incubated for 1 h at 4 °C on ice with anti-FLAG rabbit antibody (Sigma Cat #7425) diluted 1:1000 in starvation buffer.
7. Each well is then washed three times in DMEM starvation media and stimulated with or without 100 μ*M* SLIGKV (PAR2-specific agonist peptide) diluted in prewarmed starvation DMEM buffer for various times at 37 °C.
8. After stimulation, cells are placed on ice, and each well is washed twice with cold PBS, fixed with 1 ml of 4% PFA for 10 min on ice, washed with cold PBS and permeabilized with 1 ml of ice-cold 100% methanol for 30 s on ice, and then washed twice with PBS.
9. Coverslips are then removed and placed on a flat surface with cells facing upward and washed three times with 300 μl of quench buffer (PBS containing 1% (w/v) nonfat dry milk and 0.15 *M* sodium acetate, pH 7) with each wash incubated for 5 min.
10. Coverslips are then washed three times with 300 μl of wash buffer (PBS with 1% (w/v) nonfat dry milk). Each wash is incubated for 5 min at RT.
11. Coverslips are then incubated with 300 μl Alexa Fluor® 594 goat anti-rabbit antibody diluted in wash buffer at 1:1000 for 1 h at RT in the dark, washed three times with PBS, and then mounted in 20 μl of FluorSave™ Reagent as described in Steps 10 and 11 of Section 2.1.

3.2. Detection of endogenous β-arrestin on endosomes by fluorescence microscopy

We have adapted this method from Marchese et al. (Malik & Marchese, 2010) to examine the intracellular localization of endogenous β-arrestins with thrombin-activated PAR1–PAR2 heterodimer in HeLa cells and endothelial cells (Lin & Trejo, 2013). This method can be used to image endogenous β-arrestin recruitment to endosomes following thrombin-activated PAR1–PAR2 heterodimer or activated PAR2 protomer.

1. Coverslips (12 mm circular No. 1) are prepared and coated with fibronectin as described in Step 1 of Section 2.1 and placed in 24-well dishes.
2. HeLa cells expressing PAR1 and PAR2 are seeded at 3×10^4 cells/well of 24-well dishes. EA.hy926 endothelial cells expressing endogenous PAR1 and low levels of PAR2 are seeded 1.5×10^5 cells/well of 24-well dishes. To increase expression of PAR2, endothelial cells can be pretreated with 10 ng/ml TNF-α (PeproTech Inc. Cat. #300-01A) for 18 h.
3. After 48 h, each well is washed twice with DMEM starvation buffer and then incubated for an additional 3 h at 37 °C in starvation media.
4. Cells are then stimulated with 10 n*M* α-thrombin, which transactivates PAR1–PAR2 heterodimer or 100 μ*M* SLIGKV (PAR2 agonist peptide) diluted in DMEM starvation buffer for various times at 37 °C.
5. Cells are placed on ice, and each well is then washed twice with cold PBS, fixed with 4% PFA for 10 min at RT, washed twice with PBS, and then permeabilized with 500 μl of 0.05% (w/v) saponin diluted in PBS for 10 min at RT.
6. Coverslips are then incubated with 500 μl of blocking buffer (PBS containing 0.05% (w/v) saponin and 5% goat serum) for 30 min at 37 °C.
7. Coverslips are then placed on a flat surface and incubated with 50 μl of anti-β-arrestin rabbit antibody generously provided by Dr. Jeffrey Benovic (Thomas Jefferson University, Philadelphia, PA) diluted at 1:50 in blocking buffer for 1 h at 37 °C in a moist chamber.
8. After incubation, coverslips with adherent cells are placed back into 24-well plates and washed five times with 500 μl of 0.05% (w/v) saponin diluted in PBS with the last wash incubated for 15 min at 37 °C.
9. Coverslips are incubated with Alexa Fluor® 594 conjugated goat anti-rabbit antibody diluted at 1:200 in blocking buffer for 30 min at 37 °C in a moist chamber.
10. Coverslips are then washed five times with 0.05% (w/v) saponin diluted in PBS and mounted in 14 μl of FluorSave™ Reagent as described in Steps 10 and 11 of Section 2.1.

4. RAB5 Q79L EXPANSION OF EARLY ENDOSOMES

Rab5 is a small GTP-binding protein that regulates vesicle budding, transport, tethering, and fusion (Zerial & McBride, 2001). Ectopic expression of the Rab5 Q79L constitutively active mutant perturbs vesicle fusion and results in enlarged endosomes (Stenmark et al., 1994). The expansion of the early endosomal compartment can be used to enhance visualization of p38, ERK1/2, and PAR1 localized on endosomes. A similar strategy can be used for other PARs:

1. Glass coverslips are prepared as described in Step 1 of Section 2.1. HeLa cells expressing PAR1 or PAR2 are then seeded at 4×10^5 cells/well of 24-well plates to achieve ~40% confluency and grown overnight in DMEM supplemented with 10% FBS. Note that a higher confluency will reduce transfection efficiency.
2. After 24 h, media is removed and replaced with 500 μl of antibiotic-free DMEM containing 10% FBS.
3. For each well of a 24-well dish, the following is prepared. An aliquot 0.6 μl of PEI at 1 mg/ml is added to 50 μl of reduced serum Opti-MEM, gently mixed, and incubated for 10 min at RT. Then, 100 ng of cDNA plasmid encoding Rab5 Q79L tagged with GFP is added to the PEI–Opti-MEM solution, gently mixed, and incubated at RT for 20 min. Transfection of plasmids encoding Rab5 wild type and/or GFP only should be performed in parallel as controls.
4. The final PEI and plasmid mixture diluted in Opti-MEM is then added dropwise to each well.
5. Cells in suspension can be used to increase the transfection efficiency. In this method, 50 μl of the PEI, plasmid, and Opti-MEM mixture is added directly to $\sim 8 \times 10^4$ cells diluted in 500 μl of antibiotic-free DMEM containing 10% FBS and incubated for 5 min at RT and then plated on fibronectin-coated glass coverslips in 24-well plates.
6. After transfections, cells are then treated with various agonists, and p38, ERK1/2, β-arrestins, and PAR1 can be detected by immunofluorescence microscopy as described earlier.

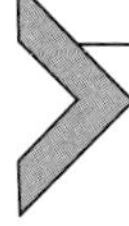

5. IMMUNOPRECIPITATION OF PAR1 SIGNALING COMPLEXES

Activation of PAR1 results in rapid internalization with the majority of the receptor localized to early endosomes after 10 min of agonist stimulation (Dores et al., 2012; Paing et al., 2006). Immunoprecipitation of internalized PAR1 can be used as a method to determine the potential coassociation of signaling effectors using HeLa cells stably expressing PAR1 or endothelial cells expressing endogenous PAR1. The detection of signaling effectors associated with activated PAR1 can then be assayed by immunoblotting. Similar approaches can be used to examine other PAR family members:

1. HeLa cells expressing PAR1 are seeded at 1×10^6 cells per 10 cm^2 dish coated with 4 ml of 0.33 μg/ml of fibronectin. Endothelial EA.hy926 cells are plated at 1.4×10^6 cells per 10 cm^2 dish without fibronectin coating. Cells are grown for 48 h in normal growth media.
2. Cells are then incubated with DMEM BSA starvation (HeLa cells) or DMEM supplemented with 0.2% FBS (EA.hy926 cells) for 24 h at 37 °C.
3. Cells are washed twice with 8 ml of prewarmed DMEM starvation buffer, incubated for additional 3 h at 37 °C, and then stimulated with 10 n*M* α-thrombin for various times at 37 °C.
4. After agonist stimulation, cells are placed on ice, washed with 10 ml of cold PBS supplemented with 1 m*M* $CaCl_2$ (added just prior to use), and then lysed with 750 μl of ice-cold lysis buffer (50 m*M* Tris–HCl, pH 7.4, 1% Triton™ X-100, 150 m*M* NaCl, 50 m*M* NaF, 10 m*M* NaPP, 25 m*M* β-glycerophosphate containing freshly added protease inhibitors including aprotinin 2 μg/ml, leupeptin 1 μg/ml, pepstatin A 1 μg/ml, benzamidine 10 μg/ml, and soybean trypsin inhibitor 1 μg/ml).
5. Cells lysates are transferred to a 1.5 ml microcentrifuge tube, passed 10 times through a 22.5 gauge needle and syringe, and gently rocked for 20 min at 4 °C.
6. Cell lysates are centrifuged at 16,000 × *g* for 30 min at 4 °C and transferred to a new 1.5 ml centrifuge tube, and 50 μl is removed to determine protein concentrations determined using the Pierce™ BCA Protein Assay Kit (Thermo Scientific Cat. #23225). Note that it is important to save ~50 μg of lysates to use as controls for expression of proteins in total cell lysates.
7. Protein A–Sepharose CL-4B (GE Health Biosciences Cat. #17-0780-01) beads are prepared by washing twice using lysis buffer and centrifuged at 450 × *g* for 3 min. Protein A–Sepharose beads are then preincubated with 1 mg/ml BSA to block nonspecific-binding sites 1 h at 4 °C and then washed with twice with lysis buffer.

8. For each sample, 12.5 μl of Protein A–Sepharose blocked with BSA is preincubated with 0.4 μg of anti-PAR1 WEDE mouse antibody or IgG antibody control for 2 h at 4 °C and then washed four times with lysis buffer before addition of cell lysates.
9. To preclear, add ~1000 μg of cell lysates from each sample to 1.5 ml microcentrifuge tubes containing 12.5 μl of Protein A–Sepharose beads blocked with BSA and gently rock for 1 h at 4 °C.
10. Cell lysates are then centrifuged at 450 × *g* for 5 min at 4 °C and supernatant is transferred to a new tube containing 12.5 μl of antibody-immobilized Protein A–Sepharose beads from Step 8 and gently rocked overnight at 4 °C.
11. Samples are centrifuged at 450 × *g* and 4 °C, the supernatant is removed, and the beads are washed three more times.
12. After the final wash, lysis buffer is completely removed from the beads using 27-gauge needle and syringe and then 30 μl of 2 × sample buffer (125 m*M* Tris–HCl, pH 6.5, 5% SDS, 20% glycerol, 0.003% bromophenol blue diluted in H_2O) containing 200 m*M* dithiothreitol (added just prior to use) is added immediately. Samples are heated to 50 °C for 10 min (to reduce sample/protein aggregation) and then to 95 °C for 5 min.
13. Samples are centrifuged at 450 × *g* for 5 min at RT and ~15 μl is loaded on SDS–PAGE, transferred to PVDF membranes, and immunoblotted with various antibodies. Membranes can be stripped with Restore Western Blot Stripping Buffer (Thermo Scientific Cat. # 46430) and then reprobed with an anti-PAR1 rabbit polyclonal antibody to detect PAR1. Total cell lysates (10 μg) from Step 6 in the preceding text can also be examined by immunoblotting to ensure that equivalent amounts of proteins are present in cell lysates before immunoprecipitations.

6. SUMMARY

PARs have important functions in vascular biology and cancer progression (Arora, Ricks, & Trejo, 2007; Coughlin, 2005). The regulation of PAR signaling is critical for the fidelity of thrombin signaling (Trejo, Hammes, & Coughlin, 1998) and dysregulation of PAR signaling has been implicated in pathophysiological disease processes (Arora et al., 2007; Leger, Covic, & Kuliopulos, 2006). While it is clear that activation of PARs at the plasma membrane results in coupling to heterotrimeric G proteins, the signaling elicited by internalized receptors from endosomes has yet to be fully elucidated and represents a significant gap in our knowledge that is critical to understand.

ACKNOWLEDGMENTS

This work was supported by the National Institutes of Health Grants GM090689 and HL073328 (J. T.). N. G. is supported by an American Heart Association Postdoctoral Fellowship.

REFERENCES

Arora, P., Ricks, T. K., & Trejo, J. (2007). Protease-activated receptor signalling, endocytic sorting and dysregulation in cancer. *Journal of Cell Science, 120*, 921–928.

Chen, C. H., Paing, M. M., & Trejo, J. (2004). Termination of protease-activated receptor-1 signaling by β-arrestins is independent of receptor phosphorylation. *The Journal of Biological Chemistry, 279*, 10020–10031.

Coughlin, S. R. (2005). Protease-activated receptors in hemostasis, thrombosis and vascular biology. *Journal of Thrombosis and Haemostasis, 3*, 1800–1814.

DeFea, K. A., Zalevski, J., Thoma, M. S., Dery, O., Mullins, R. D., & Bunnett, N. W. (2000). β-Arrestin-dependent endocytosis of proteinase-activated receptor-2 Is required for intracellular targeting of activated ERK1/2. *The Journal of Cell Biology, 148*, 1267–1281.

Dery, O., Thoma, M. S., Wong, H., Grady, E. F., & Bunnett, N. W. (1999). Trafficking of proteinase-activated receptor-2 and β-arrestin-1 tagged with green fluorescent protein. *The Journal of Biological Chemistry, 274*, 18524–18535.

Dores, M. R., Chen, B., Lin, H., Soh, U. J. K., Paing, M. M., Montagne, W. A., et al. (2012). ALIX binds a YPX3L motif of the GPCR PAR1 and mediates ubiquitin-independent ESCRT-III/MVB sorting. *The Journal of Cell Biology, 197*, 407–419.

Edgell, C. J., McDonald, C. C., & Graham, J. B. (1983). Permanent cell line expressing human factor VIII-related antigen established by hybridization. *Proceedings of the National Academy of Sciences of the United States of America, 12*, 3734–3737.

Lefkowitz, R. J., & Shenoy, S. K. (2005). Transduction of receptor signals by β-arrestins. *Science, 308*, 512–517.

Leger, A. J., Covic, L., & Kuliopulos, A. (2006). Protease-activated receptors in cardiovascular diseases. *Circulation, 114*(10), 1070–1077.

Lin, H., & Trejo, J. (2013). Transactivation of the PAR1-PAR2 heterodimer by thrombin elicits beta-arrestin endosomal signaling. *The Journal of Biological Chemistry, 288*, 11203–11215.

Malik, R., & Marchese, A. (2010). Arrestin-2 interacts with the endosomal sorting complex required for transport machinery to modulate endosomal sorting of CXCR4. *Molecular Biology of the Cell, 21*(14), 2529–2541.

Paing, M. M., Johnston, C. A., Siderovski, D. P., & Trejo, J. (2006). Clathrin adaptor AP2 regulates thrombin receptor constitutive internalization and endothelial cell resensitization. *Molecular and Cellular Biology, 28*, 3231–3242.

Paing, M. M., Stutts, A. B., Kohout, T. A., Lefkowitz, R. J., & Trejo, J. (2002). β-Arrestins regulate protease-activated receptor-1 desensitization but not internalization or down-regulation. *The Journal of Biological Chemistry, 277*, 1292–1300.

Ricks, T., & Trejo, J. (2009). Phosphorylation of protease-activated receptor-2 differentially regulates desensitization and internalization. *The Journal of Biological Chemistry, 284*, 34444–34457.

Stalheim, L., Ding, Y., Gullapalli, A., Paing, M. M., Wolfe, B. L., Morris, D. R., et al. (2005). Multiple independent functions of arrestins in regulation of protease-activated receptor-2 signaling and trafficking. *Molecular Pharmacology, 67*, 1–10.
Stenmark, H., Parton, R. G., Steele-Mortimer, O., Lutcke, A., Gruenberg, J., & Zerial, M. (1994). Inhibition of rab5 GTPase activity stimulates membrane fusion in endocytosis. *The EMBO Journal, 13*, 1287–1296.
Trejo, J., Hammes, S. R., & Coughlin, S. R. (1998). Termination of signaling by protease-activated receptor-1 is linked to lysosomal sorting. *Proceedings of the National Academy of Sciences of the United States of America, 95*, 13698–13702.
Zerial, M., & McBride, H. (2001). Rab proteins as membrane organizers. *Nature Reviews Molecular Cell Biology, 2*, 107–118.

CHAPTER TWENTY-THREE

Investigating Signaling Consequences of GPCR Trafficking in the Endocytic Pathway

Roshanak Irannejad, Sarah J. Kotowski, Mark von Zastrow[1]

Department of Psychiatry, University of California School of Medicine, San Francisco, California, USA
Department of Cellular & Molecular Pharmacology, University of California School of Medicine, San Francisco, California, USA
[1]Corresponding author: e-mail address: mark.vonzastrow@ucsf.edu

Contents

Abstract

Ligand-dependent regulation of adenylyl cyclase by the large family of seven-transmembrane G protein-coupled receptors (GPCRs) represents a deeply conserved and widely deployed cellular signaling mechanism. Studies of adenylyl cyclase regulation by catecholamine receptors have led to a remarkably detailed understanding of the basic biochemistry of G protein-linked signal transduction and have elaborated numerous mechanisms of regulation. Endocytosis of GPCRs plays a significant role in controlling longer-term cellular responses, such as under conditions of prolonged or repeated receptor activation occurring over a course of hours or more. It has been more challenging to investigate regulatory effects occurring over shorter time intervals, within the minutes to tens of minutes spanning the time course of many acute cyclic AMP (cAMP)-mediated signaling processes. A main reason for this is that biochemical methods used traditionally to assay changes in cytoplasmic cAMP concentration are

Methods in Enzymology, Volume 535
ISSN 0076-6879
http://dx.doi.org/10.1016/B978-0-12-397925-4.00023-7

limited in spatiotemporal resolution and typically require perturbing cellular structure and/or function for implementation. Recent developments in engineering genetically encoded cAMP biosensors linked to optical readouts, which can be expressed in cells or tissues and detected without cellular disruption or major functional perturbation, represent a significant step toward overcoming these limitations. Here, we describe the application of two such cAMP biosensors, one based on enzyme complementation and luminescence detection and another using Förster resonance energy transfer and fluorescence detection. We focus on applying these approaches to investigate cAMP signaling by catecholamine receptors and then on combining these analytical approaches with manipulations of receptor endocytic trafficking.

1. INTRODUCTION

Much of what is presently known about cellular G protein-coupled receptor (GPCR) signaling has been inferred from analysis of semi-intact cells, cell extracts, or isolated membrane fractions using biochemical methods. Such approaches have provided extremely powerful mechanistic insight and have led to the appreciation of a complex set of regulatory processes that affect GPCR signaling activity over a wide temporal range. Endocytic membrane trafficking processes have been recognized for many years to impact GPCR-mediated signaling responsiveness after prolonged or repeated exposure to agonist ligands and drugs, typically over a period of hours or more. With increasing interest in more rapid regulatory effects, and toward elucidating the subcellular localization and dynamics of particular protein interactions mediating GPCR function and regulation, there is a need to investigate signaling processes in intact cells and with spatiotemporal resolution exceeding that typically available using conventional biochemical assays.

Considerable recent progress has been made in engineering genetically encoded biosensors to specifically detect a wide variety of metabolites and signaling mediators in unperturbed, or minimally perturbed, cells and tissues. A number of useful biosensors of cytoplasmic cyclic AMP (cAMP) are now available, most based on linking specific cAMP-binding domains to conformation-dependent readouts such as enzyme complementation (Fan et al., 2008) or Förster resonance energy transfer (FRET) (Lohse, Nuber, & Hoffmann, 2012; Zhou, Herbst-Robinson, & Zhang, 2012). This chapter will discuss application of one example of each approach to investigate regulation of cytoplasmic cAMP by catecholamine receptors in cultured cells. We then discuss experimental manipulations of the endocytic pathway and of specific GPCR engagement with the endocytic pathway, which can be combined with optical biosensor technology to investigate the impact of receptor endocytic trafficking on the cellular cAMP response. Studies using this combination of experimental approaches have revealed a previously unanticipated role of endocytic membranes in supporting canonical GPCR–Gs–adenylyl cyclase signaling and suggest that endocytosis may significantly impact acute and longer-term G protein-linked cellular signaling responses (Calebiro et al., 2009; Feinstein et al., 2011; Ferrandon et al., 2009; Kotowski, Hopf, Seif, Bonci, & von Zastrow, 2011; Mullershausen et al., 2009; Werthmann, Volpe, Lohse, & Calebiro, 2012; Irannejad et al., 2013).

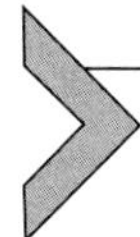

2. LUMINESCENCE-BASED ASSAY OF ACUTE cAMP REGULATION IN CELL POPULATIONS

Intramolecular enzyme complementation has emerged as a powerful approach for detecting many signaling mediators and metabolites in intact cells. Split luciferase linked to various AMP-binding domains can provide a convenient way to detect increases in cytoplasmic cAMP concentration in a cell population (Fan et al., 2008). Typically, these sensors are engineered so that binding of cAMP stabilizes a conformational change that complements the active site and results in increased luciferase activity. A problem with early versions of such biosensors was poor reversibility, limiting temporal resolution and thus obscuring the regulation of acute signaling effects. We have had good experience with a commercially marketed sensor (pGloSensor-20F, Clontech), based on cyclic-permuted split luciferase fused to a cAMP-binding domain modified from the RIIβB regulatory domain of protein kinase A (Fan et al., 2008). We have found that the cAMP-stimulated luciferase activity produced by this sensor can be reversed within tens of seconds in intact cells (J. Tomshine, unpublished). We describe a relatively economical system, taking advantage of an existing electron-multiplying charged coupled device (EMCCD) camera that is used otherwise in the laboratory for fluorescence microscopy, to detect luminescence changes representing regulation of cytoplasmic cAMP concentration in cell populations aliquoted into multiwell plates. A conventional photomultiplier tube (PMT)-based plate luminometer could be used as well, provided that a unit with sufficient sensitivity and acquisition speed is available. In our experience, the EMCCD approach is advantageous for detecting changes in cytoplasmic cAMP accumulation simultaneously in multiple samples and with temporal resolution sufficient to reliably report increases in cytoplasmic cAMP occurring on the order of seconds.

2.1. Materials

2.1.1 Cell culture

1. HEK293 cell passage 20–50 (ATCC:CRL-1573)
2. 10 cm and 24-well tissue culture plates (Costar)
3. Dulbecco's modified Eagle's medium with high glucose (DMEM) supplemented with 10% fetal bovine serum (Sigma)
4. DMEM phenol red-free Imaging media supplemented with 30 m*M* HEPES (Invitrogen)
5. Lipofectamine 2000 (Invitrogen)
6. Opti-MEM (Invitrogen)
7. Poly-D-Lysine (Sigma P0899)
8. (−)-Isoproterenol (Sigma or RBI)
9. pGloSensor-20F (Promega)
10. Luciferin (Biogold)
11. Forskolin (Sigma-Aldrich).

2.1.2 Imaging equipment and settings

1. Electron multiplying CCD sensor (Hamamatsu C9100-13) fitted with 8.5 mm f/1.3 portrait lens (Edmund Scientific)
2. Light-proof gel documentation cabinet with internal heater to achieve 37 °C internal temperature (we use a simple resistive heater and thermostatic controller purchased from Omega Scientific, mounted well outside of the camera's field of view to reduce glare in the luminometry image due to infrared emission that is generated by the heater and can be detected by the camera)
3. PC running image acquisition software interfaced to camera (we use Micromanager, micro-manager.org, running as a plug-in to ImageJ, rsb.info.nih.gov/ij/).

2.2. Methods

2.2.1 Cell preparation

1. Plate HEK293 cells on 10 cm dishes.
2. Transfect pGloSensor-20F (Fig. 23.1A) using Lipofectamine 2000, following manufacturer's protocol, 24 h prior to luminescence assay.
3. On the day of luminescence assay, dissolve poly-D-lysine in sterile water (50 mg/ml) and place 1 ml in each 24-well culture dish for 15 min at room temperature. Wash away poly-D-lysine with sterile water (three washes) and dry the culture dishes.

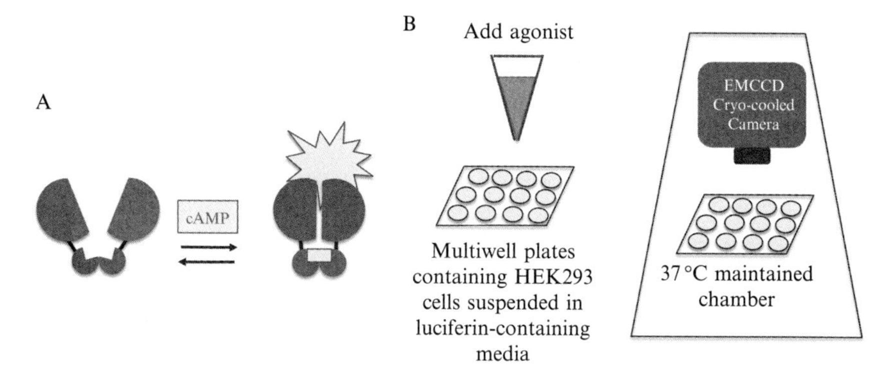

Figure 23.1 Schematic summarizing luminescence-based method used to measure cAMP accumulation in intact cells. (A) Increased cytoplasmic cAMP (cAMP) concentration is detected by increased enzyme (luciferase) activity of the genetically encoded biosensor that is caused by intramolecular protein complementation. (B) Multiwell plates containing HEK293 cells suspended in luciferin-containing media are placed on a temperature-controlled cabinet (37 °C) after agonist application and images are taken every 10 s using EMCCD camera (Hamamatsu C9100-13). (For color version of this figure, the reader is referred to the online version of this chapter.)

4. Plate ~200,000 cells/well in 500 μl DMEM onto the freshly coated wells and let them seed for 5 h.
5. Equilibrate cells for 1 h in the presence of 250 μg/ml luciferin (Biogold) in 250 μl DMEM without phenol red and no serum supplemented with 30 m*M* HEPES (Invitrogen) (Fig. 23.1B).

2.2.2 Imaging

1. Turn on the camera and the temperature-controlled light-proof cabinet (37 °C) 30 min before data acquisition.
2. Place the plates containing cells from Section 2.2.1 (Step 5) and focus the plate on the EMCCD sensor, with EM gain set to zero, using room light and video rate readout.
3. Close the light-proof cabinet, increase EM gain to 100, and record basal frames by sequential 10 s exposure. Close camera shutter and open cabinet to add agonist as desired to particular wells (e.g., varying concentrations of isoproterenol) diluted in 250 μl DMEM without phenol red and no serum supplemented with 30 m*M* HEPES (Invitrogen).
4. In each multiwell plate, measure a reference value of luminescence in the absence of any ligand addition (basal) and in the presence of 5 μ*M* forskolin (forskolin-stimulated signal). In our HEK293 cells, this stimulates a moderate amount of receptor-independent activation of

adenylyl cyclase that is equivalent (in peak luminescence intensity) to approximately two-thirds of the signal elicited by full activation (i.e., in the presence of a saturating isoproterenol concentration) mediated by the endogenous complement of β2-adrenergic receptors (β2ARs) present in these cells.

5. Close the light-proof chamber and continue to collect sequential luminescence images every 10 s. Acquisition times and EM gain can be adjusted as needed for the experiment and to empirically optimize signal to noise of the data. With our camera and for 10 s acquisition, we typically use an EM gain of 100.
6. Using ImageJ, determine mean luminescence values as a function of time for each well (Fig. 23.1B).

2.2.3 Analysis

1. Calculate integrated luminescence intensity detected from each well after background subtraction. Our camera-based system also requires correction of intensity values for vignetting, an optical artifact caused by the limited aperture afforded by the (relatively inexpensive) lens used to focus the image on the EMCCD sensor. This is determined based on the position of each well relative to the center of field, computed using the imaging toolbox of Matlab (MathWorks). A better solution would be to use a wider-aperture lens or a sufficiently sensitive plate-based luminometer, if available, for which vignetting is either less of an issue or not an issue at all.
2. Normalize the average luminescence value measured across duplicate wells to the forskolin-stimulated value measured on that plate (Fig. 23.2).
3. We typically use the forskolin-normalized luminescence signal as the cAMP readout directly and find that this value is reliable across experiments. In principle, this arbitrary readout scale could be calibrated to an absolute cAMP concentration, but we have not found a practical way to accomplish this without introducing significant additional experimental error.

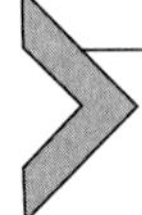

3. FRET IMAGING OF ACUTE cAMP REGULATION IN INDIVIDUAL CELLS

cAMP biosensors are also available that couple a conformational change stabilized by AMP binding to a change in FRET between linked

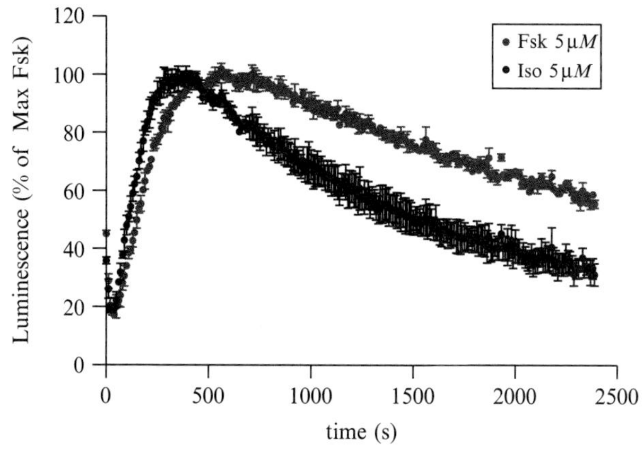

Figure 23.2 Kinetics of the cAMP response detected using the luminescence assay. Representative example of luminescence imaging from triplicate wells after application of 5 μ*M* isoproterenol (blue) or forskolin (red) at $t = 0$. Data shown are normalized to the percentage of maximum forskolin response. (For interpretation of the references to color in this figure legend, the reader is referred to the online version of this chapter.)

protein labels. Here, we describe application of a FRET-based cAMP sensor, Epac1-cAMPs, which is based on a cAMP-binding domain derived from Epac1 fused to spectrally shifted green fluorescent protein variants (Calebiro et al., 2009). cAMP binding to this biosensor results in a decreased FRET signal, and, in our hands, this approach offers slightly better temporal resolution than the luminometry-based biosensor described in the preceding text. Epac1-cAMPs also allow single-cell analysis of cAMP changes using a conventional epifluorescence microscope. We describe application of this FRET-based cAMP biosensor for single-cell analysis of FRET using a basic sensitized emission fluorescence microscopy setup.

3.1. Materials

3.1.1 Cell culture

1. HEK293 cell passage 20–50 (ATCC:CRL-1573)
2. 10 cm conventional tissue culture dishes (Costar) and 3.5 cm glass bottom dishes (MatTek)
3. DMEM supplemented with 10% fetal bovine serum (Sigma)
4. DMEM phenol red-free imaging media supplemented with 30 m*M* HEPES (Invitrogen)
5. Effectene (Qiagen)

6. Poly-D-Lysine (Sigma P0899)
7. Plasmid encoding Epac1-cAMPs, a FRET-based cAMP sensor based on the Epac1-binding domain cloned into pcDNA3 (Calebiro et al., 2009)
8. Plasmid encoding the donor fluorophore (pECFP, Clontech)
9. Plasmid encoding the acceptor fluorophore (pEYFP, Clontech).

3.1.2 Imaging equipment

1. Inverted epifluorescence microscope (we use a Nikon TE2000) fitted with 20 × NA0.4 objective, electronically shuttered mercury arc lamp, JP4 PC dichroic mirror (Chroma 104947), and motorized excitation and emission filter wheels (Lambda 10 system, Sutter Instruments) containing the following dichroic bandpass filters: excitation filter wheel—S436/10 (CFP exciter, Chroma 51232) and S500/20 (YFP exciter, Chroma 51724) and emission filter wheel—S470/30 (CFP emitter, Chroma 52479) and S535/30 (YFP emitter, Chroma 51706)
2. Temperature-/CO_2-controlled imaging chamber (we use a homemade unit but there are various commercial vendors)
3. Cooled CCD camera (Roper CoolSnap or equivalent)
4. PC running image acquisition software (e.g., Micromanager) interfaced to the Lambda 10 controller and camera.

3.2. Methods

3.2.1 Cell preparation

1. Plate HEK293 cells on 10 cm dishes.
2. Transfect Epac1-cAMPs plasmid using Effectene, following manufacturer's protocol, 48 h prior to assay. Also transfect cells singly with pECFP and pEYFP for direct excitation and bleedthrough controls (see in the succeeding text). 24 hours before assay, plate transfected cells onto MatTek dishes (10^4–10^5/cm^2).
3. On the day of experiment, transfer cells to phenol red-free imaging medium.

3.2.2 Imaging

1. Turn on microscope and temperature-/CO_2-controlled environment to achieve 37 °C and 5% CO_2 prior to imaging.
2. Place a 3.5 cm dish of Epac1-cAMPs transfected HEK293 cells on the stage, focus on a field with transfected cells and empty space.

3. Collect sequential images at the desired frequency for the desired duration. Acquire the following three images at every time point:
   a. FRET image (for determination of I_{FRET}): CFP exciter, YFP emitter.
   b. CFP image (for determination of I_{CFP}): CFP exciter, CFP emitter.
   c. YFP image (for determination of I_{YFP}): YFP exciter, YFP emitter.
   Note that optimal camera and arc lamp intensity settings will vary depending on equipment. Images must be collected in the linear range of camera detection and at the lowest practical illumination intensity and exposure settings to minimize photobleaching and toxicity.
4. Agonist can be added to the dish at any point in the time-lapse sequence, though acquisition of at least two baseline (untreated) frames is recommended.
5. Steps 2–4 can be repeated for multiple dishes of Epac1-cAMPs transfected HEK293 cells on a given day of imaging.
6. Place a 3.5 cm dish of pECFP (only) transfected HEK293 cells on the stage; focus on a field with transfected cells and empty space.
7. Collect one set of the following images to determine a factor to correct for bleedthrough of donor (CFP) fluorescence emission into the YFP emission bandpass (BT_{DONOR}):
   a. $I_{FRET(CFP\ ONLY)}$: CFP exciter, YFP emitter.
   b. $I_{CFP(CFP\ ONLY)}$: CFP exciter, CFP emitter.
   Intensity and acquisition settings should be the same as those used in the preceding text.
8. Place a 3.5 cm dish of pEYFP (only) transfected HEK293 cells on the stage; focus on a field with transfected cells and empty space.
9. Collect one set of the following images to determine a factor to correct for direct excitation of acceptor by the CFP excitation ($DE_{ACCEPTOR}$):
   a. $I_{FRET(YFP\ ONLY)}$: CFP exciter, YFP emitter.
   b. $I_{YFP(YFP\ ONLY)}$: YFP exciter, YFP emitter.
   Intensity and acquisition settings should be the same as those used in the preceding text.

3.2.3 Analysis

1. Using ImageJ, create a stack of I_{FRET} images for a given time-lapse series.
2. Draw an ROI around each cell and measure integrated intensity at each time point to obtain a numerical value for I_{FRET}.

3. Move the ROI to an area on the image stack where no cells are present, and measure integrated intensity at each time point to obtain a numerical value for background (BG_{FRET}).
4. Repeat this for all cells within the field.
5. Repeat Steps 1–5 with I_{CFP} and I_{YFP} image series to obtain numerical values for I_{CFP}, BG_{CFP}, I_{YFP}, and BG_{YFP}.
6. Open the $I_{FRET(CFP\ ONLY)}$ and $I_{CFP(CFP\ ONLY)}$ images. Create a stack.
7. Draw an ROI around a cell and determine BT_{DONOR} constant values by measuring the ratio of integrated FRET intensity to integrated CFP intensity in cells expressing only CFP ($BT_{DONOR} = I_{FRET(CFP\ ONLY)}/I_{CFP(CFP\ ONLY)}$).
8. Open the $I_{FRET(YFP\ ONLY)}$ and $I_{YFP(YFP\ ONLY)}$ images. Create a stack.
9. Draw an ROI around a cell and determine $DE_{ACCEPTOR}$ constant values by measuring the ratio of integrated FRET intensity to integrated YFP intensity in cells expressing only YFP ($DE_{ACCEPTOR} = I_{FRET(YFP\ ONLY)}/I_{YFP(YFP\ ONLY)}$).
10. Obtain a normalized FRET value for each cell at each time point using the following equation (Fig. 23.3):

$$n\text{FRET} = [(I_{FRET} - BG_{FRET}) - (I_{CFP} - BG_{CFP})BT_{DONOR} - (I_{YFP} - BG_{YFP})DE_{ACCEPTOR})]/I_{CFP}$$

4. EXPERIMENTAL MANIPULATION OF GPCR ENDOCYTIC TRAFFICKING

The biosensors described in the preceding text have many potential applications and are not specific to examining the effects of endocytosis. However, these biosensors are compatible with various experimental manipulations of GPCR endocytic trafficking and thus facilitate studies of the signaling consequences of GPCR endocytic trafficking. Some experimental manipulations that we have found useful are briefly described in the succeeding text.

4.1. Receptor mutation

For some GPCRs, such as the β2AR, we know enough about structural determinants engaging the endocytic machinery that it is possible to manipulate receptor trafficking properties selectively by mutation and without affecting other cargoes. One concern with receptor mutation is the effect

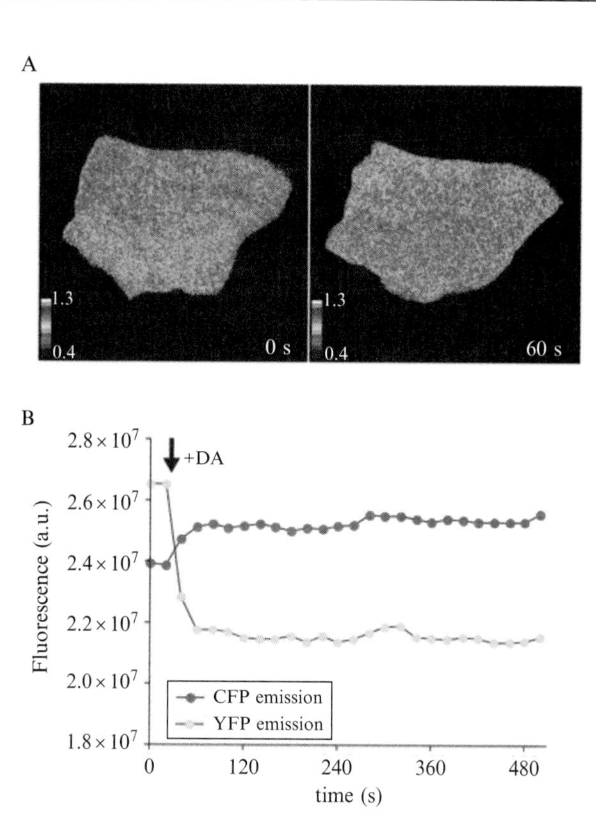

Figure 23.3 FRET-based detection of D1R-mediated cAMP accumulation. (A) Pixel-by-pixel calculation of *n*FRET in a representative HEK293 cells expressing Flag-tagged D1R before (left) and 60 s after (right) application of 10 μ*M* dopamine. (B) Time course of integrated whole-cell CFP and YFP fluorescence intensity changes on which the *n*FRET calculation is based. (For color version of this figure, the reader is referred to the online version of this chapter.)

of specific mutations on other (nonendocytic) receptor functions. A second concern with the use of receptor mutation is that it requires study of recombinant rather than endogenous receptors, with associated potential complications of overexpression. We usually address this by generating stably transfected cell clones selected for a defined level of receptor expression (we typically strive for 100–300 pmol/mg cell protein, determined by saturating radioligand-binding assay) and taking care to compare mutations in multiple cell clones all selected for comparable expression. A third concern with the use of recombinant receptors is resolving the signaling consequences of their activation from the effects of endogenous receptors present in the same cells. Our HEK293 express endogenous β2ARs but not D1 dopaminergic receptors (D1Rs). Therefore, one approach is to focus on GPCRs such as the D1R in HEK293 cells, where the recombinant receptor can be examined unambiguously using D1R-selective agonists. Another approach is to use a knockdown/replacement strategy. Based on our experience using HEK293 cell clones expressing recombinant β2ARs at >100 pmol/mg, we find that recombinant receptors become the dominant determinant of cellular cAMP accumulation. This makes it feasible to assess mutational effects by simply neglecting the endogenous pool, as a first approximation, but caution is advised and this approach is probably not sufficient for detailed examination of mutational effects. In this case, one should search for cells that do not endogenously express the receptor of interest, or devise a sufficient knockout or knockdown strategy to deplete the endogenous receptor.

4.1.1 Inhibiting β2AR endocytosis

Regulated endocytosis of β2ARs can be inhibited by mutating a small cluster of phosphorylatable residues in the receptor's proximal cytoplasmic tail (Hausdorff et al., 1991). This strongly inhibits β2AR internalization but does so by reducing receptor recruitment of β-arrestins. Accordingly, signaling effects of such mutation are not necessarily specific to endocytosis and could represent other effects such as inhibited desensitization and/or arrestin-linked signaling.

4.1.2 Manipulating β2AR trafficking after endocytosis

β2AR membrane trafficking after endocytosis can also be manipulated by mutation. Lysine mutation of the β2AR, presumably by preventing receptor ubiquitination and association with ubiquitin-dependent sorting proteins on the endosome membrane (Henne, Buchkovich, & Emr, 2011), has been reported to impair long-term downregulation of receptors (Shenoy, McDonald, Kohout, & Lefkowitz, 2001). A C-terminal PDZ motif present in the distal cytoplasmic tail of the β2AR is required for efficient recycling of receptors after endocytosis and, when mutated, internalized receptors are effectively rerouted for lysosomal downregulation (Cao, Deacon, Reczek, Bretscher, & von Zastrow, 1999). There are also caveats of these manipulations. For example, lysine mutation may produce pleiotropic effects or affect other ubiquitin-dependent regulatory processes (Shenoy, 2007), and there is evidence that the β2AR PDZ motif functions in other receptor signaling and regulatory processes distinct from endocytic sorting (Romero, von Zastrow, & Friedman, 2011).

4.2. Genetic manipulation of the endocytic pathway

Manipulations of the endocytic machinery allow GPCR trafficking to be altered without requiring structural modification of the receptor itself. This has advantages with respect to avoiding the various potential complications of receptor mutation but has potential disadvantages because the trafficking of many other cellular proteins is also affected.

4.2.1 Inhibiting clathrin-dependent endocytosis

Many GPCRs undergo regulated endocytosis primarily via clathrin-coated pits. The GTP-binding protein dynamin is an essential component of this machinery in animal cells, and mutations of dynamin isoforms have been described that inhibit clathrin-dependent endocytosis. For example, K44A mutant dynamin 1 produces a dominant negative effect by blocking the scission of clathrin-coated pits from the plasma membrane (van der Bliek et al., 1993). K44A mutant dynamin has been used successfully to block regulated endocytosis of a number of GPCRs including the β2AR (Zhang, Ferguson, Barak, Menard, & Caron, 1996). Depletion of cellular clathrin heavy chain by RNA interference is an alternative strategy that blocks endocytosis by disrupting formation of clathrin-coated pits (Motley, Bright, Seaman, & Robinson, 2003). For example, clathrin knockdown strongly inhibits regulated endocytosis of the D1 dopamine receptor (Kotowski et al., 2011) and the β2AR (J. Tomshine, unpublished).

4.2.2 Inhibiting recycling

Some GPCRs engage specific recycling machinery that can be targeted by RNA interference. The β2AR, for example, undergoes rapid recycling by PDZ motif-directed sorting on the endosome membrane by a mechanism that requires sorting nexin 27 (SNX27) and the retromer complex. Depleting either SNX27 or the retromer component VPS35 inhibits β2AR recycling after regulated endocytosis and effectively reroutes internalized receptors to lysosomes (Lauffer et al., 2010; Temkin et al., 2011). β1-Adrenergic receptors are also sensitive to these manipulations, but it is not yet clear how many other GPCRs are sorted by this machinery and some (such as the D1R) are not. Therefore, manipulations of the recycling mechanism by knockdown of such specific trans-acting components need to be validated on a case-by-case basis.

4.3. Acute chemical inhibition of endocytosis

A significant concern of using genetic manipulations of endocytosis is the relatively prolonged time course (typically several days) over which they are imposed. This increases the probability of secondary effects impacting cellular signaling responses and thus adds caveats to interpretation. It is possible to circumvent this using temperature-sensitive alleles to produce more rapid induction of genetic inhibition (Damke, Baba, Warnock, & Schmid, 1994). However, a number of chemical inhibitors of endocytosis are now available and are being used more commonly. Several small molecule inhibitors of clathrin-mediated endocytosis have been described (Hill et al., 2009; Macia et al., 2006; von Kleist et al., 2011). We presently use Dyngo-4a, a chemical inhibitor of dynamin, to impose acute endocytic inhibition. This compound is commercially available (Abcam Biochemicals). It is dissolved in dry DMSO at 30 m*M* (aliquots can be stored under dry nitrogen or argon at −80 °C) and delivered to cells by 1:1000 dilution from this DMSO stock into the culture medium. Cells should be washed and equilibrated in serum-free medium prior to application of Dyngo-4a because this drug binds avidly to serum proteins. Typically, we find that 15 min preincubation with 30 μ*M* Dyngo-4a is sufficient to achieve strong endocytic inhibition, and we identify effects by comparison to exposing cells to 0.1% DMSO under the same conditions (vehicle control). Potential disadvantages of chemical endocytic inhibition include off-target effects, which are still poorly understood and may vary across cell types or experimental conditions (Irannejad et al., 2013).

5. SUMMARY

The availability of improved cAMP biosensors, combined with advances in manipulating the endocytic machinery and specific GPCR engagement with this machinery, is revealing previously unexplored spatial and temporal features of GPCR signaling as they occur in intact, living cells. One intriguing hypothesis that has emerged from these studies is that endosomes may represent a membrane surface from which GPCRs are able to elicit classical G protein-mediated signaling. The strategy of perturbing GPCR-trafficking processes and then observing associated effects on cellular cAMP accumulation, as described in this chapter, suggests the occurrence of such signaling from endosomes but remains an indirect approach. A logical next step is to develop biosensors that can directly detect the operation of signaling machineries in intact cells and allow precise subcellular resolution of their location.

ACKNOWLEDGMENTS

We thank present and former members of the laboratory for contributions, Drs. Martin Lohse (University of Würzburg), Jin Zhang (Johns Hopkins University), and Phil Robinson (University of Queensland) for generously providing advice and reagents. We also thank Dr. Kurt Thorn of the UCSF Nikon Imaging Center for critical advice and assistance in experimental design and interpretation. R. I. is supported by a postdoctoral fellowship from the American Heart Association. S. J. K. received support from a National Science Foundation predoctoral fellowship. Research in the authors' laboratory is supported by the National Institutes of Health.

REFERENCES

Calebiro, D., Nikolaev, V. O., Gagliani, M. C., de Filippis, T., Dees, C., Tacchetti, C., et al. (2009). Persistent cAMP-signals triggered by internalized G-protein-coupled receptors. *PLoS Biology*, *7*, e1000172.

Cao, T. T., Deacon, H. W., Reczek, D., Bretscher, A., & von Zastrow, M. (1999). A kinase-regulated PDZ-domain interaction controls endocytic sorting of the beta2-adrenergic receptor. *Nature*, *401*, 286–290.

Damke, H., Baba, T., Warnock, D. E., & Schmid, S. L. (1994). Induction of mutant dynamin specifically blocks endocytic coated vesicle formation. *Journal of Cell Biology*, *127*, 915–934.

Fan, F., Binkowski, B. F., Butler, B. L., Stecha, P. F., Lewis, M. K., & Wood, K. V. (2008). Novel genetically encoded biosensors using firefly luciferase. *ACS Chemical Biology*, *3*, 346–351.

Feinstein, T. N., Wehbi, V. L., Ardura, J. A., Wheeler, D. S., Ferrandon, S., Gardella, T. J., et al. (2011). Retromer terminates the generation of cAMP by internalized PTH receptors. *Nature Chemical Biology*, *7*, 278–284.

Ferrandon, S., Feinstein, T. N., Castro, M., Wang, B., Bouley, R., Potts, J. T., et al. (2009). Sustained cyclic AMP production by parathyroid hormone receptor endocytosis. *Nature Chemical Biology*, *5*, 734–742.

Hausdorff, W. P., Campbell, P. T., Ostrowski, J., Yu, S. S., Caron, M. G., & Lefkowitz, R. J. (1991). A small region of the beta-adrenergic receptor is selectively involved in its rapid regulation. *Proceedings of the National Academy of Sciences of the United States of America*, *88*, 2979–2983.

Henne, W. M., Buchkovich, N. J., & Emr, S. D. (2011). The ESCRT pathway. *Developmental Cell*, *21*, 77–91.

Hill, T. A., Gordon, C. P., McGeachie, A. B., Venn-Brown, B., Odell, L. R., Chau, N., et al. (2009). Inhibition of dynamin mediated endocytosis by the dynoles—Synthesis and functional activity of a family of indoles. *Journal of Medicinal Chemistry*, *52*, 3762–3773.

Irannejad, R., Tomshine, J. C., Tomshine, J. R., Chevalier, M., Mahoney, J. P., Steyaert, J., et al. (2013). Conformational biosensors reveal GPCR signalling from endosomes. *Nature*, *495*, 534–538.

Kotowski, S. J., Hopf, F. W., Seif, T., Bonci, A., & von Zastrow, M. (2011). Endocytosis promotes rapid dopaminergic signaling. *Neuron*, *71*, 278–290.

Lauffer, B. E., Melero, C., Temkin, P., Lei, C., Hong, W., Kortemme, T., et al. (2010). SNX27 mediates PDZ-directed sorting from endosomes to the plasma membrane. *Journal of Cell Biology*, *190*, 565–574.

Lohse, M. J., Nuber, S., & Hoffmann, C. (2012). Fluorescence/bioluminescence resonance energy transfer techniques to study G-protein-coupled receptor activation and signaling. *Pharmacological Reviews*, *64*, 299–336.

Macia, E., Ehrlich, M., Massol, R., Boucrot, E., Brunner, C., & Kirchhausen, T. (2006). Dynasore, a cell-permeable inhibitor of dynamin. *Developmental Cell*, *10*, 839–850.

Motley, A., Bright, N. A., Seaman, M. N., & Robinson, M. S. (2003). Clathrin-mediated endocytosis in AP-2-depleted cells. *Journal of Cell Biology*, *162*, 909–918.

Mullershausen, F., Zecri, F., Cetin, C., Billich, A., Guerini, D., & Seuwen, K. (2009). Persistent signaling induced by FTY720-phosphate is mediated by internalized S1P1 receptors. *Nature Chemical Biology*, *5*, 428–434.

Romero, G., von Zastrow, M., & Friedman, P. A. (2011). Role of PDZ proteins in regulating trafficking, signaling, and function of GPCRs: Means, motif, and opportunity. *Advances in Pharmacology*, *62*, 279–314.

Shenoy, S. K. (2007). Seven-transmembrane receptors and ubiquitination. *Circulation Research*, *100*, 1142–1154.

Shenoy, S. K., McDonald, P. H., Kohout, T. A., & Lefkowitz, R. J. (2001). Regulation of receptor fate by ubiquitination of activated beta 2-adrenergic receptor and beta-arrestin. *Science*, *294*, 1307–1313.

Temkin, P., Lauffer, B., Jager, S., Cimermancic, P., Krogan, N. J., & von Zastrow, M. (2011). SNX27 mediates retromer tubule entry and endosome-to-plasma membrane trafficking of signalling receptors. *Nature Cell Biology*, *13*, 715–721.

van der Bliek, A. M., Redelmeier, T. E., Damke, H., Tisdale, E. J., Meyerowitz, E. M., & Schmid, S. L. (1993). Mutations in human dynamin block an intermediate stage in coated vesicle formation. *Journal of Cell Biology*, *122*, 553–563.

von Kleist, L., Stahlschmidt, W., Bulut, H., Gromova, K., Puchkov, D., Robertson, M. J., et al. (2011). Role of the clathrin terminal domain in regulating coated pit dynamics revealed by small molecule inhibition. *Cell*, *146*, 471–484.

Werthmann, R. C., Volpe, S., Lohse, M. J., & Calebiro, D. (2012). Persistent cAMP signaling by internalized TSH receptors occurs in thyroid but not in HEK293 cells. *FASEB Journal*, *26*, 2043–2048.

Zhang, J., Ferguson, S. S., Barak, L. S., Menard, L., & Caron, M. G. (1996). Dynamin and beta-arrestin reveal distinct mechanisms for G protein-coupled receptor internalization. *Journal of Biological Chemistry*, *271*, 18302–18305.

Zhou, X., Herbst-Robinson, K. J., & Zhang, J. (2012). Visualizing dynamic activities of signaling enzymes using genetically encodable FRET-based biosensors from designs to applications. *Methods in Enzymology*, *504*, 317–340.

CHAPTER TWENTY-FOUR

MICAL-Like1 in Endosomal Signaling

Ahmed Zahraoui[1]
Phagocytosis and Bacterial Invasion Laboratory, INSERM U.1016-CNRS UMR8104, Institut Cochin, Université Paris Descartes, Paris, France
[1]Corresponding author: e-mail address: ahmed.zahraoui@inserm.fr

Contents

Abstract

Small GTPase Rabs are required for membrane protein sorting/delivery to precise membrane domains. Rab13 regulates tight junction assembly and polarized membrane transport in epithelial cells. Using yeast two-hybrid screen, we identified MICAL-like1 (MICAL-L1), a protein that interacts with GTP-bound Rab13 and shares a similar domain organization with MICAL protein family. MICAL-L1 has a calponin homology, Lin11, Isl-1 & Mec-3 (LIM), proline-rich, and coiled-coil domains. It is associated with late and recycling endosomes. Time-lapse video microscopy shows that GFP–Rab7 and

Methods in Enzymology, Volume 535
ISSN 0076-6879
http://dx.doi.org/10.1016/B978-0-12-397925-4.00024-9

cherry–MICAL-L1 are present within vesicles that move rapidly in the cytoplasm. Depletion of MICAL-L1 by short hairpin RNA does not alter the distribution of tight junction proteins, but affects the trafficking of epidermal growth factor receptor (EGFR). Overexpression of MICAL-L1 leads to the accumulation of EGFR in late endosomal compartments. In contrast, knocking down MICAL-L1 results in the distribution of internalized EGFR in vesicles spread throughout the cytoplasm and promotes its degradation. Our data show that MICAL-L1 inhibits EGFR degradation, suggesting that MICAL-L1 is involved in sorting/targeting the receptor to the recycling pathway. They provide novel insights into MICAL-L1/Rab protein complex that can regulate EGFR trafficking/signaling.

ABBREVIATIONS

CH calponin homology
EGFR epidermal growth factor receptor
GST glutathione *S*-transferase
MDCK Madin–Darby canine kidney
PRD proline-rich domain
WT wild type

1. INTRODUCTION

Endocytosis is a process whereby cells internalize membrane proteins such as receptors and solutes, from the extracellular space by engulfing them within plasma membrane vesicles. It becomes clear that endocytosis regulates cell signaling by controlling receptor trafficking. The endocytosis of many signaling receptors is stimulated by ligand-induced activation. Activated receptor such as epidermal growth factor receptor (EGFR) undergoes rapid endocytosis through clathrin-coated pits (Sigismund et al., 2008; Sorkin & von Zastrow, 2009). Signaling is regulated not only at the level of the plasma membrane but also at the level of the endocytic compartments. The balance between endocytic uptake and recycling controls the composition of the plasma membrane and contributes to diverse cellular processes, such as nutrient uptake, cell adhesion, cell migration, cell polarity, and signal transduction (Hsu, Bai, & Li, 2012).

In eukaryotic cells, endocytosis involves the capture of transmembrane proteins and their ligands into cytoplasmic membrane bound vesicles. When coated with clathrin, these pits invaginate inward with the help of several accessory proteins and pinch off to form a clathrin-coated vesicle in a process that requires the GTPase dynamin. Many clathrin-independent pathways of endocytosis also exist. Membrane protein traffic between endosomes is controlled by small GTPase Rab proteins. Upon their activation, each GTP-bound Rab protein associates with a particular subcellular membrane domain and functions by coordinating the recruitment of a set of specific effector proteins. Endocytic vesicles derived from clathrin-dependent and clathrin-independent pathways lose their coat and fuse with a Rab5 early endosomes (EE). Receptors can rapidly recycle back to the plasma membrane by a Rab4- and Rab35-dependent mechanism. EE maturation/transition to multivesicular bodies (MVBs)/late endosomes (LE) requires Rab7. Cargo vesicles traffic to the Rab11-positive recycling endosomes (RE) or remain in MVBs/LE. Fusion of LE/MVBs with lysosomes carrying proteases results in protein degradation (Fig. 24.1).

Compelling evidence shows that endocytic membrane trafficking and signaling are strictly interconnected and that not only early/recycling but also LE/MVBs can act as sorting/signaling platforms (Abou-Zeid et al., 2011;

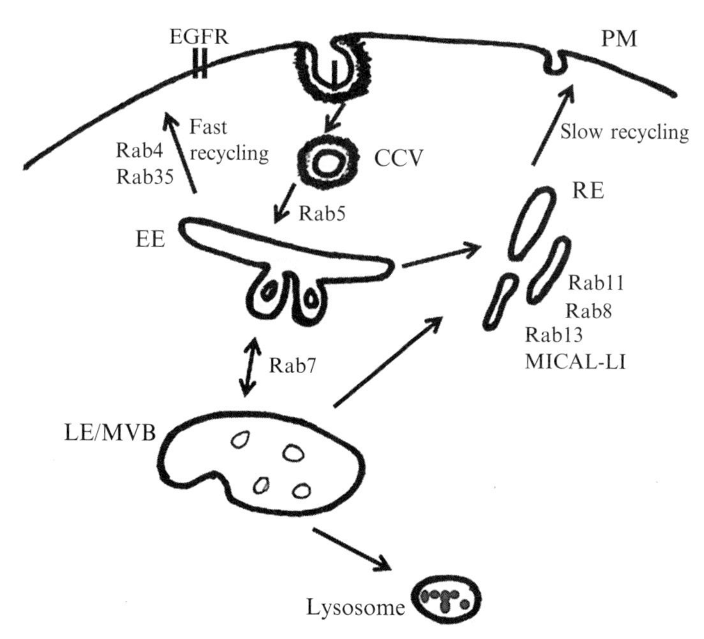

Figure 24.1 Scheme of endocytic traffic from plasma membrane (PM) to endosomes. Early endosomes (EE) receive proteins and lipids from clathrin-coated vesicles (CCV) and deliver them either to recycling endosomes (RE) or to late endosomes/multivesicular bodies (LE/MVBs) and lysosomes. Similarly to EE, LE/MVBs may constitute a sorting station for delivery of proteins either to lysosomes for degradation or to the RE. MICAL-L1 and several small GTPases Rab are indicated.

Dobrowolski & De Robertis, 2012; Le Roux et al., 2012; Seto, Bellen, & Lloyd, 2002). The imaging techniques established that endosomes are a mosaic of different membrane subdomains. Rab proteins, through the recruitment of specific effectors, help organize membrane subdomains required for cargo progression along the endocytic/recycling pathways (Gruenberg, 2001; Miaczynska & Zerial, 2002). An essential issue is how membrane trafficking through endosomes is coordinated and how regulatory elements such as small GTPase Rab proteins and their effectors control membrane trafficking/signaling.

Rab proteins, by their virtue to organize membrane microdomains, may establish platforms for the recruitment of molecular machineries required for the coordination of specific membrane transport events including vesicle formation, sorting/targeting, and docking/fusion. They cycle between GDP-bound (guanosine 5′ diphosphate) inactive and GTP-bound (guanosine 5′ triphosphate) active conformations. Under GTP-bound state, Rab proteins promote the recruitment of protein effectors to regulate endocytosis as well as exocytosis (Schwartz, Cao, Pylypenko, Rak, & Wandinger-Ness, 2007; Zahraoui, Louvard, & Galli, 2000; Zerial & McBride, 2001). They are also implicated in signaling to the nucleus. Rab5, a key regulator of endocytosis, controls nucleocytoplasmic shuttling of APPL1, a protein involved in chromatin remodeling and gene expression (Bucci & Chiariello, 2006; Miaczynska et al., 2004; Zhu et al., 2007). Rab27a protein has been implicated in the delivery to the cell surface of exosomes, small vesicles that form by inward budding of the limiting membrane of endocytic compartments, known as multivesicular bodies (Ostrowski et al., 2009; Thery, Ostrowski, & Segura, 2009). Recent evidence from the literature highlights new pathways of recruitment and fusion of LE and lysosomes with the plasma membrane. Lysosomes' fusion with the plasma membrane in fibroblasts plays a role in membrane repair, parasite entry, and tumor cells invasiveness. In some specialized cells, lysosome-related organelles, such as melanosomes, dense granules of platelets, or lytic granules of cytotoxic T cells, fuse with the plasma membrane (Andrews, 2000; Laulagnier et al., 2011; Raposo, Marks, & Cutler, 2007; Reddy, Caler, & Andrews, 2001; Tolmachova, Abrink, Futter, Authi, & Seabra, 2007). Numerous observations link Rab proteins to the actin cytoskeleton (Seabra & Coudrier, 2004). Rab proteins cooperate with cytoskeleton motors to regulate membrane trafficking (Goud, 2002; Zerial & McBride, 2001). The activated Rab27a on melanosomes binds melanophilin, which in turn recruits myosin Va allowing association and docking of melanosomes to cortical actin (Chen, Samaraweera, Sun, Kreibich, & Orlow, 2002; Fukuda, Kuroda, & Mikoshiba, 2002; Tolmachova et al., 2007). Similarly, Rab11 associates with RabFIP2 and Myosin Vb (Fan, Lapierre, Goldenring, Sai, & Richmond, 2004; Horgan, Hanscom, Jolly, Futter, & McCaffrey, 2010), and this platform is involved in sorting from RE and cargo docking/fusion with the plasma membrane through interaction with cortical actin (Gidon et al., 2012; Uzan-Gafsou et al., 2007). The small GTPase Rab8, which is closely related to Rab10 and Rab13 and to the yeast Sec4, is required for the localization of apical proteins in intestinal epithelial cells (Sato et al., 2007). Interestingly, Rab11, Rab8, and Rabin8 (a guanine nucleotide exchange factor for Rab8 and an effector of Rab11) are part of a platform regulating vesicular trafficking during primary ciliogenesis. Defects in primary cilia formation have been implicated in a number of genetic disorders (Knodler et al., 2010). Depletion of Rab8 promotes cell–cell adhesion and actin stress fibers formation (Hattula et al., 2006). Rab13 inhibits the recruitment/recycling of membrane proteins to cell–cell junctions (Marzesco et al., 2002; Morimoto et al., 2005). Moreover, we showed that Rab13 directly binds to the alpha catalytic subunit of protein kinase A (PKA) and reversibly inhibits PKA-dependent phosphorylation of VASP (vasodilator-stimulated phosphoprotein), a key actin cytoskeleton remodeling protein. The inhibition of PKA activity abolishes the targeting of VASP, claudin-1, and ZO-1 to cell–cell junction. These data provide the first direct link between activation of small GTPases and the recruitment of cytoskeleton modulator into cell–cell contact (Kohler, Louvard, & Zahraoui, 2004). Our work clearly implicates Rab13 in the recruitment/recycling of membrane proteins to the plasma membrane. It strongly suggests a regulatory role of Rab–effector complexes in protein cargo sorting from endosomes and their subsequent docking with the plasma membrane. To understand how Rab13 regulates endosomal sorting/signaling, we used two-hybrid screen and identified MICAL-like1 (MICAL-L1), a protein that interacts with Rab13 (Abou-Zeid et al., 2011). MICAL-L1 shares a similar domain organization to MICAL protein family that has been implicated in cortical actin remodeling underneath the plasma membrane (Giridharan, Rohn, Naslavsky, & Caplan, 2012). MICAL-L1 has a calponin homology (CH), LIM, proline-rich, and coiled-coil domains. The CH domain of MICAL-L1 shares high similarity to the CH domains identified in various actin-associated proteins like α-actinin. The presence of a CH domain in MICAL-L1 emphasizes a potential role of MICAL-L1 in actin cytoskeleton. Interestingly, MICAL-L1 distribution is F-actin-dependent (Abou-Zeid et al., 2011). MICAL-L1-PRD region encompasses putative proline-rich, PxxP, sequences. It was reported that the proline-rich region of MICAL interacts with the SH3 domain of CasL, a protein required for β1 integrin-induced signal transduction and actin filament organization (Terman, Mao, Pasterkamp, Yu, & Kolodkin, 2002). Similarly, MICAL-L1 proline-rich motives could mediate interaction with SH3 domains of signaling proteins implicated in the regulation of membrane trafficking.

MICAL-L1 is associated with late/MVBs and RE. It mediates EGFR trafficking and impairs its degradation, probably, by targeting the EGFR cargo to the recycling pathway (Abou-Zeid et al., 2011). Our work on Rab13/MICAL-L1 proteins strengthens the notion that the function of MVBs/LE is not only an intermediate that delivers membrane proteins for lysosomal degradation but also a sorting/signaling platform that may be common to diverse receptor signaling pathways.

2. METHODS

2.1. Yeast two-hybrid screen

The development of the yeast two-hybrid screen revolutionized the way protein interactions could be detected. Yeast two-hybrid screen was designed to enable detection of protein–protein interactions and has been improved to decrease false-positives. It is based on the reconstitution of a functional transcription factor when two proteins of interest interact. This takes place in genetically modified yeast strains, in which the transcription of a reporter gene leads to grow on a selective medium lacking histidine.

Two fusions are generated between each protein of interest, the DNA-binding domain (DBD) and the activation domain (AD) of the yeast transcription factor. The protein fused to the DBD is referred to as the "bait," and the protein fused to the AD as the "prey." Upon interaction between the bait and the prey, the DBD and AD are brought in close proximity leading to the reconstitution of the functional transcription factor. This promotes the transcription of the HIS_3 reporter gene, which allows yeast cells to grow on a selective medium lacking histidine.

2.2. Identification of MICAL-L1

The coding region of Rab13Q67L mutant (GTP-bound active Rab13) is cloned in frame with a C-terminal LexA DBD in the yeast pVJL12 two-hybrid vector. The yeast reporter strain L40 (Matα) is transformed with pVJL12 plasmid encoding DBD-Rab13Q67L by electroporation. Several transformants are checked by immunoblot in order to determine the expression levels of the fusion proteins. The clone with the highest expression level is used as a bait for the two-hybrid screen. A library derived from human placenta is cloned into the yeast pGAD1318 vector followed by transformation into the strain Y187 (Matα). 10×10^9 transformants are conjugated to approximately 100×10^9 L40 cells by 8 h mass mating in liquid culture. The mating is controlled by placing aliquots under a light microscope in order to determine the percentage of diploids. After 8 h, yeast cells are plated on synthetic medium lacking leucine, tryptophan, and histidine. Thirty-five positive clones are isolated. Plasmid DNAs are prepared and inserts sequenced. Plasmids derived from positive clones are again transformed into Y187 cells and another mating assay is performed. Diploids that could grow in selective medium are tested for β-galactosidase activity. Out of the 35 positive clones, 18 contained overlapping fragments of 600–850 bp length. These fragments are sequenced and translated in frame to the N-terminal DBD of LexA used for the two-hybrid screen. Thereby, a partial sequence encoding a putative open reading frame could be detected. This sequence is named Rab13-binding domain (RBD). By sequence comparison, several EST cDNAs could be identified that contained the RBD sequence. The longest EST containing an insert of about 5800 bp length is purchased from the IMAGE consortium (IMAGE clone 2262785) and sequenced. Sequence analysis revealed an open reading frame of 2586 bp encoding a protein of 863 amino acids. The results of the two-hybrid assay indicated that RBD of MICAL-L1 interacts with the GTP-bound Rab13Q67L. MICAL-L1-RBD (coiled-coil) also interacts with Rab7, Rab8, and Rab11 (Table 24.1). Using a pull-down assay, we confirmed that MICAL-L1-RBD interacts with Rabs 7, 8, 11, and 13.

2.3. cDNA constructs and cloning

For overexpression experiments, a 3000 bp fragment encoding the full-length MICAL-L1 is cut out of IMAGE clone 2262785 by a XhoI–XbaI digest and ligated into the corresponding sites of a pEGFP-N2 vector (Clontech, Palo Alto, CA), thereby replacing the cutout green fluorescent protein (GFP). This plasmid now contained the MICAL-L1 open reading frame, 68 bp of the 5′-untranslated region, and 200 bp of the 3′-untranslated region. MICAL-L1 construct encoding residues 1–863 is amplified with Taq polymerase using sense and antisense primers containing BamH1and

Table 24.1 Yeast two-hybrid interaction of MICAL-L1-RBD with several small GTPases; wild type, WT; Q67L(GTP-Rab13), T22N(GDP-Rab13)

Rab protein	Interaction with MICAL-L1-RBD
Rab13 WT	+
Rab13 Q67L	+
Rab13 T22N	−
Rab7 WT	+
Rab8 WT	+
Rab11 WT	+
Rab5 WT	−
Rab6 WT	−
CDC42 WT	−
Arf6 WT	−

MICAL-L1-RBD domain interacts with the WT and the active form (Q67L) of Rab13, but not with the inactive form (T22N) of Rab13. It also interacts with Rab7, Rab8, and Rab11, but not with other small GTPases.

EcoR1 restriction sites. The amplified construct is fused to the C-terminus of the enhanced GFP or to the monomeric cherry fluorescent protein (Ch) and cloned into the pGFP-C3 or cherry vector (Clontech Inc., Palo Alto, CA). All constructions are verified by sequencing. Transient transfections are performed by using the Lipofectamine 2000 reagent according to the manufacturer's instruction (Invitrogen).

2.4. Expression and purification of recombinant GST–RBD

The glutathione *S*-transferase (GST) fusion protein is generated by inserting a BamHI–XhoI fragment containing RBD of MICAL-L1 (amino acid 652–845) into the corresponding sites of pGEX-4T bacterial expression vector. *E. coli* strain BL21 is transformed with GST–RBD construct. A single colony transformant is picked and cultured overnight in a temperature-controlled orbital shaking incubator at 37° in 10 ml of sterile LB (Luria–Bertani broth) containing 0.1 mg/ml ampicillin (Sigma). Next morning, this culture is diluted 1/50 into 500 ml LB broth plus 0.1 mg/ml ampicillin and cultured at 37° with agitation until the OD600 nm reaches 0.7. Protein production is then induced with 0.1 m*M* isopropyl-b-D-thiogalactopyranoside for 1.5 h at 37°. Bacteria are pelleted by centrifugation at 8000 × *g* for 10 min at 4°. The pellet is washed with ice-cold phosphate-buffered saline (PBS, 137 m*M* NaCl, 2.7 m*M* KCl, 1.8 m*M* KH_2PO_4, 8.1 m*M* Na_2 HPO_4, pH 7.0) and resuspended in 10 ml of PBS containing a protease inhibitor cocktail (Sigma-Aldrich). Bacteria are lysed by sonication and Triton X100 is added to a final concentration of 1%. They are then incubated at 4° for 20 min. While bacteria are being lysed, 50% (v/v) glutathione Sepharose slurry (GE Healthcare, Uppsala, SE) is equilibrated in PBS–1% Triton X100 buffer with protease inhibitor cocktail. The lysate is clarified by centrifugation at 15,000 × *g* for 15 min at 4°. The supernatant is incubated with 0.2 ml of glutathione Sepharose at room temperature for 30 min. The resulting beads with bound GST–RBD recombinant protein are centrifuged at 1000 × *g* for 2 min. Beads are washed three times at 4° with the same PBS buffer and stored at 4°. An aliquot (2–5 μl) is analyzed by SDS-PAGE (10%) and stained with Coomassie blue to assess the purity and the relative amount of the fusion protein. The GST fusion protein is used in subsequent pull-down experiments.

2.5. Antibodies

The purified GST–RBD protein is injected into rabbits to generate polyclonal antibodies. The resulting antiserum was consecutively affinity purified on a GST column and an RBD column. GFP monoclonal antibody is purchased from Boehringer Mannheim (Indianapolis, IN). Mouse monoclonal anti-EGFR is purchased from Calbiochem (Darmstadt, Germany). Polyclonal anti-Rab13 is obtained from Sigma-Aldrich. Polyclonal anti-EGFR is obtained from Santa Cruz Biotechnology (Santa Cruz, CA).

2.6. Pull-down experiments and immunoblotting

Our data suggest that MICAL-L1 inhibits Rab13 effect on EGFR trafficking (AZ unpublished data). Although, MICAL-L1 does not encompass a TBC (tree/Bub2/Cdc16) domain, a GTPase-activating protein (GAP) domain for Rab proteins, we decided to verify whether MICAL-L1 inactivates Rab13 and acts as a GAP. We used the RBD of MICAL-L1 to perform a GTP-Rab13 pull-down assay. This method has been successfully used to identify putative GAP activities (Ishibashi, Kanno, Itoh, & Fukuda, 2009; Itoh & Fukuda, 2006). Because the MICAL-L1-RBD specifically recognizes the GTP-bound Rab13, the amount of the active form of Rab13 can be estimated by quantitatively determining the Rab13 bound to the MICAL-L1-RBD. Briefly, lysates from Madin–Darby canine kidney

(MDCK) cells stably expressing GFP or GFP–MICAL-L1 are incubated with 20 μg of GST–MICAL-L1-RBD beads in 1.5 ml centrifuge tube for 4 h at 4°. Beads are then washed three times with PBS, and the bound GTP-Rab13 is analyzed by immunoblotting with anti-Rab13 antibodies as described (Kohler et al., 2004). Immunoblot detection is done by ECL according to manufacturer's protocols (Pierce, Rockford, IL). Compared to GFP, expression of GFP–MICAL-L1 did not decrease the amount of GTP-Rab13 in cells, suggesting that MICAL-L1 is not a GAP for Rab13 (Fig. 24.2).

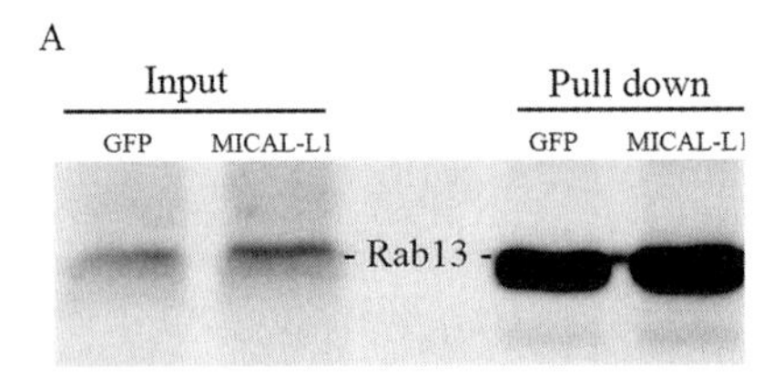

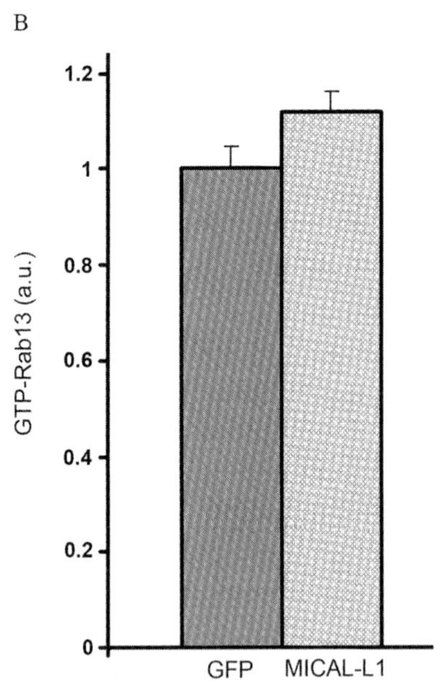

Figure 24.2 MICAL-L1 is not a GAP for Rab13 as revealed by the GTP-Rab13 pull-down assay. Equal amount of lysates from cells expressing GFP or GFP–MICAL-L1 is incubated with GST–MICAL-L1-RBD beads. The lysates used as inputs (1/40 of the reaction mixture) and proteins trapped (pull down) with RBD are analyzed by 12% SDS-PAGE and immunoblotting with anti-Rab13 antibodies. (B) Quantification of three independent experiments performed as in (A). After blotting, Rab13 bands are quantified using ImageJ program (NIH image, Rockville, MD) and Microsoft Excel software. The amount of GTP-Rab13 is calculated as the means ± SD (a.u., arbitrary units).

2.7. Short hairpin RNAs cloning

RNA interference is a powerful technology for studying gene functions in eukaryotic cells. We used InvivoGen psiRNA-h7SKGFPzeo plasmid that allows the production of short hairpin RNAs (shRNAs) within the cells. shRNAs consist of 56 mer RNA molecule with a hairpin structure, called shRNA, which is homologous to a region within the target gene. Introduction of shRNAs in mammalian cells induces strong and specific suppression of the target gene.

2.8. Vector preparation for Bbs I cloning (InvivoGen)

The plasmid contains two cloning sites, Bbs I/Bbs I (although these sites are recognized by the same enzyme, they are different avoiding self-ligation of the plasmid).

Digest overnight 5 μg of psiRNA-h7SKGFPzeo plasmid with 50 U of Bbs I in 50 μl buffer G (Fermentas).

Gel purify the vector onto a 0.8% agarose gel. Do not dephosphorylate the vector.

2.9. Hairpin design and cloning

Design the hairpin on the web http://www.sirnawizard.com/.

Order one pair of 56 base oligos for each construct.

I order them usually HPLC purified. shRNA sequences that efficiently inhibited MICAL-L1 expression are shRNA oligo sens1, 5′ACCTCGTCG CAGTATTACAACCACTT*TCAAGAG*AAGTGGTT GTAATACT-GCGACTT-3′; shRNA oligo-antisens1, 5′-CAAAAAGTC GCAGTATTACAACCACTT*CTCTTGA*AAGTGGTTGTA ATACTG CGACG-3′; shRNA oligo sens2: 5′ ACCTCGACCTACGTGTCGCAGT ATTA*TCAAGAG*TAATACTGCGACACGTAGGTCTT-3′; and shRNA oligo-antisens2, 5′-CAAAAAGACCTACGTGTCGCAGTATTA*CTCTT GA*TAATACTGCGACACGTAGGTCG-3′.

Scramble oligo-sense and antisense are 5′-ACCTCGCTGTTCC TACTCGCAAATAA*TCAAGAG*TTATTTGCGAGTAGGAACAGC-TT-3′ and 5′-CAAA AAGCTGTTCCTACTCGCAAATAA*CTCTTGA* TTATTTGCGAGTAGGAACAGCG-3′, respectively. Italics indicate the seven base pairs hairpin loop. Underlined nucleotides correspond to Bbs I cloning sites.

- Avoid gel purification, which I tried once and could not clone.
- Constructs targeting two species (i.e., man and mouse) might be an advantage.

Align the two genes and select oligos targeting a conserved region.

2.10. shRNA oligonucleotides annealing

- Mix the two complementary oligos: 23 μl H_2O, 6 μl 0.5 *M* NaCl, and 0.5 μl of each oligo at 100 μ*M*.
- Boil for 2 min, and let cool down to room temperature for approximately 40 min.

Ligation

1 μl of psiRNA-h7SKGFPzeo vector digested with Bbs I.
0.5 μl of annealed oligos.
3.5 μl H_2O.
5 μl Takara ligase.
Incubate at room temperature for 5 h.

2.11. Transform GT116 competent cells (InvivoGen, San Diego, CA)

- Mix 5 μl of ligation mixture with 50 μl of GT116 competent cells in ice for 15 min.
- Incubate the tubes in a 42° water bath for exactly 30 s, and then place the tubes back in ice for 1–2 min.
- Add 0.2 ml of SOC medium to each reaction and incubate at 37° for 1 h with shaking at 200 rpm.
- Spread each transformation reaction onto agar plate prepared with Fast-Media® Zeo X-Gal (InvivoGen), to take advantage of the white/blue selection. After 24 h, this procedure yields 20–70 colonies with approximately half of them white.

2.12. Plasmid preparations

Prepare plasmid DNAs from four white colonies in 2.5 ml of Fast-Media® Zeo TB medium. Verify the sequence of the shRNA insert of four positive clones using the appropriate sequencing primers.

Plasmids DNA of positive clones are prepared in 100 ml of Fast-Media® Zeo TB medium.

2.13. Cell culture and transfection

MDCK cells (clone II) are grown in DMEM (Dulbecco's modified Eagle's medium) supplemented with 10% fetal calf serum, 2 m*M* glutamine, 100 U/ml penicillin, and 10 mg/ml streptomycin. The cells are incubated at 37 °C under 10% CO_2 atmosphere. Stable MDCK cell lines expressing MICAL-L1 wild type or shRNAs are generated by transfection using the

Lipofectamine 2000 reagent according to the manufacturer's instructions (Invitrogen). After transfection, cells are allowed to recover and are then resuspended, serially diluted, and plated in 10 cm dishes supplemented with 1.0 mg/ml of G418 (Life Technologies, Inc.) or 50 μg/ml of Zeocin (InvivoGen). The cells are allowed to grow for 15 days until colonies are observed. Individual colonies are transferred to 48-well plates and amplified. Three independent positive cell lines for each MICAL-L1 or shRNA constructs are characterized. Stable cell lines expressing GFP–MICAL-L1 are maintained under selection in 0.5 mg/ml of G418, those expressing shRNA control or shRNAs targeting MICAL-L1 in 50 μg/ml of Zeocin. The psiRNA-h7SKGFPzeo vector encodes the GFP reporter gene used as a marker for transfected cells and Zeocin resistance.

Figure 24.3 shows epithelial MDCK cells transiently transfected with shRNA targeting dog MICAL-L1 mRNA sequence (KD) or mock

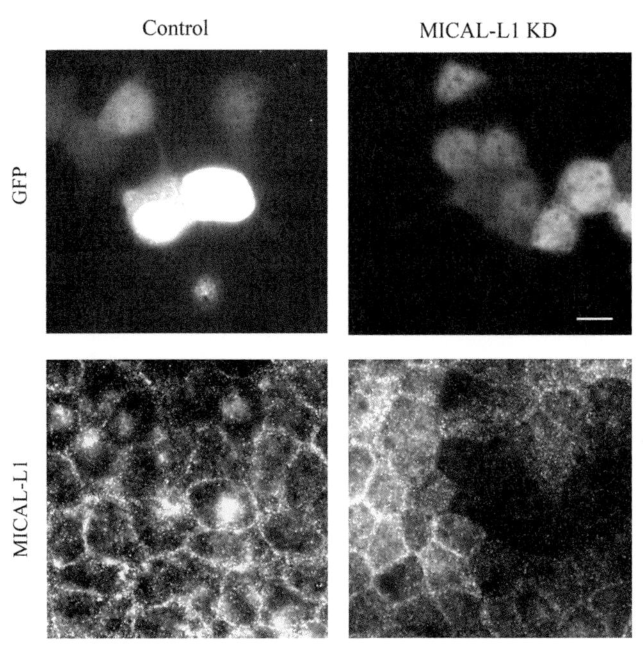

Figure 24.3 Epithelial MDCK cells are transiently transfected with shRNA targeting dog MICAL-L1 mRNA sequence indicated as MICAL-L1 KD or mock transfected (control). The cloning vector used in this study encodes the GFP reporter gene used as a marker for transfected cells. After 6 days, cells are processed by immunofluorescence with anti-MICAL-L1 antibodies. Note that MICAL-L1 is not detected in transfected "KD" cells, indicating that the expression of shRNA induces a strong depletion of MICAL-L1 protein. Bar, 10 μm.

transfected (control). After 6 days, MDCK cells are analyzed by immunofluorescence with polyclonal anti-MICAL-L1 antibodies. In transfected cells, the shRNA expression induces a strong depletion of MICAL-L1 protein (Fig. 24.3). We also generated stable MDCK cells producing different shRNAs targeting MICAL-L1 and confirmed that the shRNA expression induces MICAL-L1 depletion by blotting (Abou-Zeid et al., 2011).

2.14. Immunofluorescence microscopy

MDCK cells are grown on 10 mm round glass coverslips in 24-well plates and allowed to reach 80–90% confluency. Cells are then washed three times with PBS containing 1 m*M* $CaCl_2$ and 0.5 m*M* $MgCl_2$ and fixed in 3% paraformaldehyde-PBS for 20 min at room temperature. Immunofluorescence is performed essentially as described (Marzesco et al., 2002). Cells are incubated in blocking buffer (PBS+0.2% BSA containing 0.5% TritonX-100). Permeabilized cells are incubated for 1 h with the primary antibodies, rinsed three times for 10 min with the blocking buffer, and then incubated with affinity purified secondary antibodies raised in goat and conjugated to Cy2 or Cy3 (Jackson ImmunoResearch Laboratories, West Grove, PA). After washing, samples are analyzed with a three-dimensional (3D) deconvolution microscopy.

2.15. Image acquisition, processing, and fluorescence quantification

3D stacks with a 0.3 μm step are acquired by 3D deconvolution microscopy (Angenieux et al., 2005), adapted on an Eclipse 90i upright microscope (Nikon, S.A., France) equipped with a cool SNAP HQ2 CCD camera (Photometrics, Tucson, United States) and using a 100× CFI Plan Apo VC objective NA 1.4 (Nikon, S.A., France) controlled in the *Z*-axis by a Piezo Objective (PI, S.A.S., France). All deconvolution processes are performed automatically using an iterative and measured PSF-based algorithm (Gold-Meinel) on batches of image stacks, as a service proposed by the PICT-IBiSA imaging facility of the Curie Institute, Paris. Images are further processed with ImageJ (National Institutes of Health, Bethesda, MD) or Adobe Photoshop Software (Adobe Systems, Mountain View, CA).

The cellular distribution of GFP–MICAL-L1 is examined by immunofluorescence in epithelial MDCK cells. GFP–MICAL-L1 is localized to vesicle structures that showed significant overlap with GFP–Rab13 (Fig. 24.4).

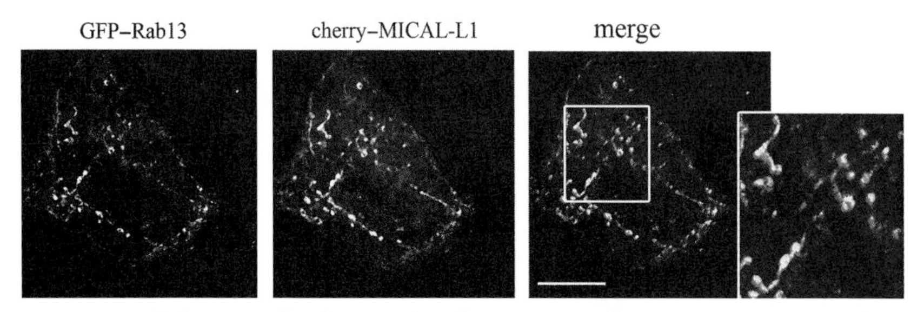

Figure 24.4 MICAL-L1 colocalizes with Rab13. MDCK cells coexpressing GFP–Rab13/mcherry–MICAL-L1 are processed for immunofluorescence. Three-dimensional (3D) stacks are acquired by 3D deconvolution microscopy. 3D projections of images are shown. High magnification shows that MICAL-L1 colocalizes with Rab13. Bar, 10 μm. (For color version of this figure, the reader is referred to the online version of this chapter.)

2.16. EGFR endocytosis

Stable MDCK cell lines (10^6 cells) expressing GFP, GFP–MICAL-L1 or shRNA (control), and shRNA–MICAL-L1 grown on 3.5 cm diameter culture plates for 1 day. They are serum-starved for 16 h, pretreated with cycloheximide (CHX, 40 μg/ml) at 37° for 60 min, and then cooled to 4°. Subsequently, 50 ng/ml EGF (Sigma-Aldrich) is applied at 4° in the presence of CHX to allow ligand binding to the receptor. Cells are then pulsed for 15 min at 37°, washed, and chased in low serum medium (0.5%) plus CHX for 3 h. To answer visualization of internalized EGFR, coverslips are washed and stripped for 3 min at 4° in 50 m*M* glycine, 100 m*M* NaCl, pH 3.0, to remove surface labeling. Cells are then subjected to immunofluorescence analysis. After a 15 min pulse, the distribution of endogenous EGFR is not affected by expression or depletion of MICAL-L1 as compared with control cells. The receptor is clearly detected as vesicular structures at perinuclear regions. After a 3 h chase, the EGFR is still accumulated at a perinuclear endosomal compartment in control cells as well as in cells expressing GFP–MICAL-L1. In contrast, MICAL-L1-depleted cells exhibited different EGFR distribution. EGFR vesicles are not detected at the perinuclear region but are found in vesicles spread throughout the cytoplasm. Moreover, we observed a significant EGFR degradation in MICAL-L1-depleted cells (Fig. 24.5).

2.17. EGFR degradation

MDCK cells are serum-starved for 16 h and incubated with 40 μg/ml CHX for 1 h. Cells are then stimulated with 50 ng/ml of EGF in presence of CHX

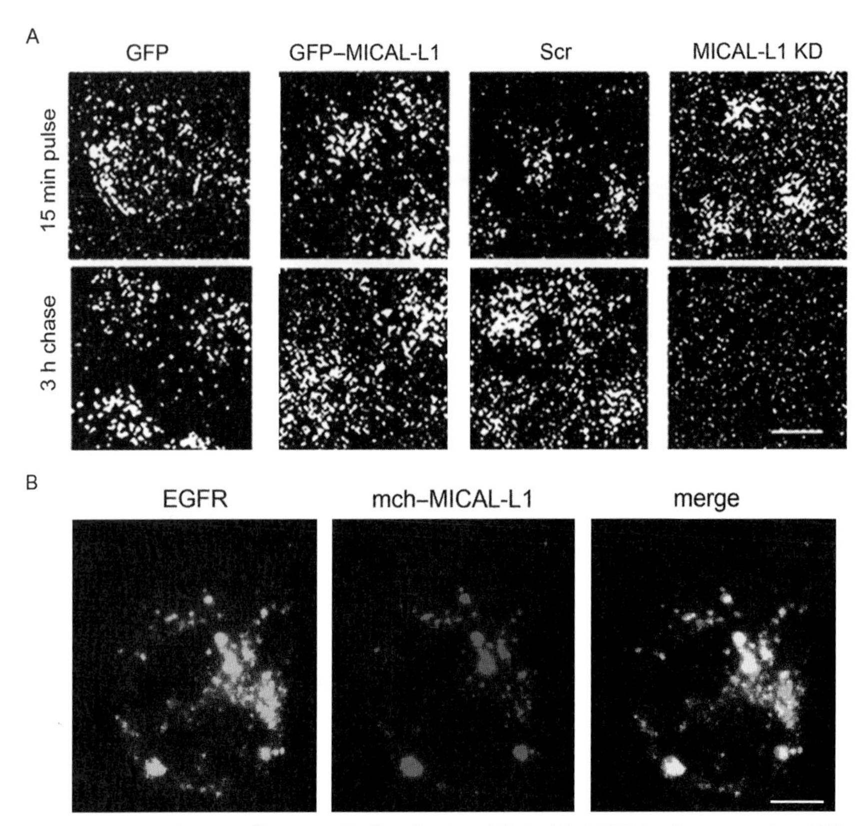

Figure 24.5 MICAL-L1 affects EGFR distribution. (A) Stable MDCK cells expressing GFP, GFP–MICAL-L1, control, MICAL-L1–shRNA (KD) are serum-starved for 16 h, pretreated with CHX, and incubated with 50 ng/ml EGF in the presence of CHX. Cells are then pulsed for 15 min at 37°, washed, and chased in low serum medium plus CHX for 3 h. They are washed and stripped with acidic buffer to remove surface labeling. Cells are then analyzed by immunofluorescence with monoclonal anti-EGFR. 3D stacks are acquired by 3D deconvolution microscopy. 3D projections of images are shown. EGFR staining is mainly detected at a perinuclear region after 15 min EGF pulse. After 3 h chase, MICAL-L1-depleted cells exhibited a different EGFR distribution. The receptor is distributed in vesicles spread throughout the cytoplasm. Bar, 10 μm. (B) Stable MDCK cells expressing cherry–MICAL-L1 are incubated with EGF and pulse/chased for 3 h as indicated in (A). Cells are stained with anti-EGFR and a Cy3 coupled secondary antibodies. They are then analyzed by 3D deconvolution microscopy. 3D projections of images are shown. Note that EGFR is accumulated in MICAL-L1 positive endosomes. Bar, 10 μm. (For color version of this figure, the reader is referred to the online version of this chapter.)

for 15 min at 37°. Cells are washed and chased in low serum medium plus CHX for 1, 2, and 3 h. They are lysed in 10 m*M* Tris–HCl, pH 7.6, 150 m*M* NaCl, 25 m*M* KCl, 1.8 m*M* $CaCl_2$, 1% Triton X-100, and a mixture of protease inhibitors. Cell extracts are then cleared by centrifugation, separated by SDS-PAGE, and immunoblotted with polyclonal anti-EGFR antibodies. Protein bands are quantified using ImageJ software (National Institutes of Health, Bethesda, MD). We assessed the biochemical EGFR degradation in MDCK cells expressing control or MICAL-L1–shRNA and found that cells depleted of MICAL-L1 exhibited a substantial EGFR degradation compared with control cells (Abou-Zeid et al., 2011).

3. CONCLUSION

MICAL-L1, a scaffold protein, regulates EGFR trafficking and inhibits its degradation. MICAL-L1 by binding Rab7, Rab8a, Rab11a, and Rab13, which regulate sequential transport steps along the endocytic/recycling pathway, may help organize membrane domains required for cargo progression between MVBs and RE. Therefore, MICAL-L1 could serve as a platform for the regulation of EGFR sorting/signaling at MVBs/RE.

ACKNOWLEDGMENTS

These investigations are supported by CNRS, INSERM, Fondation pour la recherche médicale, ANR MIME 2011, and ARC (SFI20111203516).

REFERENCES

Abou-Zeid, N., Pandjaitan, R., Sengmanivong, L., David, V., Le Pavec, G., Salamero, J., et al. (2011). MICAL-like1 mediates epidermal growth factor receptor endocytosis. *Molecular Biology of the Cell, 22*, 3431–3441.

Andrews, N. W. (2000). Regulated secretion of conventional lysosomes. *Trends in Cell Biology, 10*, 316–321.

Angenieux, C., Fraisier, V., Maitre, B., Racine, V., van der Wel, N., Fricker, D., et al. (2005). The cellular pathway of CD1e in immature and maturing dendritic cells. *Traffic, 6*, 286–302.

Bucci, C., & Chiariello, M. (2006). Signal transduction gRABs attention. *Cellular Signalling, 18*, 1–8.

Chen, Y., Samaraweera, P., Sun, T. T., Kreibich, G., & Orlow, S. J. (2002). Rab27b association with melanosomes: Dominant negative mutants disrupt melanosomal movement. *Journal of Investigative Dermatology, 118*, 933–940.

Dobrowolski, R., & De Robertis, E. M. (2012). Endocytic control of growth factor signalling: Multivesicular bodies as signalling organelles. *Nature Reviews Molecular Cell Biology, 13*, 53–60.

Fan, G. H., Lapierre, L. A., Goldenring, J. R., Sai, J., & Richmond, A. (2004). Rab11-family interacting protein 2 and myosin Vb are required for CXCR2 recycling and receptor-mediated chemotaxis. *Molecular Biology of the Cell, 15*, 2456–2469.

Fukuda, M., Kuroda, T. S., & Mikoshiba, K. (2002). Slac2-a/melanophilin, the missing link between Rab27 and myosin Va: Implications of a tripartite protein complex for melanosome transport. *Journal of Biological Chemistry, 277*, 12432–12436.

Gidon, A., Bardin, S., Cinquin, B., Boulanger, J., Waharte, F., Heliot, L., et al. (2012). A Rab11A/myosin Vb/Rab11–FIP2 complex frames two late recycling steps of langerin from the ERC to the plasma membrane. *Traffic, 13*, 815–833.

Giridharan, S. S., Rohn, J. L., Naslavsky, N., & Caplan, S. (2012). Differential regulation of actin microfilaments by human MICAL proteins. *Journal of Cell Science, 125*, 614–624.

Goud, B. (2002). How Rab proteins link motors to membranes. *Nature Cell Biology, 4*, E77–E78.

Gruenberg, J. (2001). The endocytic pathway: A mosaic of domains. *Nature Reviews Molecular Cell Biology, 2*, 721–730.

Hattula, K., Furuhjelm, J., Tikkanen, J., Tanhuanpaa, K., Laakkonen, P., & Peranen, J. (2006). Characterization of the Rab8-specific membrane traffic route linked to protrusion formation. *Journal of Cell Science, 119*, 4866–4877.

Horgan, C. P., Hanscom, S. R., Jolly, R. S., Futter, C. E., & McCaffrey, M. W. (2010). Rab11–FIP3 links the Rab11 GTPase and cytoplasmic dynein to mediate transport to the endosomal-recycling compartment. *Journal of Cell Science, 123*, 181–191.

Hsu, V. W., Bai, M., & Li, J. (2012). Getting active: Protein sorting in endocytic recycling. *Nature Reviews Molecular Cell Biology, 13*, 323–328.

Ishibashi, K., Kanno, E., Itoh, T., & Fukuda, M. (2009). Identification and characterization of a novel Tre-2/Bub2/Cdc16 (TBC) protein that possesses Rab3A-GAP activity. *Genes to Cells, 14*, 41–52.

Itoh, T., & Fukuda, M. (2006). Identification of EPI64 as a GTPase-activating protein specific for Rab27A. *Journal of Biological Chemistry, 281*, 31823–31831.

Knodler, A., Feng, S., Zhang, J., Zhang, X., Das, A., Peranen, J., et al. (2010). Coordination of Rab8 and Rab11 in primary ciliogenesis. In: *Proceedings of the National Academy of Sciences of the United States of America, 107*, 6346–6351.

Kohler, K., Louvard, D., & Zahraoui, A. (2004). Rab13 regulates PKA signaling during tight junction assembly. *Journal of Cell Biology, 165*, 175–180.

Laulagnier, K., Schieber, N. L., Maritzen, T., Haucke, V., Parton, R. G., & Gruenberg, J. (2011). Role of AP1 and Gadkin in the traffic of secretory endo-lysosomes. *Molecular Biology of the Cell, 22*, 2068–2082.

Le Roux, D., Le Bon, A., Dumas, A., Taleb, K., Sachse, M., Sikora, R., et al. (2012). Antigen stored in dendritic cells after macropinocytosis is released unprocessed from late endosomes to target B cells. *Blood, 119*, 95–105.

Marzesco, A. M., Dunia, I., Pandjaitan, R., Recouvreur, M., Dauzonne, D., Benedetti, E. L., et al. (2002). The small GTPase Rab13 regulates assembly of functional tight junctions in epithelial cells. *Molecular Biology of the Cell, 13*, 1819–1831.

Miaczynska, M., Christoforidis, S., Giner, A., Shevchenko, A., Uttenweiler-Joseph, S., Habermann, B., et al. (2004). APPL proteins link Rab5 to nuclear signal transduction via an endosomal compartment. *Cell, 116*, 445–456.

Miaczynska, M., & Zerial, M. (2002). Mosaic organization of the endocytic pathway. *Experimental Cell Research, 272*, 8–14.

Morimoto, S., Nishimura, N., Terai, T., Manabe, S., Yamamoto, Y., Shinahara, W., et al. (2005). Rab13 mediates the continuous endocytic recycling of occludin to the cell surface. *Journal of Biological Chemistry, 280*, 2220–2228.

Ostrowski, M., Carmo, N. B., Krumeich, S., Fanget, I., Raposo, G., Savina, A., et al. (2009). Rab27a and Rab27b control different steps of the exosome secretion pathway. *Nature Cell Biology, 12*, 19–30, sup pp. 11–13.

Raposo, G., Marks, M. S., & Cutler, D. F. (2007). Lysosome-related organelles: Driving post-Golgi compartments into specialisation. *Current Opinion in Cell Biology, 19*, 394–401.

Reddy, A., Caler, E. V., & Andrews, N. W. (2001). Plasma membrane repair is mediated by Ca(2+)-regulated exocytosis of lysosomes. *Cell, 106*, 157–169.

Sato, T., Mushiake, S., Kato, Y., Sato, K., Sato, M., Takeda, N., et al. (2007). The Rab8 GTPase regulates apical protein localization in intestinal cells. *Nature, 448*, 366–369.

Schwartz, S. L., Cao, C., Pylypenko, O., Rak, A., & Wandinger-Ness, A. (2007). Rab GTPases at a glance. *Journal of Cell Science, 120*, 3905–3910.

Seabra, M. C., & Coudrier, E. (2004). Rab GTPases and myosin motors in organelle motility. *Traffic, 5*, 393–399.

Seto, E. S., Bellen, H. J., & Lloyd, T. E. (2002). When cell biology meets development: Endocytic regulation of signaling pathways. *Genes & Development, 16*, 1314–1336.

Sigismund, S., Argenzio, E., Tosoni, D., Cavallaro, E., Polo, S., & Di Fiore, P. P. (2008). Clathrin-mediated internalization is essential for sustained EGFR signaling but dispensable for degradation. *Developmental Cell, 15*, 209–219.

Sorkin, A., & von Zastrow, M. (2009). Endocytosis and signalling: Intertwining molecular networks. *Nature Reviews Molecular Cell Biology, 10*, 609–622.

Terman, J. R., Mao, T., Pasterkamp, R. J., Yu, H. H., & Kolodkin, A. L. (2002). MICALs, a family of conserved flavoprotein oxidoreductases, function in plexin-mediated axonal repulsion. *Cell, 109*, 887–900.

Thery, C., Ostrowski, M., & Segura, E. (2009). Membrane vesicles as conveyors of immune responses. *Nature Reviews Immunology, 9*, 581–593.

Tolmachova, T., Abrink, M., Futter, C. E., Authi, K. S., & Seabra, M. C. (2007). Rab27b regulates number and secretion of platelet dense granules. In: *Proceedings of the National Academy of Sciences of the United States of America, 104*, 5872–5877.

Uzan-Gafsou, S., Bausinger, H., Proamer, F., Monier, S., Lipsker, D., Cazenave, J. P., et al. (2007). Rab11A controls the biogenesis of Birbeck granules by regulating Langerin recycling and stability. *Molecular Biology of the Cell, 18*, 3169–3179.

Zahraoui, A., Louvard, D., & Galli, T. (2000). Tight junction, a platform for trafficking and signaling protein complexes. *Journal of Cell Biology, 151*, F31–F36.

Zerial, M., & McBride, H. (2001). Rab proteins as membrane organizers. *Nature Reviews Molecular Cell Biology, 2*, 107–117.

Zhu, G., Chen, J., Liu, J., Brunzelle, J. S., Huang, B., Wakeham, N., et al. (2007). Structure of the APPL1 BAR-PH domain and characterization of its interaction with Rab5. *EMBO Journal, 26*, 3484–3493.

AUTHOR INDEX

Note: Page numbers followed by "*f*" indicate figures and "*t*" indicate tables.

D

E

G

H

L

M

N

Q

R

S

W

SUBJECT INDEX

Note: Page numbers followed by "*f*" indicate figures and "*t*" indicate tables.

F

G

H

N

O

P

S

U

V

W

Y

Z

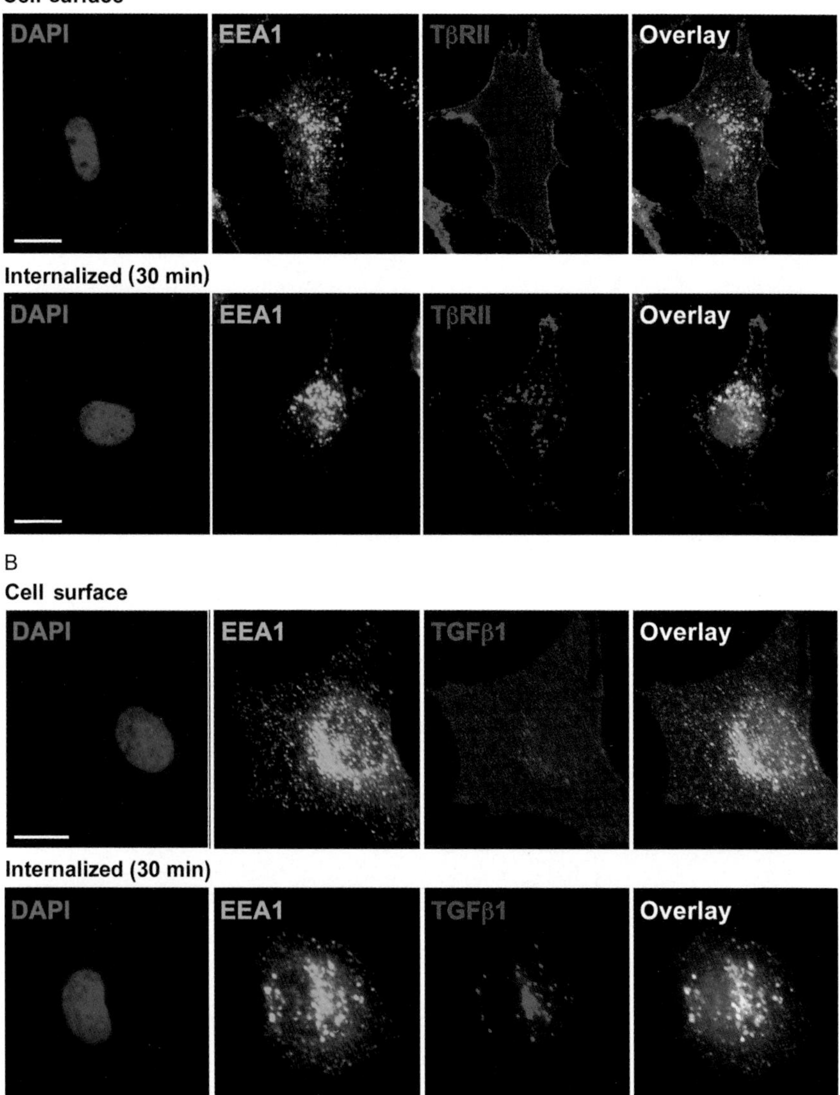

Sarah McLean and Gianni M. Di Guglielmo, Figure 3.2 *TGFβ receptor trafficking to the early endosome*. (A) Mv1Lu cells stably overexpressing HA–TβRII were labeled at 4 °C with anti-HA antibodies. Following incubation of Cy3-labeled anti-rabbit antibodies at 4 °C, cells were fixed (cell surface) or allowed to internalize for 30 min at 37 °C to permit receptor internalization. Standard immunofluorescence staining was used to visualize EEA1, a marker for the early endosome, and nuclei (DAPI staining). (B) Mv1Lu cells stably overexpressing HA–TβRII were labeled at 4 °C with biotinylated TGFβ1. Following incubation of Cy3-labeled streptavidin at 4 °C, cells were fixed or incubated at 37 °C as described in panel A. Standard immunofluorescence staining was used to visualize EEA1 and nuclei (DAPI staining). Receptor complex localization with the early endosomal marker results in a yellow overlay.

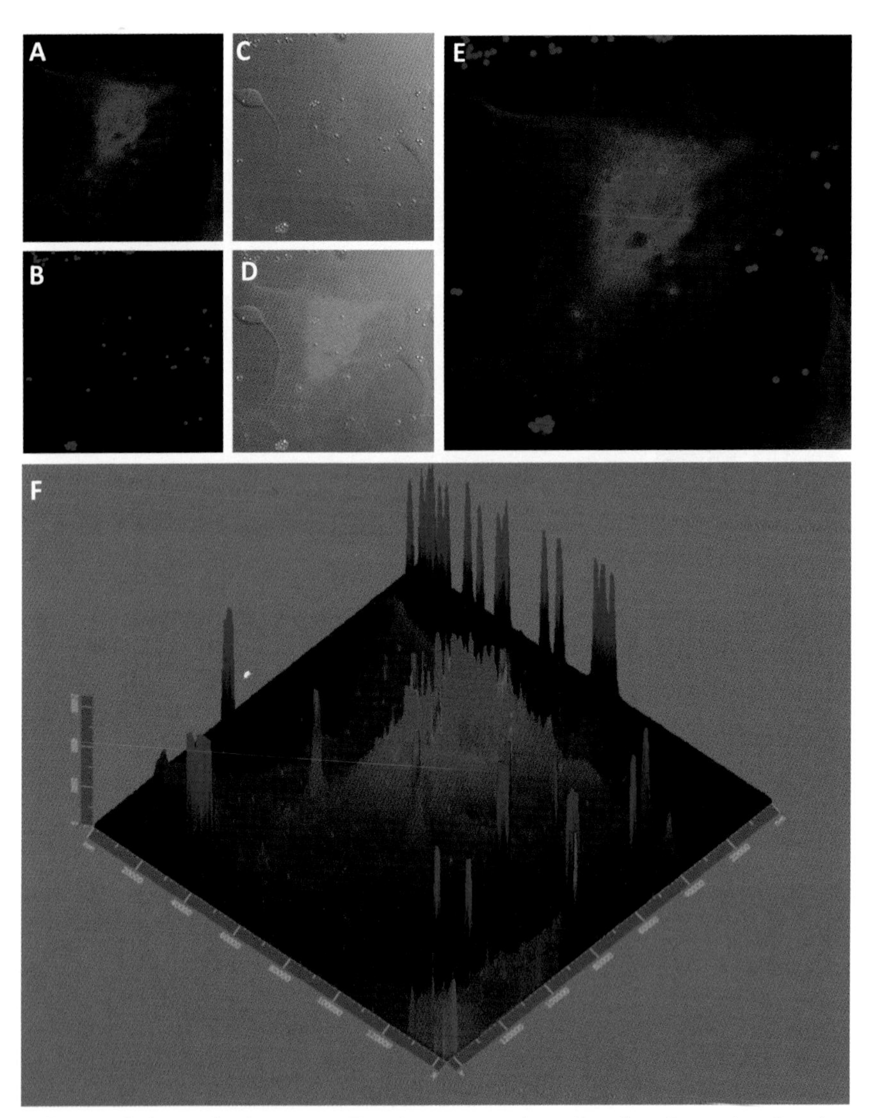

Helen R. Clark *et al.*, Figure 7.2 Transient expression of endocytic marker fused to mCherry in mammalian cell line BEAS-2B colocalizing with internalized fluorescent polystyrene beads. Transfected cells were incubated with polystyrene beads for 9 h, washed with dPBS, and stored in formalin until imaging by confocal microscopy. (A) mCherry fluorescent channel. (B) Polystyrene fluorescent channel. (C) Light image. (D) Total overlay. (E) Fluorescent overlay. (F) 2.5 D fluorescent channel, *x*- and *y*-axes indicate spatial position of fluorescence; *z*-axis indicates intensity at given position. An increase in intensity around an internalized fluorescent bead in comparison to other parts of the cells is indicative of colocalization.

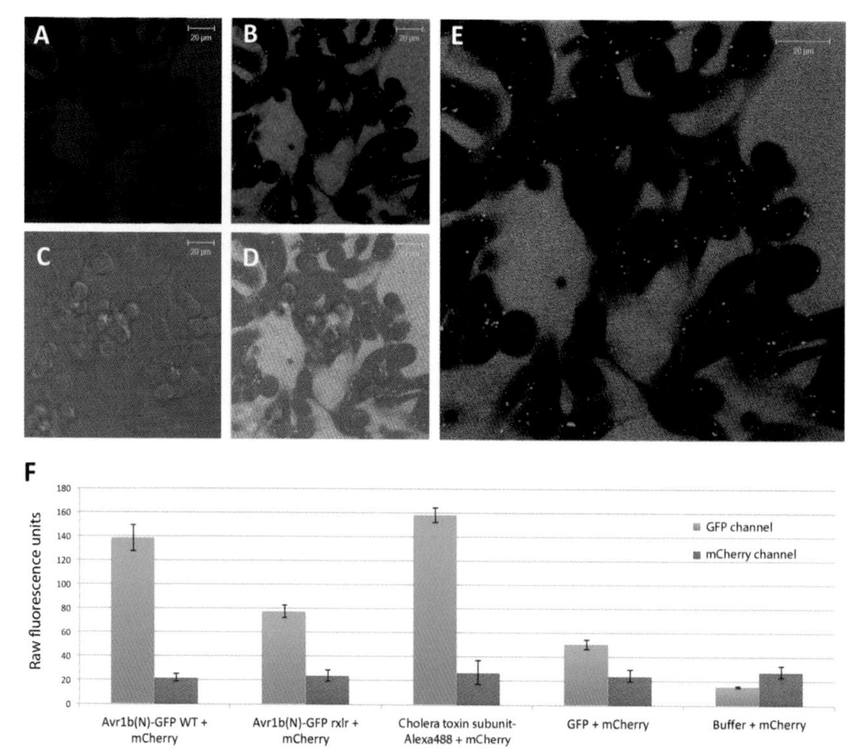

Helen R. Clark *et al.*, Figure 7.3 Qualitative and quantitative measurements of cell entry of Avr1b(N)–GFP in BEAS-2B mammalian cells. (A) Fluorescence from control mCherry samples. (B) Fluorescence from Avr1b(N)–GFP. (C) Light image. (D) Complete overlay. (E) Fluorescence overlay. (F) Bar graphs represent the mean raw signal of six replicates from fluorescent microplate reader assay from one experiment. In both cases, cells were incubated with construct of interest and control mCherry. Avr1b(N)–GFP WT refers to the wild-type protein purified in Section 4. Avr1b(N)–GFP RxLR refers to the same fusion protein except that the RxLR motif has been replaced with four alanines. Internalized fluorescent protein levels are compared using Duncan's multiple-range test. Control cholera toxin beta subunit was conjugated with Alexa Fluor 488 (1 m*M* of cholera toxin = 5 m*M* Alexa Fluor 488).

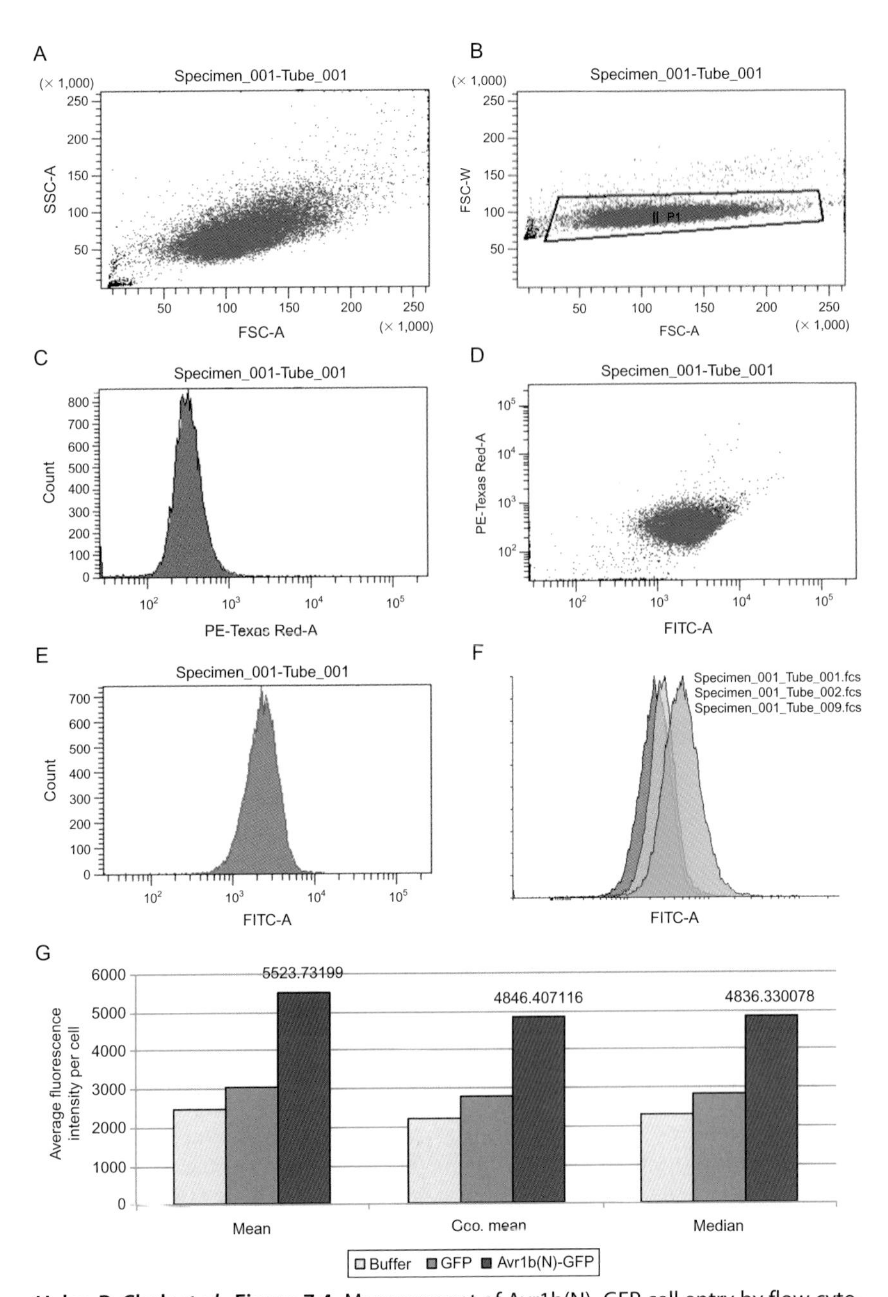

Helen R. Clark *et al.*, Figure 7.4 Measurement of Avr1b(N)–GFP cell entry by flow cytometry for one biological replicate. An example given is the gating of 20,000 cells positive for singleton status and PI-negative. (A–E) Gating a setup for analysis prior to beginning cytometry. (A) Forward scatter area (FSC-A) versus side scatter area (SSC-A). (B) FSC-A versus FSC width (W) to separate singletons from doublets. Singleton population is gated, shown in red. (C) Histogram display of propidium iodide fluorescence (log scale). (D) Dot plot of cellular fluorescence associated with propidium iodide and GFP (FITC). Histogram display of GFP (FITC channel) fluorescence. Logarithmic histogram overlay of FITC channel fluorescence for control (red, left), GFP (green, middle), and Avr1b(N)–GFP (right). (E) Presentation of different methods to determine average fluorescence among population. Utilization of mean results in artificial skewing. Populations of cells that respond homogeneously and have parametric distribution, such as these, have similar values for geometric means, medians, and modes (not shown).

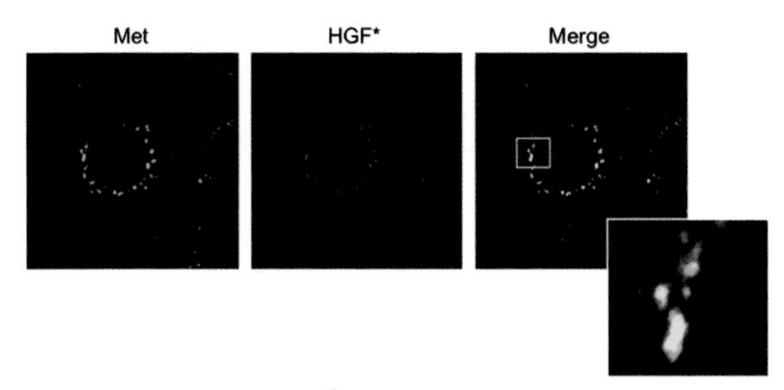

Rachel Barrow *et al.*, Figure 8.1 HGF* colocalizes with Met immunostaining. Confocal sections of cells stimulated with HGF* (red) for 60 min and then stained with an antibody against Met (green) and with DAPI (blue).

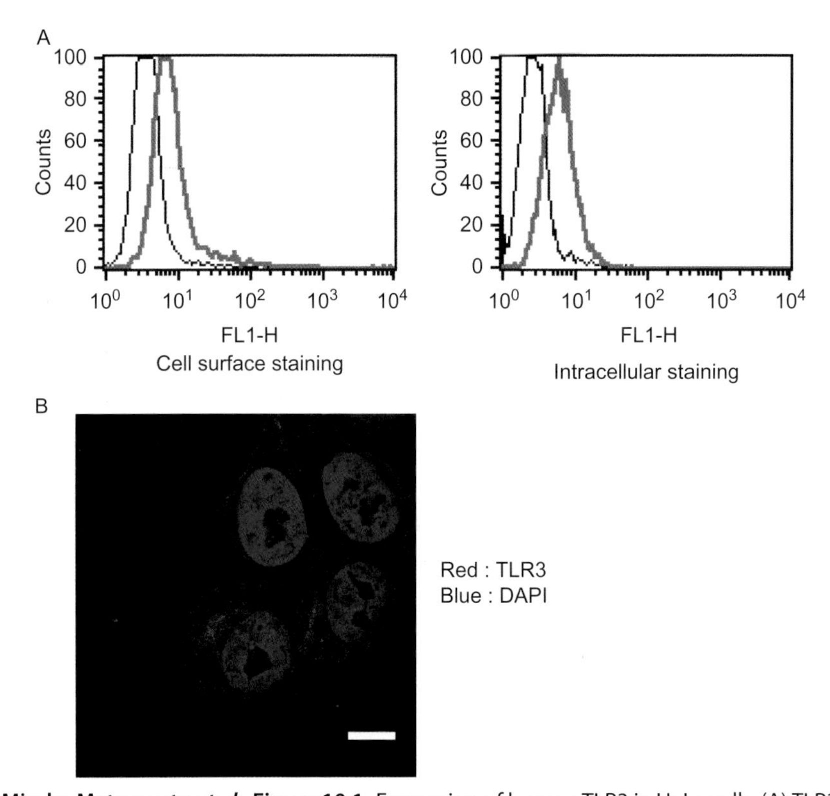

Misako Matsumoto *et al.*, Figure 10.1 Expression of human TLR3 in HeLa cells. (A) TLR3 is expressed on the cell surface and inside the cells in HeLa cells. Cell-surface (left) staining and intracellular (right) staining were performed using the TLR3.7 mAb and analyzed by flow cytometry. The black line indicates control mouse IgG staining. The red line indicates TLR3.7 staining. (B) TLR3 is localized to intracellular vesicles in HeLa cells. Cells were fixed and permeabilized, and endogenous TLR3 was stained with the TLR3.7 mAb. The red signal indicates endogenous TLR3 and the blue signal indicates DAPI staining. Scale bar, 10 μm.

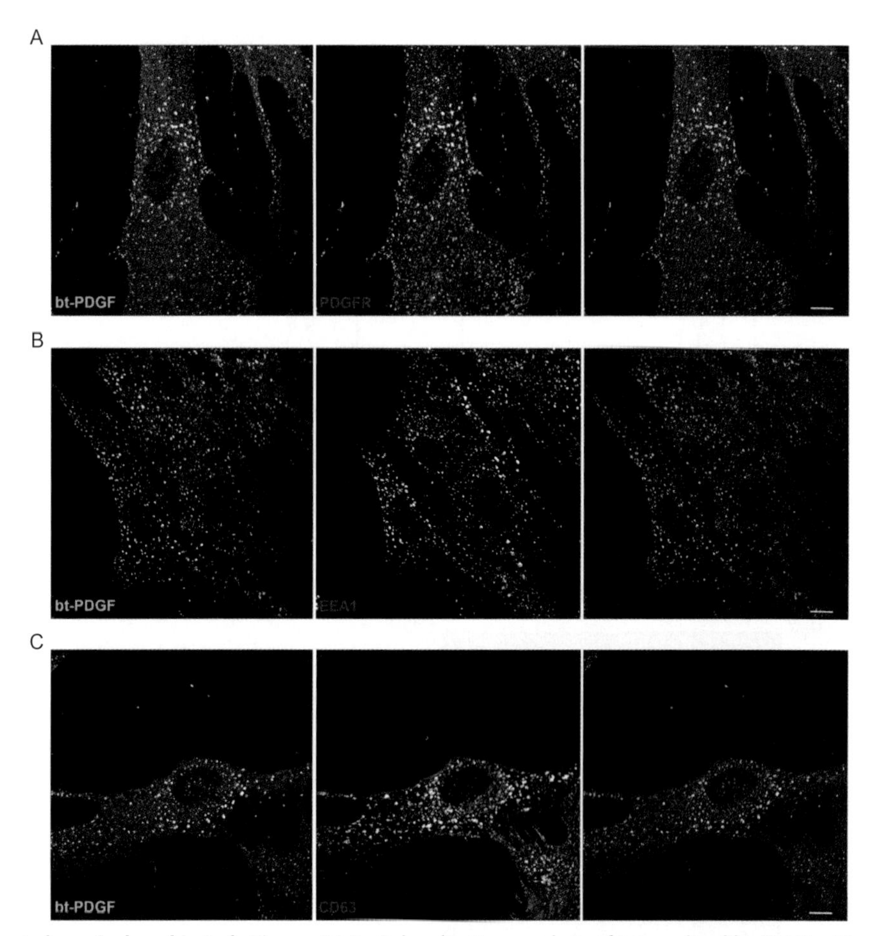

Łukasz Sadowski *et al.*, Figure 11.3 *Colocalization analysis of internalized bt-PDGF with PDGFR and endocytic markers.* CCD-1070Sk human foreskin fibroblasts were stimulated with bt-PDGF for 40 min (A), 30 min (B), or 60 min (C). Cells were immunostained for biotin (in order to detect bt-PDGF) and either PDGFRβ (A), EEA1 (marker of early endosomes; B), or CD63 (marker of multivesicular bodies; C). Scale bar 10 μm.

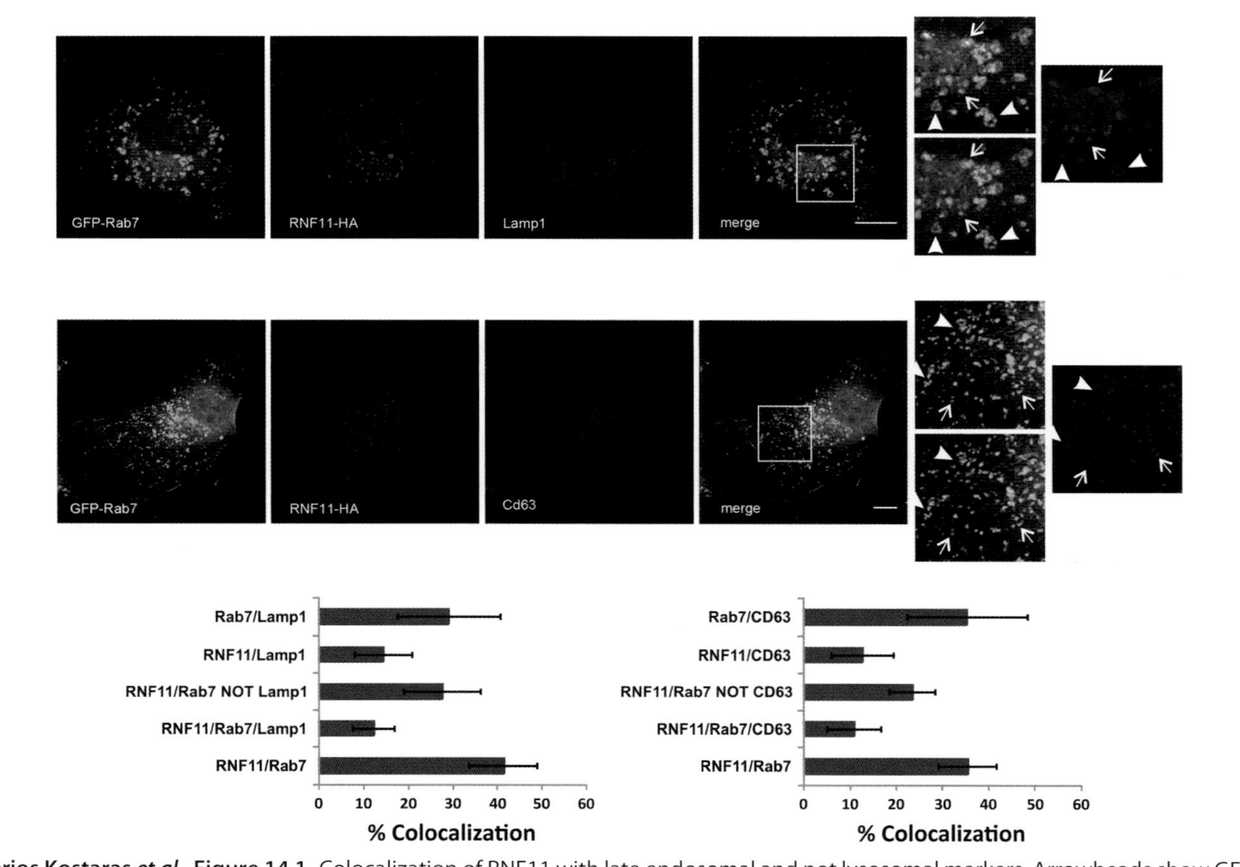

Eleftherios Kostaras *et al.*, Figure 14.1 Colocalization of RNF11 with late endosomal and not lysosomal markers. Arrowheads show GFP-Rab7 vesicles positive for Lamp1 or CD63 but RNF11-HA-negative. Arrows indicate that the RNF11-positive Rab7 endosomes are Lamp1- or CD63-negative. Scale bars, 10 μm. Quantitation of the triple colocalization was calculated using Kalaimoscope motionTracker and error bars depict the standard deviation of the mean. *From Kostaras et al. (2012).*

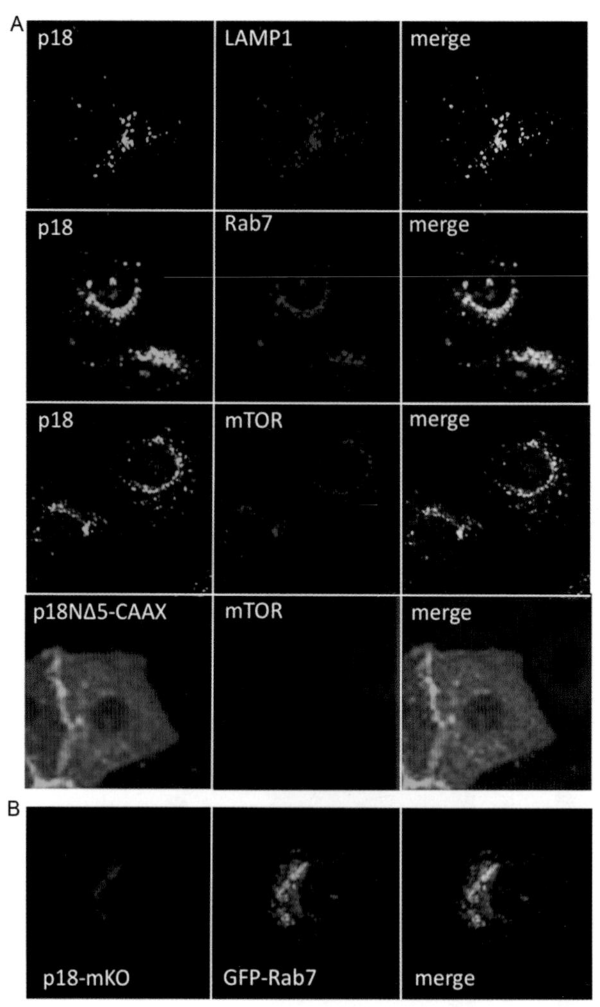

Shigeyuki Nada *et al.*, Figure 15.3 (A) Intracellular localization of p18. HA-tagged wild-type p18 or p18N△5-CAAX was expressed in p18$^{-/-}$ cells, and the cells were stained with anti-HA antibody. p18N△5-CAAX is localized to the plasma membrane as a result of deletion of the N-terminal acylation sites as well as addition of a CAAX motif at its C-terminus. LAMP-1 is a marker of both late endosomes and lysosomes. Rab7 is used as a specific marker of late endosomes. The results show that mTOR is localized to LE/Lys in a manner that depends on p18. (B) Analysis of p18 localization by expression of fluorescent fusion proteins (p18-mKO and GFP-Rab7).

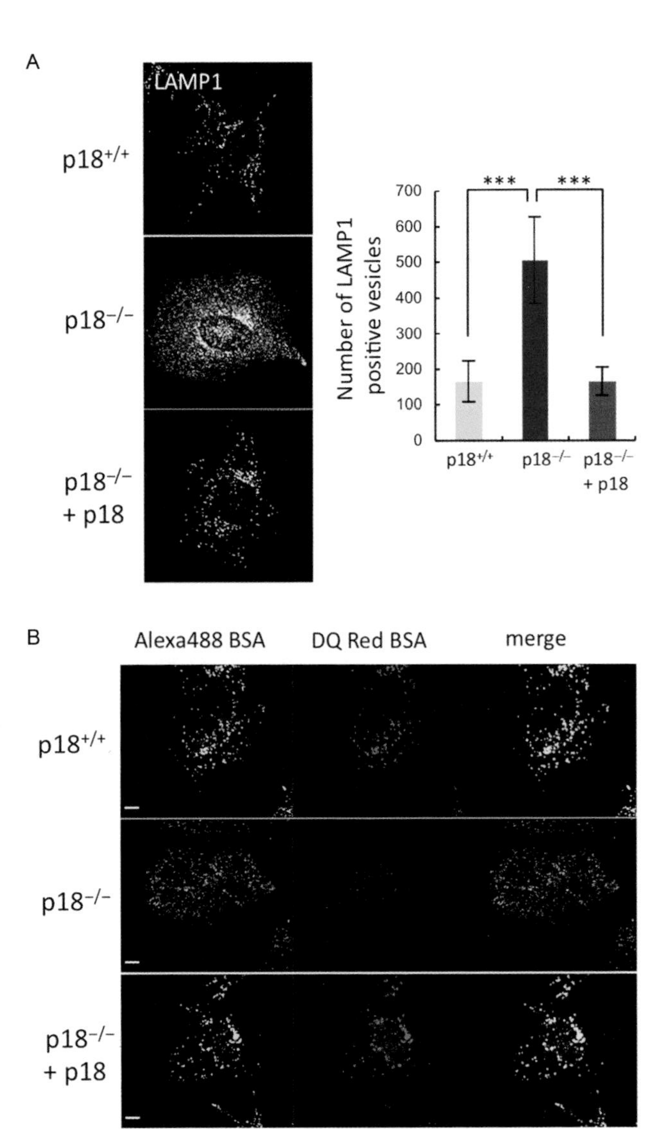

Shigeyuki Nada *et al.*, Figure 15.5 (A) Changes in distribution, size, and numbers of LE/Lys upon loss of p18, investigated using immunostaining for LAMP-1. Smaller-sized LE/Lys are diffusely distributed in the cytoplasm of p18$^{-/-}$ cells. (B) The indicated cells were cultured in the presence of Alexa488-BSA and DQ Red BSA, and localization of Alexa488-BSA and the fluorescent degradation products of DQ Red BSA were imaged. Degradation of DQ Red BSA was suppressed by loss of p18.

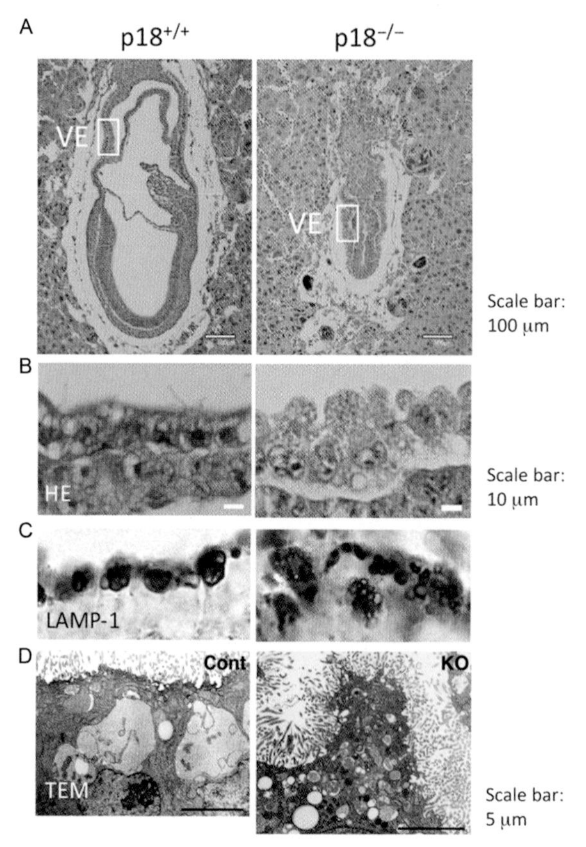

Shigeyuki Nada *et al.*, Figure 15.6 Phenotypes of $p18^{-/-}$ embryos. (A) HE staining of $p18^{+/+}$ and $p18^{-/-}$ embryos. Locations of visceral endoderm are indicated by white boxes. (B) Magnified views of VE. (C) Immunohistochemical analysis for LAMP-1. (D) Transmission electron microscopy (TEM) analysis.

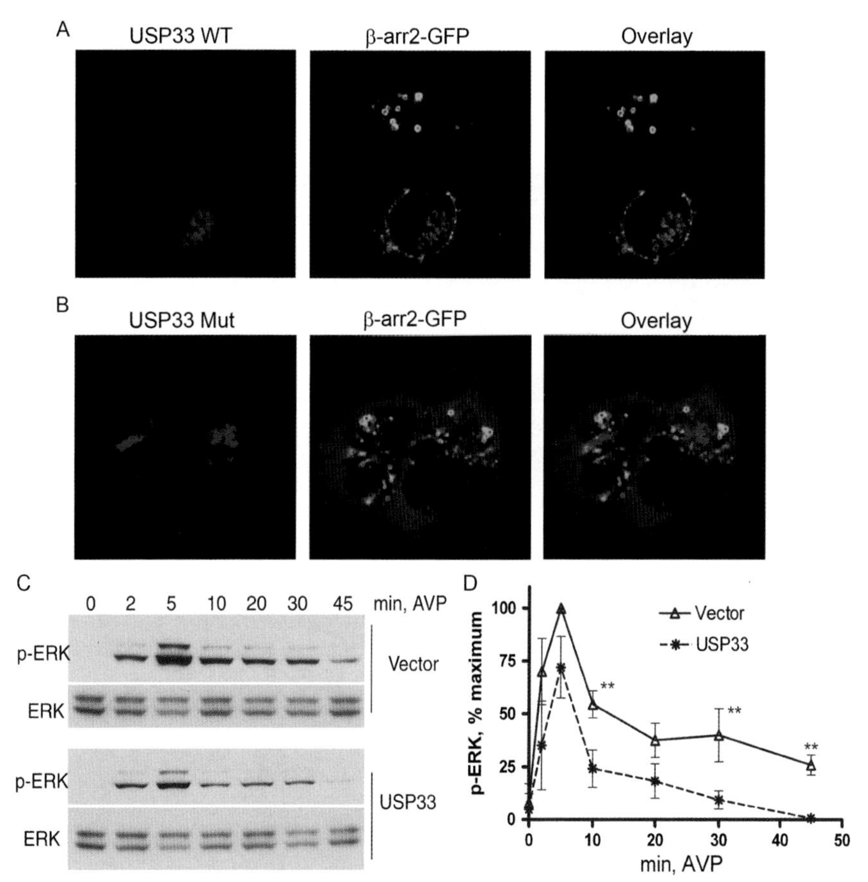

Sudha K. Shenoy, Figure 20.3 (A) HEK-293 cells with stable HA-V_2R expression were transiently transfected with β-arrestin2-GFP and RFP-USP33 and stimulated with 1 μM AVP. The panel shows two cells, one with no detectable RFP-USP33 that has the expected endosomal recruitment of β-arrestin2 and the other cell that expresses USP33 in which β-arrestin recruitment to the endosomes is inhibited. (B) Confocal micrographs show normal endosomal distribution of β-arrestin2-GFP upon AVP stimulation in cells overexpressing RFP-USP33$^{CYS:HIS}$. (C) HEK-293 cells transiently expressing HA-V_2R with either vector or USP33 were stimulated with 1 μM AVP for the indicated times. Equivalent amounts of whole cell lysates were analyzed for pERK and ERK content. (D) The graph represents quantification of pERK bands from four independent experiments. $^{**}p < 0.01$ two-way ANOVA, $n = 4$. *Adapted from Shenoy et al. (2009).*

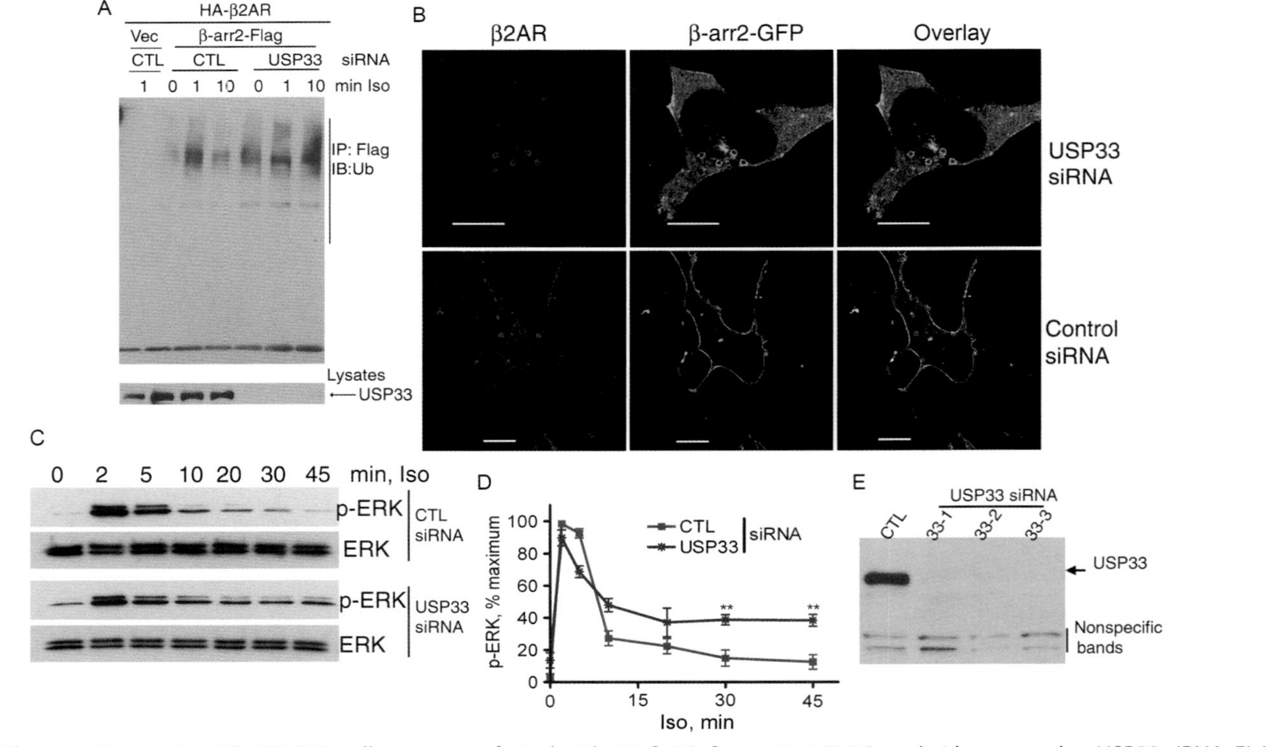

Sudha K. Shenoy, Figure 20.4 (A) HEK-293 cells were transfected with HA-β_2AR, β-arrestin2-FLAG, and either control or USP33 siRNA. FLAG immunoprecipitates were isolated after Iso stimulation for the indicated times and probed with an antiubiquitin antibody. The lower panel shows USP33 levels in control and USP knockdown samples. (B) Confocal micrographs show distribution of FLAG-β_2AR (red) and β-arrestin2-GFP (green) in USP33-depleted cells (top row) and control cells (bottom row) in agonist-stimulated HEK-293 cells. Scale bars, 10 μm. (C) Western blots of pERK and ERK in response to a time course of Iso (100 n*M*) activation, detected in HEK-293 cells with stable β_2AR expression (2 pmol/mg protein) and with indicated siRNA transfections. (D) This graph represents quantification of pERK signals from three independent experiments. $^{**}p < 0.01$, control versus USP33, two-way ANOVA. (E) A representative blot for USP33 levels is displayed showing western blot analysis of 25 μg of lysate protein from each siRNA transfection. *Adapted from Shenoy et al. (2009).*

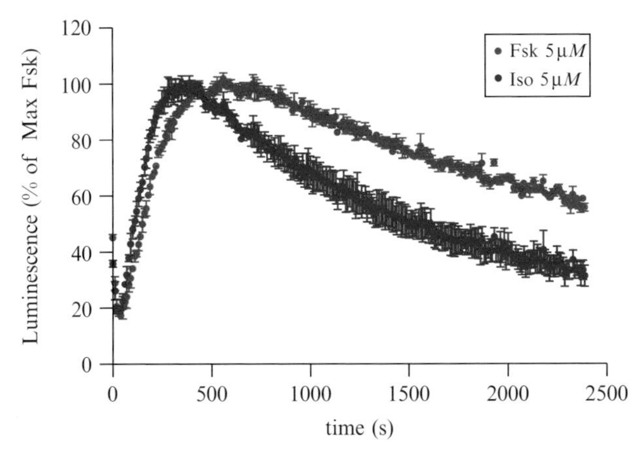

Roshanak Irannejad *et al.*, Figure 23.2 Kinetics of the cAMP response detected using the luminescence assay. Representative example of luminescence imaging from triplicate wells after application of 5 μ*M* isoproterenol (blue) or forskolin (red) at $t = 0$. Data shown are normalized to the percentage of maximum forskolin response.

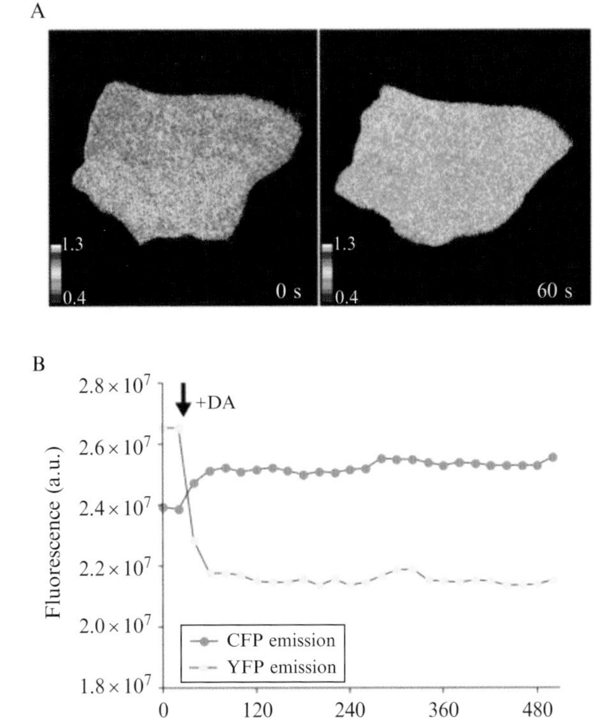

Roshanak Irannejad *et al.*, Figure 23.3 FRET-based detection of D1R-mediated cAMP accumulation. (A) Pixel-by-pixel calculation of *n*FRET in a representative HEK293 cells expressing Flag-tagged D1R before (left) and 60 s after (right) application of 10 μ*M* dopamine. (B) Time course of integrated whole-cell CFP and YFP fluorescence intensity changes on which the *n*FRET calculation is based.

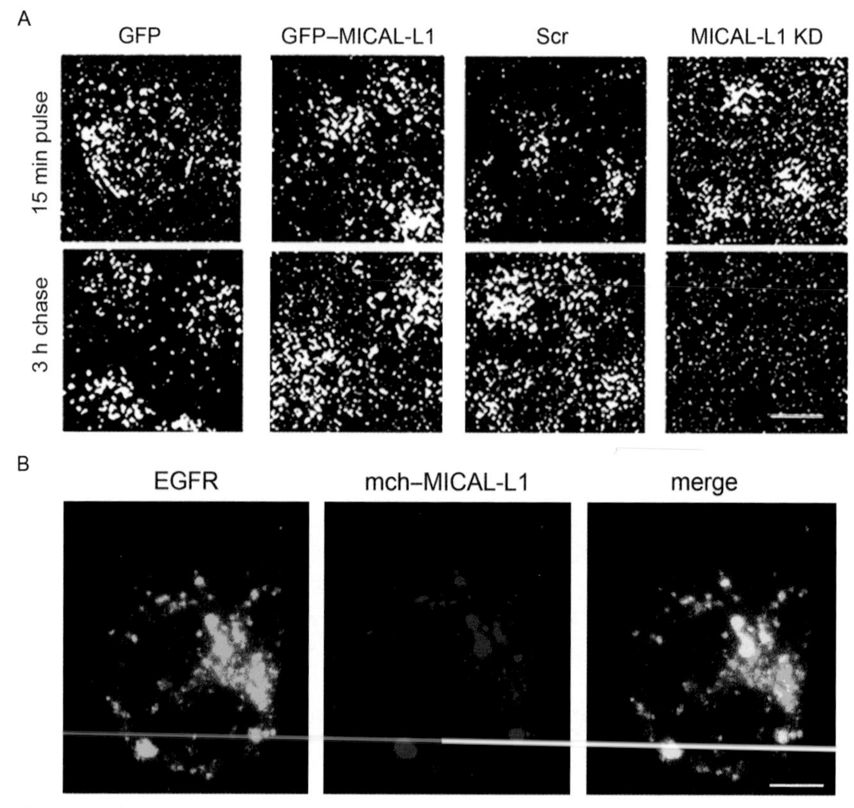

Ahmed Zahraoui, Figure 24.5 MICAL-L1 affects EGFR distribution. (A) Stable MDCK cells expressing GFP, GFP–MICAL-L1, control, MICAL-L1–shRNA (KD) are serum-starved for 16 h, pretreated with CHX, and incubated with 50 ng/ml EGF in the presence of CHX. Cells are then pulsed for 15 min at 37°, washed, and chased in low serum medium plus CHX for 3 h. They are washed and stripped with acidic buffer to remove surface labeling. Cells are then analyzed by immunofluorescence with monoclonal anti-EGFR. 3D stacks are acquired by 3D deconvolution microscopy. 3D projections of images are shown. EGFR staining is mainly detected at a perinuclear region after 15 min EGF pulse. After 3 h chase, MICAL-L1-depleted cells exhibited a different EGFR distribution. The receptor is distributed in vesicles spread throughout the cytoplasm. Bar, 10 μm. (B) Stable MDCK cells expressing cherry–MICAL-L1 are incubated with EGF and pulse/chased for 3 h as indicated in (A). Cells are stained with anti-EGFR and a Cy3 coupled secondary antibodies. They are then analyzed by 3D deconvolution microscopy. 3D projections of images are shown. Note that EGFR is accumulated in MICAL-L1 positive endosomes. Bar, 10 μm.